This book belongs to Jea[illegible]
work 385-3961 381-4989

work 232-5300
Cathy Barida
247-2498

no class
no class 11-8

'80 Dennis Cronin
381-6653

P9-DYD-278

William M. Setek, Jr.
Monroe Community College

Fundamentals of Mathematics

Second Edition

Glencoe Publishing Co., Inc.
Encino, California
Collier Macmillan Publishers
London

Copyright © 1979, 1976 by William M. Setek, Jr.

Printed in the United States of America

All rights reserved. No part of this book may be reproduced or transmitted in any form or by any means, electronic or mechanical, including photocopying, recording, or by any information storage and retrieval system, without permission in writing from the Publisher.

Glencoe Publishing Co., Inc.
17337 Ventura Boulevard
Encino, California 91316
Collier Macmillan Canada, Ltd.

Library of Congress Catalog Card Number:77-094777

1 2 3 4 5 6 7 8 9 10 83 82 81 80 79

ISBN 0-02-477690-4

To my wife, Addie, for her encouragement, understanding, assistance, and patience throughout this project, and to my sons, Scott and Joe, who helped in their own special way

CONTENTS

3 PROBABILITY

4 STATISTICS

5 AN INTRODUCTION TO THE METRIC SYSTEM

8

SETS OF NUMBERS AND THEIR STRUCTURE

9

AN INTRODUCTION TO ALGEBRA

PREFACE

TO THE INSTRUCTOR

In writing this second edition of *Fundamentals of Mathematics,* I have tried to reflect the changing approach to teaching liberal arts mathematics courses. The student population of colleges has changed; there is a greater diversity of students enrolled in the typical liberal arts mathematics course. Today, such a course may enroll students ranging from recent high school graduates to mature students with a wide variety of mathematical backgrounds. Interest and motivation vary greatly among these students, and many of them suffer from "math anxiety."

The only prerequisite for this text is a working knowledge of arithmetic. The approach is intuitive. The text contains an abundance of completely worked-out examples with systematic step-by-step solutions; there are no gaps or "magic" solutions. I have found that this type of experience provides the student with confidence and competence when doing homework or test problems.

ORGANIZATION OF THE TEXT

The text is divided into eleven chapters. Chapters 1 and 2 develop the basic ideas of sets and logic from an intuitive standpoint. Generous use is made of Venn diagrams and truth tables. Chapters 3 and 4 introduce the student to probability and statistics. Chapter 5 gives a thorough treatment of the metric system, emphasizing both metric-metric and metric-English conversions. Chapters 6, 7, and 8 are designed to broaden students' ideas about mathematics by exposing them to various mathematical systems and systems of numeration and the structure of the real number system. Chapter 9 gives the student an introduction to algebra, including experience in solving elementary linear equations, graphing equations and inequalities, and solving word problems. Chapter 10 is an intuitive introduction to basic concepts of plane geometry, and chapter 11 covers a number of mathematical topics of use to students in their role as consumers.

FEATURES OF THE SECOND EDITION

The second edition of *Fundamentals of Mathematics* reflects many improvements suggested by instructors who used the first edition. A new chapter, "Consumer Mathematics" (chapter 11), has been added as a result of growing interest in applications of mathematics to problems of everyday living. Chapter 9, "An Introduction to Algebra," has been expanded to include sections on graphing quadratic equations and linear programming, and chapter 10, "An Introduction to Geometry," has a new section on networks. The sections on applying Venn diagrams to logic have been moved from chapter 1 to chapter 2 so that they follow the basic topics of logic. Chapter 7 has been completely reorganized to make the distinctions among simple grouping, multiplicative grouping, and place-value systems of numeration clear-cut.

As with the first edition, the emphasis in this text is on encouraging the student to actively participate—to *do* mathematics by working problems and examples. (See *To the Student* following this Preface.) To this end, the more than 600 worked-out examples in the text have been completely reviewed for clarity and effectiveness, and the many exercises have been carefully reevaluated. Many new examples and exercises have been added. Furthermore, the learning objectives for each chapter have been completely rewritten, and the chapter review exercises have been rewritten and reorganized to test these objectives effectively.

To further encourage students to become active participants, each exercise set concludes with a Just for Fun problem. These problems range from serious extensions of mathematical ideas in the text to light-hearted puzzles and "brain teasers." They have been chosen primarily for their ability to capture student interest, and users of the first edition have been pleased with the results.

COURSE OUTLINES

Since liberal arts mathematics is not a well-defined course and its content varies from one school to the next, several suggested course outlines follow. Essentially, the chapters of this text are designed to be independent of one another so that the topics can be covered in any order. A student who fails to master the material in one chapter will

SUGGESTED COURSE OUTLINES

Chapters	Possible Omissions	Time Allotment
1–6	2.8–2.10, 3.8, 3.9	one semester
2–6, 11	2.8–2.10, 3.8, 3.9	one semester
3–6, 9–11	3.8, 3.9	one semester
1, 2, 5–8	2.8–2.10, 8.7	one semester
1, 2, 4, 9–11	none	one semester
1–11	chapters 1, 6, 10	two semesters

not necessarily be at a disadvantage when a new topic is begun. Many users of the first edition began their course with chapter 1 (sets), others began with chapter 3 or 4 (probability or statistics), and still others began with other chapters that suited the needs of their classes.

PEDAGOGICAL AIDS

Each chapter begins with learning objectives and a list of the symbols that will be introduced in the chapter. A generous set of exercises, graded in level of difficulty, follows each section. In addition, each chapter concludes with a summary and a set of review exercises. The review exercises are organized so that they test each learning objective in order; more challenging exercises that require the student to master several objectives are placed at the end of the review exercise set.

The answers to all odd-numbered exercises (including their multiple parts) for each section appear in Appendix B, along with all answers to the chapter review exercises. Therefore, assignments can be made with confidence in either fashion: with or without answers available.

SUPPLEMENTARY MATERIALS

A Solutions Manual for students containing detailed solutions for all exercises has been prepared by Professors Patricia I. Hooper and Linda Pulsinelli of Western Kentucky University. This manual can be a valuable source of reinforcement for the text materials.

The Instructor's Manual contains answers to the even-numbered exercises for each section, teaching suggestions, sets of more challenging exercises for each chapter, suggested projects and student activities, and lists of films and readings related to the topics in the text. It also includes a computer supplement consisting of text, examples, and problems that can be used to teach an introductory unit in BASIC programming. A Test Package containing two examinations for each chapter, a midterm, and a final is also available to the instructor.

ACKNOWLEDGMENTS

I am grateful and indebted to those users of the first edition who provided me with many valuable suggestions and constructive criticisms—in particular, my students and colleagues at Monroe Community College, and Professors Joseph Albree (Auburn University at Montgomery), Charles Ernst (St. Cloud State University), Clair Glossner (Corning Community College), and Elaine S. Johnson (Jamestown Community College).

I would like to thank those at Glencoe Publishing Company, Inc., for their enthusiastic interest and support throughout the project, especially Tanya Mink who guided me and the project from its beginning.

A special note of appreciation goes to Pamela Dretto, who typed the manuscript, and David Rogachefsky, who assisted in many ways.

I welcome any and all comments. Feel free to write and let me know your thoughts.

William M. Setek, Jr.
Monroe Community College
Rochester, New York 14623

TO THE STUDENT

This book is designed to help you learn some mathematics, regardless of your mathematical background. It is written so that you can understand, appreciate, and even enjoy areas of mathematics to which you may or may not have been exposed. But, in order for this to occur, you must use this book. Someone once said:

I hear and I forget

I see and I remember

I do and I understand

Mathematics is not a spectator subject; it is a participation sport—you must actively use the text. Read it with pencil in hand. Work the illustrative examples. There are more examples in this text than any other of this nature. Their purpose is to help you understand the material and learn by doing. Make use of the wide margins—they are designed for scratch work. A Solutions Manual, containing worked-out solutions to all of the exercises in the book, is available to help you check your homework.

The objectives, chapter summaries, and chapter review exercises are designed to highlight the contents of each chapter, and to help you check your progress. Finally, the Just for Fun problems are just that. They are provided as a change of pace. Some are relevant, some are not.

Hopefully, you will find reading and using this book a worthwhile and enjoyable endeavor. Good luck!

Fundamentals of Mathematics

1 SETS

After studying this chapter, you will be able to do the following:

1. Describe the meaning of the word **set**, and write a given set in two ways
2. Identify **well-defined sets, ill-defined sets, finite sets,** and **infinite sets**
3. Identify **equal sets, equivalent sets,** and **disjoint sets**
4. Find the **subsets** and **proper subsets** of a given set
5. Identify a **universal set** and find the **complement** of any set contained in some universal set
6. Find the **intersection** and **union** of two or more sets
7. Draw **Venn diagrams** to show the relationship between sets
8. Show a **one-to-one correspondence** between any two equivalent sets, and find the **cardinality** of sets
9. Use Venn diagrams to solve survey problems
10. Write the **Cartesian product** of any two given sets.

Symbols frequently used in this chapter

$\{ \}$	braces, used to enclose members of a set
$\in$	"is an element of"
$\notin$	"is not an element of"
$\ldots$	proceed in the indicated pattern
ϕ	the empty set, also denoted by $\{ \}$
$\subset$	"is a proper subset of"
$\subseteq$	"is a subset of"
$\not\subseteq$	"is not a subset of"
U	the universal set
A'	the complement of A
$\cap$	intersection
$\cup$	union
$n(A)$	the cardinal number of set A
$A \times B$	the Cartesian product of sets A and B
(a, b)	the ordered pair a and b
$=$	"is equal to"

Los Angeles Times

1.1 INTRODUCTION

The concept and properties of a *set* are often used in mathematics. Students study some form of set theory at all levels of mathematics, from grade school through college and graduate school. It has been said that the use of sets is the one unifying idea that unites all of the different branches of mathematics.

We do not define *set* as it is an intuitive concept. This is not unusual; you may recall that in geometry we do not attempt to define what we mean by a *point* or a *line*. A set may be thought of as a collection of objects. These objects are called **elements,** or **members,** of the set.

It is not uncommon for us to use the concept of set in everyday experiences. You have probably examined some of the following:

A set of dishes
A set of tires for a car
A set of silverware
A set of encyclopedias
A set of golf clubs

In mathematics you may have discussed some of the following:

The set of counting numbers
The set of integers
A set of points
A set of solutions for an equation
A set of ordered pairs (x, y)

It is important for you to remember that a set is an intuitive concept. But you should also be aware that there are specific properties of sets. We are going to examine some basic properties, rules, and operations that pertain to sets.

1.2 NOTATION AND DESCRIPTION

We do not have to describe a set by a name as we did in the introduction; we can also describe it by listing or naming its elements. Braces, { }, are used to enclose the members of a set when we list them. Note that the only correct fence or enclosure is braces, not parentheses, ( ), or brackets, [ ].

If we are talking about the set of vowels in the English alphabet, we may denote this set of vowels as $\{a, e, i, o, u\}$. The listing of the elements in a different order does not change the set: we could also write the set as $\{i, o, u, e, a\}$, and it would still be the set of vowels in the English alphabet. When we list the elements in a set, as we have just done, it is called the **roster form** of the set.

Sets are usually denoted by capital letters such as A, B, or C. Therefore, we could write

$$A = \{a, e, i, o, u\}$$

to indicate that set A contains the elements a, e, i, o, u.

We shall use $\in$ to indicate that elements are members of a set. The symbol $\in$ is read "is a member of" or "is an element of," and the notation $\notin$ is read "is *not* an element of." Using our previous example $A = \{a, e, i, o, u\}$, we may say that $a \in A$, $e \in A$, $u \in A$, $2 \notin A$, and $z \notin A$.

If a set contains many elements, we often use three dots, . . . , called an **ellipsis,** to indicate that there are elements in the set that have not been written down. The following are some examples of sets where we list some elements and then use an ellipsis to indicate that the pattern is to be continued indefinitely.

$$N = \{1, 2, 3, 4, \ldots\}, \qquad C = \{5, 10, 15, \ldots\}, \qquad D = \{2, 4, 6, 8, \ldots\}$$

The set N is called the set of **counting** numbers (or **natural** numbers). Using the sets N, C, and D, we may say that $10 \in N$, $10 \in C$, $10 \in D$, $99 \in N$, $99 \notin C$, and $99 \notin D$.

We can also use an ellipsis in another manner when listing elements in a set, to indicate that some elements are missing in the listing. Consider the set K consisting of those counting numbers from 1 through 100. We can make use of the ellipsis to list the set K as

$$K = \{1, 2, 3, \ldots, 98, 99, 100\}$$

This notation tells us the first element in the set, some of the succeeding elements, and the last element in the set.

When we use three dots in the roster form of listing elements, we must list some of the elements so that the pattern can be determined. Remember that the ellipsis means that the listing of the elements will continue in the indicated pattern.

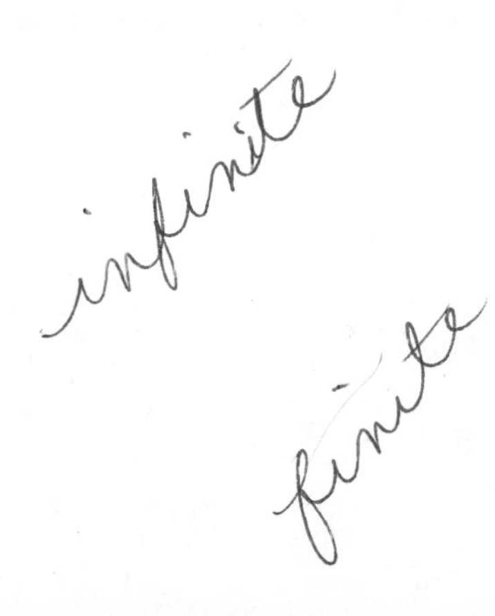

If we use the three dots to indicate that the pattern continues indefinitely, as in $N = \{1, 2, 3, 4, \ldots\}$, then we have what is known as an **infinite set** because the set has an unlimited number of elements. The pattern is unending, and the list of elements goes on and on. If we have a set like $A = \{1, 2, 3, \ldots, 10\}$—where the ellipsis shows an indicated pattern, and we know the last element of the pattern and how many elements are in the set—then we have a **finite** set. An infinite set has an unlimited number of elements; a finite set has a last element, *and* we can count the number of elements in the set.

EXAMPLE 1

Is set $A = \{1, 2, 3, 4, \ldots\}$ an infinite set or a finite set?

Solution

Set $A = \{1, 2, 3, 4, \ldots\}$ is an infinite set. It has an unlimited number of elements.

EXAMPLE 2

Is set $B = \{a, b, c, d, \ldots, x, y, z\}$ an infinite set or a finite set?

Solution

Set $B = \{a, b, c, d, \ldots, x, y, z\}$ is a finite set. From the pattern indicated, we might assume that set B is the set of letters in the alphabet, and that there are 26 elements in the set.

We should be able to determine whether or not any given element is a member of any given set; that is, any set that we consider should be **well defined.** There should not be any ambiguity as to whether an element belongs to a set. The following are some sets that are not well defined. Why?

The set of interesting courses you can take
The set of nice people in your class
The set of good instructors in your school

None of these sets is well defined, because there is no common agreement as to what is meant by "interesting courses," "nice people," or "good instructors."

A set is either well defined or it is not. If a set is not well defined, then it is an **ill-defined** set.

American Stock

Is the set of contented cows a well-defined set?

EXAMPLE 3

Is the set of big people a well-defined set?

Solution

No, it is an ill-defined set. What is meant by big people—height, power, money? The word *big* is ambiguous.

EXAMPLE 4

Given $A = \{2, 4, 6, 8, \ldots\}$, is set A a well-defined set?

Solution

Yes, set A is a well-defined set. Set A is the set of even counting numbers beginning with 2 and proceeding on. From the given description we can ascertain that $100 \in A$. Note that set A is a well-defined infinite set.

It is sometimes cumbersome to write a word description of a set, and it also is sometimes awkward to describe a set by listing all of its elements in roster form. There is another method that we can use to describe a set, called **set-builder** notation.

Consider set $A = \{2, 4, 6, 8, \ldots\}$. We have described set A as the set of even counting numbers. We could say A = the set of x's such that x is an even counting number. We can refine this to

$$A = \{x\text{'s such that } x \text{ is an even counting number}\}$$

In set notation we use a vertical line ($|$) to stand for "such that." Hence, we now have $A = \{x | x \text{ is an even counting number}\}$. This is read as "$A$ is the set of all x such that x is an even counting number." The set-builder notation is commonly used in mathematics when discussing sets.

EXAMPLE 5
What does $A = \{x | x \text{ is a Great Lake}\}$ mean?

Solution
This set is described in set-builder notation. Set A is the set of all x such that x is a Great Lake. The vertical line after the x stands for "such that." The x after the first brace tells us that we are considering all x's, that is, all of the Great Lakes.

EXAMPLE 6
List the elements of $\{x | x \text{ is a vowel in the word } \textit{Westhampton}\}$.

Solution
We have the set of all x such that x is a vowel in the word *Westhampton.* The vowels are a, e, i, o, u, and those that appear in *Westhampton* are e, a, and o.

A set does not have to contain elements that are related. It may be that the only thing that the elements in a set have in common is that they are in the same set. For example, consider the sets $A = \{\triangle, \square, a, 2\}$ and $B = \{\text{red, blue, 1, 1000, XII}\}$.

Sets may contain a definite number of elements or an unlimited number of elements. It is also true that a set may contain *no* elements. If a set does not contain any elements, it still contains a definite number of elements, namely, zero elements. Consider the set of lobsters that live in Lake Ontario. This set has no elements:

there are no lobsters living in Lake Ontario. If we were to list the elements for this set, we would have to put nothing between the braces. We would have { }. This is an **empty set,** also called the **null set.** The empty set is usually denoted by { }, but another common symbol is $\emptyset$.

When you denote the empty set you may use either symbol, but do not use both together.

{ } or $\emptyset$ correct
{$\emptyset$} WRONG!

The notation {$\emptyset$} is incorrect because the set is not empty: the symbol $\emptyset$ is inside the braces. It is false that there is nothing inside the braces; hence the set cannot be empty.

EXAMPLE 7
List the elements in each of the following sets.

a. The set of months containing 33 days

b. The set of months whose name contains the letter q

Solution

a. There are no months that have 33 days, and therefore there are no elements in the set. Hence the set of months containing 33 days is the empty set, or null set, denoted by { } or $\emptyset$.

b. None of the months has a name that contains the letter q, so there are no elements in the set. The solution is the empty set, { } or $\emptyset$.

Two sets that contain exactly the same elements are said to be **equal** or **identical** sets. If we are given $A = \{a, e, i, o, u\}$ and $B = \{i, o, u, a, e\}$, then we can say that $A = B$. These two sets contain exactly the same elements and therefore they are equal.

Two sets that contain exactly the same number of elements are **equivalent** sets. If we are given $A = \{a, e, i, o, u\}$ and $B = \{1, 2, 3, 4, 5\}$, then we can say that A is equivalent to B. Both sets contain five elements and hence they are equivalent; but they are not equal.

EXAMPLE 8
Are the following sets equal?

a. $A = \{d, a, b\}$ $B = \{b, a, d\}$

b. $C = \{1, 2, 3, 4, \ldots\}$ $D = \{5, 10, 15, \ldots\}$

Solution

a. Yes, sets A and B are equal sets because they contain exactly the same elements. The order of the listing of the elements does not change the set.

b. No, sets C and D are not equal sets. They do not contain exactly the same elements. For example, from the set descriptions we know that $6 \in C$, but $6 \notin D$, because set D contains only those counting numbers that are multiples of 5.

EXAMPLE 9

Are the following sets equivalent?

a. $A = \{d, a, b\}$ $\quad B = \{b, a, d\}$

b. $C = \{1, 2, 3, \ldots\}$ $\quad D = \{l, o, v, e\}$

Solution

a. Sets A and B are equivalent because they each contain 3 elements.

b. Set C contains an unlimited number of elements while D contains 4 elements. Hence sets C and D are not equivalent.

EXERCISES FOR SECTION 1.2

1. Are the following statements true or false?
   a. $2 \in \{1, 2, 3, 4\}$
   b. $8 \in \{2, 4, 6, \ldots\}$
   c. $\{m, o, r, e\} = \{r, o, m, e\}$
   d. $\{1, 2, 3\}$ is equivalent to $\{4, 5, 6\}$.
   e. $12 \in \{x \mid x \text{ is a counting number}\}$
   f. $0 \in \{\ \}$

2. Are the following statements true or false?
   a. $4 \notin \{1, 2, 3, 4, 5\}$
   b. $M \in \{a, b, c, d, \ldots, z\}$
   c. $\{\ \} = \emptyset$
   d. $\{1, 2, 3\}$ is both equivalent to and equal to $\{3, 1, 2\}$.
   e. March $\in \{x \mid x \text{ is a month of the year}\}$
   f. $\{1, 2\} = (1, 2)$

3. Are the following statements true or false?
   a. $\{\emptyset\}$ is a finite set.
   b. $\{\emptyset\}$ is an empty set.
   c. $\{0\}$ is an empty set.
   d. The set of students enrolled in this course is a finite set.
   e. The set of small cars manufactured in the United States is a well-defined set.
   f. Equivalent sets are equal sets.

4. List the elements of each set.
   a. The set of Great Lakes
   b. The set of states whose names begin with the word *New*
   c. The set of states whose names begin with the letter *A*
   d. The set of states that border New York State
   e. The set of states whose names begin with the letter *W*

5. List the elements of each set.
   a. The set of days of the week
   b. The set of days with names containing the letter *s*

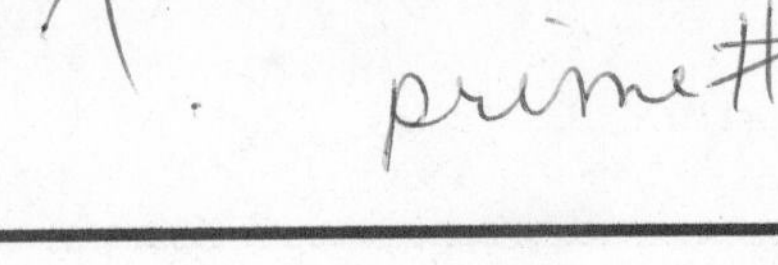

c. The set of days with names containing the letter x
d. The set of months containing 31 days
e. The set of months containing 32 days

6. List the elements of each set.
   a. $\{x \mid x$ is a letter in the English alphabet$\}$
   b. $\{x \mid x$ is a Great Lake$\}$
   c. $\{x \mid x$ is the capital of New York$\}$
   d. $\{x \mid x + 2 = 2\}$
   e. $\{x \mid x + 2 = x\}$
7. List the elements of each set.
   a. $\{x \mid x$ is a counting number less than $10\}$
   b. $\{x \mid x$ is a counting number greater than $5\}$
   c. $\{x \mid x$ is a counting number greater than 5 and less than $10\}$
   d. $\{x \mid x$ is a month whose name contains an $r\}$
   e. $\{x \mid x$ is a month whose name does not contain an $r\}$
8. Write each set in set-builder notation.
   a. {Monday, Tuesday, Wednesday, Thursday, Friday, Saturday, Sunday}
   b. $\{a, e, i, o, u\}$
   c. $\{1, 3, 5, 7, \ldots\}$
   d. $\{2, 3, 5, 7, 11, 13, 17, \ldots\}$
9. Write the following in set-builder notation.
   a. $\{a, b, c, d, \ldots, z\}$
   b. $\{1, 2, 3, \ldots\}$
   c. $\{2, 4, 6, 8, \ldots\}$
   d. $\{5, 10, 15, \ldots\}$
10. Is the empty set a well-defined set? Explain your answer.

Just For Fun

We have considered the set $\{a, e, i, o, u\}$ in our discussion. Can you find a word that contains all of these vowels in the order they are listed? (The vowels may be separated by other letters.)

© 1965 United Feature Syndicate, Inc.

1.3 SUBSETS

Many times two or more sets contain some, but not all, of the same elements. Consider the set of positive even whole numbers, $A = \{2, 4, 6, 8, \ldots\}$, and the set of positive whole numbers, $B = \{1, 2, 3, 4, \ldots\}$. We can see that $4 \in A$ and $4 \in B$; similarly, we note that

$10 \in A$ and $10 \in B$. In fact, every element that is in set A is also contained in set B. Therefore, we can say that set A is **contained in** set B, or, symbolically, we can write

$$A \subseteq B$$

When a set A is contained in another set B, we say that A is a *subset* of B.

> **Given any two sets A and B, if every element in A is also an element in B, then A is a subset of B.**

Since every set is a subset of itself, we have to be careful in our notation. A subset of a given set that is *not* the set itself is called a *proper subset*. If set A is a proper subset of set B, then two conditions must be satisfied: first, A must be a subset of set B; second, set B must contain at least one element that is not found in set A. If A is a proper subset of B, then we say that A is **properly contained in** B, and we write

$$A \subset B$$

> **If A is a subset of B, and there is at least one element in B not contained in A, then A is a proper subset of B.**

If A is a subset of B, but not necessarily a proper subset of B, then we denote this by

$$A \subseteq B$$

The notation $A \not\subset B$ means that A is not a proper subset of B.

Consider the sets $A = \{1, 2, 3\}$ and $B = \{1, 2, 3, 4\}$. We can say that $A \subset B$ since each element in A is also an element in B, and there is at least one element in B not contained in A. But we cannot say that $B \subset A$. Why? Set B is not a subset of A because $4 \in B$, but $4 \notin A$. Hence $B \not\subset A$, but $A \subset B$. In order to show that B was not a subset of A, we found an element in B that was not in A.

Consider the empty set, $\{\ \}$. The empty set has no elements. This means that it is impossible to find an element in the empty set that is not in set A. Since the empty set has no elements there are none that can fail to be elements of A. Hence the empty set is a subset of A. By the same reasoning the empty set is a subset of the set B. In fact, the empty set is a subset of all sets.

EXAMPLE 1

Determine all the possible subsets of the set $\{1, 2\}$.

Solution
From our discussion on subsets, we know that every set is a subset of itself: thus {1, 2} is a subset. We also know that the empty set is a subset of all sets, so we also have { }. Are there any others? Yes; namely, {1} and {2}. It appears that a complete list of subsets of {1, 2} is

{1, 2}, { }, {1}, {2}

EXAMPLE 2
Determine all the possible subsets of $\{a, b, c\}$.

Solution
We know that we can list the set itself, $\{a, b, c\}$, and the empty set, { }. But what about the others? If we are to proceed in a manner that has some order, then we should probably first consider the subsets that would be obtained by taking the elements one at a time, then two at a time, and so on. Note that taking zero elements at a time gives us the empty set, and the empty set is a subset of all sets. Since the set $\{a, b, c\}$ contains three elements, we would have the following:

zero at a time	one at a time	two at a time	three at a time
{ }	$\{a\}$	$\{a, b\}$	$\{a, b, c\}$
	$\{b\}$	$\{a, c\}$	
	$\{c\}$	$\{b, c\}$	

There are eight subsets for the given set. Remember that, since $\{a, b\}$ is the same set as $\{b, a\}$, we do not list both of these, because we would not wish to list the same subset twice.

The sets in the two examples we just considered contained 2 and 3 elements, respectively, and it was not too difficult to determine all of the subsets for each set. In order to determine the number of subsets for any given set with n elements we may use the following rule.

If a set contains n elements, then it has 2^n subsets.

The set {1, 2} contains 2 elements, so it has 2^2, or $2 \times 2 = 4$, possible subsets. The set $\{a, b, c\}$ contains 3 elements, so it has 2^3, or $2 \times 2 \times 2 = 8$, possible subsets. Note that this rule also works for the empty set { }. The empty set contains zero elements, so we have 2^0, which is equal to 1. The subsets of the empty set are the set

Camerique

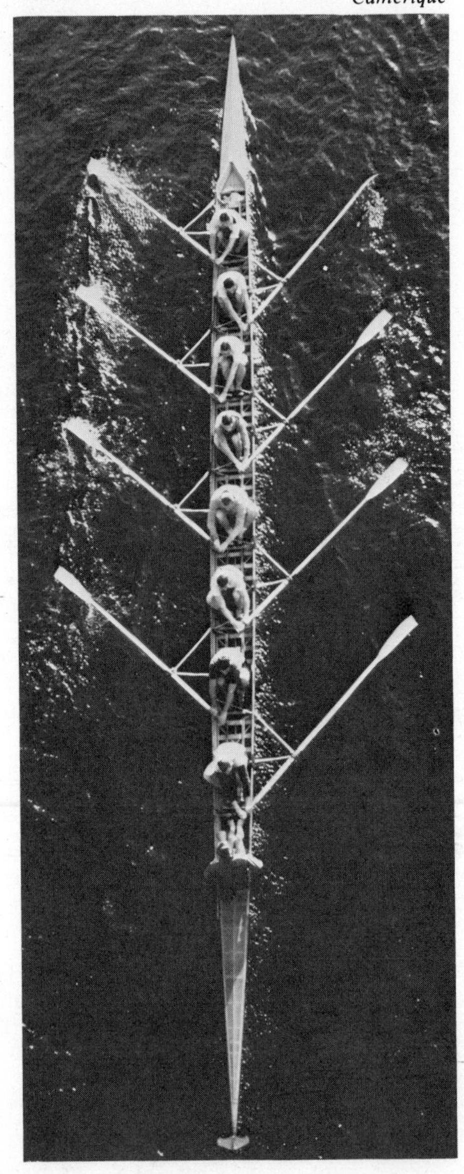

How many subsets does the set of rowers in this boat have?

itself, and the empty set, which is again the set itself; hence we only list it once, and have only 1 subset.

When we discuss sets, we often refer to some general set that contains all of the other sets under consideration. If we are discussing a set of numbers, this set could be generated from many other sets—the set of positive numbers, the set of whole numbers, the set of negative numbers, and so on. To avoid confusion, it is necessary to know what general set the elements are taken from. This general set is called the **universal set,** and it contains all of the elements being considered in the given discussion or problem. The universal set can change from problem to problem, depending upon the nature of the sets being discussed. We usually denote the universal set in any problem by the capital letter U.

For example, the universal set $U = \{0, 1, 2, 3, 4, 5, 6, 7, 8, 9\}$ contains the digits 0 through 9. In a discussion using this universal set, we would only consider those sets whose elements are members of U. For example, $A = \{0, 2, 4\}$ might be discussed, but $C = \{2, b, c\}$ would not be, because not all of the elements of C are elements of U.

The **complement** of a set A is the set of all the elements in the given universal set, U, that are not in the set A. The notation for the complement of A is A'. Some texts use the notation $\bar{A}$ for the complement of A, but we shall use the "prime" notation. In order to find the complement of a set, we must be given a universal set U.

EXAMPLE 3
Given $U = \{a, e, i, o, u\}$ and $A = \{i, o, u\}$, find A'.

Solution
The set A', the complement of A, is the set of elements that are in U, but not in A. These elements are a and e. Hence we have $A' = \{a, e\}$.

EXAMPLE 4
Given $U = \{a, e, i, o, u\}$, $B = \{a, i\}$, and $C = \emptyset$, find—

a. B' b. C'

Solution

a. The complement of B, B', is the set of elements that are in U, but not in B. These elements are e, o, and u. Therefore $B' = \{e, o, u\}$.

b. Set C is the empty set and therefore contains no elements, and its complement is the set of elements in U that are not in C. In this case, the complement of C is all the elements in the universal set. Therefore, $C' = \{a, e, i, o, u\} = U$.

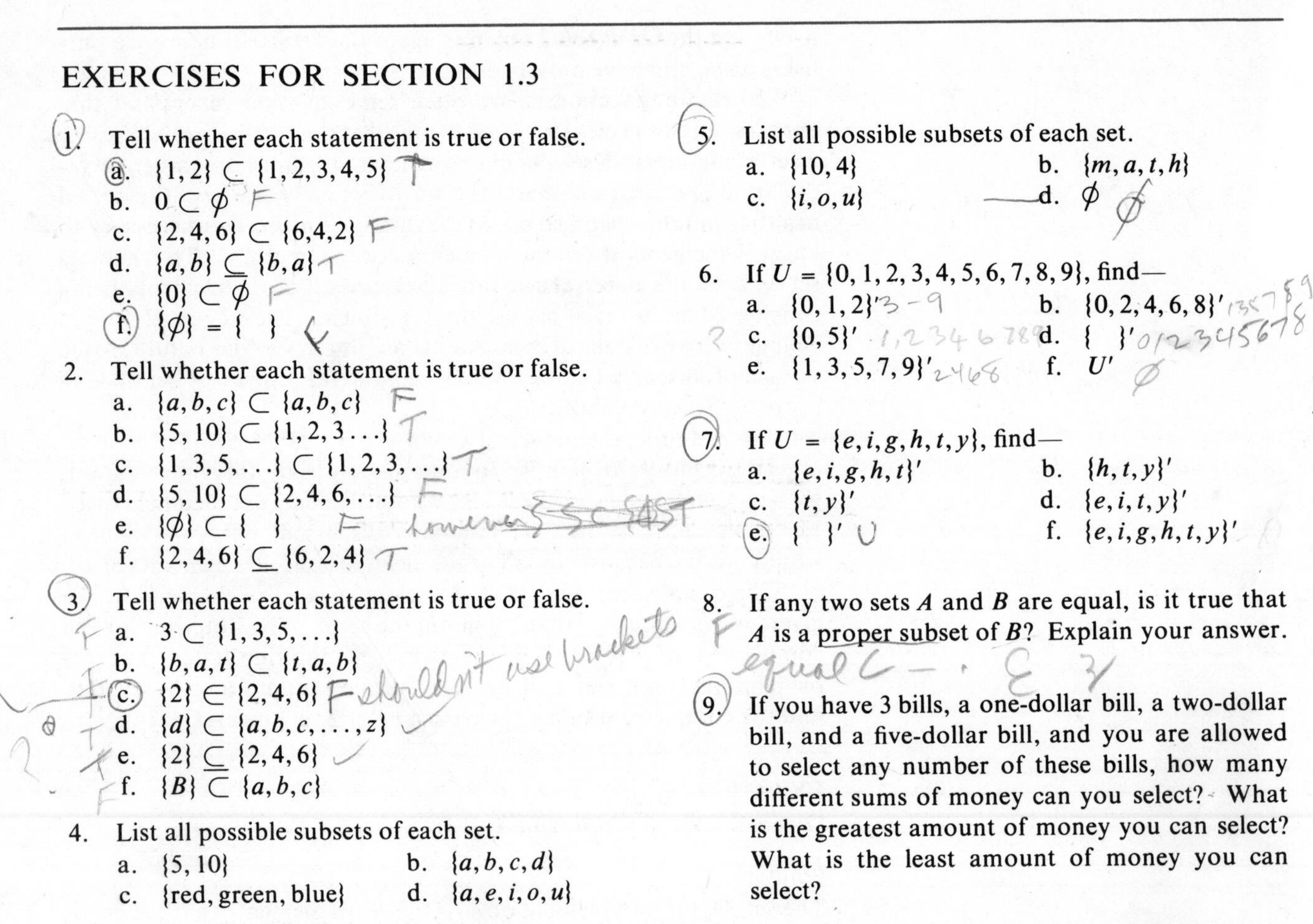

EXERCISES FOR SECTION 1.3

1. Tell whether each statement is true or false.
   a. $\{1, 2\} \subseteq \{1, 2, 3, 4, 5\}$
   b. $0 \subset \phi$
   c. $\{2, 4, 6\} \subset \{6, 4, 2\}$
   d. $\{a, b\} \subseteq \{b, a\}$
   e. $\{0\} \subset \phi$
   f. $\{\phi\} = \{\ \}$

2. Tell whether each statement is true or false.
   a. $\{a, b, c\} \subset \{a, b, c\}$
   b. $\{5, 10\} \subset \{1, 2, 3 \ldots\}$
   c. $\{1, 3, 5, \ldots\} \subset \{1, 2, 3, \ldots\}$
   d. $\{5, 10\} \subset \{2, 4, 6, \ldots\}$
   e. $\{\phi\} \subset \{\ \}$
   f. $\{2, 4, 6\} \subseteq \{6, 2, 4\}$

3. Tell whether each statement is true or false.
   a. $3 \subset \{1, 3, 5, \ldots\}$
   b. $\{b, a, t\} \subset \{t, a, b\}$
   c. $\{2\} \in \{2, 4, 6\}$
   d. $\{d\} \subset \{a, b, c, \ldots, z\}$
   e. $\{2\} \subseteq \{2, 4, 6\}$
   f. $\{B\} \subset \{a, b, c\}$

4. List all possible subsets of each set.
   a. $\{5, 10\}$
   b. $\{a, b, c, d\}$
   c. {red, green, blue}
   d. $\{a, e, i, o, u\}$

5. List all possible subsets of each set.
   a. $\{10, 4\}$
   b. $\{m, a, t, h\}$
   c. $\{i, o, u\}$
   d. ϕ

6. If $U = \{0, 1, 2, 3, 4, 5, 6, 7, 8, 9\}$, find—
   a. $\{0, 1, 2\}'$
   b. $\{0, 2, 4, 6, 8\}'$
   c. $\{0, 5\}'$
   d. $\{\ \}'$
   e. $\{1, 3, 5, 7, 9\}'$
   f. U'

7. If $U = \{e, i, g, h, t, y\}$, find—
   a. $\{e, i, g, h, t\}'$
   b. $\{h, t, y\}'$
   c. $\{t, y\}'$
   d. $\{e, i, t, y\}'$
   e. $\{\ \}'$
   f. $\{e, i, g, h, t, y\}'$

8. If any two sets A and B are equal, is it true that A is a proper subset of B? Explain your answer.

9. If you have 3 bills, a one-dollar bill, a two-dollar bill, and a five-dollar bill, and you are allowed to select any number of these bills, how many different sums of money can you select? What is the greatest amount of money you can select? What is the least amount of money you can select?

Just For Fun

Using four 4s and the various operations, see how many whole numbers you can write, starting with zero. Here are some examples:

$$\frac{4 - 4}{4 + 4} = 0 \qquad \frac{4 + 4}{4 + 4} = 1 \qquad \frac{4 \times 4}{4 + 4} = 2$$

1.4 SET OPERATIONS

In the preceding sections, we discussed sets to some extent and considered various properties of sets and subsets. Now we are ready to examine set operations which will enable us to combine sets.

In arithmetic, we have operations such as addition and subtraction that enable us to combine numbers. In this section, we shall consider the intersection and union of sets, and we shall also do some more work with the complement of a set.

If we have two sets A and B, then the intersection of A and B is a set of elements that are members of both A and B. The notation for A intersection B is $A \cap B$.

In other words, the *intersection* of sets A and B results in another set, denoted by $A \cap B$. This resulting set is composed of elements which are common to both A and B.

EXAMPLE 1

Given $A = \{1, 2, 3, 4, 5\}$ and $B = \{2, 4, 6, 8\}$, find $A \cap B$.

Solution

The elements that are in A and also in B are 2 and 4. Hence $A \cap B = \{2, 4\}$.

EXAMPLE 2

Given $A = \{a, e, i, o, u\}$ and $B = \{i, o, u\}$, find $A \cap B$.

Solution

The elements in both A and B are i, o, and u. Therefore, $A \cap B = \{i, o, u\}$. Note that here we can say that $A \cap B = \{i, o, u\}$, or that $A \cap B = B$.

EXAMPLE 3

Given $A = \{1, 2, 3, \ldots, 10\}$ and $B = \{2, 4, 6, \ldots\}$, find $A \cap B$.

Solution

We see that set A contains the numbers 1 through 10, while set B contains the positive even whole numbers. The elements common to both sets are 2, 4, 6, 8, and 10, and therefore $A \cap B = \{2, 4, 6, 8, 10\}$.

EXAMPLE 4

Given $A = \{2, 4, 6, 8\}$ and $B = \{a, e, i, o, u\}$, find $A \cap B$.

Solution

Examining sets A and B, we see that there are no elements common to both. Therefore, the intersection of these two sets is the empty set. Hence $A \cap B = \{\ \}$, or $A \cap B = \phi$.

Two sets whose intersection is the empty set are said to be **disjoint.** Disjoint sets have no elements in common.

disjoint sets

EXAMPLE 5

Given $U = \{a, e, i, o, u\}$, $A = \{a, e, o\}$, and $B = \{e, i, o\}$, find $(A \cap B)'$.

Solution

We are asked to find $(A \cap B)'$. As in arithmetic and algebra, when we have parentheses, we first do the work inside the parentheses, and then perform the operation(s) outside the parentheses. So we must find $A \cap B$, and then find the complement of our answer. $A \cap B = \{e, o\}$. The complement of $\{e, o\}$ is all the elements that are in U, but not in $A \cap B$. Therefore the complement of $\{e, o\}$ is $\{a, i, u\}$, and

$$(A \cap B)' = \{e, o\}' = \{a, i, u\}$$

EXAMPLE 6

Given $A = \{1, 2, 4, 5, 6, 7\}$, $B = \{1, 3, 5, 7, 9\}$, and $C = \{2, 4, 6, 7, 8\}$, find $(A \cap B) \cap C$.

Solution

First we find $A \cap B$, then find the intersection of that with C. Since $A \cap B = \{1, 5, 7\}$,

$$(A \cap B) \cap C = \{1, 5, 7\} \cap C$$

$$= \{1, 5, 7\} \cap \{2, 4, 6, 7, 8\} = \{7\}.$$

The *union* of sets A and B is the set of all the elements that are members of either set A or set B, or both. When we list the elements in the union of two sets, we list all of the elements in set A and all of the elements in set B, but if an element is in both sets, we list it only once. Therefore, the union of sets A and B is the set of elements that are elements of at least one of the two sets. The notation for A union B is $A \cup B$.

If we have two sets A and B, then the *union* of A and B, $A \cup B$, is the set of elements that are members of A, or members of B, or members of both A and B.

EXAMPLE 7

Given $A = \{1, 2, 3, 4, 5\}$ and $B = \{2, 4, 6, 8\}$, find $A \cup B$.

Solution

The elements of A are 1, 2, 3, 4, 5. The elements of B are 2, 4, 6, 8. The union of A and B is the set of all these elements, because all of these elements belong either to set A or to set B, or to both. Therefore, $A \cup B = \{1, 2, 3, 4, 5, 6, 8\}$.

Note that 2 and 4 are members of both sets, but we list each only once in our solution. Also, it is not necessary to list the elements in order, but it does provide a means of checking that all elements are included.

EXAMPLE 8
Given $A = \{2, 4, 6, 8\}$ and $B = \{a, e, i, o, u\}$, find $A \cup B$.

Solution
The two given sets have no elements in common, but their union is the set containing all of these elements. Hence $A \cup B = \{2, 4, 6, 8, a, e, i, o, u\}$. Note that each element in $A \cup B$ is either a member of A or a member of B.

EXAMPLE 9
Given $A = \{1, 3, 5, \ldots\}$ and $B = \{2, 4, 6, \ldots\}$, find $A \cup B$.

Solution
We have the set of odd counting numbers and the set of even counting numbers. Hence the union of these two sets is the set of counting numbers: $A \cup B = \{1, 2, 3, 4, 5, 6, \ldots\}$.

EXAMPLE 10
Given $U = \{a, e, i, o, u\}$, $A = \{a, e, o\}$, and $B = \{e, i, o\}$, find $A' \cup B'$.

Solution
We want to find the union of two sets, A' and B'. First we must find the complement of each set, and then the union of the two complements. $A' = \{i, u\}$ and $B' = \{a, u\}$. Therefore $A' \cup B' = \{i, u, a\}$.

EXAMPLE 11
Given $U = \{a, e, i, o, u\}$, $A = \{a, e, o\}$, and $B = \{e, i, o\}$, find $(A \cup B)'$.

Solution
This problem is a little different than the previous example. This time we want $(A \cup B)'$, which is the complement of $A \cup B$. Recall that we must first perform the operation inside the parentheses, and then perform the operation outside the parentheses. First we find $A \cup B = \{a, e, i, o\}$; then we find the complement of $\{a, e, i, o\}$, which is $\{u\}$. Hence $(A \cup B)' = \{a, e, i, o\}' = \{u\}$.

Note that examples 10 and 11 show that $A' \cup B' \neq (A \cup B)'$. It is important to remember this, and to exercise care in reading problems and computing solutions. It is also true that $A' \cap B' \neq$

$(A \cap B)'$. As we shall see later, $(A \cap B)' = A' \cup B'$, and $(A \cup B)' = A' \cap B'$.

EXAMPLE 12

If $U = \{a, e, i, o, u\}$, $A = \{a, i, u\}$, $B = \{a, o, u\}$, and $C = \{e, i\}$, find $(A \cap B) \cup C'$.

Solution

This is a more complex problem than we have been considering, but we can do it if we do the operations in an orderly manner. Because of the parentheses, the first thing we should do is find $A \cap B$. $A \cap B = \{a, u\}$. Next we should find C' so that we can form the union with $A \cap B$. $C' = \{a, o, u\}$, and $\{a, u\} \cup \{a, o, u\} = \{a, o, u\}$. Therefore $(A \cap B) \cup C' = \{a, u\} \cup \{a, o, u\} = \{a, o, u\}$.

There are many different combinations of operations in set theory. However, if you do each operation as it is indicated, you will be able to do any problem involving set operations.

It should be noted that the words *and, or,* and *not* correspond to operations in set theory. The intersection of two sets A and B is the set of elements in set A *and* set B. The union of two sets A and B is the set of elements in set A *or* set B *or* both. The complement of set A is the set of elements in the universal set U, but *not* in set A.

EXERCISES FOR SECTION 1.4

1. For each pair of sets, find $A \cap B$ and $A \cup B$.
   a. $A = \{2, 4, 6, 8\}$, $B = \{1, 3, 4, 6, 7\}$
   b. $A = \{a, b, c\}$, $B = \{d, e, f\}$
   c. $A = \{a, e, i, o, u\}$, $B = \{a, e, i\}$
   d. $A = \{g, i, a, n, t, s\}$, $B = \{j, e, t, s\}$
   e. $A = \{5, 10, 15, \ldots\}$, $B = \{10, 20, 30, \ldots\}$
   f. $A = \{1, 3, 5, 7, \ldots\}$, $B = \{2, 4, 6, 8, \ldots\}$

2. Find $A \cap B$ and $A \cup B$ for each pair of sets.
   a. $A = \{\text{dog, cat, pig}\}$, $B = \{\text{fox, dog, cow}\}$
   b. $A = \{1, 2, 3, \ldots\}$, $B = \{2, 4, 6, 8, \ldots\}$
   c. $A = \{\$, ?, !\}$, $B = \{\$, ¢, ?\}$
   d. $A = \{a, b, c\}$, $B = \{x, y, z\}$
   e. $A = \{1, 3, 5\}$, $B = \emptyset$
   f. $A = \{m, a, t, h\}$, $B = \{e, a, s, y\}$

3. Given the sets $U = \{1, 2, 3, 4, 5, 6, 7\}$, $A = \{2, 4, 6, 7\}$, and $B = \{1, 3, 5, 6, 7\}$, find—
   a. A'
   b. B'
   c. $A' \cap B'$
   d. $A' \cup B'$
   e. $(A \cap B)'$
   f. $(A \cup B)'$

4. Given the sets $U = \{a, e, i, o, u\}$, $A = \{a, e, u\}$, and $B = \{i, o, u\}$, find—
   a. A'
   b. B'
   c. $A' \cap B'$
   d. $A' \cup B'$
   e. $(A \cap B)'$
   f. $(A \cup B)'$

5. If $U = \{0, 1, 2, 3, 4, 5, 6, 7, 8, 9\}$, $A = \{0, 2, 3, 4, 5\}$, $B = \{4, 5, 6, 7, 8\}$, and $C = \{4, 6, 7, 8, 9\}$, find—
   a. $A' \cap B'$
   b. $B' \cup C'$
   c. $(A \cap B) \cap C$
   d. $(A \cap B)' \cap C$
   e. $(A' \cup B') \cup C$
   f. $A \cup (B \cap C)$
   g. $(A' \cap B')'$
   h. $A \cup B$

6. If $U = \{0, 1, 2, 3, 4, 5, 6, 7, 8, 9\}$, $A = \{1, 3, 4, 5, 7\}$, $B = \{2, 3, 4, 5, 6\}$, and $C = \{0, 2, 4, 6, 8, 9\}$, find—
   a. $(A \cap B)'$ b. $A' \cap B'$
   c. $(A \cup B)'$ d. $A' \cup B'$
   e. $(A \cap B) \cup C$ f. $(A \cup B) \cap C$
   g. $A' \cap (B \cup C)$ h. $(A \cup B)' \cap C'$

7. If $U = \{a, e, i, o, u\}$, $A = \{i, o, u\}$, $B = \{e, i, o\}$, and $C = \{a, i, o\}$, is each of the following statements true or false?
   a. $A \cap B = C$ b. $A \cup B = U$
   c. $B \cup C = U$ d. $A' \cap B' = \phi$
   e. $(A \cap B) \subset C$ f. $(B \cap C) \subseteq A$

Just For Fun

A snake is stuck at the bottom of a 30-foot well. It can climb up 3 feet every hour, but at the end of each hour it stops to rest and slips back 2 feet. At this rate, how long will it take for the snake to get out of the well?

1.5 PICTURES OF SETS (Venn Diagrams)

It is sometimes useful to represent sets and relationships between sets by means of a picture or diagram. This is almost always done by using a rectangle to represent the universal set and a circle or circles inside the rectangle to represent the set or sets being considered in the discussion. It is understood that the elements in the set are inside the circle that represents the set.

EXAMPLE 1

Let $U = \{a, e, i, o, u\}$ and $A = \{a, e, i\}$. Make a picture to show the relationship between A and U.

Solution

The diagram that represents the given information is figure 1.

FIGURE 1 Note that the elements of the universal set that are in A are inside the circle, while those not in A are inside the rectangle, but outside the circle.

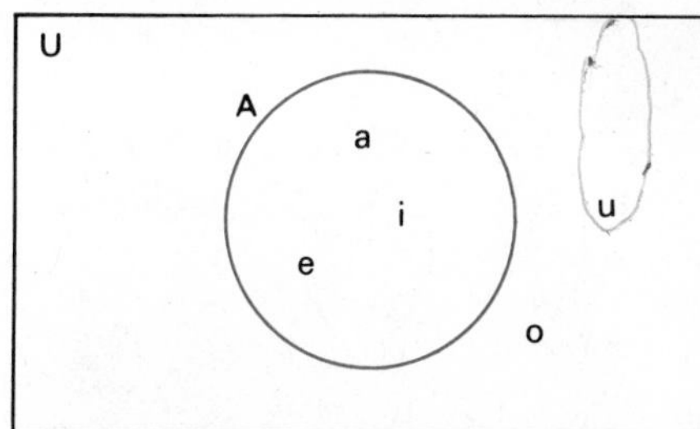

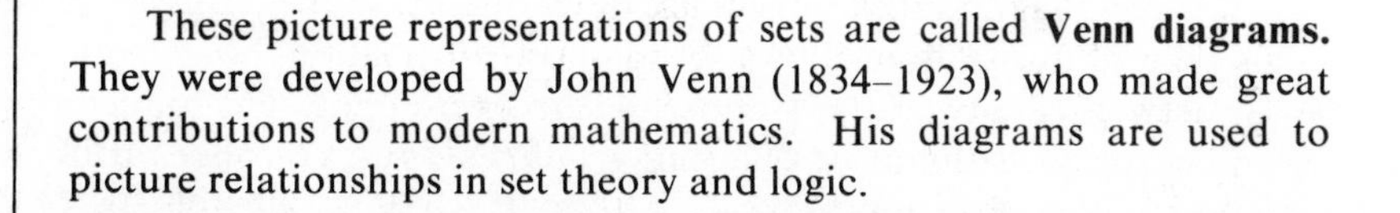

These picture representations of sets are called **Venn diagrams.** They were developed by John Venn (1834–1923), who made great contributions to modern mathematics. His diagrams are used to picture relationships in set theory and logic.

EXAMPLE 2

Use a Venn diagram to show the relationship $B \subset A$.

Solution
First, the sets under consideration are understood to be in a universal set U. More importantly, $B \subset A$. This means that all of the elements contained in B are also in A; hence our solution is the diagram in figure 2.

FIGURE 2 $B \subset A$ (*Note:* This is more correctly called an *Euler diagram,* as Leonhard Euler (1707–1783) first used circles to represent sets. Venn later added refinements to produce what we shall use here.)

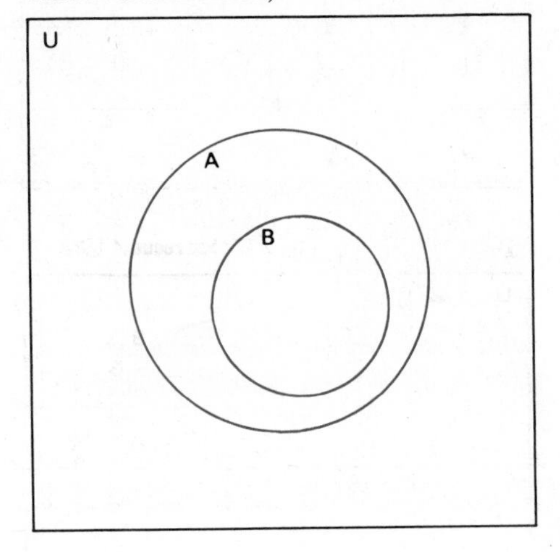

If we are considering two sets A and B, there are two other distinct pictures we could use to show a relationship between A and B. Suppose that $U = \{a, b, c, d, e\}$, $A = \{a, b, c\}$, and $B = \{d, e\}$. We see that A and B are disjoint sets. We illustrate this in figure 3.

But suppose that $A = \{a, b, c\}$ and $B = \{c, d, e\}$; now B and A have an element, c, in common. This is shown in figure 4.

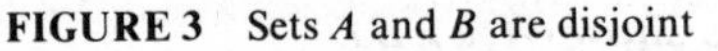
FIGURE 3 Sets A and B are disjoint

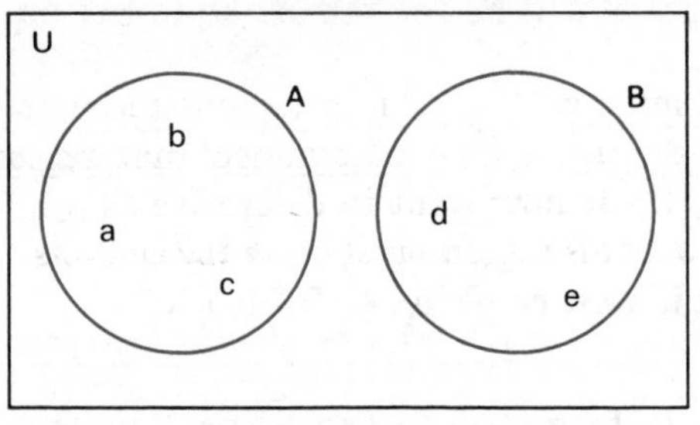

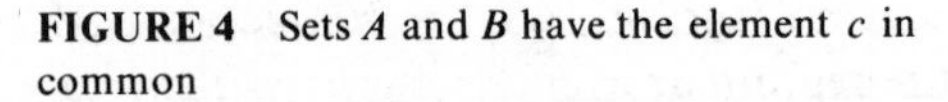
FIGURE 4 Sets A and B have the element c in common

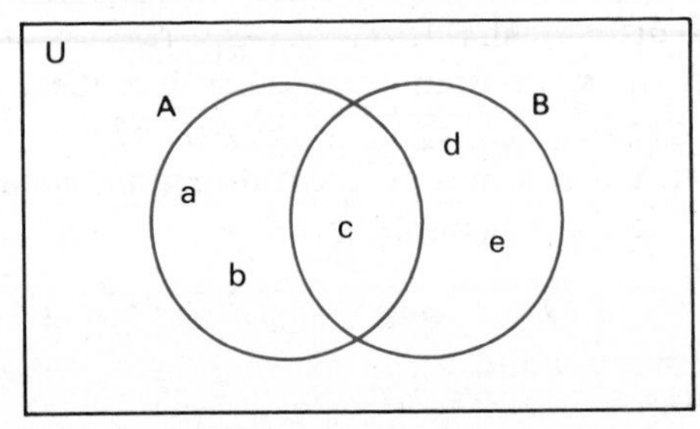

FIGURE 5

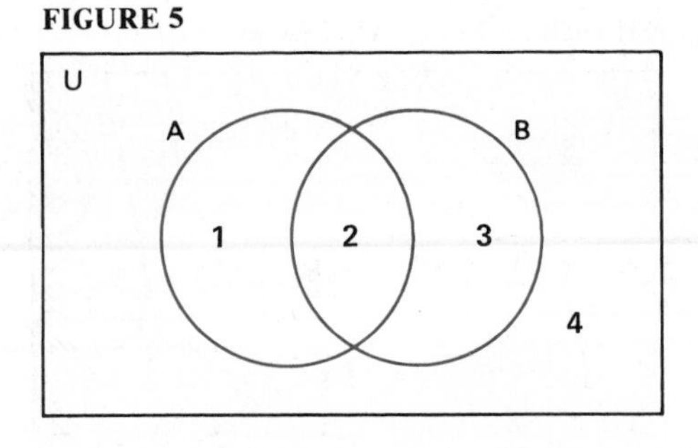

Note that the two sets overlap, which allows us to illustrate that c is an element common to both set A and set B.

A Venn diagram that we will use quite often is shown in figure 5. This type of illustration allows us to show all relations that might exist between the two sets A and B. The overlapping technique is used because it is the most efficient; it allows us to consider all of the possibilities that might exist. Although we have shown sets A and B overlapping, it does not mean that the two sets intersect.

We have assigned numbers to each region in the diagram in figure 5. This enables us to easily discuss the diagram and its various

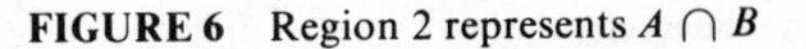
FIGURE 6 Region 2 represents $A \cap B$

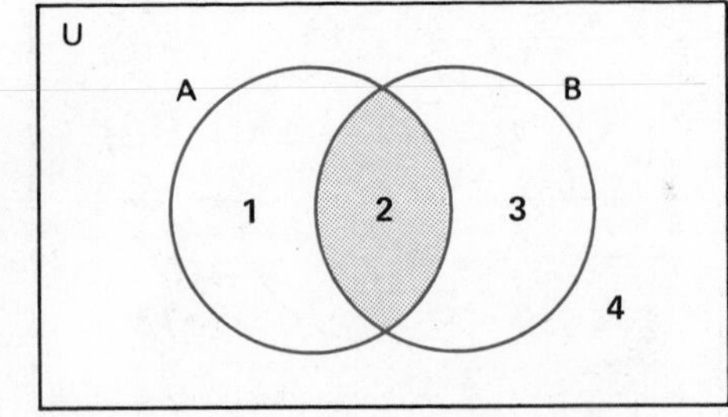

FIGURE 7 Region 2 represents $A \cap B$

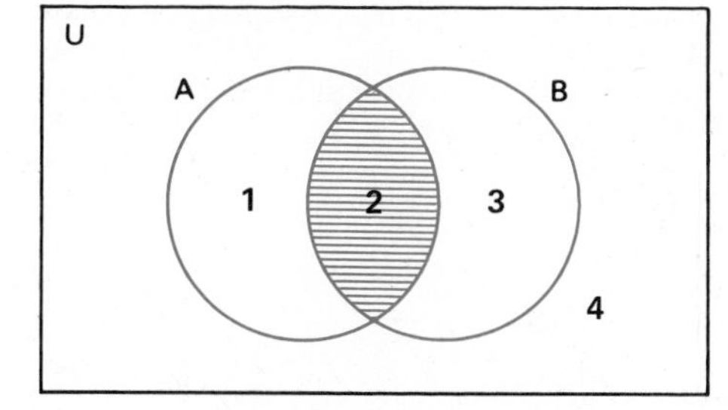

FIGURE 8 Regions 1, 2, 3 represent $A \cup B$

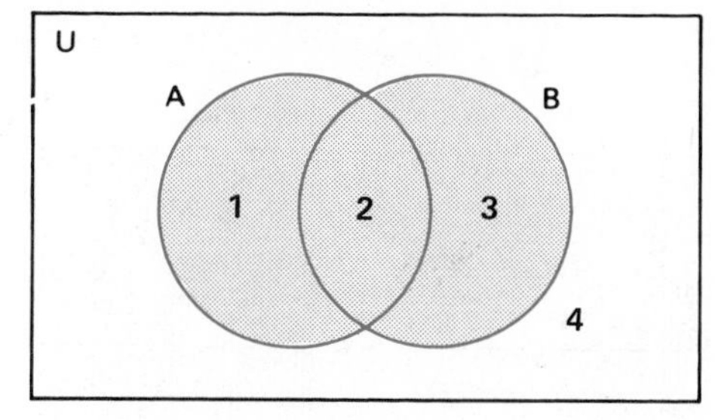

FIGURE 9 Region 4 represents $(A \cup B)'$

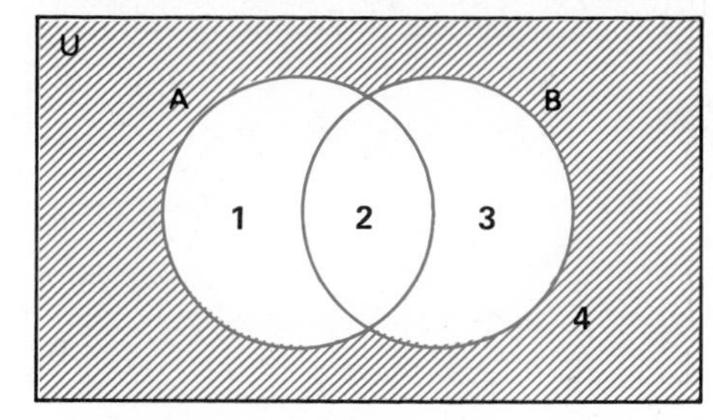

FIGURE 10 Region 2 represents $A \cap B$

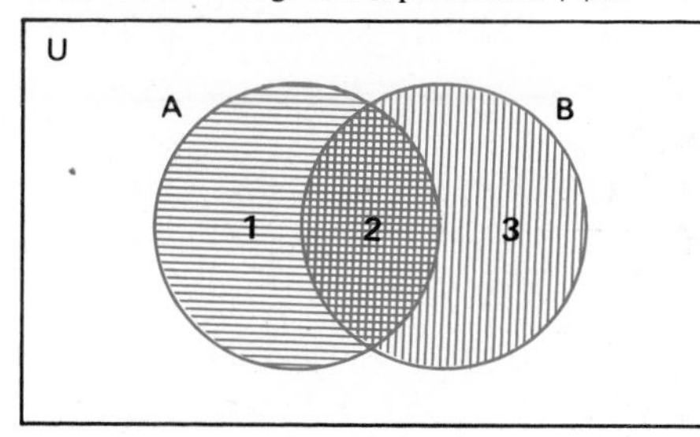

parts. Region 2, for example, represents the intersection of A and B, $A \cap B$. In using the overlapping diagram, we are not concerned whether sets A and B are disjoint or not; we just want to identify the region that represents $A \cap B$. The two most common methods of doing this are shading the region and using stripes (a type of shading) to distinguish the region. See figures 6 and 7.

We know that region 2 represents $A \cap B$, but what region or regions represent $A \cup B$? Since $A \cup B$ is the set of elements that are elements of A or B or both, we must have all the elements in both sets. Therefore the Venn diagram showing $A \cup B$ would have regions 1, 2, and 3 shaded, as in figure 8.

EXAMPLE 3

Use a Venn diagram to show $(A \cup B)$.'

Solution

We must determine what region or regions represent $(A \cup B)'$. In the preceding discussion, we determined that regions 1, 2, and 3 represent $A \cup B$; we now want to determine $(A \cup B)'$, the complement of $A \cup B$. This region must be in the universal set, but not in $A \cup B$; hence it must be region 4. See figure 9.

Note that in figure 9 we used stripes to identify a particular region in our Venn diagram. We are not concerned about what elements are in the region, or even whether there are any elements in it; we just want to distinguish it from the other regions in some way. If we have to identify more than one region in a diagram, we should use a different type of shading for each part of the problem. Each set of stripes should run in a different direction (such as horizontal or vertical), as shown in example 4.

EXAMPLE 4

Use a Venn diagram to show $A \cap B$.

Solution

In figure 10, regions 1 and 2 are in A; we shade these with horizontal stripes. Set B is made up of regions 2 and 3; we shade set B with vertical stripes. Since we are looking for the intersection of A and B, we want the region with stripes that are common to both A and B. Note that this occurs in region 2, which is shaded with both horizontal and vertical stripes. Therefore, region 2 represents $A \cap B$.

In example 4, the regions of A and B were shaded separately, and the intersection of the two sets was the shading common to both sets. For the union of A and B, $A \cup B$, we would have done the problem in the same manner, but we would have taken all of the shaded areas for our answer.

FIGURE 11 Regions 1, 3, 4 represent $(A \cap B)'$

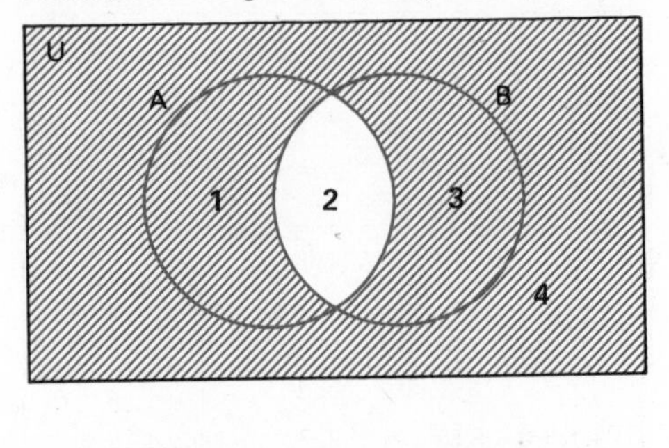

EXAMPLE 5

Use a Venn diagram to show that $(A \cap B)' = A' \cup B'$.

Solution

In order to illustrate that the statement is true, we must use two Venn diagrams. We will let one diagram represent the left side of the equation and the second diagram represent the right side, and then show that the final results for each diagram have the same regions shaded.

The set $(A \cap B)'$ is the complement of $A \cap B$. The region that would satisfy the stated problem must be in U, but not in $A \cap B$. The region satisfying this consists of regions 1, 3, and 4 in figure 11.

The set A' is represented by regions 3 and 4 in figure 12; we shade this with a vertical set of lines. The set B' is represented by regions 1 and 4, and we shade this with a horizontal set of lines.

The set $A' \cup B'$ is determined by shading the region for A' vertically, shading the region for B' horizontally, and then taking all of the shaded areas. See figure 12.

Now note that the shaded region for $(A \cap B)'$ in figure 11 is the same as the shaded region for $A' \cup B'$ in figure 12. Hence we have shown that $(A \cap B)' = A' \cup B'$.

FIGURE 12 Regions 1, 3, 4 represent $A' \cup B'$

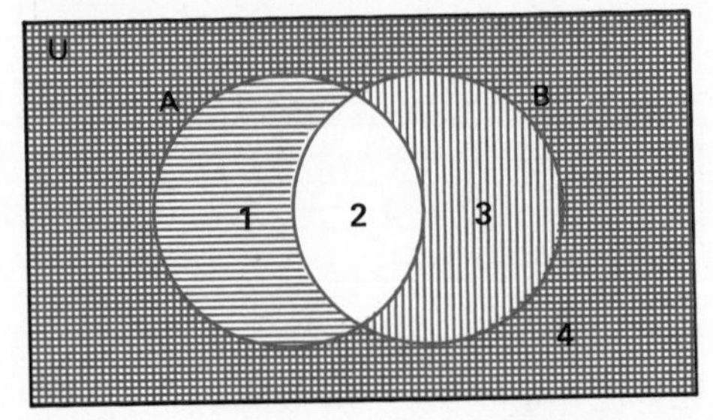

In section 1.4, we discussed relationships between two sets and three sets. Venn diagrams may also be used when working with three sets. Figure 13 is a typical Venn diagram involving three sets A, B, and C. We show the three sets overlapping as this allows us to see what relationships might exist between A and B, B and C, A and C, and so on. Since there are many possible combinations for such a problem, we have to use care in shading the proper region. For example, consider the following sets: $A' \cap (B \cup C)$ and $(A \cap B) \cup C'$. Before we attempt to illustrate such sets, let us agree that we shall number the separate regions as shown in figure 13. Note that we used Roman numerals in numbering the regions. We could have used Arabic numerals $(1, 2, 3, \ldots)$, but, as you will see later, the use of Roman numerals enables us to avoid some confusion in discussing Venn diagrams. We shall use Roman numerals from now on.

Examine the Venn diagram shown in figure 13: What regions represent $A \cup B$? We see that the regions comprising $A \cup B$ are I, II, III, IV, V, and VI. If we are asked to shade $A \cup B$, then these are the regions we should shade.

FIGURE 13

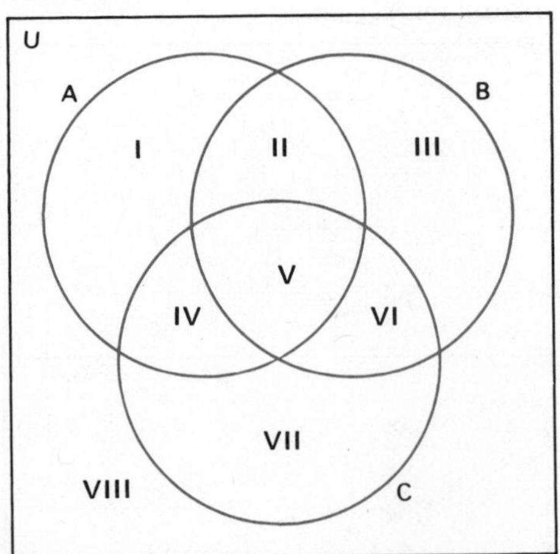

FIGURE 14 Regions II, IV, V represent $A \cap (B \cup C)$

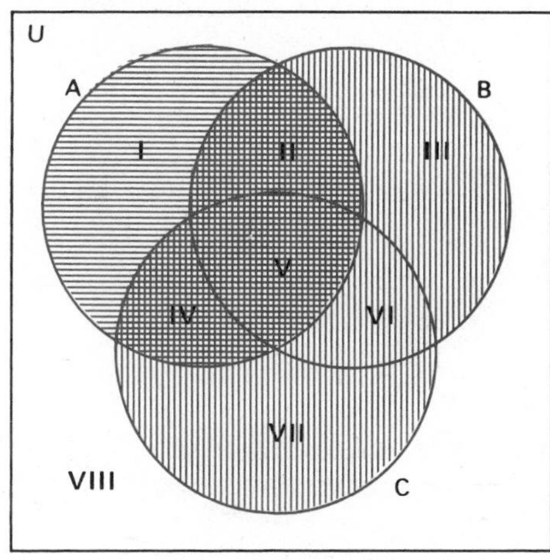

Regions II and V represent $A \cap B$; regions IV and V represent $A \cap C$; and regions V and VI represent $B \cap C$. Now, what region represents $(A \cap B) \cap C$? That is, what region is common to all three sets? Examining figure 13, we see that it is region V.

What does region VIII represent? It is the region inside U, but outside of all three sets. Hence, it must be the complement of the union of all three sets. Therefore region VIII is the complement of $A \cup B \cup C$; we write this as $(A \cup B \cup C)'$.

Now that we are somewhat familiar with the different regions, let us consider some examples using shading.

EXAMPLE 6

Use a Venn diagram to show $A \cap (B \cup C)$.

Solution

We shall first shade $(B \cup C)$ with vertical lines. In figure 14, $B \cup C$ is regions II, III, V, VI, IV, and VII. We then shade A with horizontal lines. This problem asks for $A \cap (B \cup C)$, the intersection of A with the union of B and C. Therefore our answer appears in the regions with double lines, namely, regions II, IV, and V.

EXAMPLE 7

Use a Venn diagram to show $A \cup (B \cap C)$.

Solution

We shall shade $(B \cap C)$ with horizontal lines. In figure 15, the regions shaded for $B \cap C$ are V and VI. Set A is shaded with vertical lines. Since we want the union of A with $(B \cap C)$, our answer is regions I, II, IV, V, and VI.

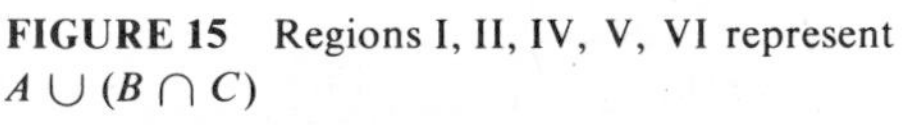

FIGURE 15 Regions I, II, IV, V, VI represent $A \cup (B \cap C)$

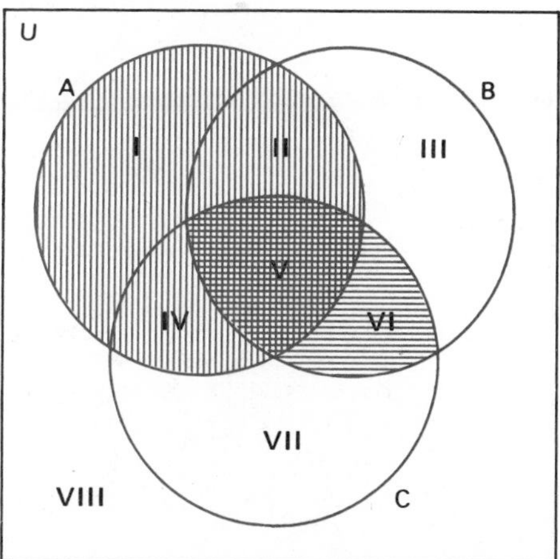

EXAMPLE 8

Show that $A \cup (B \cap C) = (A \cup B) \cap (A \cup C)$.

Solution

We shall show that $A \cup (B \cap C) = (A \cup B) \cap (A \cup C)$ by comparing the corresponding Venn diagrams. Recall that if they both contain the same regions for a final answer, then the sets that the regions represent are equal. We have already determined the regions for $A \cup (B \cap C)$ (see figure 15); they are I, II, IV, V, and VI.

Now we must represent $(A \cup B) \cap (A \cup C)$ by means of a Venn diagram. We first shade $(A \cup B)$ with horizontal lines and then shade $(A \cup C)$ with vertical lines, as shown in figure 16. We want the intersection of these two, so our answer is regions I, II, IV, V, and VI. These are precisely the same regions that we got for $A \cup (B \cap C)$. Therefore we have shown that $A \cup (B \cap C) = (A \cup B) \cap (A \cup C)$.

The reason we used Roman numerals to designate the regions of a set is that many times we may want to use Arabic numerals to tell how many members are in a set. This is called the **cardinality** of a set. A **cardinal number** tells us "how many," as opposed to an **ordinal number,** which tells us "what position." (Examples of ordinal numbers are first, second, third, fourth, fifth, and so on.)

A cardinal number gives the number of elements in a set. The empty set contains no elements, so its cardinal number is zero. We may say that the *cardinality* of the empty set is zero. The set $A = \{a, e, i, o, u\}$ has 5 elements; hence, its cardinality is 5. But suppose set A is in a Venn diagram—how could we tell how many members it has? One solution is to list the elements in the circle. Another solution is to write the Arabic numeral 5 to indicate that set A contains 5 elements. See figures 17 and 18.

While you may feel that neither technique is better than the other, suppose set A contained all the letters of the alphabet. Would you list all 26 letters separately in the circle? Carrying this idea a bit further, suppose the set contains all the students in your school. If we are concerned only with the cardinality of this set, then it is much more efficient to write the Arabic numeral for that number than to list all of the elements.

FIGURE 16 Regions I, II, IV, V, VI represent $(A \cup B) \cap (A \cup C)$

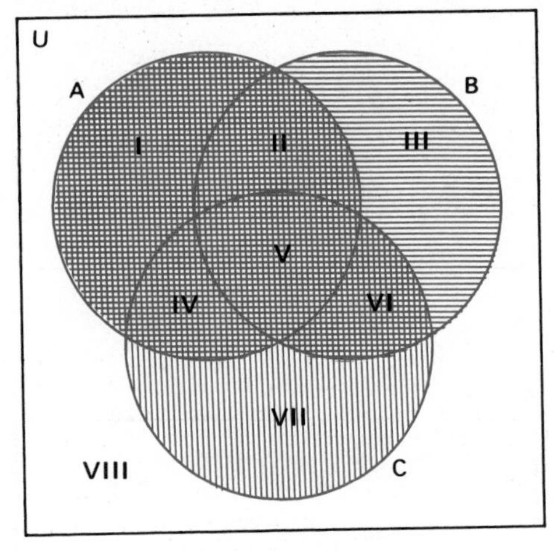

EXAMPLE 9

Use figure 19 to find the cardinality of set A.

Solution

In figure 19, set A is composed of regions I and II. The number of elements in region I is 14, and the number of elements in region II is 7. The total number of elements in set A is 21. Therefore, the cardinality of A is 21.

We shall use the notation n(A) to stand for the cardinality of A.

FIGURE 17

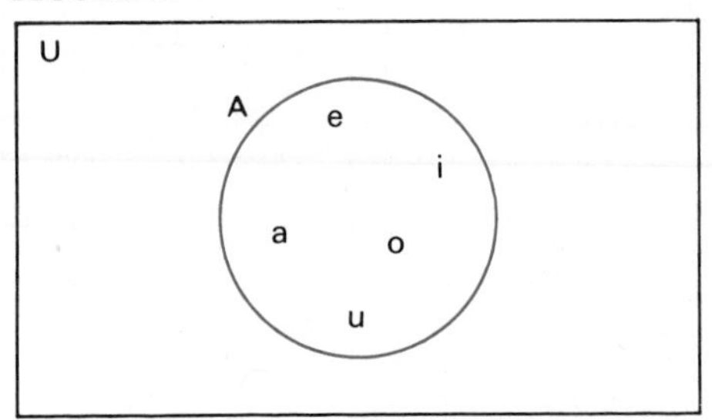

EXAMPLE 10

Use figure 20 to find the cardinalities of $A \cap B$, $B \cup C$, and the universal set.

Solution

In figure 20, we see that $A \cap B$ is composed of regions II and V. The total number of elements in these two regions is 6. Therefore the cardinality of $A \cap B$ is 6, and we can say $n(A \cap B) = 6$.

For $B \cup C$, we use regions II, III, V, VI, IV, and VII. The total number of elements in all of these regions is 39. Recall that in doing set union problems, we only list an element once even if it appears in two sets. Similarly, we need only count the elements once

FIGURE 18

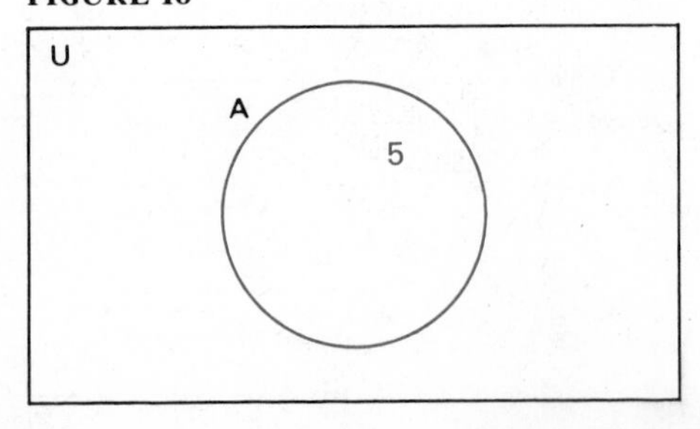

FIGURE 19

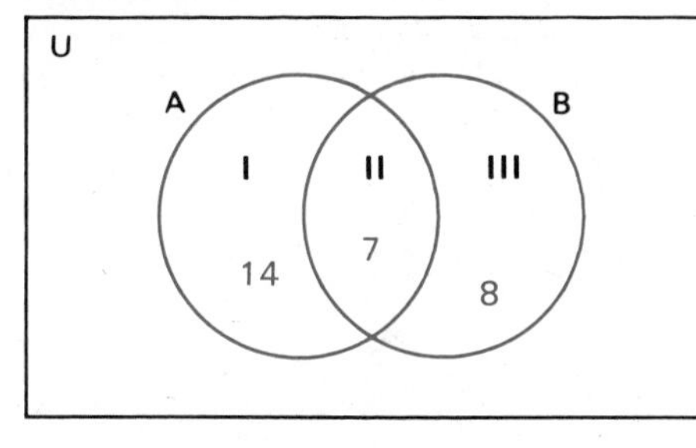

FIGURE 20

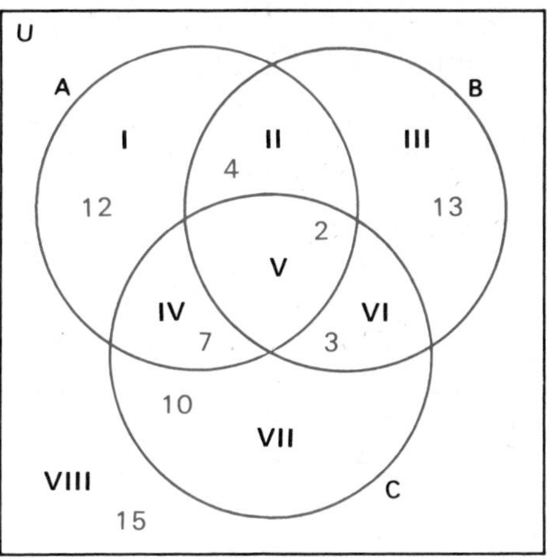

in a region even if the region is in both sets. Hence, the cardinality of $B \cup C = 39$, or $n(B \cup C) = 39$.

To find the cardinality of the universal set, we must find the total number of elements in all eight regions. Adding these numbers, we find that the cardinality of U is 66, or $n(U) = 66$.

Recall that two sets that contain the same number of elements are said to be equivalent sets. If two sets are equivalent, then they have the same cardinality, and the elements of one set can be paired with the elements of the other set. Consider the following two sets A and B.

$$A = \{a, b, c\}$$
$$B = \{\text{bat}, \text{cat}, \text{rat}\}$$

The two sets have the same cardinality, so we can match the elements of the two sets with each other. We could pair a with *bat*, b with *cat*, and c with *rat*. We can show this by

$$\begin{array}{ccc} \{a, & b, & c\,\} \\ \updownarrow & \updownarrow & \updownarrow \\ \{\text{bat}, & \text{cat}, & \text{rat}\} \end{array}$$

This is not the only way that we could have paired the elements of the two sets; there are others. For example: a with *rat*, b with *bat*, and c with *cat*. A pairing of the elements of two sets is called a **one-to-one correspondence** between the sets. When two sets are in a one-to-one correspondence, it means that each element of the first set is paired with just one element of the second set, and each element of the second set is paired with just one element of the first set. Note that two sets that are in one-to-one correspondence must be equivalent sets and have the same cardinality.

EXAMPLE 11

In how many ways can a one-to-one correspondence be established between $A = \{a, b, c\}$ and $B = \{\text{bat}, \text{cat}, \text{rat}\}$?

Solution

We have already discussed two different one-to-one correspondences for these two sets, but we shall start from the beginning. The letter a can be paired with any of the three words *bat*, *cat*, or *rat* in set B. Now, if a is paired, then b can be paired with any of the *two* remaining words. With a and b paired, c has to be paired with the remaining word. Therefore, the total number of one-to-one correspondences is

$$3 \times 2 \times 1 = 6$$

Here are the six different one-to-one correspondences.

{a, b, c} ↕ ↕ ↕ {bat, cat, rat}	{a, b, c} ↕ ↕ ↕ {cat, bat, rat}	{a, b, c} ↕ ↕ ↕ {rat, bat, cat}
{a, b, c} ↕ ↕ ↕ {bat, rat, cat}	{a, b, c} ↕ ↕ ↕ {cat, rat, bat}	{a, b, c} ↕ ↕ ↕ {rat, cat, bat}

EXAMPLE 12

Can the two sets $A = \{a, e, i, o, u\}$ and $B = \{2, 4, 6, 8\}$ be placed in a one-to-one correspondence?

Solution

No, the two sets do not have the same cardinality, because the cardinality of set A is 5, while the cardinality of set B is 4. If we tried pairing the elements of the two sets, we would have an element left over, and it would have to be paired with an element that had already been matched.

{a, e, i, o, u}
↕ ↕ ↕ ↕↙
{2, 4, 6, 8} *not* one-to-one

If two sets have the same cardinality, then a one-to-one correspondence may be established between the two sets; and if a one-to-one correspondence can be established between two sets, then the two sets have the same cardinality.

EXERCISES FOR SECTION 1.5

1. Illustrate each of the following with a Venn diagram. Number the regions in your diagram as shown in examples 3–5 and list the regions that make up your answer.
   a. $A \cap B$
   b. $A' \cap B'$
   c. $A \cup B$
   d. $A' \cup B'$
   e. $(A' \cap B')'$
   f. $(A' \cup B')'$

2. Illustrate each of the following with a Venn diagram. Number the regions in your diagram as shown in examples 3–5 and list the regions that make up your answer.
   a. $A \cap B'$
   b. $A' \cup B$
   c. $A' \cap B'$
   d. $(A \cup B)'$
   e. $(A \cap B)'$
   f. $A' \cup B'$

3. Use a Venn diagram to illustrate each of the following. Number the regions in your diagram as shown in examples 6–10 and list the regions that make up your answer.
   a. $A \cap (B \cap C)$
   b. $A \cup (B \cup C)$
   c. $(A \cap B) \cup C$
   d. $(A \cap B) \cup (A \cap C)$
   e. $(A \cup B) \cap (A \cup C)$
   f. $(A' \cap B') \cup C$

4. Use a Venn diagram to illustrate each of the following. Number the regions in your diagram as shown in examples 6–10 and list the regions that make up your answer.
   a. $(A \cap B) \cap C$
   b. $(A \cup B) \cup C$
   c. $(A \cup B) \cap C$
   d. $(A \cap B) \cap (B \cap C)$
   e. $(B \cup C) \cap (A \cup C)$
   f. $(C' \cup A') \cap B$

5. Use Venn diagrams to show that each of the following statements is true.
   a. $A \cap B = (A' \cup B')'$
   b. $A \cup B = (A' \cap B')'$
   c. $(A \cup B)' = A' \cap B'$
   d. $(A \cap B)' = A' \cup B'$
   e. $A \cap (B \cup C) = (A \cap B) \cup (A \cap C)$
   f. $A \cup (B \cap C) = (A \cup B) \cap (A \cup C)$

6. Use figure 21 to find the following cardinalities.
   a. $n(A)$ b. $n(B)$
   c. $n(A \cap B)$ d. $n(A \cup B)$
   e. $n(A')$ f. $n(B')$
   g. $n(A' \cap B')$ h. $n(A' \cup B')$

FIGURE 21

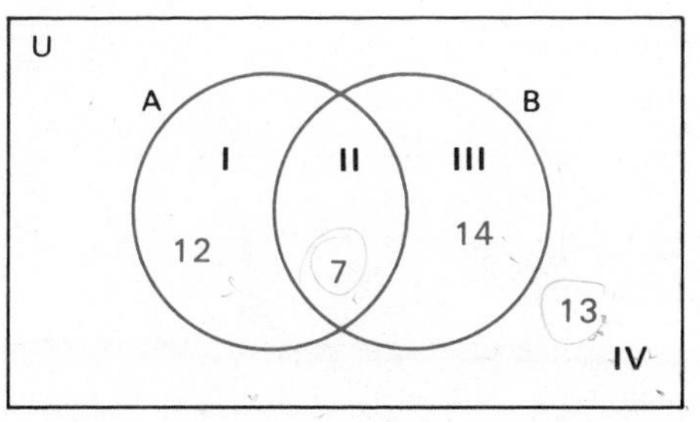

7. Use figure 22 to find the following cardinalities.
   a. $n(A)$ b. $n(B)$
   c. $n(C)$ d. $n(U)$
   e. $n(B \cup C)$ f. $n(B \cap C)$
   g. $n(A \cap B \cap C)$ h. $n(A \cup B \cup C)$

8. Show a one-to-one correspondence between $A = \{5, 10\}$ and $B = \{15, 20\}$. In how many ways can you do this?

FIGURE 22

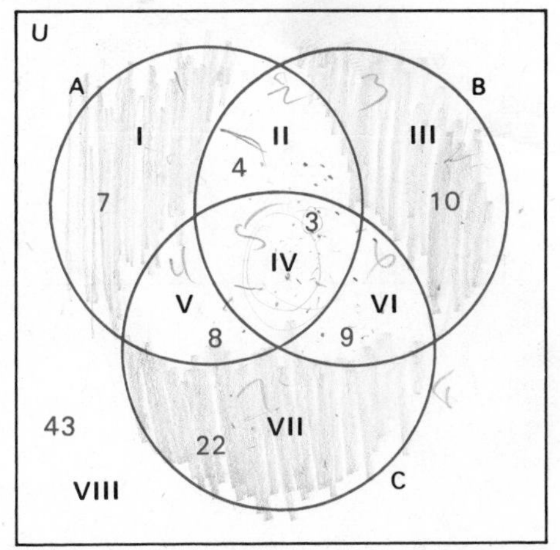

9. In how many ways can one show a one-to-one correspondence between $A = \{i, o, u\}$ and $B = \{x, y, z\}$? Show one of these correspondences.

10. Show a one-to-one correspondence between $A = \{\text{Bob, Joe, Cy, Ted}\}$ and $B = \{5, 10, 15, 20\}$. In how many ways can you do this?

11. In how many ways can a one-to-one correspondence be established between $A = \{m, a, t, h\}$ and $B = \{f, u, n\}$? Explain your answer.

12. For each of the following diagrams, use set notation to describe the situation shown. For example, figure 23 shows $A \cap B$.

FIGURE 23

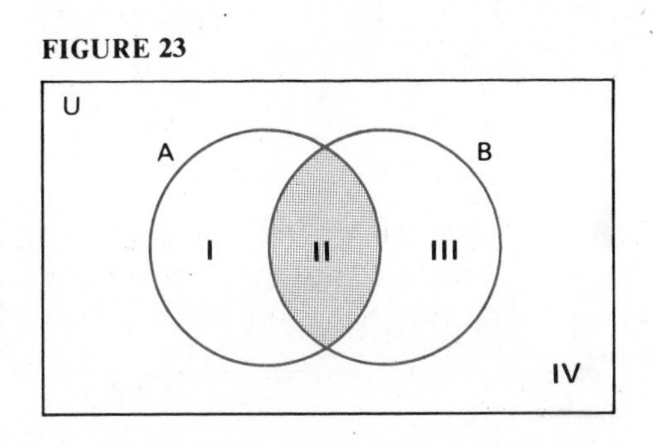

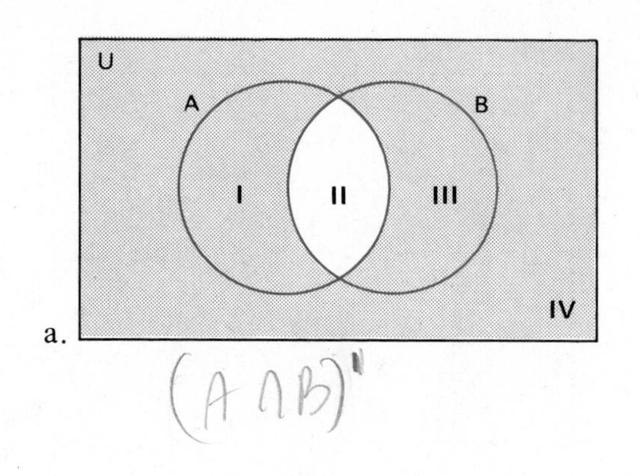
U
A
B
I
II
III
IV
a.

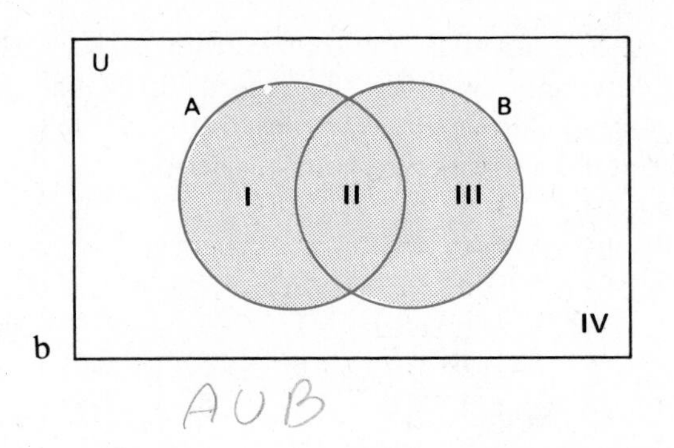
U
A
B
I
II
III
IV
b

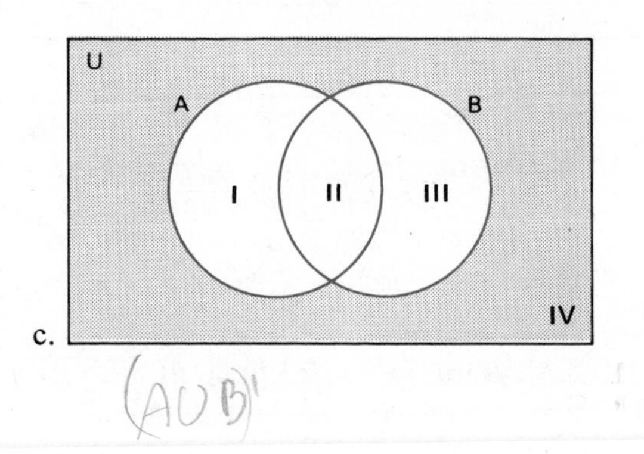
U
A
B
I
II
III
IV
c.

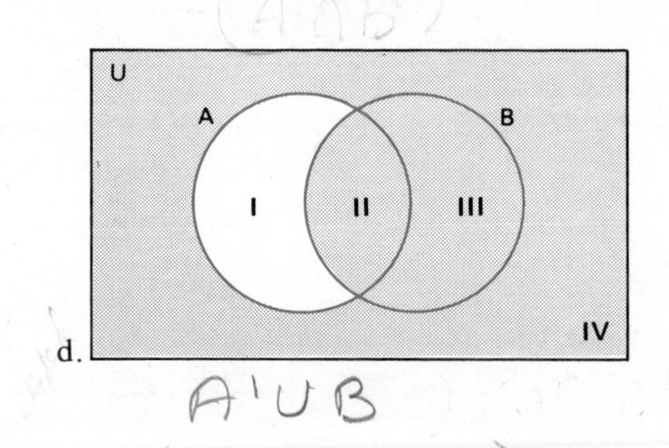
U
A
B
I
II
III
IV
d.

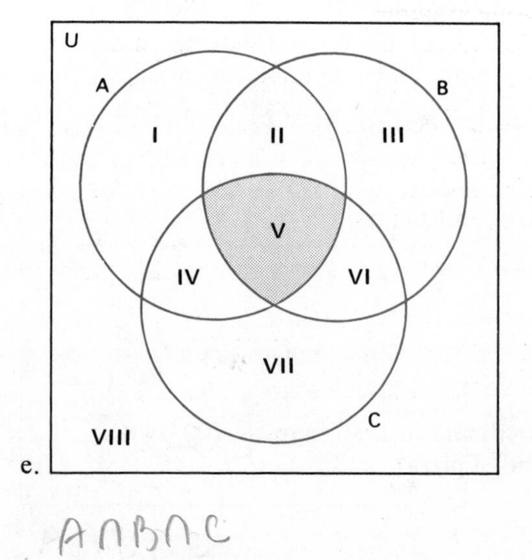
U
A
B
I
II
III
V
IV
VI
VII
C
VIII
e.

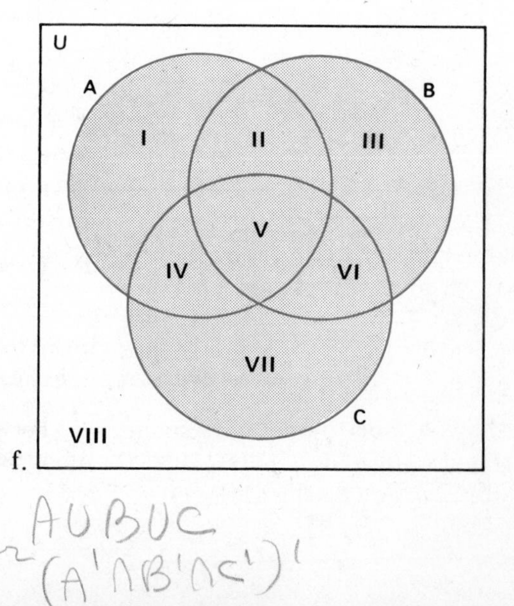
U
A
B
I
II
III
V
IV
VI
VII
C
VIII
f.

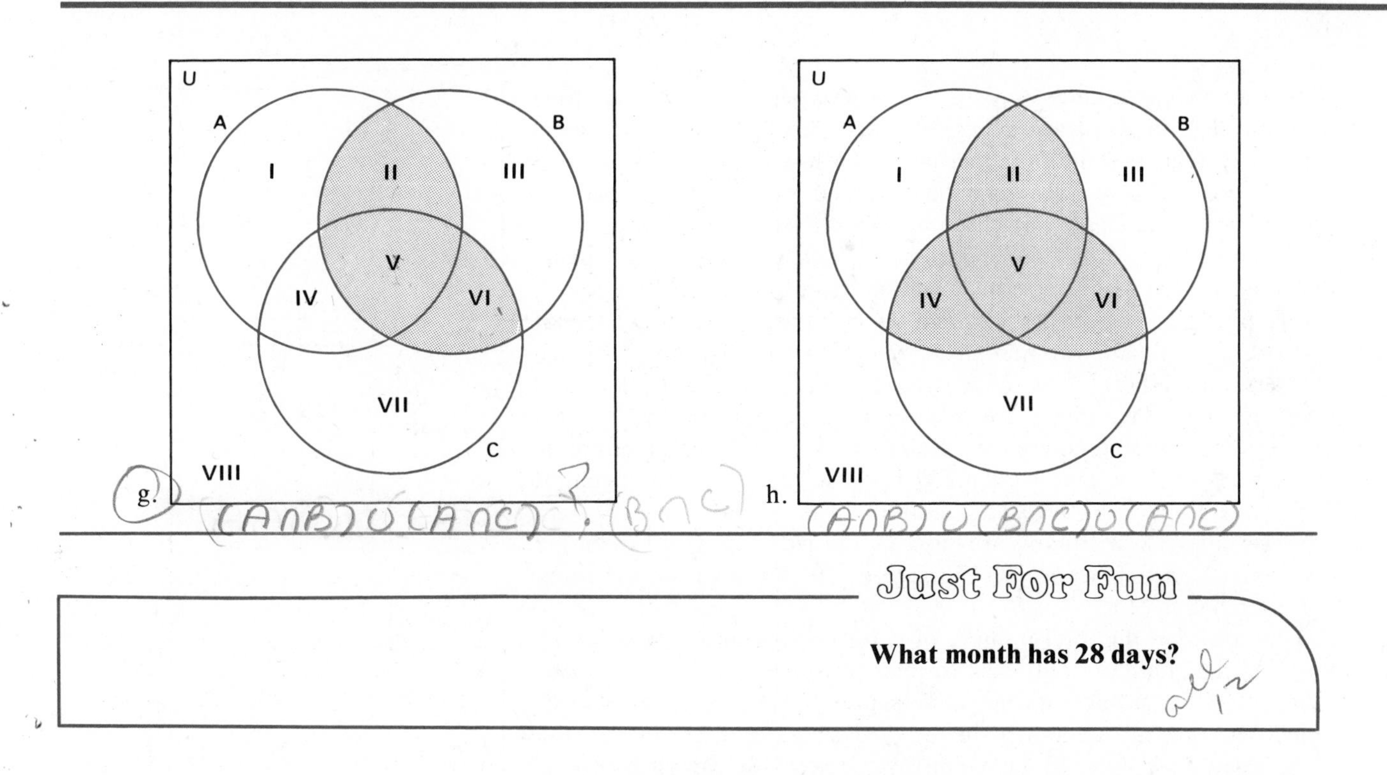

1.6 AN APPLICATION OF SETS AND VENN DIAGRAMS

Our knowledge of sets and Venn diagrams is particularly useful in solving problems involving overlapping sets of data on individuals. Consider the following example.

EXAMPLE 1

In a certain group of 100 customers at Phil's Pizza Palace, 60 customers ordered cheese and pepperoni on their pizza. But altogether, 80 customers ordered a pizza with cheese on it, and 72 customers ordered pizza with pepperoni on it.

a. How many customers ordered cheese on their pizza, but no pepperoni?

b. How many customers ordered pepperoni on their pizza, but no cheese?

c. How many customers in the group of 100 customers did not order either type of pizza?

Solution

Since we have two kinds of pizza, cheese and pepperoni, we can draw a Venn diagram illustrating these two sets. See figure 24a. We label the sets C (cheese) and P (pepperoni) and also number the regions.

FIGURE 24a

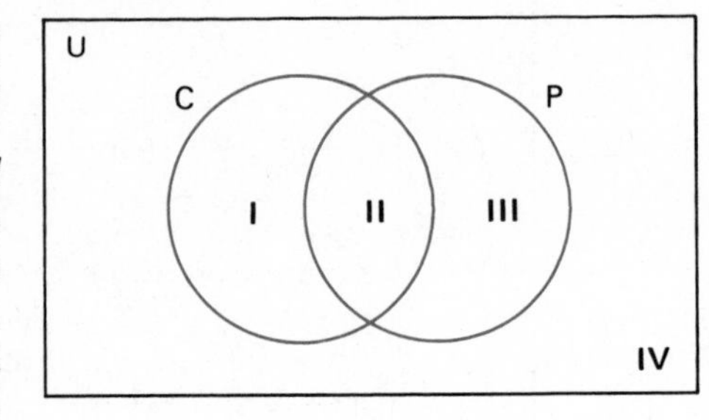

There were 60 customers that ordered pizza with cheese *and* pepperoni. The *and* is our clue that these customers belong in the intersection of C and P, namely, region II. Altogether 80 customers ordered a pizza with cheese on it, but we cannot place 80 in region I. If we put 80 in region I, since we already have 60 in region II, that would give us 140 customers in set C—and there are only 100 customers altogether. The 60 customers that ordered cheese and pepperoni on their pizza are also counted as part of the 80 customers that ordered cheese on their pizza. If altogether there are 80 customers in set C, and we already have 60 customers in region II (part of C), then we have 20 in region I.

Seventy-two customers are in set P, and we know 60 of them are in region II; hence, we write 12 in region III. We now have the diagram in figure 24b.

FIGURE 24b

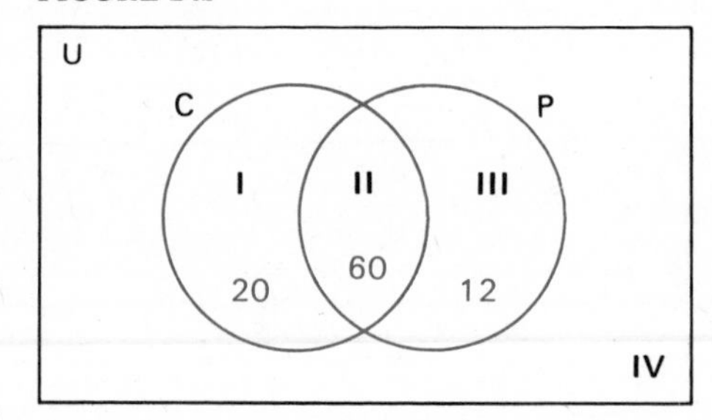

Adding up the total number of people in regions I, II, and III, we see that there are only 92 customers in our diagram. We had a total of 100 customers in our universal set. Where are the other 8? Since they are not contained in set P or set C, but are in the universal set, they must be in region IV. Therefore we place 8 customers in region IV. The completed diagram is shown in figure 24c.

Now that we have our completed diagram, we can answer the original questions.

a. How many of these customers ordered cheese on their pizza, but no pepperoni? These are the customers in region I and therefore the answer is 20.

b. How many of these customers ordered pepperoni on their pizza, but no cheese? These customers are in region III and the answer is 12.

c. How many customers in this group of 100 customers did not order either type of pizza? These customers are in region IV and we see that the answer is 8.

FIGURE 24c

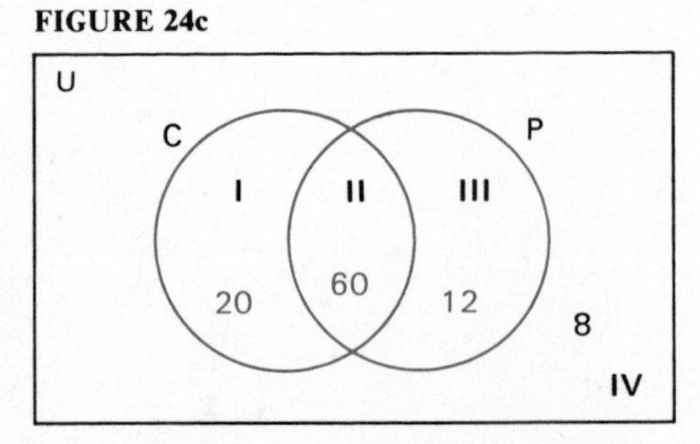

Note that in determining the numbers that went in the different regions of the diagram in example 1, we started with the most specific piece of information. We first placed the 60 customers that were in the intersection of the two sets, and then proceeded with the completion of the diagram. Once a diagram is completed, we can examine it for information and answer the questions that are asked. But we must use care in entering the data in our Venn diagram. If the data is entered correctly, then we can answer the questions by reading directly from the Venn diagram.

EXAMPLE 2

In a certain group of 75 students, 16 students are taking psychology, geology, and English. Twenty-four students are taking psychology and geology, 30 students are taking psychology and English, and 22 students are taking geology and English. Seven students are taking only psychology, 10 students are taking only geology, and 5 students are taking only English.

a. How many of these students are taking psychology?

b. How many of these students are taking psychology and English, but not geology?

c. How many students in this group are not taking any of the three subjects?

FIGURE 25a

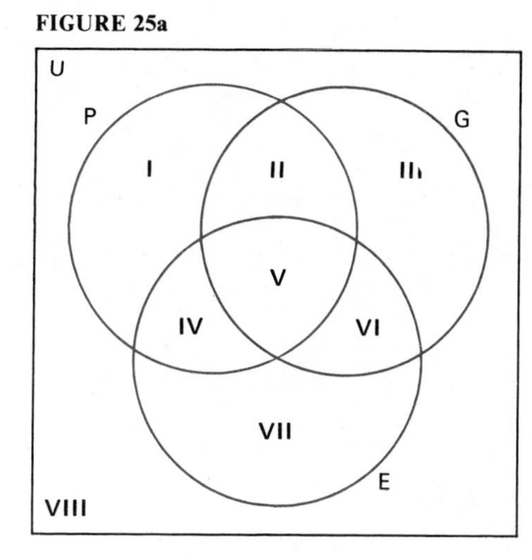

FIGURE 25b

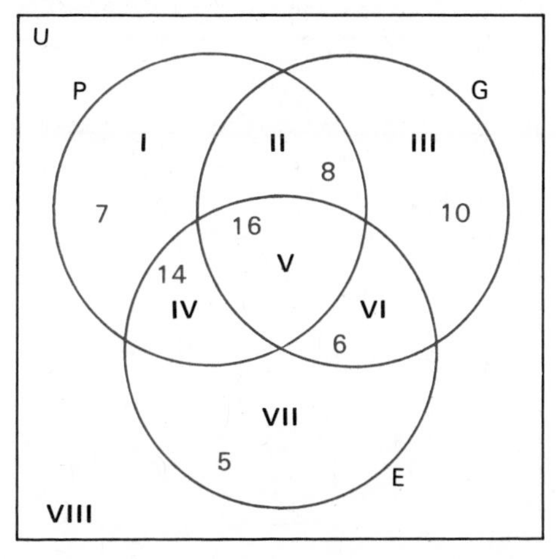

Solution

This problem involves overlapping sets of data, so we may represent the data by means of a Venn diagram. There are three subjects, so we will have three circles. (See figure 25a.) We label the sets P (psychology), G (geology), and E (English) and number the regions.

It is best to start with the most specific piece of information and work backwards from it. The most specific information we have is that 16 students are taking all three subjects (the intersection of all three sets). These students would have to appear in region V. If 24 students are taking psychology and geology (the intersection of P and G), and we already know that 16 are in region V, then we need to place 8 of them in region II. Why not put 24 there? If we did that, then we would have 24 in region II and 16 in region V, which would give us a total of 40 students taking psychology and geology. That is not the case!

Since 30 students are in English and psychology, we place 14 in region IV. Similarly, we place 6 in region VI. Since 7 students are taking only psychology, they belong in region I; similarly, we place 10 students in region III and 5 students in region VII. We now have the diagram in figure 25b.

Counting the total number of students in the circles, we get 66. But there are 75 students in our group and therefore we place 9 students in region VIII to complete our diagram. See figure 25c.

Now we can readily answer the questions by reading directly from the Venn diagram.

a. How many of these students are taking psychology? These stu-

dents are in set P, so we add the numbers of students in regions I, II, IV, and V. The answer is 45.

b. How many of these students are taking psychology and English, but not geology? These are the students in region IV, and the answer is 14.

c. How many students in this group are not taking any of the three subjects? These students are in region VIII, and the answer is 9.

EXAMPLE 3

At a meeting of 50 car dealers, the following information was obtained: 12 dealers sold Buicks, 15 dealers sold Oldsmobiles, 16 dealers sold Pontiacs; 4 dealers sold both Buicks and Oldsmobiles, 6 dealers sold both Oldsmobiles and Pontiacs, 5 dealers sold both Buicks and Pontiacs; and 1 dealer sold all three brands of cars.

a. How many dealers sold Buicks and neither of the other two brands?

b. How many of the dealers at the meeting did not sell any of these cars?

Solution

We may obtain answers to these questions by completing a Venn diagram and reading our answers directly from the figure. See figure 26. It is important to remember to start with the most specific piece of information, that which belongs in the intersection of all three sets (in this case, the one dealer who sold all three brands of cars). After using this information, we proceed to the data that belong in the intersection of two sets, and continue working backwards.

In our diagram, we labeled the circles Buick, Oldsmobile, and Pontiac in that order, but we could have changed the order if we had wanted to. The resulting information would still be the same, and we would still be able to answer the questions.

Reading the information from the Venn diagram, we can answer the questions.

a. How many dealers sold Buicks and neither of the other two brands? These are the dealers in region I, and the answer is 4.

b. How many of the dealers at the meeting did not sell any of these cars? These are the dealers in region VIII; they are not members of any of the three sets, and the answer is 21.

We can also use Venn diagrams to sort out other kinds of information. Consider the following problem.

FIGURE 25c

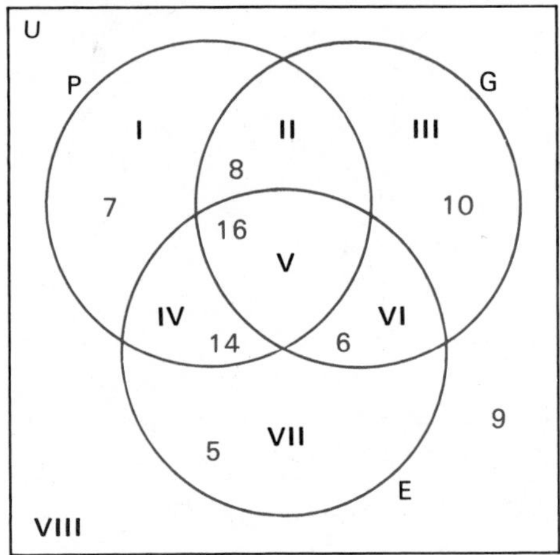

FIGURE 26 B = The set of Buicks, O = The set of Oldsmobiles, and P = The set of Pontiacs

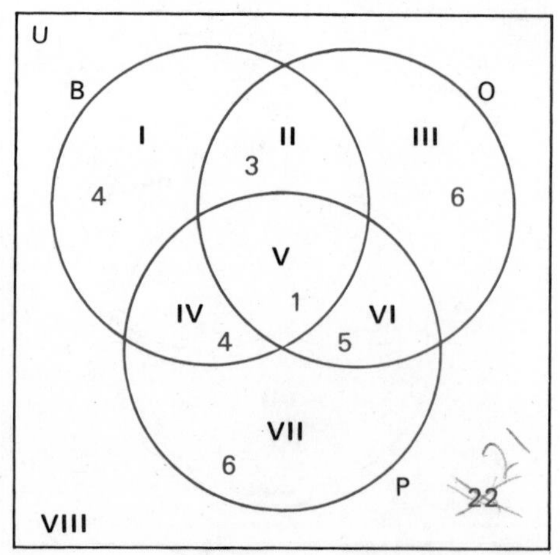

EXAMPLE 4

Benny, Nelson, and Laura made the following predictions regarding the outcomes of four selected football games for a certain Sunday. Benny picked the Bills, Giants, Jets, and Lions to win. Nelson picked the Browns, Jets, Rams, and Bills. Laura picked the Lions, Bills, Bears, and Rams. No one picked the Broncos to win, although the Broncos participated in one of the four selected games. Determine the teams that played each other.

Solution

The way to solve this problem is to put the information into a Venn diagram and read the answers directly from the diagram. First, we list three sets consisting of the teams that each person picked to win:

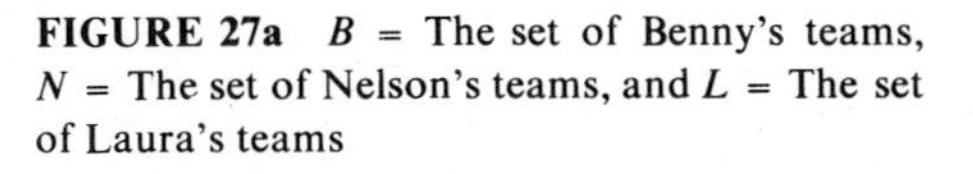

FIGURE 27a B = The set of Benny's teams, N = The set of Nelson's teams, and L = The set of Laura's teams

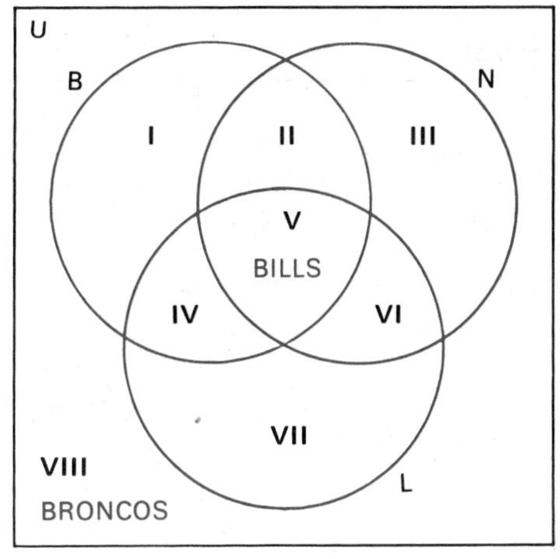

{Bills, Giants, Jets, Lions} = The set of teams picked by Benny
{Browns, Jets, Rams, Bills} = The set of teams picked by Nelson
{Lions, Bills, Bears, Rams} = The set of teams picked by Laura

Now we want to place this information in a Venn diagram consisting of three sets. Note that each person picked the Bills to win, so we must place this team in region V. Why? It is common to all three sets, so it must be in the intersection of all three sets. See figure 27a.

We placed the Broncos in region VIII because no one picked them. The fact that Benny, Nelson, and Laura all picked the Bills and none of them picked the Broncos tells us that one of the games is Bills vs. Broncos. Note that both Benny and Nelson picked the Jets, so we placed the Jets in region II. Benny and Laura picked the Lions; hence we place the Lions in region IV. What team belongs in region VI? We place the Rams in region VI because both Nelson and Laura picked the Rams. The Giants belong in region I because this is the only team in Benny's set that has not been placed. No one else picked them. Similarly, we place the Browns in region III and the Bears in region VII. See figure 27b.

FIGURE 27b

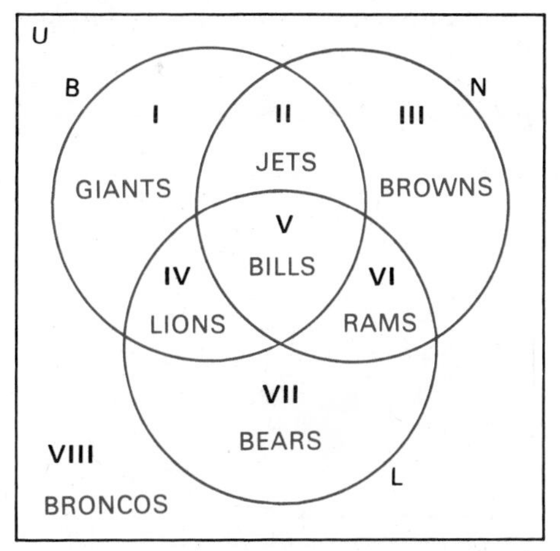

Now we have our completed diagram and it satisfies all of the given information. But we must still determine the pairings (the teams that played each other). We have already ascertained that one game was Bills vs. Broncos, so we will not consider them in the rest of our analysis. Let's figure out whom the Jets played. The Jets are in the same circle as the Giants and Lions, so they couldn't have played them. The Jets are in another circle with the Browns and the Rams, so they can't be paired with them either. Hence, the Jets must be paired with the Bears. This is the only team not in the same set as the Jets (except the Broncos, and we have already paired them with the Bills). Similarly, the Lions must be paired with the Browns, and the Rams must be paired with the Giants. The four games are:

Bills vs. Broncos
Jets vs. Bears
Lions vs. Browns
Rams vs. Giants

Note that in solving the Venn diagram problem in example 4, we again started with the most specific piece of information, and then worked backwards to the data that belong in the intersection of two sets, and then continued working backwards to the most general information.

EXERCISES FOR SECTION 1.6

1. In a survey of 75 students at a large university, the following data were collected: There were 27 students taking calculus, 26 taking political science, and 41 taking statistics. Twelve students were taking calculus and political science, 13 students were taking calculus and statistics, and 17 students were taking political science and statistics. Four students were taking all three courses.
   a. How many students were taking only political science?
   b. How many students in the group were not taking any of the three subjects?
   c. How many students were taking calculus and statistics, but not political science?
   d. How many students were taking exactly one of these subjects?

2. In a certain Las Vegas casino, a survey of 125 gamblers was taken and the following data were collected: 71 gamblers played roulette, 72 gamblers played poker, and 80 gamblers played blackjack, while 33 gamblers played roulette and poker, 42 gamblers played roulette and blackjack, and 47 gamblers played poker and blackjack. Eleven gamblers played all three games.
   a. How many of these gamblers played only blackjack?
   b. How many of these gamblers played poker and blackjack, but not roulette?
   c. How many gamblers in this group did not play any of these three games?
   d. How many gamblers did not play poker?

3. In a survey of 78 residents in a certain dormitory, the following data were collected: 35 residents had radios in their rooms, 33 residents had record players in their rooms, and 31 residents had tape players in their rooms. Sixteen residents had both tape players and record players in their rooms, 14 residents had record players and radios in their rooms, and 14 residents had radios and tape players in their rooms. Five residents had radios, record players, and tape players in their rooms.
   a. How many residents had none of these three items in their rooms?
   b. How many residents did not have record players in their rooms?
   c. How many residents had only radios in their rooms?
   d. How many residents did not have radios in their rooms?

4. In analyzing the scoring for college football teams in a particular conference, the following facts were gathered: 70 players had scored touchdowns, 44 players had scored points after touchdowns (pat's), and 32 players had scored field goals, while 19 players had scored both touchdowns and pat's, 16 players had scored touchdowns and field goals, and 21 players had scored both pat's and field goals. Six players had scored in all three ways.
   a. How many players scored only touchdowns?
   b. How many players scored field goals and pat's, but not touchdowns?
   c. How many players scored touchdowns or field goals?
   d. How many players did not score a pat?

5. At a subway stop in New York City, 125 people were asked what newspaper they read. Forty-six people read the *Times,* 43 people read the *Post,* and 65 people read the *News.* Nineteen people read the *Times* and the *Post,* 18 people read the *Post* and the *News,* and 11 people read the *Times* and the *News.* Seven people read all three papers.
   a. How many people read only the *News*?
   b. How many people did not read any of the three papers?
   c. How many people read the *Times* or the *News*?
   d. How many people read the *Post* and the *News*, but not the *Times*?

6. At a local bank, 100 customers were asked what bank services they used. Fifty-two people in the group had savings accounts, 52 people had checking accounts, and 57 people had bank charge cards, while 23 people had savings and checking accounts, 25 people had checking accounts and bank charge cards, and 24 people had savings accounts and bank charge cards. Eleven people used all three types of services—checking accounts, savings accounts, and charge cards.
   a. How many of these people had only checking accounts?
   b. How many of these people did not have savings accounts?
   c. How many of these people did not have bank charge cards?
   d. How many of these people had savings accounts or checking accounts?

7. A used-car dealer must complete an inventory of the cars on his lot. He notes that he has 22 compact, two-door, standard-transmission cars. Of the 50 standard-transmission cars on the lot, 28 are classified as compact, while 30 are two-door. Also, of the 47 two-door cars on the lot, 31 are classified as compact. The dealer also notes that he has 44 compact cars on his lot and 15 large, four-door, automatic-transmission cars.
   a. How many cars are there on the lot altogether?
   b. How many compact cars have standard transmission, but are not the two-door type?
   c. How many of the two-door cars with standard transmission are not compact?
   d. How many of the standard-transmission cars are not compact?

8. At a clambake, a survey of 100 people was taken to determine their preferences in shellfish. Sixty-eight people in the group liked lobsters, 60 people liked clams, and 56 people liked scallops, while 36 people liked lobsters and clams, 38 people liked clams and scallops, and 37 people liked lobsters and scallops. Twenty-four people liked all three kinds of shellfish.
   a. How many of the people liked only scallops?
   b. How many of the people liked lobsters and clams, but not scallops?
   c. How many of the people liked lobsters or clams?
   d. How many of the people did not like clams?

9. Many fish markets also sell clams. Usually these stores sell three different sizes of clams: cherrystones, littlenecks, and chowders. During a certain week, the following data were collected regarding the sale of these size clams: 40 customers purchased cherrystone clams, 47 cus-

tomers purchased littleneck clams, and 32 customers purchased chowder clams, while 18 customers purchased cherrystones and littlenecks, 14 customers purchased littlenecks and chowders, and 9 customers purchased cherrystones and chowders. Four customers purchased all three types of clams.

a. How many customers purchased clams during the week?
b. How many customers purchased only littlenecks?
c. How many customers purchased cherrystones or littlenecks?
d. How many customers purchased only one type of clam?

10. A statistician reported to her employer that she had gathered the following information: In a survey of 40 households in a certain tract, 36 households had a color television, 36 had two cars, and 21 owned a camper, while 22 households had both a color television and two cars, 19 had a color television and a camper, and 17 had a camper and two cars. Six households had a color television, two cars, and a camper. The statistician was promptly fired by her employer. Why?

11. An independent survey agency was hired by the Metropolitan Transit Authority to find out how people commuted to their jobs. The agency interviewed 1000 commuters and submitted the following report: 631 people came to work by car, 554 people came to work by bus, and 759 came to work by subway. Also, 373 people came to work by a combination of car and bus, 301 people came to work by bus and subway, and 268 people came to work by car and subway, while 231 people used all three means of transportation to get to work. The Transit Authority refused to accept the report, stating that it was inaccurate. Why?

12. A group of people, Barbara, Carol, Alice, Sue, Bob, Ted, Joe, and Don, are going to play bridge. The players are divided into three different sets according to their likes and dislikes. The sets are

$$A = \{\text{Ted, Sue, Bob, Barbara}\}$$
$$B = \{\text{Ted, Don, Sue, Carol}\}$$
$$C = \{\text{Ted, Joe, Bob, Don}\}$$

By means of a Venn diagram, determine the pairings of bridge partners that will result from the given information.

13. The sports writers from the *Post,* the *News,* and the *Tribune* picked the following teams to win in the quarterfinals of a certain basketball tournament. The sports writer from the *Post* picked Maryland, USC, Notre Dame, and UCLA. The sports writer from the *News* picked St. John's, Maryland, North Carolina State, and UCLA. The sports writer from the *Tribune* picked Ohio State, USC, North Carolina State, and UCLA. None of the three sports writers picked Kentucky to win. These were the eight teams in the quarterfinals. Determine the teams that played each other.

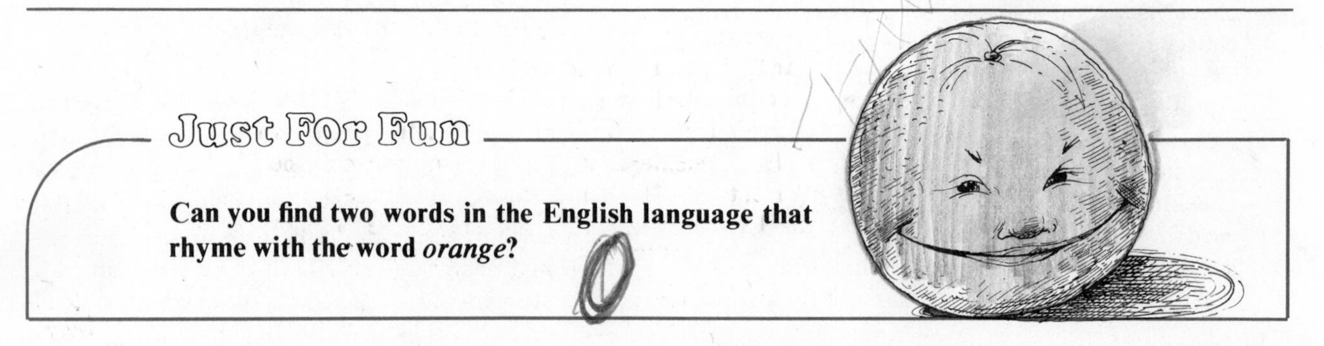

Just For Fun
Can you find two words in the English language that rhyme with the word *orange*?

1.7 CARTESIAN PRODUCTS

As the last set operation in this chapter, we shall consider the *Cartesian product* of two sets (named after René Descartes, a seventeenth-century French mathematician and philosopher). The Cartesian product is unique among set operations in that it produces new elements that are not members of the universal set. These new elements are called ordered pairs.

An **ordered pair** is a pair of objects where one element is considered first and the other element is considered second. Which element is first and which element is second is important. If we have a pair of socks, it doesn't matter which sock we put on first. A pair of socks is *not* an ordered pair. But if we have a shoe and a boot, it does matter which we put on first: try putting a shoe on over a boot!

In mathematics, if we wish to discuss an ordered pair of elements consisting of a and b, we use the notation (a, b). This notation tells us that we want to consider a first and b second. Suppose a stands for "walk a blocks north" and b stands for "walk b blocks west." Now consider the ordered pair $(3, 2)$. This would mean "walk 3 blocks north and then walk 2 blocks west." This is certainly different from $(2, 3)$, which would mean "walk 2 blocks north and then walk 3 blocks west." We would not arrive at the same place with these two ordered pairs.

Given two sets A and B, the Cartesian product of A and B, denoted by $A \times B$ (read "A cross B"), is the set of all possible ordered pairs such that the first element of the ordered pair is an element of A and the second element of the ordered pair is an element of B.

EXAMPLE 1

Given $A = \{\text{Joe, Scott}\}$ and $B = \{p, q\}$, find $A \times B$, the Cartesian product of A and B.

Solution

The Cartesian product $A \times B$ is the set of all possible ordered pairs such that the first element is an element of A and the second element is an element of B. We must pair the elements in A, Joe and Scott, with the elements in B, p and q. Joe may be paired with p or q; hence, the possible ordered pairs with Joe are (Joe, p), (Joe, q). Now we do the same thing for Scott; the possible ordered pairs are (Scott, p) and (Scott, q). Therefore we have

$$A \times B = \{(\text{Joe}, p), (\text{Joe}, q), (\text{Scott}, p), (\text{Scott}, q)\}$$

Note that $A \times B$ gives us a set where the elements are ordered pairs. These are not members of the universal set, since the universal set consists of single elements such as p, q, Joe, and Scott.

EXAMPLE 2
Given $A = \{4, 8\}$ and $B = \{a, b, c\}$, find $A \times B$.

Solution
The first element in our ordered pairs must come from A. Therefore we pair 4 with a, then 4 with b, etc. Then we do the same for 8. Hence we have

$$A \times B = \{(4, a), (4, b), (4, c), (8, a), (8, b), (8, c)\}$$

EXAMPLE 3
Given $A = \{4, 8\}$ and $B = \{a, b, c\}$, find $B \times A$.

Solution
This is similar to the last example, but here we want to find $B \times A$. Remember $B \times A$ is not the same as $A \times B$. In $B \times A$, we want the set of all possible ordered pairs such that the first element is an element of B and the second element is an element of A. Here we take the elements in B, a, b, and c, and pair them with the elements in A, 4 and 8. Therefore

$$B \times A = \{(a, 4), (a, 8), (b, 4), (b, 8), (c, 4), (c, 8)\}$$

By comparing the results of examples 2 and 3, we see that $A \times B \neq B \times A$. The two sets contain different ordered pairs, as $(4, a)$ is not the same as $(a, 4)$.

Note that in example 1 the number of elements in $A \times B$ was 4—that is, $n(A \times B) = 4$—and in example 2, $n(A \times B) = 6$. If we want to determine the number of elements in a Cartesian product $A \times B$ (the cardinality of the set), we take the number of elements in A and multiply it by the number of elements in B. If set A has m elements in it and set B has n elements in it, then the number of elements in $A \times B$ is $m \times n$. In other words,

$$n(A \times B) = n(A) \times n(B)$$

This provides us with a handy check in computing $A \times B$, because we can use the cardinality to check to see if we have all of the possible ordered pairs.

EXAMPLE 4
Find $A \times A$ if $A = \{1, 2, 3\}$.

Solution
To find $A \times A$, we pair each element in A with every element in A. Therefore we have

$$A \times A = \{(1,1), (1,2), (1,3), (2,1), (2,2), (2,3), (3,1), (3,2), (3,3)\}$$

Since there are 3 elements in A, we should have $3 \times 3 = 9$ elements in $A \times A$. Checking our answer, we see that we do have 9 different ordered pairs in $A \times A$.

The Cartesian product of two sets, $A \times B$, gives us a set of ordered pairs. This set of ordered pairs may be pictured by means of an **array** or **lattice.** Consider the following example.

EXAMPLE 5
Find $A \times B$ if $A = \{a, b, c\}$ and $B = \{d, e\}$.

Solution
There are 3 elements in A and 2 elements in B; hence there are $3 \times 2 = 6$ ordered pairs in the Cartesian product $A \times B$. Pairing each element of A with every element in B, we have

$$A \times B = \{(a,d), (a,e), (b,d), (b,e), (c,d), (c,e)\}$$

FIGURE 28

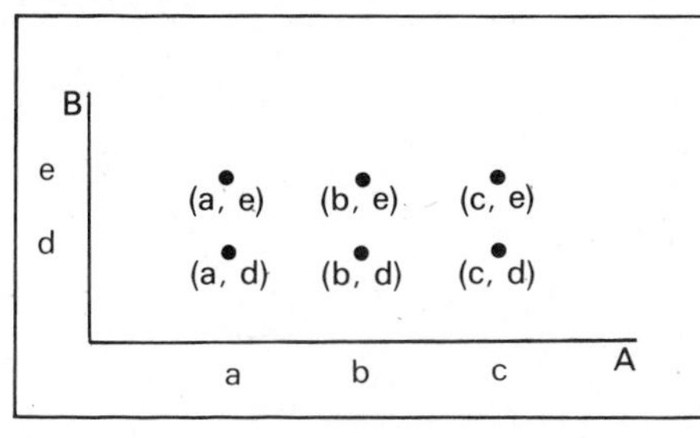

Figure 28 shows the array of ordered pairs for this example. In the lattice, a dot represents an ordered pair.

Note that the vertical axis represents set B with the elements d and e. The horizontal axis represents set A with elements a, b, and c. It is traditional to use the horizontal axis to represent the first element in an ordered pair.

One other way of picturing the formation of a Cartesian product is a *tree diagram.* A **tree diagram** consists of a number of branches that illustrate the possible pairings in $A \times B$, as in figure 29.

EXERCISES FOR SECTION 1.7

1. If $A = \{a, b, c\}$ and $B = \{10, 20\}$, find $A \times B$, $B \times A$, and $n(A \times B)$.
2. If $A = \{?, \&, !\}$ and $B = \{a, b, c\}$, find $A \times B$, $B \times A$, and $n(A \times B)$.
3. If $C = \{2, 4, 6\}$ and $D = \{1, 3, 5\}$, find $C \times D$, $D \times C$, and $n(C \times D)$.
4. If $V = \{a, e, i, o, u\}$ and $Z = \{x, y, z\}$, find $V \times Z, Z \times V$, and $n(V \times Z)$.
5. Let $T = \{t, f\}$. Find $T \times T$ and $n(T \times T)$.
6. Let $A = \{c, d, e\}$ and $B = \{1, 2, 3\}$. Make a lattice showing $A \times B$.
7. Let $A = \{4, 5, 6\}$ and $B = \{x, y\}$. Make a lattice showing $A \times B$.
8. Let $A = \{a, b, c, d\}$ and $B = \{x, y, z\}$. Make a tree diagram showing $B \times A$.

FIGURE 29

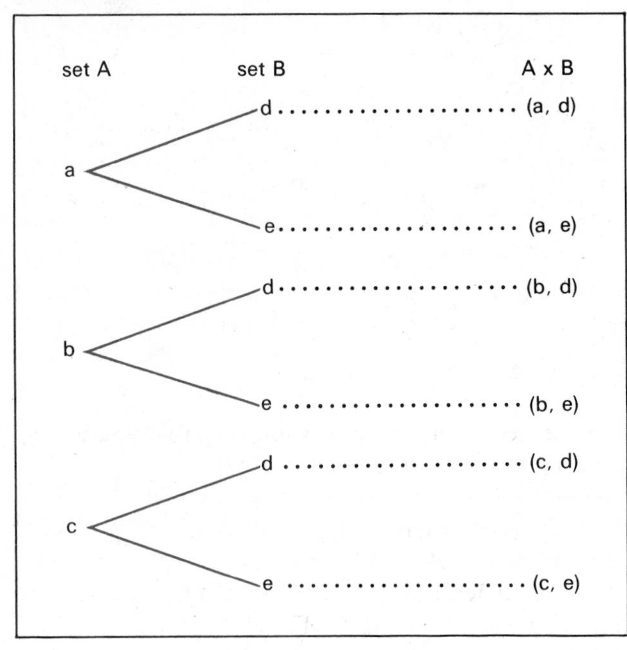

Just For Fun

Can you number the rest of the squares in figure 30 so that each row, column, and diagonal totals 15? No number may be used more than once.

FIGURE 30

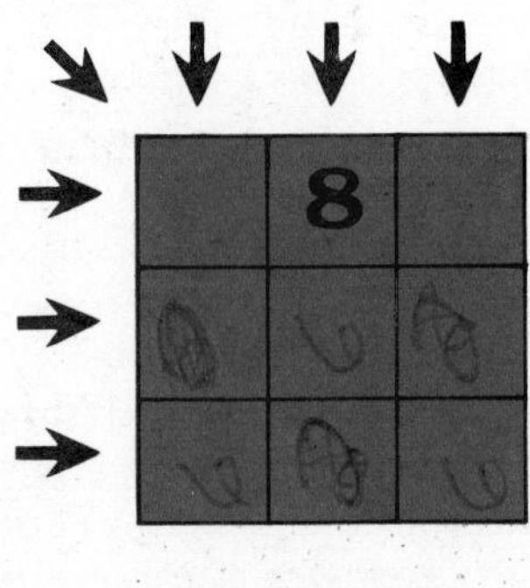

9. If $A = \{x, y\}$ and $B = \{a, e, i, o, u\}$, make a tree diagram showing $A \times B$.

10. If $U = \{1, 2, 3, 4\}$, find $U \times U$ and $n(U \times U)$.

11. Find each of the following, given $A = \{a, b\}$, $B = \{b, c, d\}$, and $C = \{c, d, e\}$.
    a. $A \times B$
    b. $A \times C$
    c. $n(B \times C)$
    d. $(A \cap B) \times C$
    e. $C \times (A \cup B)$
    f. $A \times (B \cap C)$

12. If $A = \{1, 2, 3\}$, $B = \{3, 4, 5, 6\}$, and $C = \{4, 5, 6, 7\}$, find each of the following:
    a. $A \times B$
    b. $n(B \times C)$
    c. $(A \cap C) \times B$
    d. $(B \cup C) \times A$
    e. $(B \cap C) \times A$
    f. $(A \cap B) \times (B \cap C)$

13. Given that $A = \{1, 2\}$ and $B = \{3, 4\}$, does $A \times B = B \times A$? Why or why not?

1.8 SUMMARY

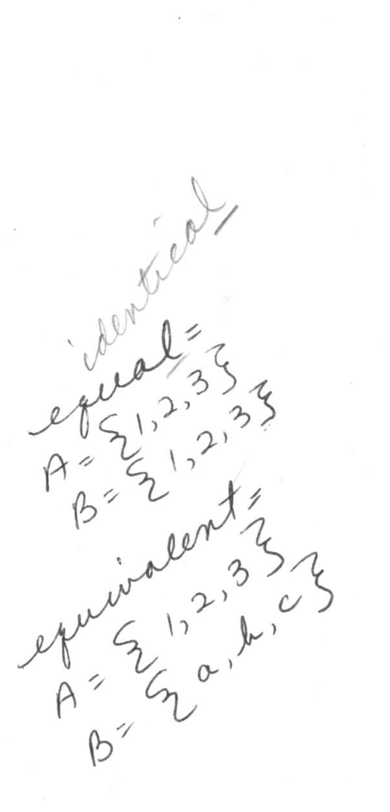

The introduction presented in this chapter should enable you to understand the properties and operations of sets and Venn diagrams.

You should be aware that *set* is an intuitive concept, but there exist special properties of sets. Given any two sets A and B, A is a *subset* of B if every element in A is also an element in B. If A is a subset of B and there is at least one element in B that is not contained in A, then A is a *proper subset* of B. The *intersection* of A and B, $A \cap B$, is a set of elements that are elements of both A and B. The *union* of sets A and B, $A \cup B$, is a set of all elements that are elements of A or B or both. The *complement* of set A, A', is the set of all the elements in the given universal set, U, that are not in set A. The *Cartesian product* of sets A and B, $A \times B$, is the set of all possible ordered pairs such that the first element of the ordered pair is an element of A, and the second element of the ordered pair is an element of B.

Equal sets are sets that contain exactly the same elements, while *equivalent sets* are sets that have the same number of elements, that is, the same cardinality. If two sets are equivalent, then they may be placed in a *one-to-one correspondence.*

Venn diagrams are picture representations of sets. The universal set is represented by a rectangle, and circles inside the rectangle represent the sets being considered. In a Venn diagram, we can show the sets overlapping one another, so that all possible combinations may be considered. Set operations, such as $A \cap B$, are represented by the intersection of the circles in a Venn diagram. We can determine if different set expressions, such as $(A \cap B)'$ and $A' \cup B'$, are equivalent by examining the corresponding Venn diagrams to see if both expressions determine the same regions.

Venn diagrams may also be used to answer questions when we are given certain data and wish to determine more information from the data. A typical example is the survey type of question discussed in section 1.6.

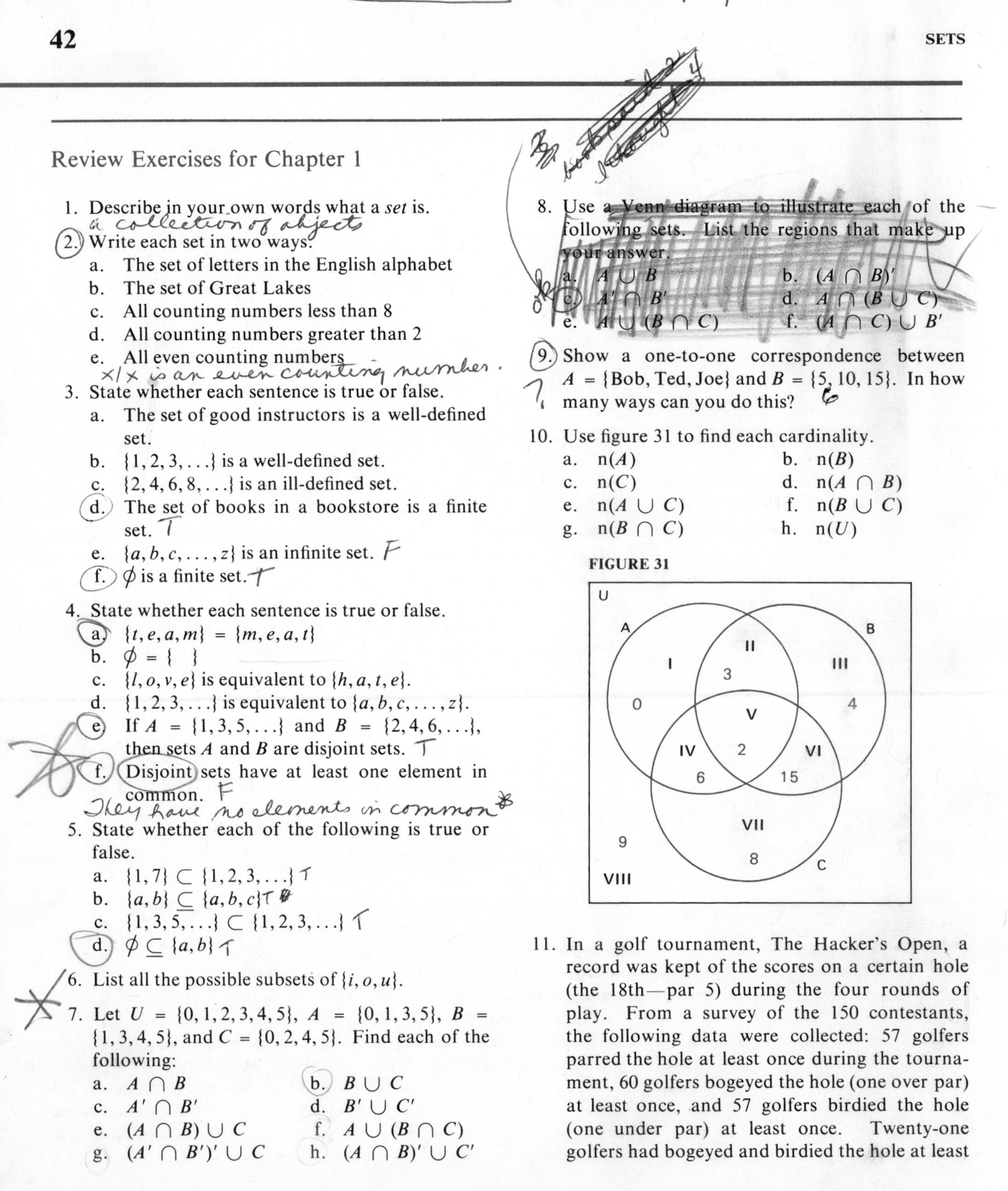

Review Exercises for Chapter 1

1. Describe in your own words what a *set* is.
2. Write each set in two ways.
   a. The set of letters in the English alphabet
   b. The set of Great Lakes
   c. All counting numbers less than 8
   d. All counting numbers greater than 2
   e. All even counting numbers
3. State whether each sentence is true or false.
   a. The set of good instructors is a well-defined set.
   b. $\{1, 2, 3, \ldots\}$ is a well-defined set.
   c. $\{2, 4, 6, 8, \ldots\}$ is an ill-defined set.
   d. The set of books in a bookstore is a finite set.
   e. $\{a, b, c, \ldots, z\}$ is an infinite set.
   f. ϕ is a finite set.
4. State whether each sentence is true or false.
   a. $\{t, e, a, m\} = \{m, e, a, t\}$
   b. $\phi = \{\ \}$
   c. $\{l, o, v, e\}$ is equivalent to $\{h, a, t, e\}$.
   d. $\{1, 2, 3, \ldots\}$ is equivalent to $\{a, b, c, \ldots, z\}$.
   e. If $A = \{1, 3, 5, \ldots\}$ and $B = \{2, 4, 6, \ldots\}$, then sets A and B are disjoint sets.
   f. Disjoint sets have at least one element in common.
5. State whether each of the following is true or false.
   a. $\{1, 7\} \subset \{1, 2, 3, \ldots\}$
   b. $\{a, b\} \subseteq \{a, b, c\}$
   c. $\{1, 3, 5, \ldots\} \subset \{1, 2, 3, \ldots\}$
   d. $\phi \subseteq \{a, b\}$
6. List all the possible subsets of $\{i, o, u\}$.
7. Let $U = \{0, 1, 2, 3, 4, 5\}$, $A = \{0, 1, 3, 5\}$, $B = \{1, 3, 4, 5\}$, and $C = \{0, 2, 4, 5\}$. Find each of the following:
   a. $A \cap B$ b. $B \cup C$
   c. $A' \cap B'$ d. $B' \cup C'$
   e. $(A \cap B) \cup C$ f. $A \cup (B \cap C)$
   g. $(A' \cap B')' \cup C$ h. $(A \cap B)' \cup C'$
8. Use a Venn diagram to illustrate each of the following sets. List the regions that make up your answer.
   a. $A \cup B$ b. $(A \cap B)'$
   c. $A' \cap B'$ d. $A \cap (B \cup C)$
   e. $A \cup (B \cap C)$ f. $(A \cap C) \cup B'$
9. Show a one-to-one correspondence between $A = \{\text{Bob, Ted, Joe}\}$ and $B = \{5, 10, 15\}$. In how many ways can you do this?
10. Use figure 31 to find each cardinality.
    a. $n(A)$ b. $n(B)$
    c. $n(C)$ d. $n(A \cap B)$
    e. $n(A \cup C)$ f. $n(B \cup C)$
    g. $n(B \cap C)$ h. $n(U)$

FIGURE 31

11. In a golf tournament, The Hacker's Open, a record was kept of the scores on a certain hole (the 18th—par 5) during the four rounds of play. From a survey of the 150 contestants, the following data were collected: 57 golfers parred the hole at least once during the tournament, 60 golfers bogeyed the hole (one over par) at least once, and 57 golfers birdied the hole (one under par) at least once. Twenty-one golfers had bogeyed and birdied the hole at least

once, 24 golfers had parred and birdied the hole at least once, and 27 golfers had bogeyed and parred the hole at least once. Fifteen golfers had at least one par, one birdie, and one bogey on the hole.

a. How many of the 150 contestants did not score a par, birdie, or bogey on the 18th hole during the tournament?
b. How many golfers birdied and bogeyed, but did not par, the 18th hole?
c. How many golfers did not par the 18th hole?
d. How many golfers did not bogey the 18th hole?

12. A group of tennis enthusiasts, Jim, Joe, Jon, Cal, Bob, Al, Ted, and Don are going to play doubles. The instructor divided the group into three sets according to ability and compatibility. The sets are—

A = {Jim, Joe, Cal, Bob}
B = [Joe, Jon, Al, Bob}
C = {Cal, Al, Ted, Bob}

Use a Venn diagram to determine the doubles partners that result from the given information.

13. Let $A = \{1, 2, 3\}$ and $B = \{a, b, c\}$.
a. Find $A \times B$.
b. Find $n(A \times B)$.
c. Make an array or lattice showing $A \times B$.
d. Make a tree diagram showing $A \times B$.

14. Let $A = \{m, a, t, h\}$ and $B = \{e, a, s, y\}$.
a. Show a one-to-one correspondence between A and B.
b. In how many ways can you do this?
c. List all possible subsets of A.
d. Find $A \times B$.

15. When are the following statements true?
a. $A \cap B = A$
b. $A \cup B = A \cap B$
c. $(A \cap B)' = A' \cup B'$
d. $B \subseteq \phi$
e. $(A \cup B)' = A' \cap B'$

Just For Fun

Mr. Apple and Mr. Jack have an 8-gallon container full of cider. They also have two empty containers whose capacities are 5 gallons and 3 gallons, respectively. How can the cider be poured, using all three containers, so that Mr. Apple and Mr. Jack each has 4 gallons of cider?

J. Chapman
Sculp.

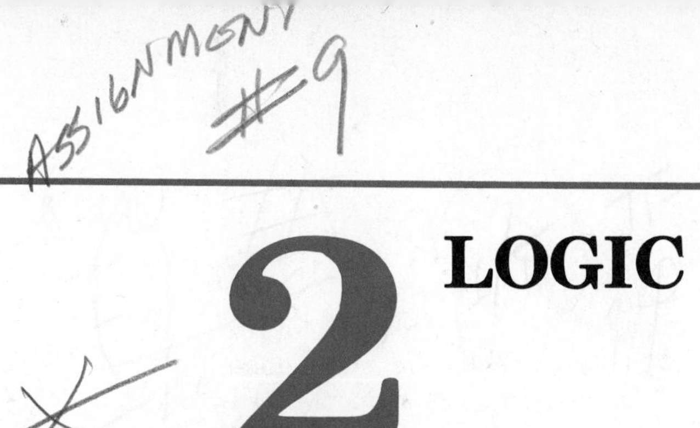

2 LOGIC

After studying this chapter, you will be able to do the following:

1. Distinguish between **simple statements** and **compound statements,** and identify a compound statement as a **negation, conjunction, disjunction, conditional,** or **biconditional**
2. Write compound verbal statements in **symbolic form** by using letters for the simple statements and the proper logical connectives, and write symbolic statements as English sentences
3. Construct a **truth table** for a given compound statement containing up to three different variables
4. Determine whether two statements are **logically equivalent** by means of truth tables or De Morgan's law
5. Determine whether an argument is **valid** or **invalid** by means of truth tables using up to three variables
6. Use Venn diagrams to determine whether two statements are **consistent** or **inconsistent,** and also use them to determine whether an argument is valid or invalid
7. Identify the various forms of the **conditional statement—converse, inverse,** and **contrapositive**—and write a conditional statement in any of these forms
8. Symbolize verbal **conditional** and **biconditional statements** stated in the only if, necessary, sufficient, necessary and sufficient, and if and only if forms.

Notation frequently used in this chapter

$P \wedge Q$	P and Q
$P \vee Q$	P or Q
$P \veebar Q$	P or Q, but not both
$P \rightarrow Q$	if P, then Q
$P \leftrightarrow Q$	P if and only if Q
$\sim P$	not P it is false that P it is not the case that P
iff	if and only if
$\equiv$	is the same as

Smithsonian Institution

2.1 INTRODUCTION

What is logic? It is often defined as the science of thinking and reasoning correctly. Many times false assumptions are made about things or people because some of us misinterpret statements. The reader or listener understands something other than what the writer or speaker has written or said. An understanding of logic and its uses will help us to avoid these pitfalls and increase our skills in analytical thinking.

We should always phrase our statements so that they express our meaning exactly. An old proverb states, "Say what you mean and mean what you say." All too often statements are misconstrued, sometimes deliberately. The following are some examples of statements that have hidden implications or may be misinterpreted:

It's good to see you here on time!
I am glad to see you sober today!
Four dentists out of five said they used "Smiley" toothpaste.
One thousand heroin addicts admitted that they had used marijuana.

Symbolic logic enables us to make some complex problems manageable. The symbols in logic, as those in other areas of mathematics (algebra, for example), do much of the thinking for us. Many large corporations, such as insurance companies, have begun to use symbolic analysis to check contracts and policies for loopholes and inconsistencies. But the major use of logic is still in mathematics, where it has had a powerful influence on the development of ideas.

Logic was first studied extensively by Aristotle in approximately 400 B.C., which means traditional logic is over two thousand years old. Some other famous contributors to the development of logic where Leonhard Euler (1707–1783), George Boole (1815–1864), John Venn (1834–1923), and Bertrand Russell (1872–1970). These dates show that modern logic is relatively young; it is only a little over 200 years old.

2.2 STATEMENTS AND SYMBOLS

In preparing for any task, we must first equip ourselves with the proper tools and the "know-how" required to do the job. The first thing that we must be able to do is identify and symbolize sentences. In logic we concern ourselves only with those sentences that are either true or false, but not both. A **statement** is a declarative sentence which is either true or false (but not both true and false).* We shall not concern ourselves with sentences that cannot be assigned a true or false value. (Sentences of this nature are usually questions or commands.)

Note that it is not possible to assign a true or false value to the following:

Is your homework finished?
Hand in your paper.
Is it raining?
Stop the bubble machine!
Shut the door when you leave.

The following statements are either true or false:

February has 30 days.
3 + 2 = 1.
Gerald Ford was president of the United States.
Phoenix is the capital of Arizona.
Tomorrow is Saturday.

There are other types of sentences that cannot be assigned a true or false value. The sentence "I am lying to you" is one example. Suppose it is true that I am lying to you; then—if I am lying—the sentence is false. On the other hand, assume that the sentence is false. If that is the case, then I am not lying, so the sentence is true. This is known as a paradox.

Another example of a paradox is "All rules have exceptions." This rule negates itself. It says that the rule itself must have exceptions and, therefore, cannot be true. Many people like the paradox about the little boy who is concerned about God. He has been told that God can do anything. The boy then asks, "If that is the case, then can God make a stone so big that he can't move it?"

Remember that in logic we concern ourselves with statements, that is, sentences that are true or false, but not both. The basic type of statement in logic is called a *simple* statement. A **simple statement** is a complete sentence that conveys one thought with no connecting words. The following are examples of simple statements:

Five is a counting number.
The Raiders have won the Super Bowl.
Sally was late for class.

*Many texts distinguish between a *proposition* and a *statement*. A **proposition** is defined to be a statement that is either true or false, but not both. We shall use both terms interchangeably as this is the only type of statement we shall consider.

Today is Monday.
Jimmy Carter won the 1976 presidential election.

Now, if we take simple statements and put them together using connecting words, we form sentences that are known as **compound** or **complex** statements. The basic connectives are *and, or, if . . . then . . . , if and only if,* and the negation *not.*

The word *not* does not connect two simple statements, but it is still thought of as a connective. It negates a simple statement. Some logicians do not like to call a negated simple statement a compound statement, but if it is no longer simple, then it must be compound. Therefore we shall think of a compound statement as a sentence that is formed by connecting one or more simple statements with a connective.

A simple statement such as "Today is Monday" is no longer simple if we say "Today is not Monday" or "It is false that today is Monday." The original simple statement has been negated, so we call the newly formed compound statement a **negation.**

When we connect two simple statements using the word *and,* we have a compound statement that is called a **conjunction.** The sentence "Today is Monday *and* tomorrow is Wednesday" is a conjunction. Remember that we are not concerned about the meaning of the sentence, only what type of statement it is. Consider:

All looms are booms and all booms are zooms.

This statement is a conjunction, even though we cannot make too much sense out of it.

Sometimes the word *but* will be used in place of *and* in a sentence.

Bonnie was early and Clyde was late

could be written as

Bonnie was early, but Clyde was late.

The connective *or* is used in forming a compound statement called a **disjunction.** The following are some examples of disjunctions:

Today is Monday or tomorrow is Wednesday.
Either you took my coat or someone stole it.
I will pass history or I will be sad.

The connective *if . . . then . . .* is used in compound statements referred to as **conditionals.** An example of a conditional is:

If you do your homework, then you will pass the exam.

The statement between the *if* and *then* ("you do your homework") is called the **antecedent** of the conditional. The part of the sentence that follows *then* ("you will pass the exam") is called the **consequent.** As with other connectives, there are variations in writing conditional statements. Two of the more common variations are illustrated by the following examples:

If someone was late, it was Benny.
We will win the game if Jackson doesn't play for them.

In the first sentence, *then* was omitted, but it is understood to be there. In the second sentence, we switched the two parts around and also omitted *then.* Nonetheless, both of these statements are conditional.

If two sides of a triangle are equal, then two angles of the triangle are equal, and if two angles of a triangle are equal, then two sides of the triangle are equal.

You may remember the above statement from the study of geometry. It is the conjunction of two conditional statements where the antecedent and consequent of the first statement have been switched in the second. This type of sentence is usually stated as:

Two sides of a triangle are equal if and only if two angles of the triangle are equal.

This type of statement is called a **biconditional.** It has the advantage of shortening the original statement. An abbreviation for *if and only if* is *iff,* and we shall sometimes use this abbreviation when we write out a biconditional statement. Remember that a biconditional statement is the conjunction of two conditional statements where the antecedent and consequent of the first statement have been switched in the second.

From this discussion, you should be able to identify a simple statement or a compound statement. If the statement is compound, then it must be one of the following: negation, conjunction, disjunction, conditional, or biconditional.

In mathematics and the English language, many statements are lengthy and cumbersome. We would all tire quickly if we had to copy this page word for word. In logic, this problem is taken care of by using symbols to represent simple statements.

It is traditional in algebra to use x, y, and z as symbols for variables and a, b, and c as symbols for constants. In logic we use the letters P, Q, R, and S, and sometimes A and B, to represent statements. Other letters may be used if needed.

Consider the statement "Today is Friday." We shall let the letter P represent this simple statement. We shall also let Q stand for the statement "I have a test." Hence we have the following:

P = *Today is Friday*
Q = *I have a test*

If we combine the two statements to form a conjunction, "Today is Friday and I have a test," we could symbolize this as

P and Q

It certainly seems odd to have the simple statements symbolized, but not the connective. However, we will find that there are symbols for each of the connectives.

Ampersand, &, is the typewriter symbol for *and.* In logic, it is more common to use an inverted V, that is, $\wedge$, to represent *and.* Some of you may be familiar with this symbol as a *caret.* Using this symbol, the statement "Today is Friday and I have a test" can be completely symbolized as

$P \wedge Q$ (conjunction)

∧ and

If we have a statement *not* P, and P stands for "Today is Friday," it is awkward to say

Not today is Friday.

We would be more comfortable if we said

It is not the case that today is Friday.

We probably would be even more comfortable if we said

Today is not Friday.

However, we must be careful here because we might change the meaning of a statement by rearranging the *not.* Consider the following:

A = *All men are not good*
B = *Not all men are good*

Statement A could be interpreted to mean that no men are good, while statement B says that it is not the case that all men are good, meaning that there exists at least one man who is not good. Therefore statements A and B do not have the same meaning. It is this type of structure that we must handle carefully; we must remember to interpret a statement exactly in the manner that we hear or read it.

When we use the word *not* in a sentence, we are negating the original statement. The logical symbol most commonly used to show negation is a *tilde,* which looks like this: $\sim$. The tilde is a dia-

~ not

critical mark used in some languages. If we let P stand for "Today is Friday," then the statement "Today is not Friday" would be symbolized as

$$\sim P \qquad \text{(negation)}$$

Remember that $\sim P$ may also be interpreted as

Not P
It is false that P
It is not the case that P

EXAMPLE 1
Let P = *Today is Monday* and Q = *I am tired.* Write each of the following statements in symbolic form.

a. Today is not Monday.

b. Today is Monday and I am tired.

c. Today is Monday and I am not tired.

d. Today is not Monday and I am tired.

e. Today is not Monday and I am not tired.

Solution
a. This statement is the negation of P, and hence it would be symbolized as $\sim P$.

b. Statement *b* is a conjunction of P and Q and therefore would be symbolized as $P \wedge Q$.

c. This is also a conjunction, but here we have Q negated. The proper symbolization is $P \wedge \sim Q$

d. This is similar to statement *c*, but now P is negated, and we have $\sim P \wedge Q$.

e. This time each part of the statement is negated. We would symbolize this as $\sim P \wedge \sim Q$.

We shall see later that statements such as $\sim P \wedge Q$ and $\sim(P \wedge Q)$ are not the same and must be interpreted differently.

A disjunction is a compound statement consisting of two statements connected by the word *or*. The symbol for this connective is $\vee$. If we let

P = *Today is Monday*
Q = *Tomorrow is Wednesday*

V or

Then the statement

Today is Monday or tomorrow is Wednesday

is symbolized as

$$P \vee Q \qquad \text{(disjunction)}$$

Consider the statement

Either two is not even or three is not odd.

In this case we would let

P = *Two is even*
Q = *Three is odd*

and the compound statement would be symbolized as

$$\sim P \vee \sim Q$$

EXAMPLE 2
Let P = *Today is Monday* and Q = *Yesterday was Sunday.* Write each of the following statements in symbolic form.

a. Either today is Monday, or yesterday was Sunday.

b. Yesterday was not Sunday, or today is Monday.

c. Either today is not Monday, or yesterday was not Sunday.

Solution
a. This statement is a disjunction since we have the connective *or;* the word *either* also tells us that we have a disjunction. The correct symbolization is $P \vee Q$.

b. This is also a disjunction, but here we have the statements interchanged and Q is negated. Therefore the statement should be symbolized as $\sim Q \vee P$.

c. Here each part of the compound statement is negated; we would symbolize this as $\sim P \vee \sim Q$.

It should be noted that the word *or* can be used in two different ways in a sentence. For example, consider the following statements:

The weather forecast calls for rain or snow.
I will get an A or B for this course.

The first statement illustrates the **inclusive** use of *or,* since it might

rain, it might snow, or it might do both. The second statement illustrates the **exclusive** use of *or,* since it is not possible for both things to occur. That is, the grade for the course is an A or a B, but not both. The symbol commonly used for the exclusive *or* is $\veebar$. Hence *P or Q but not both* is symbolized as $P \veebar Q$. Unless otherwise noted, we shall assume that *or* is used in the inclusive sense.

A conditional is a statement that implies something. The symbol used in mathematics for implication is $\rightarrow$. The statement $P \rightarrow Q$ is usually interpreted as

If P then Q

or, equivalently,

P implies Q

Consider the statement

If the Browns win the championship, then I'll eat my hat.

If we let

P = *The Browns win the championship*
Q = *I'll eat my hat*

the compound statement is symbolized as

$$P \rightarrow Q \qquad \text{(conditional)}$$

Let us examine a conditional statement where the antecedent and consequent are negated. If we let

P = *x is negative*
Q = *x is less than zero*

then $\sim P \rightarrow \sim Q$ is interpreted as

If x is not negative, then x is not less than zero.

Barbara Baker

A biconditional is the conjunction of two conditional statements where the antecedent and consequent of the first statement have been switched in the second. It may be helpful to think of a biconditional as a statement where the antecedent implies the consequent and the same consequent implies the same antecedent. The symbol for the connective in a biconditional is $\leftrightarrow$.

Consider the statement:

Skating is permitted if and only if the ice is six inches thick.

Let

P = *Skating is permitted*
Q = *The ice is six inches thick*

The compound statement is symbolized as

$$P \leftrightarrow Q \qquad \text{(biconditional)}$$

We know from the preceding discussion that this statement is the same as the conjunction of two conditionals, and therefore we are aware that if skating is permitted, then the ice is six inches thick, and if the ice is six inches thick, then skating is permitted; that is,

$$P \leftrightarrow Q \equiv (P \rightarrow Q) \wedge (Q \rightarrow P)$$

(The symbol $\equiv$ means *is the same as.*)

EXAMPLE 3

Let P = *I was late* and Q = *My car broke down.* Write each of the following statements in symbolic form.

a. If I was late, then my car broke down.

b. If my car broke down, then I was late.

c. I was late if and only if my car broke down.

Solution

a. This statement is a conditional; the key is the connective *if ... then ....* We would symbolize this as $P \rightarrow Q$.

b. This is similar to statement *a*, but here we have "my car broke down" as the antecedent, and therefore we symbolize the statement as $Q \rightarrow P$.

c. Statement *c* is a biconditional, as indicated by the phrase "if and only if," and the correct symbolization is $P \leftrightarrow Q$.

EXAMPLE 4

Let P = *Today is Monday,* Q = *Yesterday was Sunday.* Write each of the following statements in words.

a. $P \wedge Q$

b. $\sim P \vee Q$

c. $P \rightarrow \sim Q$

d. $\sim P \leftrightarrow \sim Q$

e. $\sim P \wedge \sim Q$

f. $P \rightarrow P \wedge \sim Q$

Solution

a. This is a conjunction and we may write this directly from the symbols: "Today is Monday and yesterday was Sunday."

b. This is a disjunction with the first part negated: "Today is not Monday or yesterday was Sunday."

c. This is a conditional statement with the consequent negated: "If today is Monday, then yesterday was not Sunday."

d. This statement is a biconditional with both parts negated: "Today is not Monday if and only if yesterday was not Sunday."

e. This conjunction could be written as "Today is not Monday and yesterday was not Sunday." Another correct interpretation is "Neither is today Monday, nor was yesterday Sunday." A *neither-nor* statement is a conjunction where both parts are negated.

f. This statement is a conditional whose consequent is a conjunction: "If today is Monday, then today is Monday and yesterday was not Sunday."

At this point, you should be familiar with the following types of statements and their connective symbols.

Type of statement	Connective	Symbol
negation	not	$\sim$
conjunction	and	$\wedge$
disjunction	or	$\vee$
conditional	if . . . then . . .	$\rightarrow$
biconditional	if and only if iff	$\leftrightarrow$

EXERCISES FOR SECTION 2.2

1. Identify each of the following sentences as a simple statement, compound statement, or neither. Classify each compound statement as a negation, conjunction, disjunction, conditional, or biconditional.
   a. Today is Friday.
   b. It is false that Scott is in class.
   c. You may play tennis here iff you are a member of the club.
   d. Joey is not going to Canada during vacation.
   e. If Addie went swimming, then Julia went sailing.
   f. Close the door when you leave.
   g. Both Ruth and Horace are members of the council.
   h. Either Bill is here, or he did not come to school.

2. Identify each sentence as a simple statement, compound statement, or neither. Classify each compound statement as a negation, conjunction, disjunction, conditional, or biconditional.
   a. If $3 + 4 = 8$, then $9 - 2 = 6$.
   b. When in doubt, punt!
   c. A student may take Math 176 iff he has successfully completed Math 175.
   d. Neither Hugh nor Bill is here.

e. Addie is not late, and Julia is not early.
f. Mary went swimming or cycling.
g. Yesterday was Sunday.
h. Norma was not in class.

3. Let P = *Joe is good* and Q = *Scott is good,* and let us agree that *bad* = *not good.* Write each of the following statements in symbolic form.
   a. Both Joe and Scott are good.
   b. Either Scott or Joe is good.
   c. Joe is not good or Joe is bad.
   d. If Scott is good, then Joe is bad.
   e. It is false that Joe is bad.
   f. Joe is good iff Scott is good.

4. If *Joe is good* and *Scott is good* are true statements, which of the statements in exercise 3 do you think are true?

5. Let P = *Algebra is difficult* and Q = *Logic is easy.* Write each of the following statements in symbolic form.
   a. Algebra is difficult, or logic is easy.
   b. Logic is not easy, and algebra is difficult.
   c. It is false that logic is not easy.
   d. Logic is easy iff algebra is difficult.
   e. If algebra is difficult, then logic is easy.
   f. Neither is algebra difficult nor is logic easy.

6. If *Algebra is difficult* and *Logic is easy* are true statements, which of the statements in exercise 5 do you think are true?

7. Write the following statements in symbolic form using the letters in the parentheses.
   a. Either I sink this putt, or I lose the match. (P, M)
   b. Five is greater than zero, and 5 is positive. (G, P)
   c. If you do not attend class, then you will be dropped from the course. (A, D)
   d. Either the bus is late, or my watch is not working correctly. (B, W)
   e. Two equals 1 iff 3 is greater than 4. (T, F)
   f. Smith will raise taxes if he is elected. (E, R)

8. Write the following statements in symbolic form using the letters in the parentheses.
   a. It is false that Sally did not pass math. (S)
   b. You cannot go to the concert, but I have a ticket. (C, T)
   c. Neither Don nor Bob likes football. (D, B)
   d. If you like ice cream, then you'll love sherbert. (I, S)
   e. I can go to the show iff I finish my homework. (S, H)
   f. I can go to the show if I finish my homework. (H, S)

9. Let P = *I like algebra* and Q = *I like geometry*. Write each of the following statements in words.
   a. $P \wedge Q$
   b. $P \rightarrow \sim Q$
   c. $P \vee Q$
   d. $P \vee \sim Q$
   e. $\sim P \wedge \sim Q$
   f. $P \leftrightarrow Q$

10. Let P = *Tom is tense* and Q = *Freddy is ready.* Write each of the following statements in words.
   a. $P \vee \sim P$
   b. $P \leftrightarrow \sim Q$
   c. $\sim P \vee Q$
   d. $\sim P \wedge \sim Q$
   e. $\sim Q \rightarrow P$
   f. $P \rightarrow \sim Q$

Just For Fun

Can you show that half of 13 is 8?

2.3 DOMINANCE OF CONNECTIVES

Harold M. Lambert

Up to this point in our discussion, except for some of the statements in the exercises, we have encountered few compound statements with multiple connectives. We have examined a few statements that contained a negation in addition to another connective, but that is all.

Suppose that we have a statement such as

A. I will go swimming or I will go cycling, and I will go to the movies.

This statement is a conjunction because the comma separates the sentence into two parts with the connective *and.* But if the comma were omitted, we would not know whether the statement was a disjunction or a conjunction. Some of us would interpret it as

B. I will go swimming, or I will go cycling and I will go to the movies.

while others would interpret it as it is stated in *A*.

We need punctuation marks in writing statements in order to make sense out of them. The following example points out this need: try to make sense out of this statement.

Tom is here or Jim left and Bob came late.

Unless this statement is correctly punctuated, we do not know if it is a disjunction or a conjunction.

Mathematics uses parentheses as punctuation marks and so does logic. Some statements do not need parentheses, while others do; some even need more than one set.

If we let

P = *I will go swimming*
Q = *I will go cycling*
R = *I will go to the movies*

then statement *A* would be symbolized as $(P \vee Q) \wedge R$, while statement *B* would be symbolized as $P \vee (Q \wedge R)$.

It is superfluous to use parentheses in statements such as the following:

$$\sim P, \quad P \wedge Q, \quad P \vee Q, \quad P \rightarrow Q, \quad P \leftrightarrow Q$$

When we interpret a statement, we shall interpret it exactly as written, because that will enable us to determine the symbolic form of the statement.

EXAMPLE 1

Identify each of the following statements as a negation, conjunction, disjunction, conditional, or biconditional.

a. Either the Jets won, or the Bills won and the Browns lost.

b. If John goes to college and Jack goes to art school, then their father will have to take a loan.

c. José is here or Larry is here, and it is hot.

d. It is false that if I took French II, then I had taken French I before.

Solution

a. This statement is a disjunction. The word *either* is the key.

b. Statement *b* is a conditional statement because of the word *if*.

c. Statement *c* is a conjunction, as the comma indicates.

d. The phrase *it is false that* identifies statement *d* as a negation, even though it contains a conditional. *It is false that* negates everything that follows.

When we want to translate symbolic statements into words, we need parentheses. We will also adopt the convention of the dominance of connectives. In logic, some connectives are considered more dominant than others. The following is a list of the connectives in their dominant order, as found in most logic books; the most dominant connective is listed first.

1. biconditional $\leftrightarrow$
2. conditional $\rightarrow$
3. conjunction $\wedge$, disjunction $\vee$
4. negation $\sim$

Note that conjunction and disjunction are listed together. They are of equal value. If a compound statement contains both of these connectives and no others, we must use parentheses to designate it as a conjunction or a disjunction.

EXAMPLE 2

Identify the type of each symbolic statement.

a. $P \wedge Q \leftrightarrow R$

b. $(P \vee Q) \wedge R$

c. $\sim P \vee Q \rightarrow R \wedge S$

d. $\sim(P \wedge Q)$

e. $\sim(P \rightarrow Q \vee R)$

f. $P \vee (Q \rightarrow R)$

Solution

a. The statement $P \wedge Q \leftrightarrow R$ is a biconditional because the double arrow is the dominant connective and there are no parentheses.

b. The parentheses in $(P \vee Q) \wedge R$ separate the statement at the connective; hence it is a conjunction.

c. In $\sim P \vee Q \rightarrow R \wedge S$ there are no parentheses and none are needed, as the conditional arrow is stronger than any of the other connectives in the statement. The statement is a conditional. Note that $\sim P \vee Q$ is the antecedent and $R \wedge S$ is the consequent.

d. At first glance it might appear that $\sim(P \wedge Q)$ is a conjunction, but it isn't, since the negation sign is outside the parentheses. If the statement were $\sim P \wedge Q$, it would be a conjunction, but with the parentheses it is a negation.

e. The negation sign in $\sim(P \rightarrow Q \vee R)$ takes precedence, since it is outside the parentheses and the arrow is inside. Therefore, it is a negation.

f. Since the parentheses divide the statement into two major parts connected by *or*, $P \vee (Q \rightarrow R)$, it is a disjunction.

EXAMPLE 3

Let P = *I can go*, Q = *You can go*, and R = *Lew can go*. Write each of the following symbolic statements in words.

a. $P \wedge (Q \vee R)$

b. $P \vee Q \rightarrow R$

c. $\sim(P \wedge Q)$

d. $\sim P \vee \sim Q$

e. $P \rightarrow Q \wedge R$

Solution

a. Because of the parentheses, this statement is a conjunction, so we write "I can go, and you can go or Lew can go."

b. The arrow is the dominant connective, so statement *b* is a conditional: "If I can go or you can go, then Lew can go."

c. Statement *c* is a negation; hence we have "It is false that both I can go and you can go."

d. This statement is a disjunction with each part negated. Therefore, we have "I cannot go or you cannot go."

e. As in statement *b*, the arrow is the dominant connective, and statement *e* is a conditional: "If I can go, then you can go and Lew can go."

Parentheses are used in logic to tell us what type of statement we are considering. If there are no parentheses, then we follow the convention of the dominance of connectives. The biconditional ($\leftrightarrow$) is the strongest connective, followed by the conditional ($\rightarrow$), and then conjunction ($\wedge$) and disjunction ($\vee$). The conjunction and disjunction are of equal value. The negation ($\sim$) is the weakest connective. If a compound statement contains multiple connectives, then we must sometimes use parentheses to designate the type of statement desired.

EXERCISES FOR SECTION 2.3

1. Add parentheses in each statement to form the type of compound statement indicated. If none are needed, indicate that fact.
   a. Negation: $\sim P \wedge Q \rightarrow R$
   b. Conditional: $\sim P \wedge Q \rightarrow R$
   c. Conjunction: $\sim P \wedge Q \rightarrow R$
   d. Biconditional: $P \wedge Q \leftrightarrow R$
   e. Disjunction: $\sim P \vee Q \wedge R$
   f. Conjunction: $P \wedge Q \leftrightarrow R$
2. Add parentheses in each statement to form the type of compound statement indicated. If none are needed, indicate that fact.
   a. Negation: $\sim P \rightarrow \sim Q$
   b. Biconditional: $\sim P \leftrightarrow Q \vee R$
   c. Disjunction: $P \vee Q \rightarrow R \wedge S$
   d. Conditional: $P \vee Q \rightarrow R \wedge S$
   e. Conjunction: $P \vee Q \rightarrow R \wedge S$
   f. Negation: $\sim P \wedge Q \rightarrow \sim R$
3. Let P = *Algebra is difficult,* Q = *Logic is easy,* and R = *Latin is interesting.* Use appropriate connectives and parentheses to symbolize each statement.
   a. If logic is easy and algebra is difficult, then Latin is interesting.
   b. Latin is interesting and algebra is difficult, or logic is easy.
   c. It is false that logic is easy and algebra is difficult.
   d. Either logic is easy and Latin is interesting, or algebra is difficult.
   e. Algebra is difficult iff Latin is interesting and logic is easy.
   f. Neither is algebra difficult nor is Latin interesting.
4. Let P = *Today is Monday,* Q = *Yesterday was Saturday,* and R = *Tomorrow is Wednesday.* Use appropriate connectives and parentheses to symbolize each statement.
   a. If today is Monday and yesterday was Saturday, then tomorrow is Wednesday.
   b. Either today is Monday and yesterday was Saturday, or tomorrow is Wednesday.
   c. It is false that today is Monday and yesterday was Saturday.
   d. Today is Monday iff yesterday was Saturday and tomorrow is Wednesday.
   e. Today is Monday or tomorrow is Wednesday, and yesterday was Saturday.
   f. If tomorrow is Wednesday, then it is false that today is Monday and yesterday was Saturday.
5. Using the suggested notation, symbolize each statement completely.
   a. If I pass the exam, then I will pass the course and I will graduate. (E, C, G)
   b. Either Scott took the book, or Joe took it and I am upset. (S, J, I)
   c. If you cut class, then you will miss the exam and receive a zero. (C, E, Z)
   d. It is not the case that today is not Friday. (F)
   e. Neither Sue nor Lou is a good swimmer. (S, L)

6. Using the suggested notation, symbolize each statement completely.
   a. Today is not Friday iff tomorrow is Sunday. (F, S)
   b. Paul likes strawberry ice cream and Juanita likes strawberry ice cream, or Connie made a mistake. (P, J, C)
   c. We will win, if Frank makes this free throw. (F, W)
   d. It is false that I was late and missed the test. (L, T)
   e. Today is Friday, and if today is Friday then I get paid. (F, P)

7. Let P = *Algebra is difficult*, Q = *Logic is easy*, and R = *Latin is interesting*. Write each symbolic statement in words.
   a. $P \wedge (Q \vee R)$
   b. $P \wedge Q \rightarrow R$
   c. $P \vee (Q \wedge R)$
   d. $P \wedge (Q \rightarrow R)$
   e. $\sim(P \wedge Q)$
   f. $\sim P \leftrightarrow Q \wedge \sim R$

8. Let P = *Harry studies*, Q = *Harry will pass*, and R = *Harry will succeed*. Write each symbolic statement in words.
   a. $P \rightarrow Q \vee R$
   b. $\sim(P \wedge Q)$
   c. $(P \rightarrow Q) \vee R$
   d. $Q \wedge R \leftrightarrow P$
   e. $\sim P \wedge \sim Q$
   f. $P \wedge (Q \vee \sim R)$

$F \wedge (F \rightarrow P)$

Just For Fun

Even the least-interested sports fan knows that a football field is 100 yards long (120 if we count the end zones), but do you know how wide it is?

2.4 TRUTH TABLES

Since a statement in logic is one which may be true or false, we must be able to determine the truth or falsity of a given statement under given conditions. Remember that logic is precise and we do not want to have to worry about ambiguity. If we are thorough in the study of statements now, we will have a solid foundation for the analysis of the more complicated problems that we will face later.

If we are given a simple statement P, we know that P must be either true or false, but not both. If P is a true statement, then we say that the **truth value** of P is *true;* if P is false, then its truth value is *false.* Now, what happens if we negate P? If P is true, then $\sim P$ must be false, and if P is false, then $\sim P$ must be true. This type of analysis is shown in table 1. This type of table is called a **truth table.** A truth table gives us the truth value of a compound statement for each possible combination of the truth or falsity of the simple statements within the compound statement.

TABLE 1

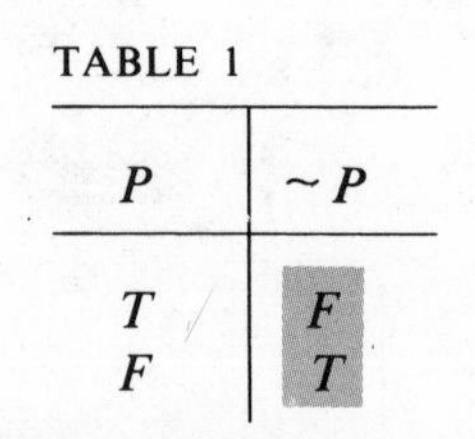

P	$\sim P$
T	F
F	T

The conjunction $P \wedge Q$ contains two simple statements. If P has two possible truth values, namely, true or false, and Q has two possible values, then how many possible values does a compound statement such as $P \wedge Q$ have? We know that P and Q could both be true, or they could both be false. That gives us two possibilities. There is also the case where one could be true and the other false, say, P true and Q false. But, remember that Q could be true and P false. Therefore we have four possibilities. We will not go through this type of reasoning every time we want to construct a truth table. If there are n simple statements in a compound statement, then there are 2^n possible true-false combinations. The statement $P \wedge Q$ has two simple statements and, therefore, we have $2^2 = 2 \times 2 = 4$ possible true-false combinations. The statement $P \wedge Q \rightarrow R$ has $2^3 = 2 \times 2 \times 2 = 8$ possible true-false combinations.

So far, a truth table for $P \wedge Q$ would look like table 2a.

TABLE 2a

P	Q	$P \wedge Q$
T	T	
T	F	
F	T	
F	F	

Now let us examine a statement that will enable us to complete the truth table for a conjunction $P \wedge Q$. Consider the statement

Today is Tuesday and I have a math class.

Let

P = *Today is Tuesday*
Q = *I have a math class*

The statement is a conjunction and is symbolized as $P \wedge Q$. When is this compound statement true? In order for the whole statement to be true, we must have both parts true. If the person speaking said the statement on any day other than Tuesday, he would not be telling the truth. This would take care of the possibilities, F—T, F—F. We have already agreed that the T—T combination (both parts true) yields a true. Now, if the person making the statement said it on Tuesday, but didn't have math class, then he would be making a false statement. Hence, the T—F possibility yields a false. The completed truth table appears in table 2b. It shows that a conjunction is only true when both parts are true; otherwise it is false.

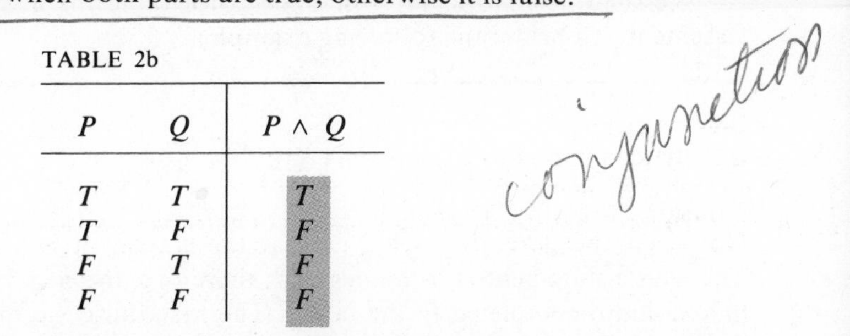

TABLE 2b

P	Q	$P \wedge Q$
T	T	T
T	F	F
F	T	F
F	F	F

When is a disjunction true? If someone says that he or she is going to go swimming or cycling, then he or she is telling the truth when he or she goes swimming and not cycling. Likewise, he or she is telling the truth when he or she goes cycling and not swimming. Suppose the person goes golfing. Then the original statement is false. Since we are using the inclusive *or* in logic, the person is also telling the truth when he or she goes both swimming and cycling. Table 3 shows the truth table for a disjunction. It shows that a disjunction is true unless both parts are false.

TABLE 3

P	Q	$P \vee Q$
T	T	T
T	F	T
F	T	T
F	F	F

Recall from section 2.2 that the exclusive *or*—that is, one event or the other, but not both—is symbolized by $P \veebar Q$. This means we want P or Q, but not both. The truth table for the exclusive *or* is shown in table 4. The statement $P \veebar Q$ is true only when exactly one part is true; otherwise it is false.

TABLE 4

P	Q	$P \veebar Q$
T	T	F
T	F	T
F	T	T
F	F	F

We shall now work out a detailed truth table for a compound statement. Consider the following example.

EXAMPLE 1
Construct a truth table for $\sim(P \vee Q)$.

Solution
Note that the parentheses tell us that the statement is a negation. The whole statement is to be negated; therefore, the negation is the final column completed in the table. The first thing we must do is list the truth values for the variables. We have numbered the columns so that you may follow, in order, the step-by-step process.

P	Q	$\sim(P$	$\vee$	$Q)$
T	T	T		T
T	F	T		F
F	T	F		T
F	F	F		F
1	2	1		2

Since $\vee$ is the connective inside the parentheses, we complete the column for this connective, column 3.

P	Q	$\sim(P$	$\vee$	$Q)$
T	T	T	T	T
T	F	T	T	F
F	T	F	T	T
F	F	F	F	F
1	2	1	3	2

Now, to complete the truth table, we must negate the statement inside the parentheses (the disjunction). To negate the statement, we negate column 3. This gives us column 4 and completes the table.

P	Q	$\sim$	$(P$	$\vee$	$Q)$
T	T	F	T	T	T
T	F	F	T	T	F
F	T	F	F	T	T
F	F	T	F	F	F
1	2	4	1	3	2

Note that we fill in the columns headed by variables before filling in those columns headed by connectives. We fill in the column of the least dominant connective before filling in the columns of the more dominant connectives.

EXAMPLE 2
Construct a truth table for $\sim(P \wedge \sim Q)$.

Solution
This problem is similar to example 1, but it is a little more involved. The statement is a negation, so that is the last column completed in

our truth table. Listing the truth values for the variables, we have the following table.

P	Q	$\sim$	$(P$	$\wedge$	$\sim$	$Q)$
T	T		T			T
T	F		T			F
F	T		F			T
F	F		F			F
1	2		1			2

The statement inside the parentheses is a conjunction, but before we can complete the column for the connective we must negate Q, because the conjunction is of P with *not* Q.

P	Q	$\sim$	$(P$	$\wedge$	$\sim$	$Q)$
T	T		T		F	T
T	F		T		T	F
F	T		F		F	T
F	F		F		T	F
1	2		1		3	2

Now, to figure out the value of the conjunction, we compare columns 1 and 3. This gives us column 4, which is the column that we negate to get column 5 and complete the table.

P	Q	$\sim$	$(P$	$\wedge$	$\sim$	$Q)$
T	T	T	T	F	F	T
T	F	F	T	T	T	F
F	T	T	F	F	F	T
F	F	T	F	F	T	F
1	2	5	1	4	3	2

The completed truth table (column 5) tells us that the statement $\sim(P \wedge \sim Q)$ is true in all cases except when P is true and Q is false.

EXERCISES FOR SECTION 2.4

In exercises 1–12, construct a truth table for each statement.

1. $\sim(P \wedge Q)$
2. $\sim P \wedge Q$
3. $\sim P \wedge \sim Q$
4. $P \vee \sim P$
5. $P \wedge \sim P$
6. $\sim P \vee \sim Q$
7. $P \vee \sim Q$
8. $\sim P \vee Q$
9. $\sim(P \vee \sim Q)$
10. $\sim(\sim P \vee \sim Q)$
11. $\sim P \vee (P \wedge \sim Q)$
12. $Q \wedge (\sim P \vee Q)$
13. Show that each statement has the same truth value as the *exclusive or*, $P \veebar Q$ (that is, truth values $F\ T\ T\ F$).
    a. $(P \vee Q) \wedge (\sim P \vee \sim Q)$
    b. $(P \wedge \sim Q) \vee (\sim P \wedge Q)$

Just For Fun

How many dimples are on a golf ball?

336

2.5 MORE TRUTH TABLES—CONDITIONAL AND BICONDITIONAL STATEMENTS

Many statements in everyday language, particularly in mathematics, deal with the conditional. We are often confronted with the *if . . . then . . .* type of statement. When is the statement $P \rightarrow Q$ true, and when is it false? We shall examine a completed truth table shortly. But, first let us try to justify the entries by means of a discussion. Try to follow the discussion closely so that the resulting table appears natural. Some of the "true" cases may seem odd, but that is because we usually do not use conditional statements where the antecedent is known to be false.

Suppose you are told the following by a counselor.

If you pass biology, then you will graduate.

Now, the statement says that if you pass biology, then you will graduate, regardless of what you may have done previously, or will do later on. Let

P = *You pass biology*
Q = *You will graduate*

Since we have two variables in the compound statement $P \rightarrow Q$, there are four possible true-false combinations. First let's examine the case where P is true and Q is true. That is, you do pass biology and you do graduate. The couselor's statement must be a true one; it is certainly not a lie. (If the counselor had lied, then the statement would be false.)

Suppose you do pass biology, but you do not graduate. This is the case where P is true and Q is false. Has the counselor told the truth? No, the original statement is not true. Therefore, the statement $P \rightarrow Q$ is false when the antecedent is true and the consequent is false.

Now, let's consider the case where P is false and Q is true. In this instance, we have the situation where you did not pass biology, but you still graduated. Examine the original statement: "If you pass biology, then you will graduate." Did the counselor lie to you? No, the counselor said what would happen if you did pass biology, not what would happen if you did not pass. There may be other ways of meeting the graduation requirements, such as taking another course. In this case, the counselor's original statement does not apply. A lie would be a false statement, but since the counselor did not lie to you, the original statement must be true! Therefore, the statement $P \rightarrow Q$ is true when P is false and Q is true.

Ewing Galloway

The last case to consider is when the antecedent and consequent are both false, that is, P is false and Q is false. You did not pass biology and you did not graduate. The original statement said, "If you pass biology, then you will graduate." In this instance, you did not pass biology and you did not graduate. Did the counselor lie to you? No, the counselor said what would happen if you *passed* biology. As we stated earlier, a lie would be a false statement; if the counselor did not lie, then the original statement must be true.

From the discussion, we are now aware that a conditional is true in all cases except one: a conditional is false when the antecedent is true and the consequent is false; otherwise it is true. The truth table for the conditional statement is shown in table 5.

TABLE 5

P	Q	$P \rightarrow Q$
T	T	T
T	F	F
F	T	T
F	F	T

Consider the following statement:

If the moon is made of green cheese, then this is a mathematics book.

Is the statement true or false? It is true! Why? The reason that the statement is true in this particular case is that the antecedent is false and the consequent is true. We see from the truth table for a conditional that F—T yields a T. Remember, the only time a conditional statement is false is when the antecedent is true and the consequent is false. Do not let the words in a statement bother you, no matter how illogical they sound. Another odd conditional statement to consider is

If $2 = 5$, *then* $6 = 9$.

Is this statement true or false? It is true. We can determine this from our truth table: since the antecedent is false, the statement has to be true.

EXAMPLE 1

Construct a truth table for $P \rightarrow \sim Q$.

Solution

The statement is a conditional; the arrow is the most dominant connective. Therefore, the last column completed in the truth table will be under the arrow. We first list the truth values for the variables.

P	Q	P	$\rightarrow$	$\sim$	Q
T	T	T			T
T	F	T			F
F	T	F			T
F	F	F			F
1	2	1			2

Next, we fill in the column for the least dominant connective, the negation.

P	Q	P	$\rightarrow$	$\sim$	Q
T	T	T		F	T
T	F	T		T	F
F	T	F		F	T
F	F	F		T	F
1	2	1		3	2

Column 3 is derived from column 2. In order to figure out the truth values for the arrow, we must compare columns 1 and 3 (the truth values for the antecedent and consequent). Thus we obtain column 4, which completes the table.

P	Q	P	$\rightarrow$	$\sim$	Q
T	T	T	F	F	T
T	F	T	T	T	F
F	T	F	T	F	T
F	F	F	T	T	F
1	2	1	4	3	2

EXAMPLE 2

Construct a truth table for $P \wedge Q \rightarrow \sim P$.

Solution

This statement is a conditional statement since there are no parentheses and the arrow is the dominant connective. The antecedent is $P \wedge Q$ and the consequent is $\sim P$. The computed truth table is

P	Q	P	$\wedge$	Q	$\rightarrow$	$\sim$	P
T	T	T	T	T	F	F	T
T	F	T	F	F	T	F	T
F	T	F	F	T	T	T	F
F	F	F	F	F	T	T	F
1	2	1	3	2	5	4	1

Column 3 is the truth value for the antecedent and column 4 is the truth value for the consequent. We compare columns 3 and 4 (using the rules for the conditional) to obtain column 5. This tells us that the statement $P \wedge Q \rightarrow \sim P$ is true in all cases except when P and Q are both true.

Recall from section 2.2 that a biconditional is the conjunction of two conditional statements where the antecedent and consequent of the first conditional have been switched in the second. Therefore, a biconditional statement $P \leftrightarrow Q$ is the same as $(P \rightarrow Q) \wedge (Q \rightarrow P)$. Using this information, let us construct the truth table for the biconditional $P \leftrightarrow Q$ by constructing a table for the statement $(P \rightarrow Q) \wedge (Q \rightarrow P)$. If we figure out the truth table for the conjunction, we will know the truth values for $P \leftrightarrow Q$, since they are equivalent. The first steps in our truth table are shown in table 6a.

TABLE 6a

P	Q	$(P$	$\rightarrow$	$Q)$	$\wedge$	$(Q$	$\rightarrow$	$P)$
T	T	T	T	T		T	T	T
T	F	T	F	F		F	T	T
F	T	F	T	T		T	F	F
F	F	F	T	F		F	T	F
1	2	1	3	2		2	4	1

So far we have worked out both sides of the conjunction. Note that each side is a conditional, but the parentheses tell us that $\wedge$ is the dominant connective. Our final step is to compare columns 3 and 4 using the rules for a conjunction. This leads to column 5, which completes the truth table. See table 6b.

TABLE 6b

P	Q	$(P$	$\rightarrow$	$Q)$	$\wedge$	$(Q$	$\rightarrow$	$P)$
T	T	T	T	T	T	T	T	T
T	F	T	F	F	F	F	T	T
F	T	F	T	T	F	T	F	F
F	F	F	T	F	T	F	T	F
1	2	1	3	2	5	2	4	1

Table 6b shows that the conjunction of $(P \rightarrow Q)$ and $(Q \rightarrow P)$ is true when P and Q are both true, and when P and Q are both false. Since this statement is equivalent to $P \leftrightarrow Q$, we also know the truth value of $P \leftrightarrow Q$. We may say that a biconditional is true when P and Q are both true and when P and Q are both false. That is, a biconditional is true when both parts have the same truth value; otherwise it is false. The truth table for the biconditional appears in table 7.

TABLE 7

P	Q	$P \leftrightarrow Q$
T	T	T
T	F	F
F	T	F
F	F	T

EXAMPLE 3
Construct a truth table for $(\sim P \vee Q) \leftrightarrow (P \rightarrow Q)$.

Solution

P	Q	$(\sim$	P	$\vee$	$Q)$	$\leftrightarrow$	$(P$	$\rightarrow$	$Q)$
T	T	F	T	T	T	T	T	T	T
T	F	F	T	F	F	T	T	F	F
F	T	T	F	T	T	T	F	T	T
F	F	T	F	T	F	T	F	T	F
1	2	3	1	4	2	6	1	5	2

As with the other truth tables, we listed the possibilities for P and Q first (columns 1 and 2). Our next step was column 3. Then we compared columns 3 and 2 to get column 4 (the disjunction). On the right side, we compared 1 and 2 to get column 5 (the conditional). Our final step was to compare columns 4 and 5 to obtain column 6. Here we used the rule for the biconditional.

Note that column 6 in example 3 contains only T's. This tells us that the compound statement $(\sim P \vee Q) \leftrightarrow (P \rightarrow Q)$ is true for all cases, regardless of the truth values of the variables P and Q. A compound statement which is true for any combination of truth values of the variables in the statement is called a **tautology.**

A statement that yields all F's, that is, a statement which is always false, is called a **contradiction.** A common example of a contradiction is $P \wedge \sim P$.

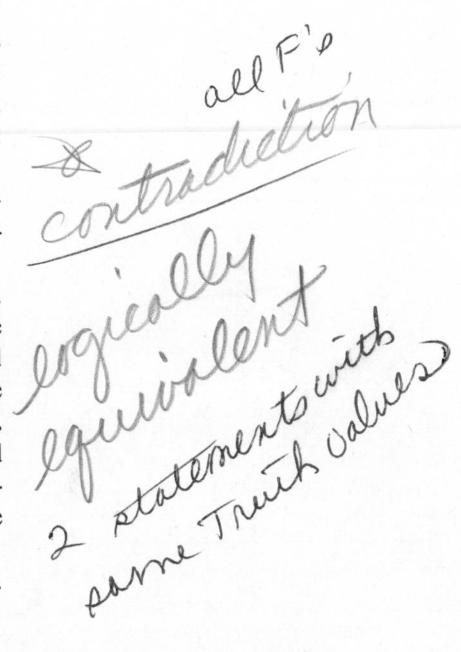

all F's
contradiction
logically
equivalent
2 statements with
same Truth values

A biconditional that is a tautology tells us something else, too. In example 3, the truth values for columns 4 and 5 are exactly the same. Columns 4 and 5 represent the truth values for the left and right members of the biconditional. Two statements that have exactly the same truth values are said to be **logically equivalent.** Therefore, a biconditional that is a tautology tells us that the left and right members are logically equivalent. In some logic texts, the biconditional is referred to as the **equivalence,** and therefore in example 3 we have a tautological equivalence.

EXAMPLE 4
How many cases have to be considered in order to construct a truth table for the statement $(P \wedge Q) \rightarrow R$?

Solution
Note that this compound statement contains three simple statements. How many true-false combinations are there? Recall that if there are n simple statements in a compound statement, then there are 2^n possible true-false combinations. So we have $2^3 = 2 \times 2 \times 2 = 8$ possible cases to consider.

EXAMPLE 5
Construct a truth table for $(P \wedge Q) \rightarrow R$.

Solution
There are 8 possible true-false combinations to consider in our trut table. We should have 8 lines in our table so we can consider possibilities for the three variables, P, Q, and R. We have to co sider "all true," "two true and one false," "one true and two fal and "all false." Here is a listing of all such possibilities for P and R.

P	Q	R
T	T	T
T	T	F
T	F	T
T	F	F
F	T	T
F	T	F
F	F	T
F	F	F

Next we complete the truth table.

P	Q	R	$(P$	$\wedge$	$Q)$	$\rightarrow$	R
T	T	T	T	T	T	T	T
T	T	F	T	T	T	F	F
T	F	T	T	F	F	T	T
T	F	F	T	F	F	T	F
F	T	T	F	F	T	T	T
F	T	F	F	F	T	T	F
F	F	T	F	F	F	T	T
F	F	F	F	F	F	T	F
1	2	3	1	4	2	5	3

EXAMPLE 6

Construct a truth table for $(P \wedge Q) \vee R$.

Solution

The completed truth table appears below. The parentheses indicate that the statement is a disjunction. Hence column 5 is the final column needed to complete our table.

P	Q	R	$(P$	$\wedge$	$Q)$	$\vee$	R
T	T	T	T	T	T	T	T
T	T	F	T	T	T	T	F
T	F	T	T	F	F	T	T
T	F	F	T	F	F	F	F
F	T	T	F	F	T	T	T
F	T	F	F	F	T	F	F
F	F	T	F	F	F	T	T
F	F	F	F	F	F	F	F
1	2	3	1	4	2	5	3

EXERCISES FOR SECTION 2.5

In exercises 1–20, construct a truth table for each symbolic statement.

1. $P \rightarrow Q$
2. $\sim(P \rightarrow Q)$
3. $\sim P \rightarrow \sim Q$
4. $P \rightarrow \sim Q$
5. $\sim P \rightarrow Q$
6. $\sim P \leftrightarrow Q$
7. $\sim P \leftrightarrow \sim Q$
8. $P \wedge Q \rightarrow P$
9. $P \vee Q \rightarrow \sim Q$
10. $P \rightarrow Q \vee \sim P$
11. $(P \rightarrow Q) \vee P \rightarrow Q$
12. $P \wedge (\sim P \vee Q) \rightarrow Q$
13. $P \wedge Q \leftrightarrow P \vee Q$
14. $\sim P \wedge \sim Q \leftrightarrow \sim(P \vee Q)$
15. $(P \vee Q) \wedge R$
16. $P \vee (Q \wedge R)$
17. $(P \wedge Q) \vee (P \wedge R)$
18. $P \wedge (Q \vee \sim R)$
19. $P \leftrightarrow Q \vee R$
20. $P \wedge \sim(Q \vee R) \rightarrow P \vee Q$
21. Determine whether $\sim(P \vee Q)$ is logically equivalent to $\sim P \wedge \sim Q$.
22. Determine whether $\sim(P \wedge Q)$ is logically equivalent to $\sim P \vee \sim Q$.
23. Are $P \wedge \sim Q$ and $\sim(\sim P \vee Q)$ logically equivalent?
24. Are $\sim P \vee Q$ and $\sim(P \wedge \sim Q)$ logically equivalent?
25. In example 3 we found that $\sim P \vee Q$ is logically equivalent to $P \rightarrow Q$. Using this equivalence, rewrite each of the following statements. For example,

    If today is Monday, then tomorrow is Tuesday

    can be rewritten as

    Today is not Monday or tomorrow is Tuesday.

a. If the tide is out, then we can go clamming.
b. If Bill drove his van, then he brought the packages.
c. If today is Wednesday, then tomorrow is not Friday.
d. Either 2 does not equal 3, or 4 equals 6.
e. Bob didn't pass the test, or he is unhappy about something else.

Just For Fun

If 3 cats kill 3 rats in 3 minutes, then how many cats will it take to kill 30 rats in 30 minutes?

2.6 DE MORGAN'S LAW AND EQUIVALENT STATEMENTS

In the preceding section, we were introduced to logically equivalent statements. Two statements that have exactly the same truth values were said to be *logically equivalent.* In example 3 of section 2.5, we showed that $P \rightarrow Q$ is logically equivalent to $\sim P \vee Q$. We can write this as

$$P \rightarrow Q \equiv \sim P \vee Q$$

We determined this by constructing the truth table for the statement $(\sim P \vee Q) \leftrightarrow (P \rightarrow Q)$, and obtaining all T's for the biconditional (a tautology). By means of this technique, we can always determine if one statement is logically equivalent to another.

Sometimes when we are given a statement we can create another statement which is logically equivalent to it. We can change a disjunction, a conjunction, or the negation of one of these to an equivalent statement by means of a rule known as **De Morgan's law.** To illustrate how this works, consider the statement

1. *Neither the Giants nor the Browns won.*

We know from previous discussions that this is the same as

2. *The Giants did not win and the Browns did not win.*

Can we restate this in still another way? Consider

3. *It is not the case that either the Giants or Browns won.*

Let P = *The Giants won* and Q = *The Browns won;* then statement 2 would be symbolized as $\sim P \wedge \sim Q$ and statement 3 would be symbolized as $\sim(P \vee Q)$. Are these two statements logically equivalent? We can determine if they are by means of a truth table. Let's examine the truth table for $(\sim P \wedge \sim Q) \leftrightarrow \sim(P \vee Q)$. If it is a tautology, then we will know that the two statements are equivalent.

TABLE 8

P	Q	$(\sim$	P	$\wedge$	$\sim$	$Q)$	$\leftrightarrow$	$\sim$	$(P$	$\vee$	$Q)$
T	T	F		F	F		T	F	T	T	T
T	F	F		F	T		T	F	T	T	F
F	T	T		F	F		T	F	F	T	T
F	F	T		T	T		T	T	F	F	F
1	2	3		5	4		8	7	1	6	2

Column 8 of table 8 shows us that the given statement is a tautology, thereby verifying the fact that statements 2 and 3 are logically equivalent.

De Morgan's law enables us to create an equivalent statement when we are given a certain type of statement. To begin with, we must have some form of a disjunction or conjunction, or the negation of one of these. In order to create an equivalent statement using De Morgan's law, we must perform the following three steps.

1. Negate the whole statement.
2. Negate each statement that makes up the disjunction or conjunction.
3. Change the conjunction to a disjunction or the disjunction to a conjunction.

EXAMPLE 1

Use De Morgan's law to create a statement equivalent to $P \wedge Q$.

Solution

We first negate the whole statement, which gives us $\sim(P \wedge Q)$. Next we negate each part, and that yields $\sim(\sim P \wedge \sim Q)$. We then make the third and final change, changing $\wedge$ to $\vee$, which gives us $\sim(\sim P \vee \sim Q)$. Hence $P \wedge Q \equiv \sim(\sim P \vee \sim Q)$.

EXAMPLE 2
Use De Morgan's law to create a statement equivalent to $\sim(\sim P \vee Q)$.

Solution
We are given $\sim(\sim P \vee Q)$. First we negate the whole statement, which makes it $\sim\sim(\sim P \vee Q)$ or $(\sim P \vee Q)$. Now we negate each part, that is, $\sim\sim P \vee \sim Q$, which leaves us with $P \vee \sim Q$. (Note that $\sim\sim P \equiv P$; $\sim\sim P$ is called a **double negation**.) We apply the third step and change $\vee$ to $\wedge$, which yields $P \wedge \sim Q$. Finally, we have $\sim(\sim P \vee Q) \equiv P \wedge \sim Q$.

EXAMPLE 3
Use De Morgan's law to create a statement equivalent to $P \vee Q$.

Solution
$\sim(\sim P \wedge \sim Q)$

EXAMPLE 4
Use De Morgan's law to write a statement equivalent to the statement "It is false that Allan abhors anatomy and Lucy loves Latin."

Solution
Let A = *Allan abhors anatomy* and L = *Lucy loves Latin;* then the given sentence may be symbolized as $\sim(A \wedge L)$. Now we create an equivalent statement using De Morgan's law: $\sim(A \wedge L) \equiv \sim A \vee \sim L$. Translating the new statement, we have "Either Allan does not abhor anatomy or Lucy does not love Latin."

EXAMPLE 5
Use De Morgan's law to write a statement equivalent to the statement "Sally did not stay and Quincy quit."

Solution
Let S = *Sally stayed* and Q = *Quincy quit.* The statement may be symbolized as $\sim S \wedge Q$. Using De Morgan's law, we have $\sim S \wedge Q \equiv \sim(S \vee \sim Q)$. Translating the equivalent statement, we have "It is not the case that Sally stayed or that Quincy did not quit."

EXAMPLE 6
Use De Morgan's law to create a statement equivalent to $\sim(P \rightarrow Q)$.

Solution
At first glance, it may seem that we cannot do this problem, because we do not have a disjunction or conjunction. But, recall that in ex-

... of section 2.5 we discovered that $P \rightarrow Q \equiv \sim P \vee Q$. Therefore, we can take the given statement and rewrite it as $\sim(\sim P \vee Q)$. Now we are ready to use De Morgan's law: $\sim(\sim P \vee Q) \equiv P \wedge \sim Q$. Therefore $\sim(P \rightarrow Q) \equiv P \wedge \sim Q$.

Note that example 6 verifies that a statement such as

It is false that if today is Friday, then I get paid

is equivalent to

Today is Friday and I did not get paid.

EXAMPLE 7

Use De Morgan's law to write a statement equivalent to the statement "Neither Mary nor Doris came to class."

Solution

Let M = *Mary came to class* and D = *Doris came to class.* The statement can be symbolized as $\sim M \wedge \sim D$. Using De Morgan's law, we have $\sim M \wedge \sim D \equiv \sim(M \vee D)$. Translating the equivalent statement, we have "It is false that Mary came to class or Doris came to class." This can also be stated as "It is false that Mary or Doris came to class."

It is interesting to note that the logical operation *and* corresponds to the set operation *intersection.* It is also the case that *or* corresponds to set *union,* and that *not* corresponds to set *complementation.* Two of the most common examples of De Morgan's law for equivalent statements are

$$\sim(P \wedge Q) \equiv \sim P \vee \sim Q$$
$$\sim(P \vee Q) \equiv \sim P \wedge \sim Q$$

The corresponding expressions in set theory are

$$(A \cap B)' = A' \cup B'$$
$$(A \cup B)' = A' \cap B'$$

These two statements can be shown to be true by means of Venn diagrams (see example 5 and exercise 5c in section 1.5). Therefore, you can see that set operations are similar to some of those in logic. The basic difference is in the notation.

Remember that in logic we can only use De Morgan's law on

some form of a disjunction, conjunction, or negation of one of these. The following is a list of logically equivalent statements. Their equivalence can be verified by means of truth tables.

Equivalent Statements	
De Morgan's law	$\sim(P \wedge Q) \equiv \sim P \vee \sim Q$
	$\sim(P \vee Q) \equiv \sim P \wedge \sim Q$
Implication	$P \to Q \equiv \sim P \vee Q$
Contraposition	$P \to Q \equiv \sim Q \to \sim P$
Biconditional	$P \leftrightarrow Q \equiv (P \to Q) \wedge (Q \to P)$
Association	$(P \wedge Q) \wedge R \equiv P \wedge (Q \wedge R)$
	$(P \vee Q) \vee R \equiv P \vee (Q \vee R)$
Distribution	$P \wedge (Q \vee R) \equiv (P \wedge Q) \vee (P \wedge R)$
	$P \vee (Q \wedge R) \equiv (P \vee Q) \wedge (P \vee R)$
Idempotent	$P \wedge P \equiv P$
	$P \vee P \equiv P$

EXERCISES FOR SECTION 2.6

In exercises 1–10, use De Morgan's law to create a statement equivalent to each given statement.

1. $\sim(\sim P \wedge Q)$
2. $\sim(P \vee \sim Q)$
3. $P \wedge Q$
4. $P \vee Q$
5. $P \wedge \sim Q$
6. $\sim P \vee \sim Q$
7. $\sim(\sim P \vee \sim Q)$
8. $\sim(\sim P \wedge \sim Q)$
9. $\sim(P \to \sim Q)$
10. $\sim[P \vee (Q \wedge R)]$

In exercises 11–22, use De Morgan's law to rewrite each statement.

11. It is not the case that John or Janie went to the party.
12. Neither Bill nor Bob is tall.
13. I did not pass the test, or I studied too much.
14. Either Sandra or Julia cut class.
15. It is false that logic is dull and not interesting.
16. Neither Sam nor Sid is a good swimmer.
17. Either the bus is late, or my watch is not working correctly.
18. You cannot go to the game and I cannot get a ticket.
19. It is not the case that x is greater than zero and x is negative.
20. It is false that today is Monday or tomorrow is Wednesday.

*21. If the wind doesn't come up, then we can't sail.

*22. If you do not attend class, then you will be dropped from the course.

Use De Morgan's law on the set expressions in exercises 23–30 to create equivalent expressions.

23. $(A \cap B)'$
24. $(A \cup B)'$
25. $A' \cap B'$
26. $A' \cup B$
27. $A \cap B$
28. $(A' \cap B)'$
29. $(A \cup B')'$
30. $(A' \cap B')'$

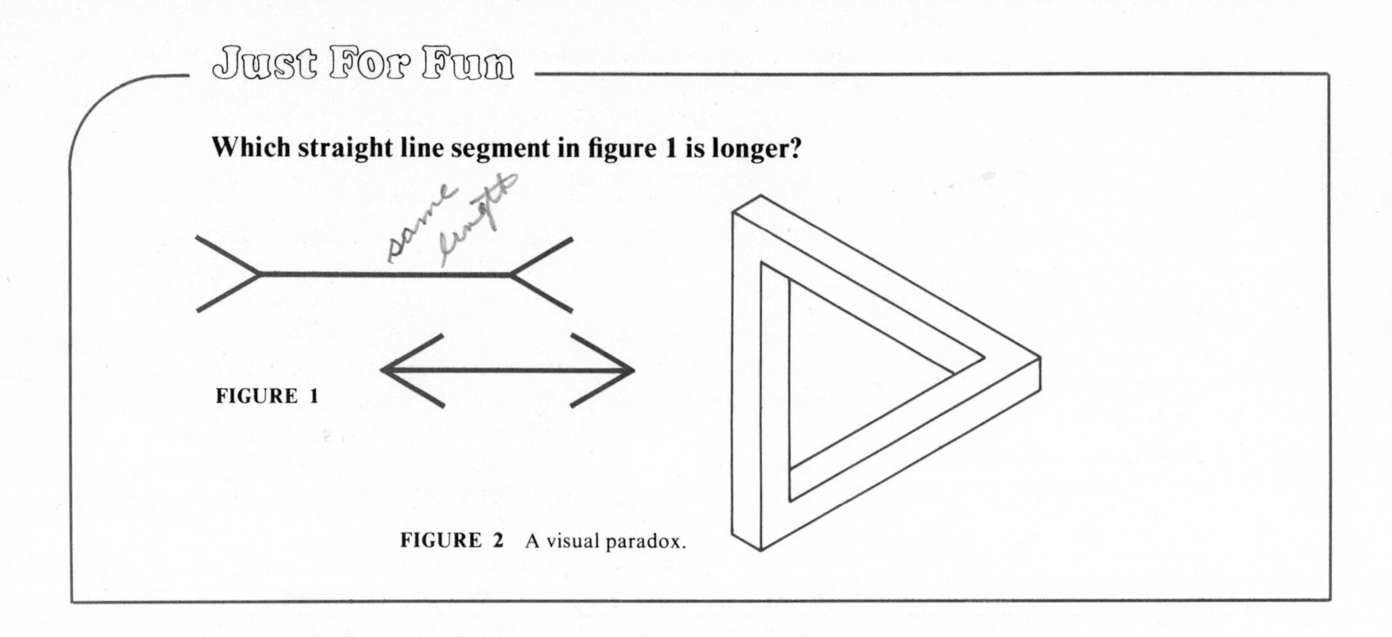

FIGURE 1

FIGURE 2 A visual paradox.

2.7 VALID ARGUMENTS

A truth table indicates all of the possible truth values for a given compound statement. We also discovered that a truth table can help us determine logical equivalence. Now we shall see another application of the truth table: by means of a truth table, we can determine if an argument is valid or not.

An **argument,** or **proof,** consists basically of two parts: the given statements, which are called premises, and the conclusion. You may recall proofs from geometry. A proof or argument is said to be **valid** if the conclusion follows logically from the premises. What does this mean? This is really another way of saying that an argument is valid if the premises imply the conclusion. We can take the premises of an argument and connect them using conjunction, and then use this compound statement as the antecedent of a conditional statement of which the conclusion is the consequent. If this conditional, or implication, is a tautology, then the argument is valid. A tautological implication tells us that the premises do indeed imply the conclusion.

Consider the following argument:

If the given figure is a square, then it is a rectangle.
(Major Premise)
The given figure is a square.
(Minor Premise)
Therefore, it is a rectangle.
(Conclusion)

Let

$$P = \textit{The given figure is a square}$$
$$Q = \textit{It is a rectangle}$$

Then the argument would be symbolized as

$$\begin{array}{l} P \rightarrow Q \\ \underline{P} \\ Q \end{array}$$

If we connect the premises using conjunction, and then imply the conclusion, the corresponding conditional statement is

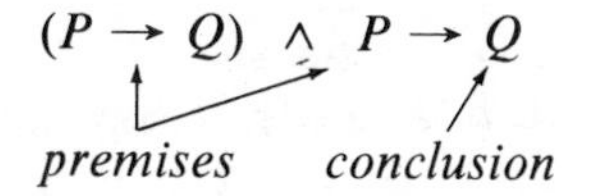

Is this argument valid? Does the conclusion follow logically from the premises? That is, is $(P \rightarrow Q) \wedge P \rightarrow Q$ true for all possible values of P and Q? Table 9 shows the truth table for $(P \rightarrow Q) \wedge P \rightarrow Q$.

TABLE 9

P	Q	$(P \rightarrow Q)$	$\wedge$	P	$\rightarrow$	Q
T	T	T	T	T	T	T
T	F	F	F	T	T	F
F	T	T	F	F	T	T
F	F	T	F	F	T	F
1	2	3	4	1	5	2

Note that we skipped a couple of steps in table 9. Our major concern is column 5, and we see that the implication is true in all cases, therefore verifying the fact that the premises imply the conclusion in all cases. Hence, the argument is valid.

Maybe you had already guessed that. However, we are trying to convey what a valid argument is, and how to determine validity. It should be noted from table 9 that a valid argument can have either a true or a false conclusion. But, if the premises are true, a valid argument has to have a true conclusion. If you have an argument in which all the premises are true, but the conclusion is false, then the argument is **invalid.**

EXAMPLE 1

Determine whether the following argument is valid or invalid.

If I study, then I will pass math.
I didn't pass math.
Therefore, I didn't study.

Solution

Let P = *I study* and Q = *I will pass math.* The argument in symbolic form would appear as

$$\begin{array}{l} P \rightarrow Q \\ \underline{\sim Q} \\ \sim P \end{array}$$

If we connect the premises using conjunction and imply the conclusion, the corresponding conditional statement is

$$\underbrace{(P \rightarrow Q) \wedge \sim Q}_{\text{premises}} \rightarrow \underbrace{\sim P}_{\text{conclusion}}$$

The truth table for this conditional statement is

P	Q	$(P \rightarrow Q)$	$\wedge$	$\sim Q$	$\rightarrow$	$\sim P$
T	T	T	F	F	T	F
T	F	F	F	T	T	F
F	T	T	F	F	T	T
F	F	T	T	T	T	T
1	2	3	6	4	7	5

Column 7 tells us that we have a tautology, and therefore the argument is valid. The premises do imply the conclusion.

The argument used in example 1 illustrates one of the basic rules of inference in logic: the **law of contraposition,** or **modus tollens.** If the statement $P \rightarrow Q$ is true, and $\sim Q$ is known to be true, then $\sim P$ must be true.

EXAMPLE 2

Determine whether the following argument is valid or invalid.

If you are healthy then you are wealthy.
You are wealthy.
Therefore, you are healthy.

Solution

Let P = *You are healthy* and Q = *You are wealthy.* The argument in symbolic form would be

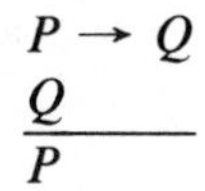

$$\begin{array}{l} P \rightarrow Q \\ \underline{Q \qquad} \\ P \end{array}$$

If we connect the premises using conjunction and imply the conclusion, the corresponding conditional statement is

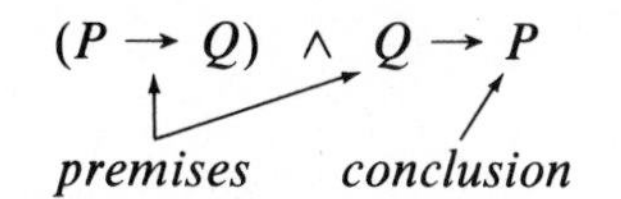

$$(P \rightarrow Q) \wedge Q \rightarrow P$$

The corresponding truth table is

P	Q	$(P \rightarrow Q)$	$\wedge$	Q	$\rightarrow$	P
T	T	T	T	T	T	T
T	F	F	F	F	T	T
F	T	T	T	T	F	F
F	F	T	F	F	T	F
1	2	3	4	2	5	1

The resulting conditional statement is not a tautology. See column 5. The premises do not imply the conclusion in all cases, and therefore the argument is invalid. Note that in the third horizontal line of the truth table we have a case where the premises are true, but the conclusion is false. A valid argument does not allow us to go from true premises to a false conclusion.

An invalid argument is one in which the conclusion does not logically follow from the premises. That is, the form of the argument does not permit only true conclusions to be logically derived from true premises. The form of the invalid argument in example 2 is

$$\begin{array}{l} P \rightarrow Q \\ \underline{Q \qquad} \\ P \end{array}$$

This form of incorrect reasoning is commonly called the **fallacy of affirming the consequent.** If the consequent Q of a conditional statement $P \rightarrow Q$, is true, it does *not* follow that the antecedent P must be true. The statement $P \rightarrow Q$ can still be true, even if the consequent Q is false.

Can an argument be valid if one or more of the premises is false and the conclusion is false? Consider the argument in example 3.

EXAMPLE 3

Determine whether the following argument is valid or invalid.

If the moon is made of green cheese, then two equals one.
The moon is made of green cheese.
Therefore, two equals one.

Solution

Let P = *The moon is made of green cheese* and Q = *Two equals one.* The argument in symbolic form would appear as

$$\begin{array}{l} P \rightarrow Q \\ \underline{P \qquad\quad} \\ Q \end{array}$$

If we connect the premises using conjunction and imply the conclusion, the corresponding conditional statement is

$$\underbrace{(P \rightarrow Q) \wedge P}_{\text{premises}} \rightarrow \underbrace{Q}_{\text{conclusion}}$$

The corresponding truth table is

P	Q	$(P \rightarrow Q)$	$\wedge$	P	$\rightarrow$	Q
T	T	T	T	T	T	T
T	F	F	F	T	T	F
F	T	T	F	F	T	T
F	F	T	F	F	T	F
1	2	3	4	1	5	2

Column 5 consists of all T's, so the conditional resulting from the argument is a tautology. Hence, the argument is valid. The symbolic argument used in this example illustrates another of the basic rules in logic. It is called the **law of detachment,** or **modus ponens.**

We have just examined an example of a valid argument, parts of which are false (refer back to the truth table in example 3). There is a difference between the validity of an argument and its truth: an argument is valid because of its form—that is, the manner in which the conclusion is derived from the premises—not because of the meaning of the statements in it.

EXAMPLE 4

Determine whether the following argument is valid or invalid.

Either the bank is closed, or it is not after three o'cloock.
It is not after three o'clock.
Therefore, the bank is not closed.

Solution

Let P = *The bank is closed* and Q = *It is after three o'clock.* The argument in symbolic form is

$$\begin{array}{l} P \vee \sim Q \\ \sim Q \\ \hline \sim P \end{array}$$

If we connect the premises using conjunction and imply the conclusion, the corresponding conditional statement is

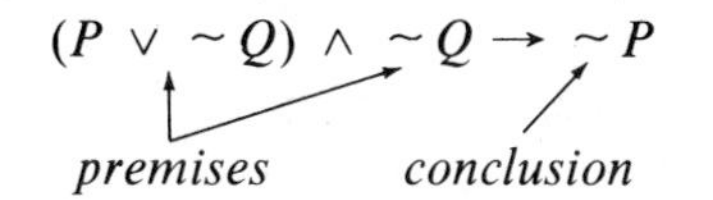

The corresponding truth table is:

P	Q	$(P \vee \sim Q)$	$\wedge$	$\sim Q$	$\rightarrow$	$\sim P$
T	T	T	F	F	T	F
T	F	T	T	T	F	F
F	T	F	F	F	T	T
F	F	T	T	T	T	T
1	2	5	6	3	7	4

The resulting conditional statement is not a tautology (see column 7). The premises do not imply the conclusion in all cases. Therefore, the argument is invalid.

In order to test whether an argument is valid or invalid, we need only consider the one conditional statement whose antecedent is the conjunction of all of the premises of the argument and whose consequent is the conclusion of the argument. If this conditional statement is a tautology, then the argument is valid. If it is not a tautology, then the argument is invalid.

EXERCISES FOR SECTION 2.7

Symbolize each of the following arguments (using the suggested notation) and, by means of a truth table, determine whether the argument is valid or invalid. State your answer.

1. If I pass the test, then I will quit coming to class. (P, Q)
   I quit coming to class.
   Therefore, I passed the test.

2. If Harry was late, then he missed the exam. (H, M)
   Harry was late.
   Therefore, Harry missed the exam.

3. If I pass math, then I will graduate. (P, G)
   I graduated.
   Therefore, I passed math.

4. If I pass math, then I will graduate. (P, G)
   I did not pass math.
   Therefore, I did not graduate.

5. I will graduate if I pass math. (G, P)
   I graduated.
   Therefore, I passed math.

6. It will be sunny or cloudy today. (S, C)
   It isn't sunny.
   Therefore, it will be cloudy.

7. Addie and Bill will be at the party. (A, B)
   Bill was at the party.
   Therefore, Addie was at the party.

8. If 2 divides 8, then 3 divides 7. (E, S)
   Three does not divide 7.
   Therefore, two does not divide eight.

9. You lost money iff you owned Sterling Homex stock. (M, S)
   You owned Sterling Homex stock.
   Therefore, you lost money.

10. If it snows, Benny will go skiing. (S, B)
    It did not snow.
    Therefore, Benny did not go skiing.

11. Pat or Sandy will bring the doughnuts. (P, S)
    Pat did not bring the doughnuts.
    Therefore, Sandy did not bring the doughnuts.

12. You can play tennis here if you are a member. (T, M)
    You play tennis here.
    Therefore, you are a member.

13. If it rains, Bobby takes his umbrella to school. (R, B)
    It did not rain.
    Therefore, Bobby did not take his umbrella to school.

14. If you finish the exam early, you may leave early. (E, L)
    You did not finish the exam early.
    Therefore, you may not leave early.

*15. Paul did not study, or he is bluffing. (P, B)
    If he is bluffing, then he will cut class. (B, C)
    Therefore, Paul did not study, or he will cut class.

*16. If I get a job, then I will not be able to study. (J, S)
It is false that I will fail history and that I will not get a job. (F, J)
Therefore, if I study, then I will not fail the course.

*17. If a is greater than b, then b is greater than c. (B, C)
If b is greater than c, then c is greater than d. (C, D)
Therefore, if a is greater than b, then c is greater than d.

*18. If the heavenly body is Mars, then it is near Venus. (M, V)
If it is not near Venus, then it is not close to Saturn. (V, S)
Therefore, either it is not close to Saturn, or the heavenly body is Mars.

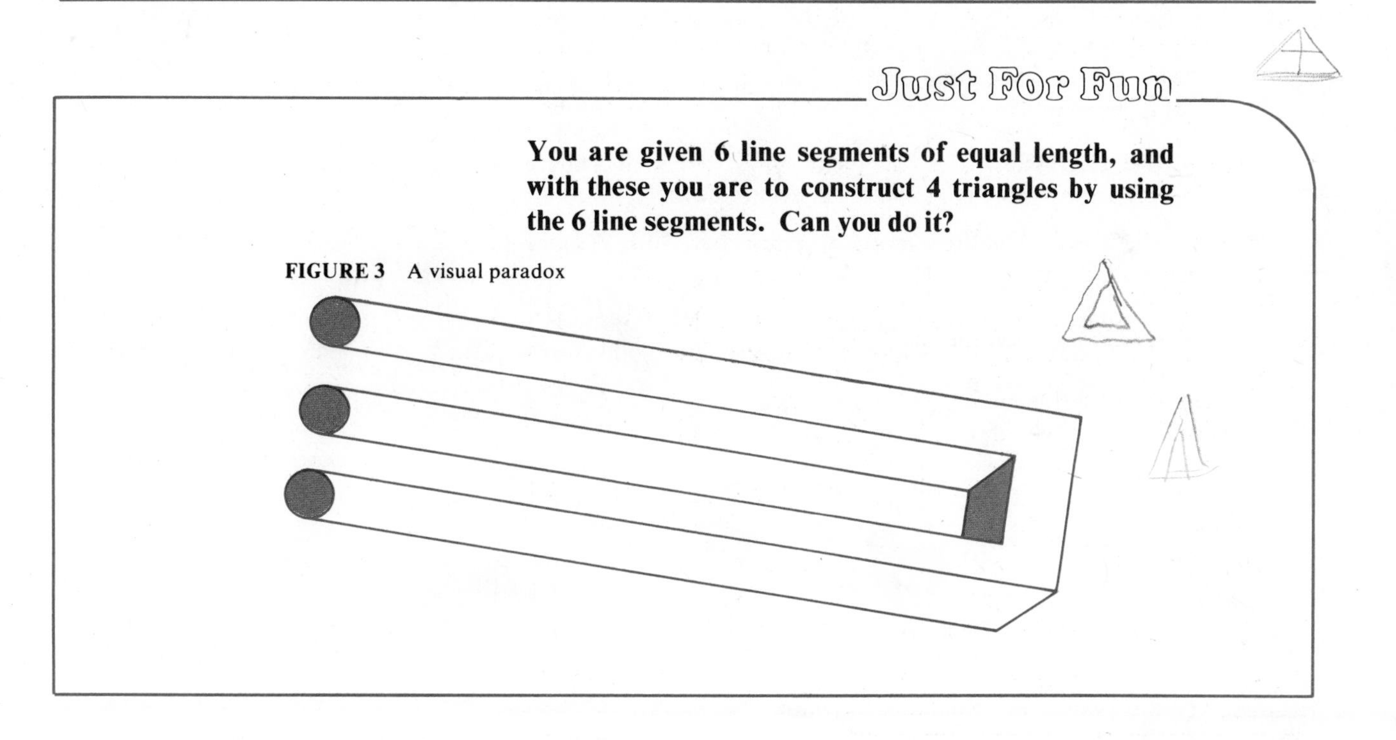

You are given 6 line segments of equal length, and with these you are to construct 4 triangles by using the 6 line segments. Can you do it?

FIGURE 3 A visual paradox

2.8 PICTURING STATEMENTS WITH VENN DIAGRAMS

Another use for Venn diagrams is to determine whether certain kinds of arguments are valid or invalid. The arguments that we shall consider here are called *syllogisms.* A **syllogism** is an argument that contains three statements: the **major premise,** the **minor premise,** and the **conclusion.**

Recall that a valid argument is one in which the conclusion follows logically from the premises. The conclusion is derived from the premises according to the laws of logic. The following is an example of a syllogism that is a valid argument. The conclusion follows from the premises.

All mathematics students are ambitious.	(Major Premise)
No ambitious people are lazy.	(Minor Premise)
Therefore, no mathematics students are lazy.	(Conclusion)

Note that each of the statements in the given example contains a quantifier such as *all* or *no*. These kinds of statements are a little different (because of the quantifiers) than the statements contained in the arguments in section 2.7. The syllogisms that we shall consider here contain quantified statements. There are four such types of statements. They are:

1. The **Universal Affirmative** statement states that "All A's are B's." For example, "All students are scholars."
2. The **Universal Negative** statement states that "No A's are B's." For example, "No students are scholars."
3. The **Particular Affirmative** statement states that "Some A's are B's." For example, "Some students are scholars."
4. The **Particular Negative** statement states that "Some A's are not B's." For example, "Some students are not scholars."

FIGURE 4 S = The set of students, and IP = The set of industrious people

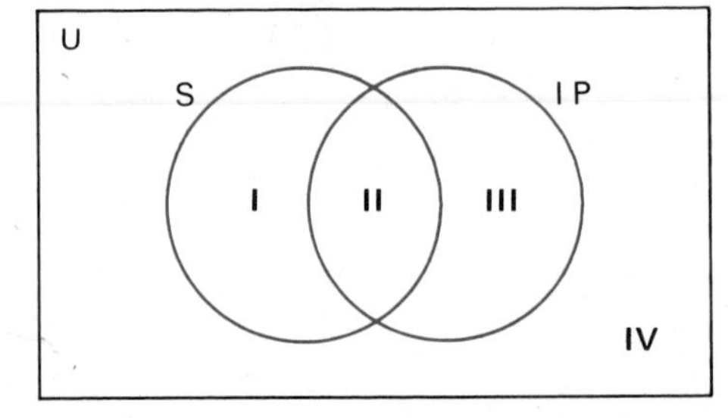

Do you think that the following syllogism is a valid argument? Does the conclusion follow from the premises?

All golfers are swingers.	(Major Premise)
No hackers are swingers.	(Minor Premise)
Therefore, no hackers are golfers.	(Conclusion)

The example syllogism above is a valid argument. You will see why shortly, but first we must be able to symbolize (picture) a statement using Venn diagrams.

Consider the statement "All students are industrious." We have a set of students and a set of industrious people. Hence, we may draw two circles and label them. See figure 4. It doesn't matter which circle is on the left, but it is convenient to take them in the order given.

FIGURE 5

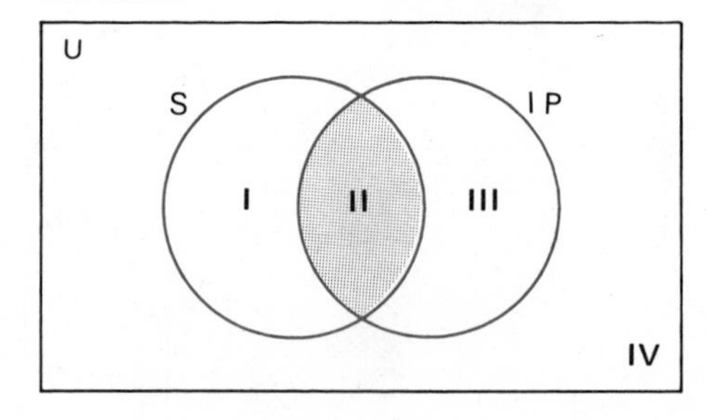

Note that we overlap the circles so that all possibilities may be considered. What do we know from the statement "All students are industrious"? We know that if a person is a student, then he is industrious; that is, the set of students is a subset of the set of indus-

FIGURE 6 P = the set of purple people-eaters, and O = The set of one-eyed objects

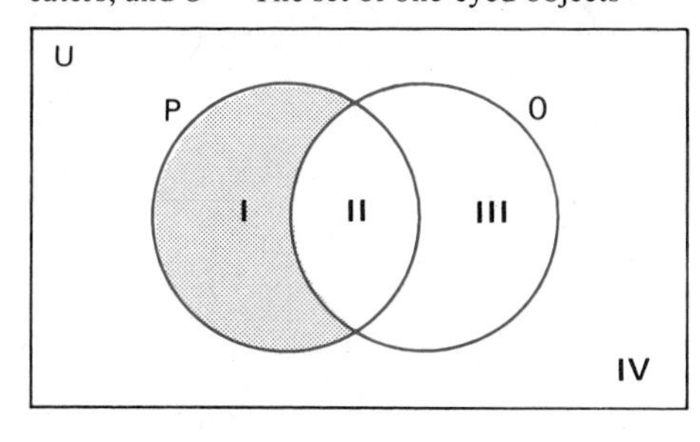

FIGURE 7 S = The set of students, and IP = The set of industrious people

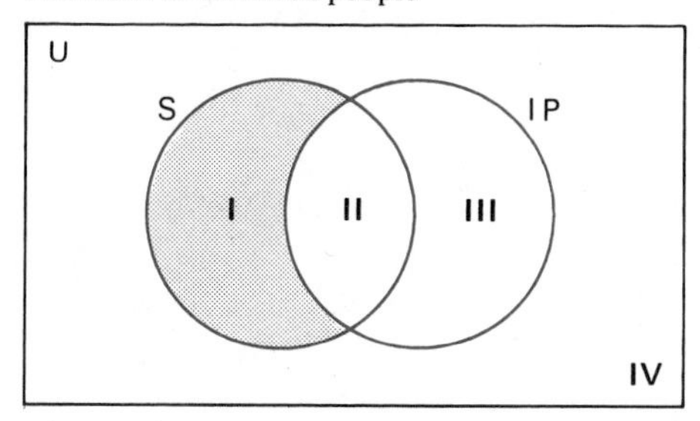

FIGURE 8

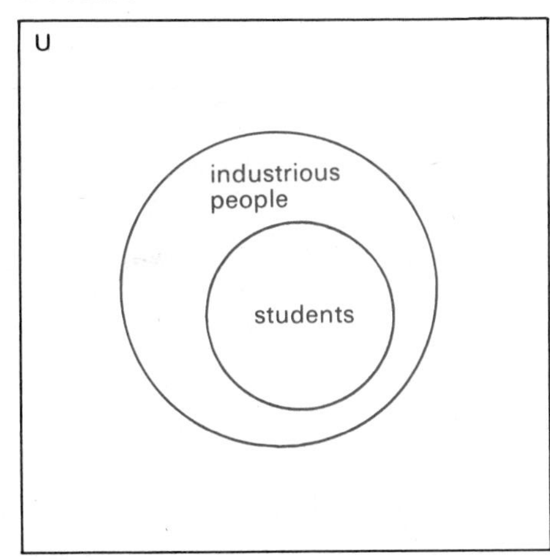

trious people. It does not follow that if a person is industrious, then he is a student.

How are we going to picture our statement? The statement tells us that all the students are in the "industrious" circle; that is, all the elements in the set containing students are in the set containing industrious people. So, they are in region II, and therefore region I is empty. Probably your first inclination is to shade region II, that is, the set $S \cap IP$, as shown in figure 5. This is because all the elements in set S are also in set IP.

However, suppose one or more of the sets we are picturing do not contain any elements. For example, consider the statement "All purple people-eaters are one-eyed." This is a Universal Affirmative statement; it is of the form "All A's are B's." To picture this statement, we must consider the set of purple people eaters, which is an empty set. The shading technique used in figure 5 would not be appropriate for the statement "All purple people-eaters are one-eyed," because shading region II would indicate that the set of purple people-eaters has members, which is not true. Therefore, we must use another way of picturing statements in Venn diagrams, since we must have a method that works for all cases.

A technique that does work for all cases is to distinguish the sets (regions) that *cannot* have elements in them, according to the given statement. For example, in constructing a picture of the statement "All purple people-eaters are one-eyed," we note that if purple people-eaters exist, then they would also all be in the set of one-eyed objects (region II in figure 6). But whether they exist or not, region I is known to be empty. According to the given statement, there are no purple people-eaters that are not one-eyed. Therefore we can eliminate region I, that is, we can cross it out. In figure 6, we have indicated that region I is empty by shading it.

Now, back to the original statement under consideration, "All students are industrious." From this statement we know that there is nothing in the set of students that is not also in the set of industrious people. That is, there are no students in set S who are not also members of set IP. Hence, region I is empty and we can eliminate it, or cross it out. The statement "All students are industrious" may be pictured as in figure 7.

Since the set of all students is contained in the set of industrious people, the set of students is a subset of the set of industrious people. You may have considered picturing the given statement as in figure 8. Here we have one set whose elements are contained in another. This picture is also correct and it may seem more reasonable to you than figure 7, but remember that we shall be picturing more than one statement in a diagram. It will be more efficient to handle them by the elimination technique than as a circle within a circle. The overlapping circles in a Venn diagram and the elimination technique will allow us to immediately determine whether an argument is valid or not.

We have discussed the diagram of the Universal Affirmative statement. Now let's examine the next one, the Universal Negative statement. Consider the statement "No students are industrious." Again we have two sets, *students* and *industrious people,* but what region do we eliminate? There are no elements in the set of students that are also in the set of industrious people. Therefore region II in figure 9 should be crossed out because it contains no elements. Hence the statement "No students are industrious" would be pictured as in figure 9.

FIGURE 9 S = The set of students, and IP = The set of industrious people

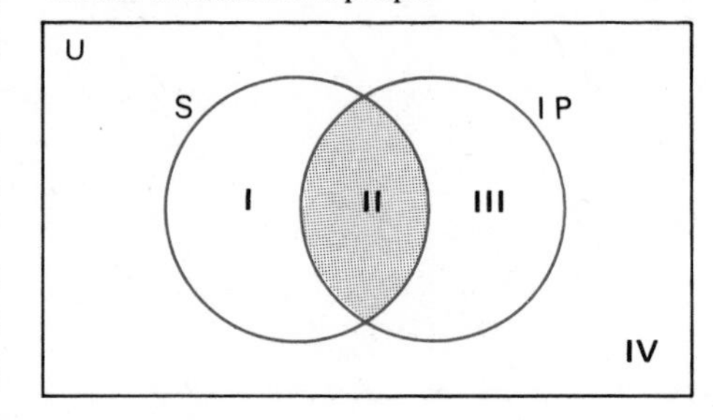

If two statements have the same diagram, the two statements are logically equivalent. The statements have the same meaning because they have the same picture in the diagram and the same region eliminated. The statement "No A's are B's" is logically equivalent to "No B's are A's," but how do "All A's are B's" and "All B's are A's" compare? Are they logically equivalent? By checking a diagram of these two statements you will see that they are not! See figure 10.

FIGURE 10 All A's are B's

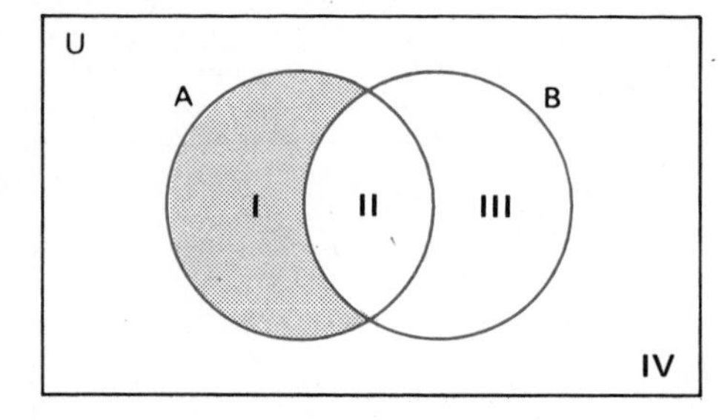

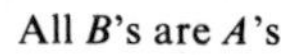
All B's are A's

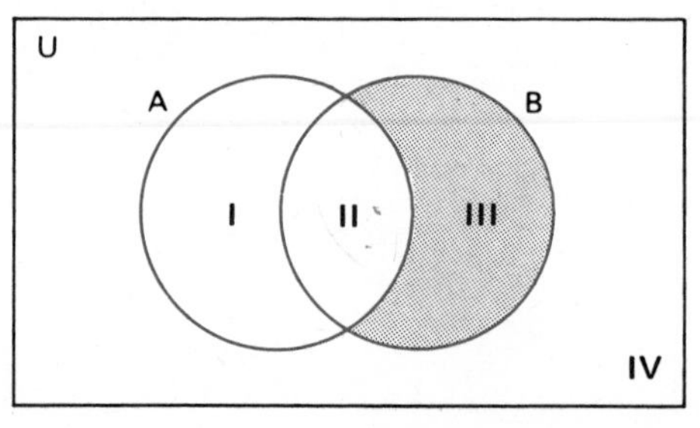

So far we have pictured the Universal Affirmative and Universal Negative statements. Consider the statement "Some students are industrious." We have the set of students and the set of industrious people, and from the given statement we know that there are some students that are industrious. We are discussing "some" students. In logic, the word *some* is interpreted to mean "There is at least one," but there could be more. The word *some* does not specify a certain number, so we just maintain that there is at least one. Since there are some students that are industrious we know that there is at least one student that is industrious. This one student is common to both sets. We diagram this by showing that the intersection of the two sets is not empty. We place something in the intersection (region II) to show that it is not empty, namely, an X. Therefore the statement "Some students are industrious" would be pictured as in figure 11. Note that when we cross out a region in a Venn diagram, then there are no elements in that region; but when we place an X in a region, then there is at least one element in that region.

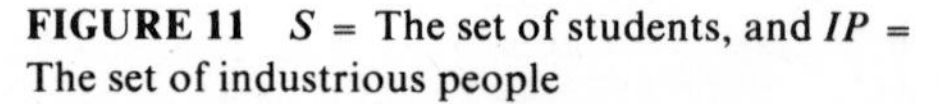
FIGURE 11 S = The set of students, and IP = The set of industrious people

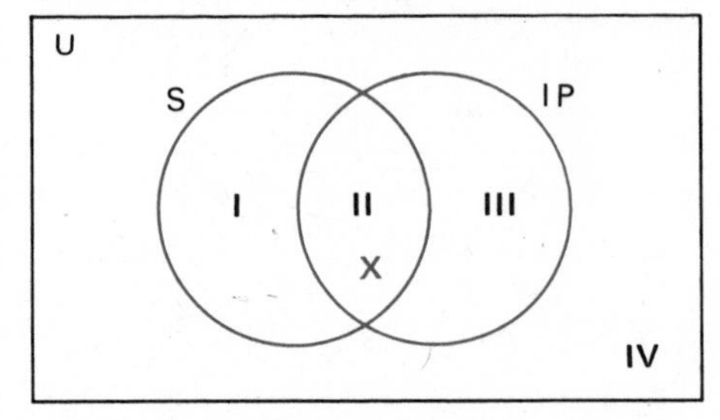

Consider the Particular Negative statement "Some students are not industrious." This statement tells us that there is at least one element in the set of students that is not in the set of industrious people. Therefore the intersection of the set of students and the set of nonindustrious people is not empty. Hence we would place an X in region I because it contains at least one element that is a student

FIGURE 12 S = The set of students, and IP = The set of industrious people

and is not industrious. The proper diagram for "Some students are not industrious" is shown in figure 12.

Using examples, we have diagrammed the four types of statements that may appear in a syllogism: Universal Affirmative, Universal Negative, Particular Affirmative, and Particular Negative. See figures 13a, 13b, 13c, and 13d.

FIGURE 13a Universal Affirmative: All A's are B's

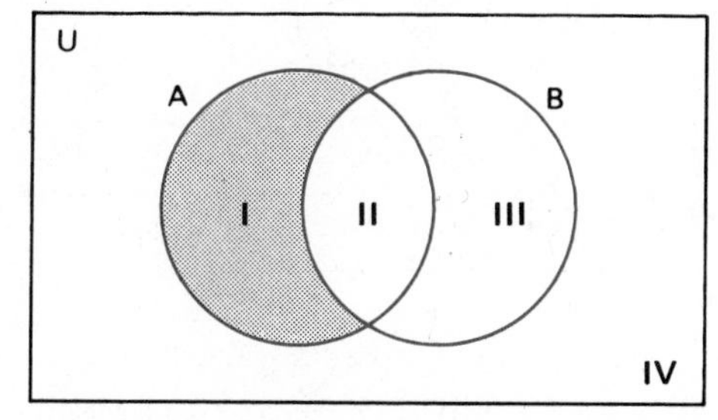

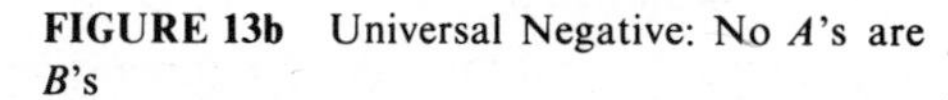

FIGURE 13b Universal Negative: No A's are B's

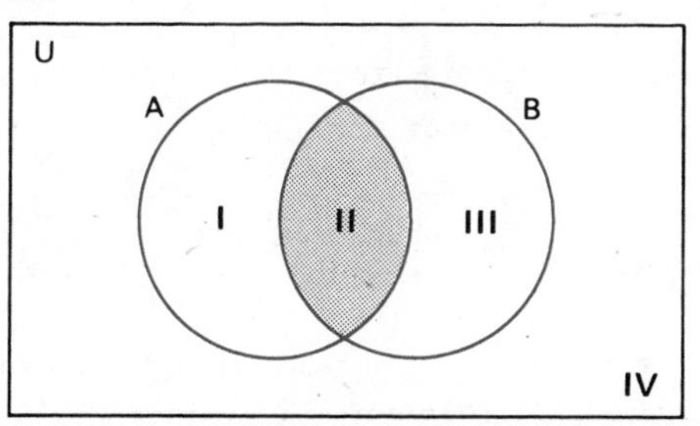

FIGURE 13c Particular Affirmative: Some A's are B's

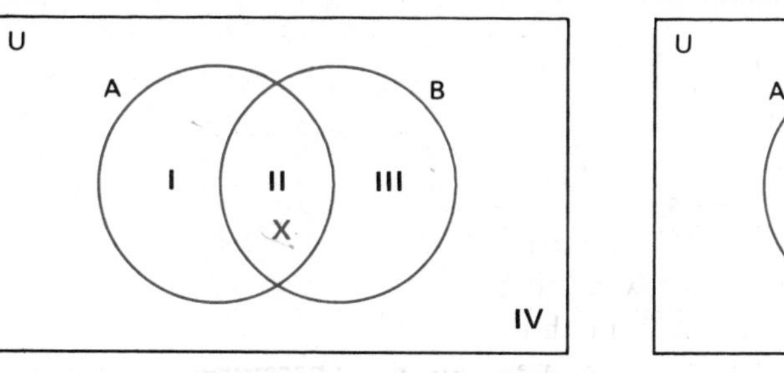

FIGURE 13d Particular Negative: Some A's are not B's

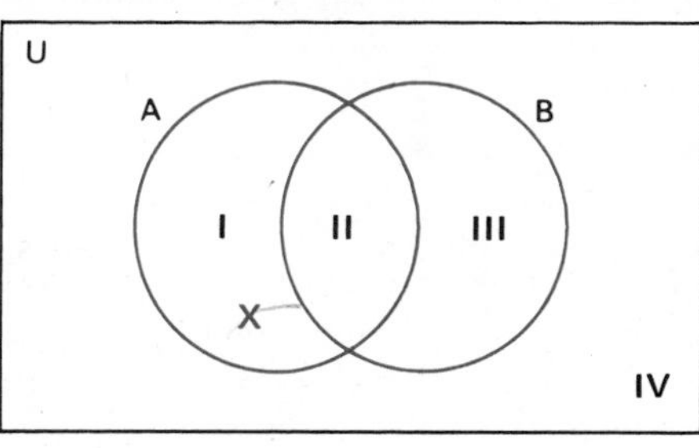

EXAMPLE 1

Identify and diagram the following:

a. All bleeps are peeps.

b. Some bleeps are peeps.

c. No bleeps are peeps.

d. Some peeps are not bleeps.

Solution
Let B = The set of bleeps, P = The set of peeps.

a. Universal Affirmative; see figure 14a.

b. Particular Affirmative; see figure 14b.

c. Universal Negative; see figure 14c.

d. Particular Negative; see figure 14d.

FIGURE 14a All bleeps are peeps

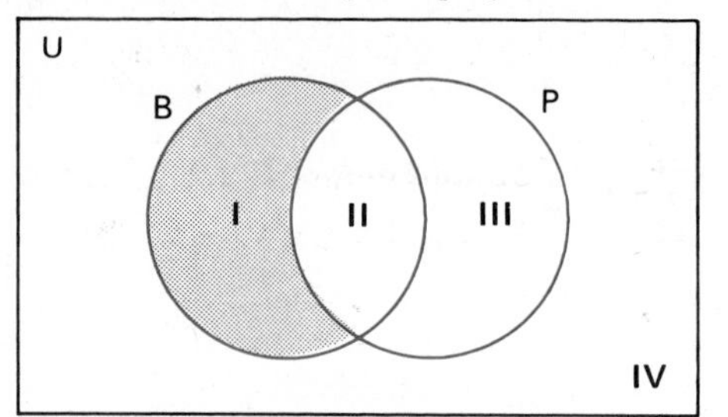

FIGURE 14b Some bleeps are peeps

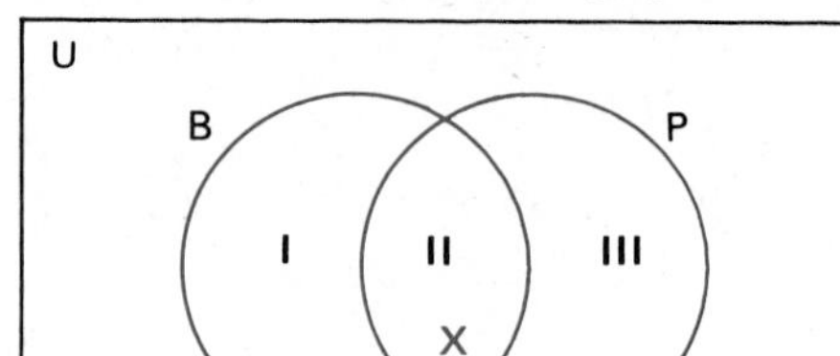

FIGURE 14c No bleeps are peeps

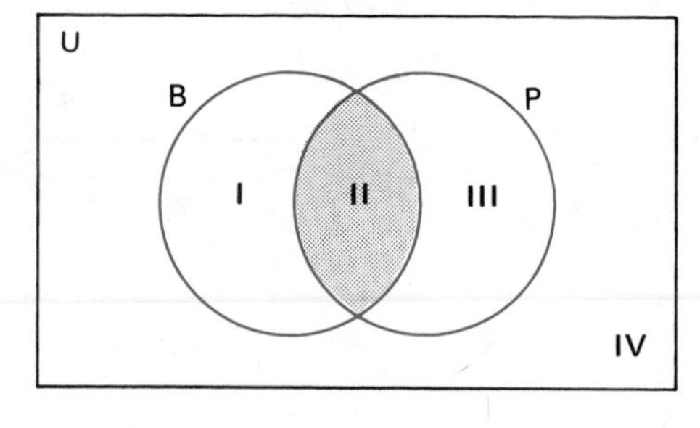

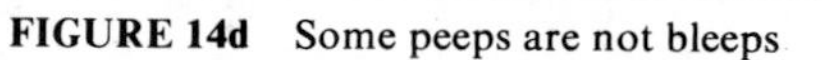

FIGURE 14d Some peeps are not bleeps

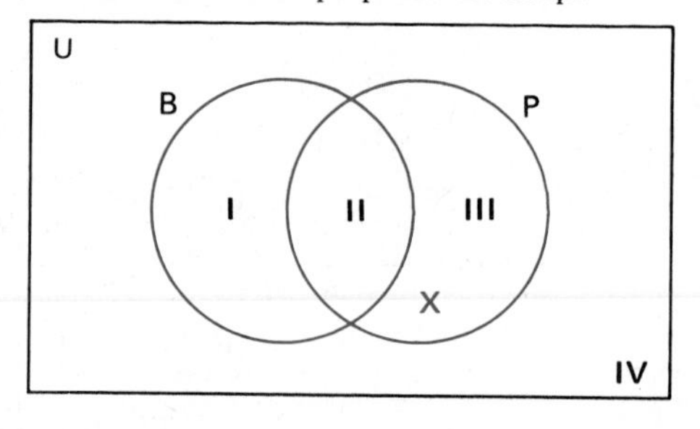

Statements that cannot be true together are said to be **inconsistent.** Given two statements, "Today is Monday" and "Today is not Monday," we see that these two statements contradict each other and cannot be true at the same time. Statements that can be true together are **consistent** statements; they do not contradict each other. Many times it is difficult to determine whether two statements are consistent or not. Let us see how we may use Venn diagrams to do this.

EXAMPLE 2
Determine whether the following pair of statements is consistent or inconsistent.

1. *Some students are lazy.*
2. *No students are lazy.*

shaded is a null set, empty set

FIGURE 15 S = The set of students, and L = The set of lazy people

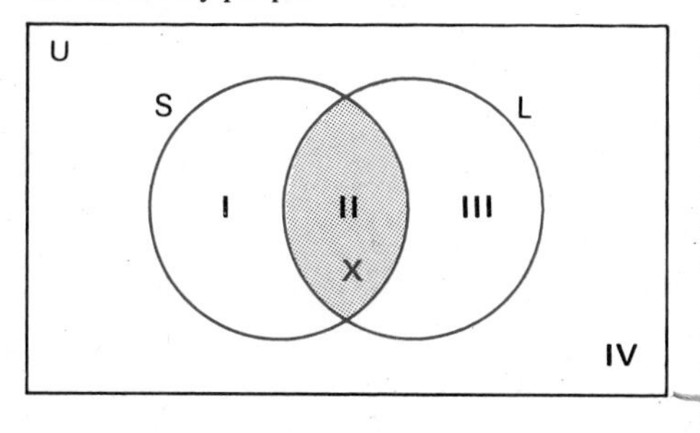

Solution
We construct a Venn diagram and diagram both statements in the same picture. See figure 15. The shading in region II represents statement 2 (*No students are lazy*), and the X in region II represents statement 1 (*Some students are lazy*). We have both crossed out a region and placed an X in the region. When we cross out a region to picture a statement, it means that there are no elements in that particular region. If we place an X in a region, then there is at least one element in that region. This is a contradiction; the two statements cannot be pictured in the same diagram. They are inconsistent. Two statements that are not consistent cannot be true at the same time.

EXAMPLE 3
Determine whether the following pair of statements is consistent or inconsistent.

1. *No politicians are honest.*
2. *Some politicians are not honest.*

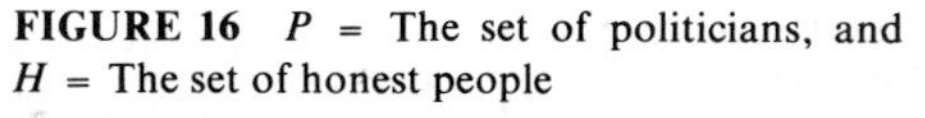

FIGURE 16 P = The set of politicians, and H = The set of honest people

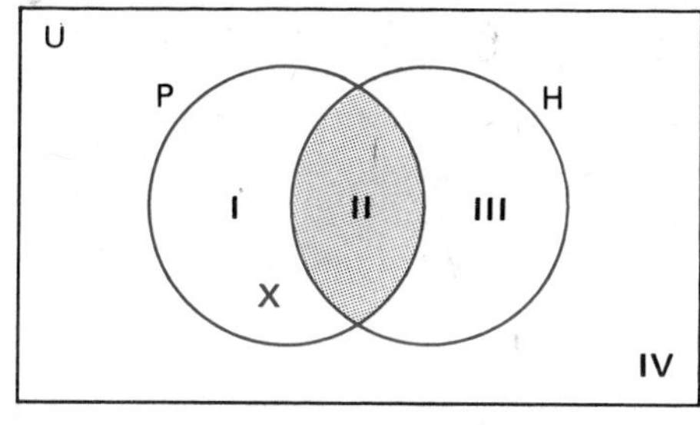

Solution
We construct a Venn diagram and diagram statements 1 and 2 in the same picture. See figure 16. The shading in region II represents statement 1 (*No politicians are honest*), and the X in region I represents statement 2 (*Some politicians are not honest*). Since we are able to picture both statements in the same diagram without any contradiction, we may conclude that the statements are consistent.

EXAMPLE 4
Determine whether the following pair of statements is consistent or inconsistent.

1. *All bleeps are peeps.*
2. *Some peeps are not bleeps.*

FIGURE 17 B = The set of bleeps, and P = The set of peeps

Solution
In testing these two statements for consistency, we construct a Venn diagram. See figure 17. Statement 1 (*All bleeps are peeps*) is pictured by crossing out region I, and statement 2 (*Some peeps are not bleeps*) is pictured by placing an X in region III. Since our diagram does not produce a contradiction, we may conclude that the given statements are consistent.

Note that we do not know what bleeps and peeps are, or whether they even exist. But, by using this technique for picturing the statements, we can determine whether the given statements are consistent or not. If the diagram does not produce a contradiction, then the statements are consistent.

EXERCISES FOR SECTION 2.8

Identify each statement as *Universal Affirmative, Universal Negative, Particular Affirmative,* or *Particular Negative,* and then diagram it by means of a Venn diagram.

1. No heros are imps.
2. All girls are heroes.
3. Some girls are not imps.
4. No cheaters are happy.
5. Some taxpayers are happy.
6. All numbers are interesting.
7. Some letters are not interesting.
8. All math courses are interesting.
9. Some math courses are challenging.
10. All politicians are sly.
11. Some politicians are not honest.
12. All Republicans are politicians.
13. No math teachers are compassionate.
14. All math teachers are compassionate.
15. Some cars are not gas eaters.
16. Some gamblers are losers.
17. Some losers are not gamblers.
18. All math teachers are dull.
19. No math teachers are boring.
20. Some tests are difficult.

Use a Venn diagram to determine whether each pair of statements is consistent or inconsistent.

21. Some kind people are clever.
    Some clever people are not kind.
22. Some math courses are interesting.
    No math courses are interesting.
23. No math teachers are compassionate.
    All math teachers are compassionate.
24. All dogs are barkers.
    Some barkers are not dogs.
25. Some gamblers are losers.
    Some losers are not gamblers.
26. Some tests are not easy.
    All tests are easy.
27. No math course is boring.
    Some math courses are not boring.
28. All students are procrastinators.
    No procrastinators are students.
29. All politicians are sly.
    No sly people are politicians.
30. Some real numbers are rational.
    Some rational numbers are real.
31. All logic students are gullible.
    No logic students are gullible.
32. No Democrats are politicians.
    Some politicians are not Democrats.
33. Some tests are long.
    Some tests are not long.
34. Some girls are not imps.
    No imps are girls.
35. Some houses are homes.
    Some homes are houses.

Just For Fun

If it takes five seconds for a clock to strike five, how long does it take to strike ten?

2.9 VALID ARGUMENTS AND VENN DIAGRAMS

An argument consists of the premises and the conclusion. An argument is said to be valid if the conclusion follows from the premises. Arguments may contain more than two premises, but here we are only going to consider syllogisms. Recall that a syllogism contains a major premise, a minor premise, and a conclusion. An argument is valid if the conclusion follows from the premises.

You should be aware of the difference between truth and validity. An argument may be valid even though your own knowledge tells you that the conclusion is false. But if the conclusion follows from the premises, then the argument is valid. On the other hand, a conclusion may be true and the argument invalid. You may know that a certain conclusion is true, but if it does not follow from the premises, the argument is not valid.

Venn diagrams are useful to determine the validity of syllogisms. An argument is valid if the conclusion follows from the premises. How do we determine if a conclusion follows from the premises? We diagram the premises in a Venn diagram, and if the conclusion is shown in the diagram of the premises without any ambiguity, then the argument is valid. If it is possible to diagram the premises without at the same time showing the conclusion, then the argument is invalid. Let's examine some examples to see how this technique is used.

Smithsonian Institution

This machine, on display at the Smithsonian Institution in Washington, D.C., can be used to evaluate the validity of syllogisms.

EXAMPLE 1

Determine whether the following argument is valid.

All students are industrious.
No dropouts are industrious.
Therefore, no dropouts are students.

Solution

First we construct a Venn diagram with three overlapping circles, one for the set of students, one for the set of industrious people, and one for the set of dropouts. See figure 18. Next we diagram each of the premises in the argument. "All students are industrious" is pictured by crossing out regions I and IV. "No dropouts are industrious" is pictured by crossing out regions V and VI. You will note that the conclusion, "No dropouts are students," is already pictured in the diagram as a result of diagramming the premises. This tells us that the argument is valid. (If the conclusion is shown in the diagram of the premises without any ambiguity, the argument is valid.) See figure 19 for the completed diagram.

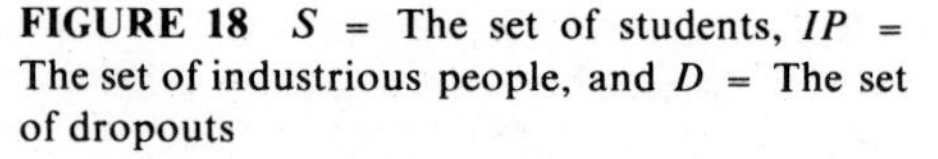
FIGURE 18 S = The set of students, IP = The set of industrious people, and D = The set of dropouts

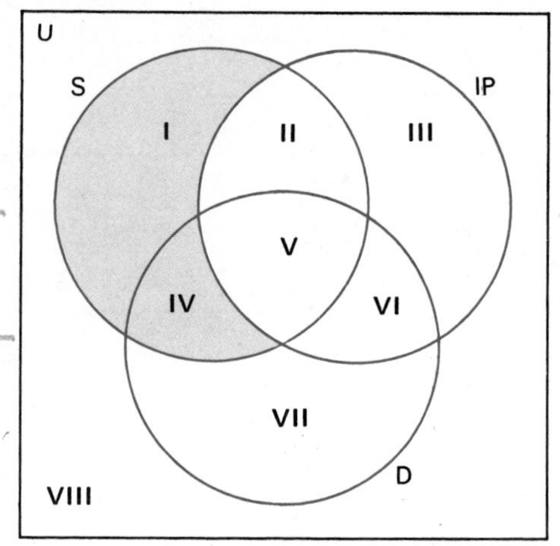

The premises of a valid argument *must* show the conclusion when they are diagrammed. We do not diagram the conclusion; its picture must be a result of diagramming the premises. If we are able to diagram the premises without showing the conclusion, then the argument is invalid.

EXAMPLE 2

Determine whether the argument is valid.

No generals are pacifists.
No pacifists are soldiers.
Hence, no generals are soldiers.

Solution

We diagram the premises in figure 20 to see if the conclusion follows from the premises. We have three circles: generals, pacifists, and soldiers. "No generals are pacifists" is pictured by eliminating region II and V. "No pacifists are soldiers" is pictured by eliminating regions V and VI. We already have crossed out region V, so we need only do region VI. The conclusion, "No generals are soldiers," is not pictured in the diagram. If it were, then regions IV and V would be crossed out—and region IV is not. The conclusion does not follow from the premises; therefore the argument is invalid.

FIGURE 19

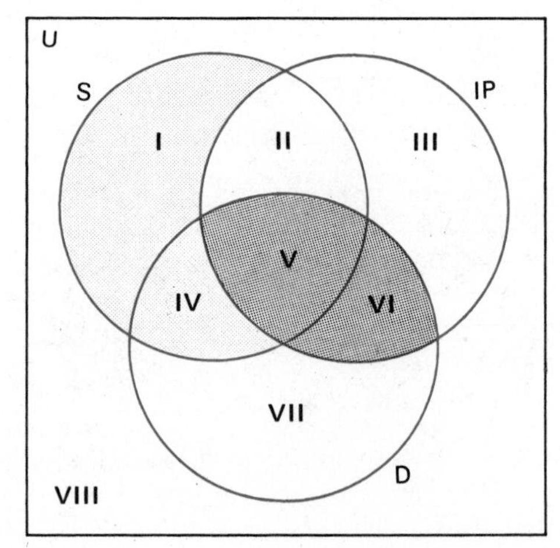

EXAMPLE 3

Determine whether the argument is valid.

All generals are soldiers.
Some generals are fighters.
Therefore, some soldiers are fighters.

Solution

We diagram the premises to see if we also obtain a picture of the conclusion. See figure 21. The statement "All generals are soldiers" is diagrammed by eliminating regions I and IV. The second premise,

FIGURE 20 G = The set of generals, P = The set of pacifists, and S = The set of soldiers

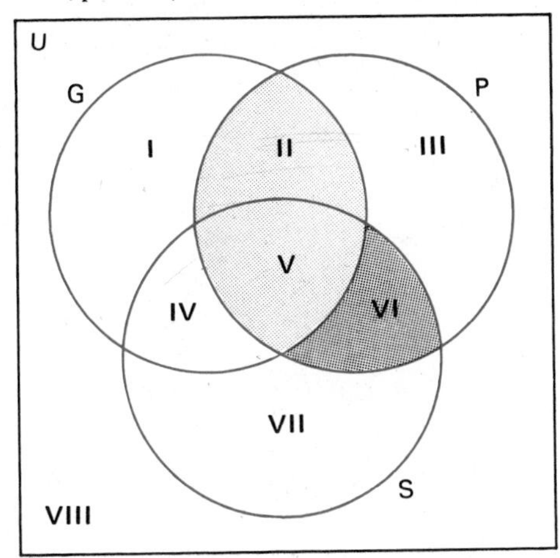

"Some generals are fighters," tells us that there are some elements in set G that are also in set F. We have a choice of placing an X in regions IV or V. We must place the X in region V, since we have already eliminated region IV in diagramming the first premise. Now we see that the conclusion, "Some soldiers are fighters," is already diagrammed in figure 21. The conclusion does follow from the given premises, so the argument is valid.

EXAMPLE 4.
Determine whether the argument is valid.

All fishermen are patient
Some fishermen are not liars.
Hence, some liars are not fishermen.

Solution
The major premise, "All fishermen are patient," is diagrammed in figure 22 by eliminating regions I and IV. "Some fishermen are not liars" tells us that there are some elements in set F that are not in set L. We must place the X in region II, since region I has already been eliminated. The conclusion, "Some liars are not fishermen," is not pictured in the diagram. This conclusion does not follow from the premises. Hence, the argument is invalid.

Note that by placing an X in region II we could have shown the statement "Some fishermen are not liars"; but that is not the desired conclusion.

EXAMPLE 5
Determine whether the argument is valid.

All generals are soldiers.
Some soldiers are not fighters.
Hence, some generals are not fighters.

FIGURE 21 G = The set of generals, S = The set of soldiers, and F = The set of fighters

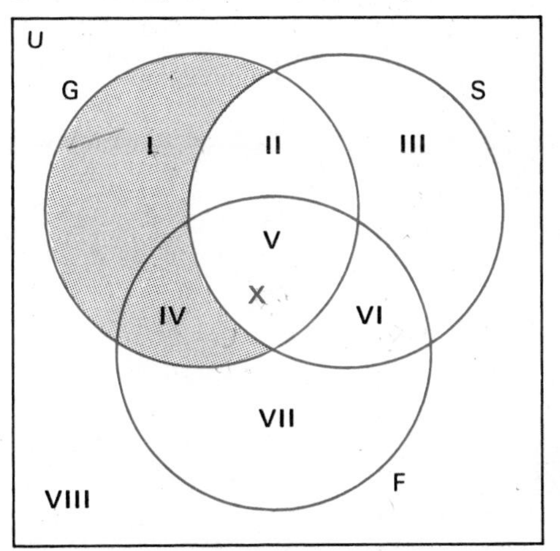

Solution
In figure 23, we diagram the premises to see if we also obtain a picture of the conclusion. "All generals are soldiers" is diagrammed by eliminating regions I and IV. The second premise, "Some soldiers are not fighters," tells us that there are some elements in set S that are not in set F. We have an option of placing an X in region II or III. Neither region has been eliminated. In order to represent exactly what is asserted by "Some soldiers are not fighters," let us place an X that is in *both* of the regions. We know that there is at least one soldier that is not a fighter. But, we do not know whether the soldier is in region II or III. The soldier may be in region II—those soldiers that are not fighters, but are generals. He may also be in region III—those soldiers that are not fighters and are not gen-

erals. The conclusion may or may not be shown. There are no conditions that say we must place the X in region II, so the premises do not guarantee the truth of the conclusion. Hence the argument is invalid.

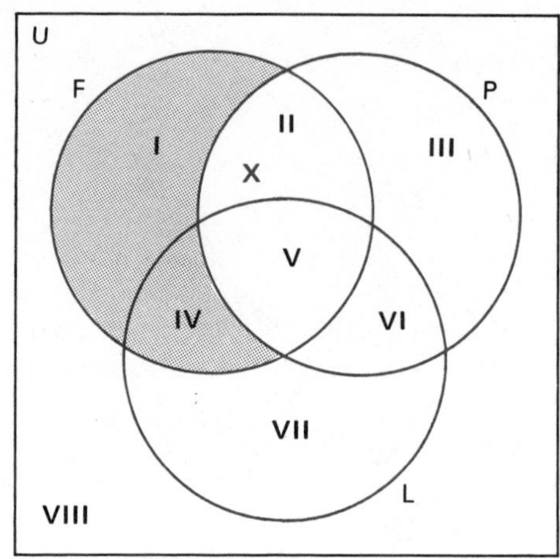

FIGURE 22 F = The set of fishermen, P = The set of patient people, and L = The set of liars

EXAMPLE 6

Determine whether the argument is valid.

All sophomores are hard workers.
All clever people are hard workers.
Therefore, all sophomores are clever.

Solution

As before, we diagram the premises to see if the conclusion is obtained. See figure 24. "All sophomores are hard workers" is diagrammed by eliminating regions I and IV. "All clever people are hard workers" is diagrammed by eliminating regions IV and VII. The conclusion, "All sophomores are clever," should be pictured with regions I and II eliminated. This is not the case, and therefore the argument is invalid.

To summarize, an argument is valid if the conclusion is shown without any ambiguity as soon as the premises are diagrammed. We do not diagram the conclusion; its picture must be a result of diagramming the premises. If it is possible to diagram the premises without at the same time showing the conclusion (as in example 5), then the argument is invalid.

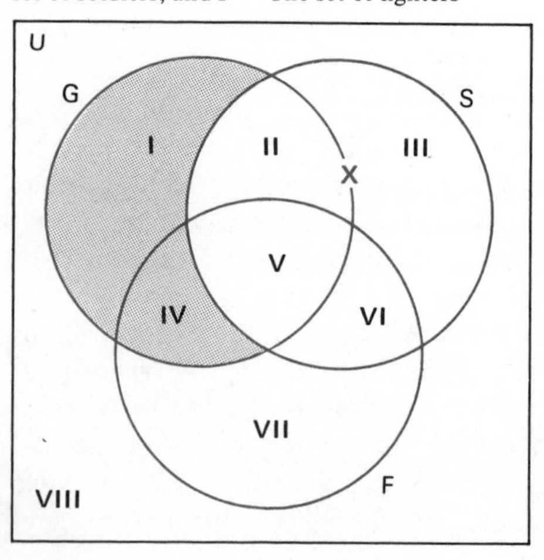

FIGURE 23 G = The set of generals, S = The set of soldiers, and F = The set of fighters

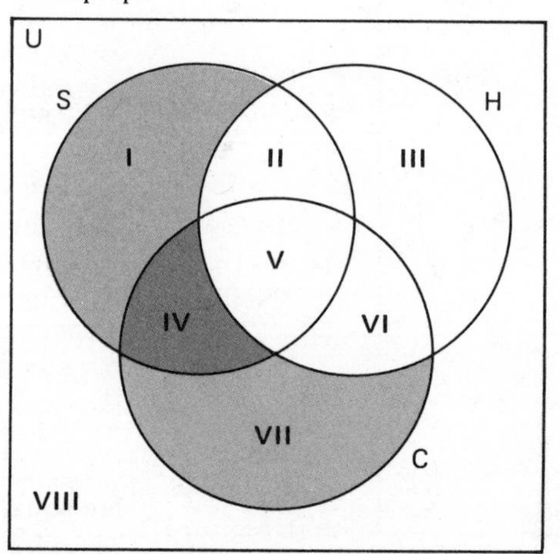

FIGURE 24 S = The set of sophomores, H = The set of hard workers, and C = The set of clever people

EXERCISES FOR SECTION 2.9

Use Venn diagrams to determine whether the following arguments are valid or invalid. State your answer.

1. No logic students are gullible.
   Some gullible people are superstitious.
   Hence, no logic students are superstitious.
2. All weightlifters are strong.
   Some football players are strong.
   So, some football players are weightlifters.
3. All coins are valuable.
   Some coins are old.
   Therefore, some old things are valuable.
4. All cars are gas eaters.
   Some gas eaters are four-wheeled.
   So, some cars are four-wheeled.
5. All cars have four wheels.
   No bikes have four wheels.
   Therefore, no cars are bikes.
6. Some jets are gigantic.
   Some gigantic things are heavy.
   Hence, some jets are heavy.
7. All books are readable.
   Some readable things are not interesting.
   So, some books are not interesting.
8. All mathematics students are clever.
   Some history students are clever.
   Therefore, some mathematics students are history students.
9. All logic students are adroit.
   Some logic students are ambitious.
   So, some ambitious people are adroit.
10. All logic students are adroit.
    Some logic students are ambitious.
    Hence, some adroit people are ambitious.
11. All logic students are adroit.
    All adroit people are ambitious.
    Therefore, some ambitious people are logic students.
12. All gamblers are losers.
    Some quarterbacks are not losers.
    So, some quarterbacks are not gamblers.
13. All gamblers are losers.
    No losers are lucky.
    Hence, no gamblers are lucky.
14. No winners are losers.
    Some gamblers are losers.
    Therefore, some gamblers are not winners.
15. All winners are gamblers.
    Some gamblers are losers.
    Hence, some gamblers are winners.
16. No skyjackers are sane.
    Some writers are sane.
    Hence, some skyjackers are not writers.
17. Some gamblers are not kibitzers.
    Some kibitzers are not winners.
    Therefore, some gamblers are not winners.
18. All politicians are leaders.
    Some politicians are lawyers.
    Hence, some lawyers are leaders.

*19. Some teachers are dull.
    All dull people are boring.
    Therefore, some teachers are boring.

*20. Some instructors are understanding.
    All understanding people are tolerant.
    So, some instructors are tolerant.

21. All warriors are brave.
    No cowards are brave.
    Hence, no cowards are warriors.
22. No circles are squares.
    No squares are triangles.
    Therefore, no circles are triangles.

23. All squares are rectangles.
    Some rectangles are parallelograms.
    So, some squares are parallelograms.

24. All handball players are agile.
    No weightlifters are agile.
    Hence, no handball players are weightlifters.

25. All actors are vain.
    Some vain people are not greedy.
    So, some actors are not greedy.

26. All vain people are greedy.
    Some actors are greedy.
    Therefore, some actors are not vain.

Just For Fun

Do you dial O for operator? No, you dial zero. What letters are missing on the standard telephone dial?

2.10 THE CONDITIONAL

The conditional statement is used a great deal in mathematics. All math students have, at one time or another, encountered rules or theorems stated as conditionals. One of the first theorems proved in high school geometry is the conditional "If two sides of a triangle are equal, then the angles opposite these sides are equal."

In this section, we shall examine the conditional statement in detail. We shall concern ourselves with statements that are equivalent to *If P then Q*, and the various forms that these statements may take.

Suppose we are given two statements, P and Q, and we are asked to construct a conditional statement using P and Q, or $\sim P$ and $\sim Q$. What are the possible variations? After some thought, you would probably come up with the following list (their names are included).

1. $P \rightarrow Q$ — statement
2. $Q \rightarrow P$ — *converse* of statement 1
3. $\sim P \rightarrow \sim Q$ — *inverse* of statement 1
4. $\sim Q \rightarrow \sim P$ — *contrapositive* of statement 1
5. $\sim(P \rightarrow Q)$ — *negation* of statement 1

Note, however, that statement 5, $\sim(P \rightarrow Q)$, is not a conditional statement, but the *negation* of a conditional statement. Therefore, we will only concern ourselves with the first four statements. By comparing their truth tables, we can discover which statements are logically equivalent.

TABLE 10

Statement			Converse			Inverse			Contrapositive		
P	$\rightarrow$	Q	Q	$\rightarrow$	P	$\sim P$	$\rightarrow$	$\sim Q$	$\sim Q$	$\rightarrow$	$\sim P$
T	T	T	T	T	T	F	T	F	F	T	F
T	F	F	F	T	T	F	T	T	T	F	F
F	T	T	T	F	F	T	F	F	F	T	T
F	T	F	F	T	F	T	T	T	T	T	T

From table 10, we see that a conditional statement is logically equivalent to its contrapositive. Hence, when we have a statement such as "If I study, then I shall pass math," we know that it is equivalent to "If I did not pass math, then I did not study." Note, too, that the inverse of a conditional is equivalent to its converse. You will probably want to memorize these variations.

The conditional statement is dangerous in the hands of the uninformed. A common error made by some people in reasoning about mathematics or everyday occurrences is that they assume the converse or inverse of a given statement has the same truth value as the given statement. This is particularly the case when the given conditional statement seems to be obviously true. Checking the truth tables in table 10, we see that the inverse and converse of a statement are not always true when the given statement is true.

Unfortunately, many people do not realize this. In some of the advertising that we see, hear, or read, the advertiser wants us to believe that the inverse or converse of a premise is true. Consider the statement.

For a cleaner wash, insist on Sudsy Soap.

This translates to "If you want a cleaner wash, then use Sudsy Soap." The advertiser wants you to use Sudsy Soap, but he or she also wants you to believe that if you do not use Sudsy Soap, then your wash will not be clean; hence, you had better buy some Sudsy Soap.

Consider another conditional statement.

If x is greater than zero, then x is not negative.

This statement is true. Its converse is:

If x is not negative, then x is greater than zero.

This statement is not true because x could be zero (zero is neither positive nor negative), and therefore x would not be greater than zero. Thus we see that an inverse or converse does not follow just because the given statement is true.

Recall that *then* is omitted in a conditional statement when it is thought to be obvious. Consider the following:

If anybody was late, it had to be Dave.

A conditional statement may be stated in a manner which is misleading to the uninformed or unobservant reader or listener. We must be careful in reading and listening. A basic understanding of the various forms of the conditional should help us to do this.

EXAMPLE 1

Write the converse, inverse, and contrapositive of the statement

If you use Sudsy Soap, then your clothes are clean.

Solution

Let P = *You use Sudsy Soap* and Q = *Your clothes are clean.* We can symbolize the given statement as $P \rightarrow Q$.

The converse of $P \rightarrow Q$ is $Q \rightarrow P$, or "If your clothes are clean, then you use Sudsy Soap." Note that even if the given statement is true, the converse is not necessarily true.

The inverse of $P \rightarrow Q$ is $\sim P \rightarrow \sim Q$, which may be stated as "If you do not use Sudsy Soap, then your clothes are not clean." This statement is also not necessarily true.

The contrapositive of $P \rightarrow Q$ is $\sim Q \rightarrow \sim P$, "If your clothes are not clean, then you do not use Sudsy Soap." If the given statement is true, then the contrapositive is also true, because it is logically equivalent to the given statement.

EXAMPLE 2

Write the converse, inverse, and contrapositive of the statement

If today is Sunday, then yesterday was not Friday.

Solution

Let P = *Today is Sunday* and Q = *Yesterday was Friday.* The given statement is symbolized as $P \rightarrow \sim Q$.

The converse is $\sim Q \rightarrow P$, "If yesterday was not Friday, then today is Sunday."

The inverse of $P \rightarrow \sim Q$ is $\sim P \rightarrow \sim\sim Q$. Since we have a double negation, $\sim\sim Q$, we may interpret it as Q; that is, $\sim\sim Q \equiv Q$. Hence the inverse $\sim P \rightarrow \sim\sim Q$ is the same as $\sim P \rightarrow Q$: "If today is not Sunday, then yesterday was Friday."

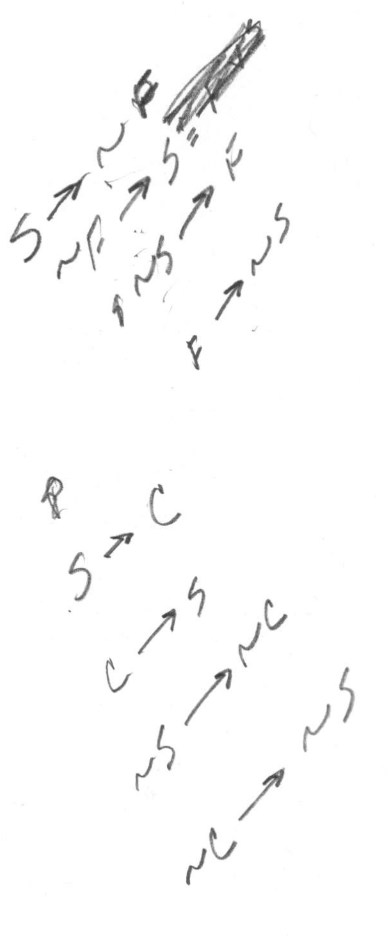

The contrapositive of $P \rightarrow \sim Q$ is $\sim\sim Q \rightarrow \sim P$, which is the same as $Q \rightarrow \sim P$: "If yesterday was Friday, then today is not Sunday."

The conditional causes a great deal of confusion and trouble for many people because they do not understand it. We shall use an example to show the various ways that a conditional may be stated. You should already be aware that $P \rightarrow Q$ can be translated as

If P, then Q
If P, Q
Q if P

In the following discussion we shall see that $P \rightarrow Q$ may also be translated in other ways. Consider the statement

If you are a citizen of Buffalo, then you are a citizen of New York State.

P = *You are a citizen of Buffalo*
Q = *You are a citizen of New York State*

Then the statement would be symbolized as $P \rightarrow Q$.

Now, suppose someone says "You are a citizen of Buffalo only if you are a citizen of New York State." What do we know if you live in New York State? We know that you live in the state of New York, but you may live in Buffalo, Rochester, Syracuse, New York City, or a thousand other places. But what do we know if you live in Buffalo? We know that you are a citizen of New York State because you are a citizen of Buffalo *only if* you are a citizen of New York State. *If* you live in Buffalo, *then* you must live in New York State.

Another statement that may help you understand *P only if Q* is:

It snows only if it is cold.

If the thermometer tells us it's cold, do we know anything? We know it's cold, but we do not know if it will snow. But suppose we look out the window and see that it is snowing. Then we can deduce that it is cold. Why? Because we know that it snows only if it is cold. That is, *if* it snows, *then* it is cold. We would conclude that *P only if Q* may be symbolized as $P \rightarrow Q$.

Consider a variation of the citizenship statement, "Being a citizen of Buffalo is sufficient for being a citizen of New York State." This statement means that in order to be a citizen of New York State, it is enough to be a citizen of Buffalo, since every citizen of the city of Buffalo is also a resident of the state of New York. Do you have to do anything else? No, because *if* you are a citizen of Buffalo, *then* you are a citizen of New York State.

Suppose an instructor tells you that doing your homework every day is sufficient for passing the course. What do you know? Do you have to do your homework every day? The instructor did not say that you had to do your homework every day; but he did say that if you did, then you would pass the course. You could pass the course some other way, perhaps by passing all of the exams. But you know that if you do your homework every day, then you will pass the course. Suppose you did not pass the course? We may conclude that you did not do your homework every day. That is, if you did not pass the course, then you did not do your homework every day. But this is the contrapositive of "If you do your homework every day, then you will pass the course." We may conclude from these discussions that *P is sufficient for Q* may be symbolized as $P \rightarrow Q$.

Another variation of "If you are a citizen of Buffalo, then you are a citizen of New York State" is the following sentence: "Being a citizen of New York State is necessary for being a citizen of Buffalo." This means that being a citizen of New York State is a necessary condition for being a citizen of Buffalo. Do you have to be a citizen of Buffalo in order to be a citizen of New York? No, but you have to be a citizen of New York State in order to be a citizen of Buffalo. Since you have to be a citizen of New York State, we might conclude that

If you are not a citizen of New York State, then you are not a citizen of Buffalo.

But that is the contrapositive of

If you are a citizen of Buffalo, then you are a citizen of New York State.

Therefore, we may conclude that *Q is necessary for P* may be symbolized as $P \rightarrow Q$.

Another example that may aid you in understanding *Q is necessary for P* is the following:

Oxygen is necessary for fire.

Given this statement, we know that we cannot have fire unless we have oxygen. That is, if we do not have oxygen, then we can't have a fire. Again, we can apply the contrapositive to this and we have the statement

If we have fire, then we have oxygen.

In summary, we have the fact that *Q is necessary for P* is symbolized as $P \rightarrow Q$.

Consider the statement

If you studied Algebra II, then you studied Algebra I.

Three other sentences that have the same meaning are:

You studied Algebra II only if you studied Algebra I.
Studying Algebra II is sufficient for studying Algebra I.
Studying Algebra I is necessary for studying Algebra II.

When we have the conditional statement $P \rightarrow Q$, it is usually interpreted as *If P, then Q*, but three other common equivalent wordings are

P only if Q
P is sufficient for Q
Q is necessary for P

The following is a list of equivalent wordings of the conditional statement, $P \rightarrow Q$. Bear in mind that the list is a sampling and not intended to be a list of all possible variations, but all statements listed are interpretations of the original conditional statement.

1. *If P, then Q*
2. *If P, Q*
3. *Q if P*
4. *P implies Q*
5. *Q is implied by P*
6. *Q whenever P*
7. *P only if Q*
8. *P is sufficient for Q*
9. *Q is necessary for P*

(all of the above) $P \rightarrow Q$

The following examples are conditional statements using some of the variations listed above.

1. If the Mets won, then the Dodgers lost.
2. If the Mets won, the Dodgers lost.
7. The Mets won only if the Dodgers lost.
8. The Mets' winning is sufficient for the Dodgers' losing.
9. The Dodgers' losing is necessary for the Mets' winning.

EXAMPLE 3

Let P = *Today is Saturday* and Q = *Tomorrow is Sunday.* Write each of the following statements in symbolic form.

a. If today is Saturday, then tomorrow is Sunday.
b. If tomorrow is not Sunday, then today is not Saturday.
c. Tomorrow is Sunday if today is Saturday.
d. Today's being Saturday is sufficient for tomorrow's being Sunday.
e. Tomorrow's being Sunday is necessary for today's being Saturday.
f. Today is not Saturday only if tomorrow is not Sunday.
g. It is false that today's being Saturday is sufficient for tomorrow's being Sunday.
h. Tomorrow's not being Sunday is sufficient for today's not being Saturday.

Solution

a. $P \rightarrow Q$ b. $\sim Q \rightarrow \sim P$ c. $P \rightarrow Q$ d. $P \rightarrow Q$
e. $P \rightarrow Q$ f. $\sim P \rightarrow \sim Q$ g. $\sim(P \rightarrow Q)$ h. $\sim Q \rightarrow \sim P$

It has probably occurred to you that since a conditional statement may be written in various ways, the same may be true for a biconditional statement. If so, then how may we rewrite it? Consider the statement

A number is positive if and only if that number is greater than zero.

From the discussion on truth tables in section 2.5, we know that this statement may be rewritten as a conjunction.

If a number is positive, then the number is greater than zero and if a number is greater than zero, then the number is positive.

Let P = *A number is positive* and Q = *A number is greater than zero.* Then the above statement would be symbolized as

$$(P \rightarrow Q) \wedge (Q \rightarrow P)$$

For the first part of the conjunction, we can say *P is sufficient for Q*, and for the second part, *P is necessary for Q*. Or we could switch them around: *P is necessary for Q* and *P is sufficient for Q*, or simply *P is necessary and sufficient for Q*.

$$P \leftrightarrow Q \quad \begin{cases} P \text{ if and only if } Q \\ P \text{ is necessary and sufficient for } Q \end{cases}$$

EXAMPLE 4
Using the suggested notation, write the following in symbolic form:

a. You may play on this course if and only if you are a member of Oak Hill. (C, O).

b. Citizens become irate whenever, and only whenever, their taxes are raised. (C, T)

c. Being a student here is a necessary and sufficient condition for obtaining a yearbook. (S, Y)

d. Quickness and coordination are necessary and sufficient for being a pole vaulter. (Q, C, P)

Solution
a. $C \leftrightarrow O$ b. $C \leftrightarrow T$ c. $S \leftrightarrow Y$ d. $(Q \wedge C) \leftrightarrow P$

EXERCISES FOR SECTION 2.10

1. Use the suggested notation to write the converse, inverse, and contrapositive for each of the following statements in symbolic form.
   a. If Jim studies, then he will pass. (S, P)
   b. We will go skating if the pond freezes over. (F, S)
   c. If the Sabres do not make the playoffs, then the fans will be sad. (S, F)
   d. If logic is easy, then geometry is not difficult. (L, G)
   e. If today isn't Thursday, then yesterday wasn't Wednesday. (T, W)

2. "If the storm is a hurricane, then its winds are traveling at least 74 miles per hour." Given that this is a true statement, which of the following must also be true? (Use your knowledge of the conditional.)
   a. If the storm is not a hurricane, then its winds are not traveling at least 74 miles per hour.
   b. If the storm's winds are traveling at least 74 miles per hour, then the storm is a hurricane.
   c. It is false that if the storm's winds are not traveling at least 74 miles per hour, then the storm is not a hurricane.
   d. If the storm's winds are not traveling at least 74 miles per hour, then the storm is not a hurricane.
   e. The storm is not a hurricane, or the storm's winds are traveling at least 74 miles per hour.
   f. The storm's winds are not traveling at least 74 miles per hour, or the storm is not a hurricane.

3. Let P = *It is a logic course* and Q = *It is interesting.* Write each of the following statements in symbolic form.
   a. It is a logic course only if it is interesting.
   b. If it isn't interesting, then it isn't a logic course.
   c. The fact that it is a logic course is sufficient for it to be interesting.
   d. Only if it is interesting is it a logic course.
   e. Being a logic course is necessary and sufficient for it to be interesting.

4. Use the suggested notation to write the following in symbolic form.
   a. A student can be successful in mathematics only if he does his homework and pays attention in class. (S, H, P)
   b. A sufficient condition for a student to flunk is that he not attend class. (S, A)
   c. A necessary condition for me to go swimming is that the water must be warm or the air temperature must be 80 degrees. (S, W, T)
   d. If a person is contributing to society, he is contributing to society only if he is helping his fellow man. (C, H)
   *e. Only exams are boring. (E, B)
5. Let P = *He works hard* and Q = *He is happy*. Write each of the following statements in symbolic form.
   a. He works hard only when he is happy.
   b. If he isn't working hard, then he isn't happy.
   c. His working hard is necessary for his being happy.
   d. He works hard iff he is happy.
   *e. He is not happy unless he works hard.
6. Sneaky Sam made the following promise to Gullible Gary: "I will help you only if you give me five dollars." Gary gave Sam five dollars and Sam walked away. Did Sam break his promise? Why or why not?
7. An instructor told a student "I will pass you only if you come to class every day." The student came to class every day and still failed the course. Does the student have a legitimate gripe? Why or why not?

Just For Fun

Let us agree that all special agents always tell the truth and all spies always lie. Once three people appeared in a court of law before a judge. The first accused person spoke, but the judge was unable to hear what the accused had said. The judge asked the second accused person what the first had said. Number 2 replied, "She said she is a spy." The judge then asked the third accused person whether number 2 was a special agent or a spy. Number 3 replied, "Number 2 is a special agent." Determine whether each accused person was a special agent or a spy.

2.11 SUMMARY

In this chapter, we were introduced to modern logic. We considered only *statements* (*propositions*) that are either true or false, but not both. For example, it is not possible to assign a true or false value to

"Stand up" or "Who are you?" Besides examining simple statements, we also studied compound statements of the following types: *negation, conjunction, disjunction, conditional,* and *biconditional.*

We learned that letters such as P, Q, R, and S, and sometimes A, B, and C, are traditionally used to symbolize simple statements in logic. The connectives $\wedge$ (*and*), $\vee$ (*or*), $\rightarrow$ (*if . . . then . . .*), $\leftrightarrow$ (*if and only if*), and $\sim$ (*not*) are used to connect the simple statements in a compound statement. Parentheses are used to tell us what type of compound statement we are considering. When parentheses are omitted, we follow the convention of the dominance of connectives. The biconditional ($\leftrightarrow$) is the most dominant connective, followed by the conditional ($\rightarrow$), and then the equally dominant conjunction ($\wedge$) and disjunction ($\vee$). The negation ($\sim$) is the least dominant connective.

Truth tables show us all possible truth values for given compound statements. A statement that is always true is called a *tautology* and a statement that is always false is called a *contradiction.* Truth tables also enable us to determine *logical equivalence* between two statements. We may also use *De Morgan's law* to create logically equivalent statements from conjunctions and disjunctions.

Truth tables are also used to determine whether an argument is *valid* or *invalid.* An argument is valid if the conclusion follows from the premises. To test whether an argument is valid, we first connect the premises by means of conjunction and then connect the resulting statement to the conclusion to form a conditional. Next we complete the truth table for this conditional, and if the truth table shows that the conditional is a tautology, then the argument is valid. If the conditional is not a tautology, then the argument is invalid.

Venn diagrams can be used to picture certain kinds of quantified statements. These statements are of four types: the *Universal Affirmative, Universal Negative, Particular Affirmative,* and *Particular Negative.* We can use Venn diagrams to determine whether two statements are *consistent.* Statements that cannot be true at the same time are *inconsistent.*

We can also use Venn diagrams to determine whether a syllogism (one type of argument) is valid. To determine if a syllogism is valid, we picture the premises in a Venn diagram, and if the conclusion is shown without any ambiguity as soon as the premises are diagrammed, then the argument is valid. If it is possible to diagram the premises without showing the conclusion at the same time, then the argument is invalid.

Section 2.10 was devoted to the conditional statement and its various forms. A conditional statement may be worded in various ways. Statements such as *P only if Q*, *P is sufficient for Q*, and *Q is necessary for P* are all logically equivalent to *If P then Q*, and may be symbolized as $P \rightarrow Q$. Logic enables us to take statements of this type and symbolize them in some form of $P \rightarrow Q$. Once we symbolize the statements, we no longer have to worry about what the words are saying and we do not have to contend with what seems "right" or what our intuition tells us. We can symbolize the statement and analyze it. A conditional statement is logically equivalent to its *contrapositive.* When a conditional statement is true, its *converse* does not have to be true; and when a conditional statement is true, its *inverse* does not have to be false.

Review Exercises for Chapter 2

1. Identify each sentence as a simple statement, compound statement, or neither. If the statement is compound, then classify it as a negation, conjunction, disjunction, conditional, or biconditional.
   a. Either Hugh sold his car, or he traded it in.
   b. Is it raining?
   c. Yesterday was Friday.
   d. Scott stayed home, but Joe went to the show.
   e. Carlos did not bring his notes to class.
   f. If Mary went swimming, then she did not study.

2. By means of the appropriate connectives and parentheses, symbolize each statement, using the given symbols for the simple statements.

   P = *Gilly is silly,* Q = *Nelson is rich,*
   R = *Cal is a pal*

   a. Gilly is silly but Nelson is rich.
   b. If Nelson is not rich, then Cal is a pal or Gilly is silly.
   c. Neither is Gilly silly, nor is Nelson rich.
   d. Gilly is silly and Nelson is rich, if Cal is a pal.
   e. Either it is false that both Gilly is silly and Nelson is rich, or Cal is a pal.
   f. Gilly is silly if and only if Nelson is rich or Cal is a pal.

3. Let P = *Sam is sulky,* Q = *Tom is tense,* and R = *Freddy is ready.* Write each statement in words.
   a. $P \wedge (Q \vee R)$
   b. $(P \wedge Q) \vee R$
   c. $P \vee Q \rightarrow R$
   d. $P \vee (Q \rightarrow R)$
   e. $R \wedge Q \rightarrow \sim P$
   f. $\sim P \leftrightarrow Q \wedge \sim R$

4. Construct a truth table for each statement.
   a. $\sim P \rightarrow Q$
   b. $P \vee \sim Q$
   c. $P \vee Q \leftrightarrow P \wedge \sim Q$
   d. $P \vee Q \rightarrow R$
   e. $P \wedge \sim Q \rightarrow Q \vee R$
   f. $\sim(P \wedge \sim Q) \rightarrow \sim(\sim P \vee Q)$

5. Use De Morgan's law to rewrite each statement.
   a. It is false that today is Monday and tomorrow is Sunday.
   b. Scott is not first or Joe is second.
   c. Hugh is painting or cutting the grass.
   d. It is not the case that Norma went to the store or Laurie went swimming.
   e. If mathematics is difficult, then logic is easy.

6. Determine whether $\sim P \vee Q$ is logically equivalent to $\sim(P \wedge \sim Q)$.

7. Use a truth table to determine whether the following arguments are valid or invalid. State your answer.
   a. If Gilly is silly or Cal is a pal, then Nelson is rich.
   It is false that Nelson is rich.
   Therefore, Gilly is not silly.

b. If Sam is sulky, then Tom is tense.
If Tom is tense, then Freddy is ready.
Therefore, if Sam is sulky, then Freddy is ready.

8. Use Venn diagrams to determine whether each pair of statements is consistent or inconsistent.
 a. No liars are virtuous.
 Some liars are virtuous.
 b. No liars are virtuous.
 All liars are virtuous.
 c. Some cars are gas eaters.
 Some gas eaters are not cars.
 d. All mathematics students are industrious.
 Some mathematics students are not industrious.

9. Use Venn diagrams to determine whether each argument is valid or invalid. State your answer.
 a. All logic students are mathematics students.
 No mathematics students are gullible.
 Hence no logic students are gullible.
 b. All logic students are mathematics students.
 Some logic students are not gullible.
 Therefore some gullible people are not mathematics students.
 c. No golfers are sane.
 All students are sane.
 Therefore no students are golfers.
 d. Some television programs are stupid.
 All stupid things are useless.
 So some television programs are useless.
 e. All commercials are deceiving.
 Some ads are deceiving.
 Hence some ads are commercials.

10. Using the suggested notation, write in symbolic form the converse, inverse, and contrapositive of each conditional statement.
 a. If I pass math, then I will graduate. (P, G)
 b. If I pass the test, then I will not come to class. (P, C)
 c. If you do not attend class, then you will not pass the course. (A, C)

11. A young man promised his girl friend, "I will marry you only if I get a job." He eventually got a job and then married someone else. Is the young man a heel? Why or why not? Explain your answer by using logic.

12. Given that "If Nixon knew about Watergate, then he is telling the truth" is a true statement.
 a. Write out another statement (in words) that you know is *true* from the above.
 b. Write out another statement (in words) that you know is *false* from the above.

13. Let P = *It is a gas eater* and Q = *It is a car.* Write each of the following statements in symbolic form.
 a. If it is a gas eater, then it is a car.
 b. Being a car is necessary for it to be a gas eater.
 c. Not being a car is sufficient for its not being a gas eater.
 d. It isn't a car only if it isn't a gas eater.
 e. It isn't a gas eater unless it is a car.

Just For Fun

Bill, the bewildered builder, discovered that a large picture window had been broken in his new big beautiful building. Bill knew that 3 workers were on the premises when the window was broken: Bob, Bart, and Barry. The workers' professions were painter, mason, and carpenter. But Bill did not know which man did which job, although he did know that one had committed the foul deed. He also knew the painter always told the truth, the mason never told the truth, and the carpenter always told one true statement and one false statement.

Barry said:	"Bart didn't do it."
	"Bob did it."
Bob said:	"I didn't do it."
	"Bart did it."
Bart said:	"I didn't do it."
	"Barry did it."

Using the true-false idea, help Bill discover the culprit's name and profession.

3 PROBABILITY

After studying this chapter, you will be able to do the following:

1. Compute the **probability** that an event A will occur
2. Compute the probability of *not* A, given the probability of A
3. Use the **fundamental counting principle** to determine how many ways two or more events can occur together
4. Construct a **sample space** showing the possible outcomes for an experiment
5. Construct a **tree diagram** for an experiment and use the tree to list the possible outcomes of the experiment
6. Compute the **odds** in favor of or against an event
7. Compute the **mathematical expectation** for an event
8. Determine whether two events are **dependent** or **independent,** and determine whether two events are **mutually exclusive**
9. Compute the probability of the event A and B and the probability of the event A *or* B, given the probability of A and of B

*10. Determine the number of **permutations** that can occur for n things taken r at a time, and the number of **combinations** that can occur for n things r at a time

*11. Compute the probabilities of events involving unordered arrangements and a large number of possible outcomes.

(*Note:* * indicates optional material)

Notation frequently used in this chapter

$P(A)$	The probability that event A will occur
$A:B$, A *to* B, $\frac{A}{B}$	Ratio of A and B
$n!$	n factorial $\quad n! = n \times (n-1) \times \cdots \times 3 \times 2 \times 1$
${}_nP_r$	the number of permutations of n things taken r at a time
${}_nC_r$	the number of combinations of n things taken r at a time

American Stock

3.1 INTRODUCTION

"I'll bet you!" This is a phrase that most of us have used at one time or another—many people have bet on something. It might have been on the results of an election or what the weather will be like tomorrow. Maybe you have bet with someone on the outcome of a World Series or Super Bowl game. Millions of people make some sort of a wager on major horse races such as the Kentucky Derby. You may have even bet with someone on how you would do on an exam.

Most people are introduced to the topic of probability through betting or games of chance. Even games that we played as small children—or play now—involve the use of a pair of dice. Even if you have never made a friendly bet or played any dice games (remember Monopoly), you have heard the weather forecaster mention that there is an "80% chance" of showers for the weekend. Probability is with us every day.

Probability has been studied by mathematicians for a long time. The concept of probability was first formally studied in the sixteenth century. At that time, it was the outgrowth of a study on gambling and games of chance.

Today, probability is still used to help people understand games of chance such as blackjack, craps, and lotteries. But there are certainly other uses of probability. Insurance companies are concerned with the probable life expectancy of their policy holders. Surveyers of public opinion use probability to determine the results of their polls: the Harris Poll and Gallup Poll arrive at their results by means of probability, as do the various television polls. Biology (genetics), astronomy, and manufacturing are some other areas that make extensive use of probability theory.

The topics covered in this chapter will provide you with a basis for understanding probability and some everyday applications.

3.2 DEFINITION OF PROBABILITY

Suppose we are given one die from a pair of dice. Upon examination, we see that it is a six-sided solid cube in which one side has 1 dot, another side has 2 dots, and so on, until the last side has 6 dots.

FIGURE 1

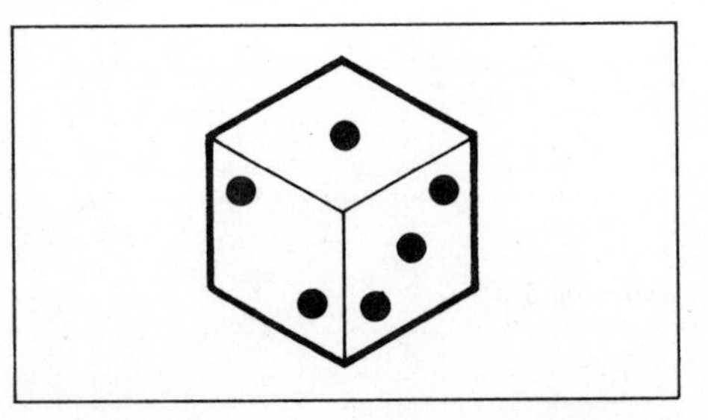

When we toss a die, we are performing an **experiment** to see which set of dots, or number, will turn face up. The number of dots that are on the top surface when we toss the die is called the **outcome** of the experiment. Each number has an equal chance of occurring, so we say that each outcome—1, 2, 3, 4, 5, or 6—is **equally likely** to occur.

The set of all possible outcomes of an experiment is called a

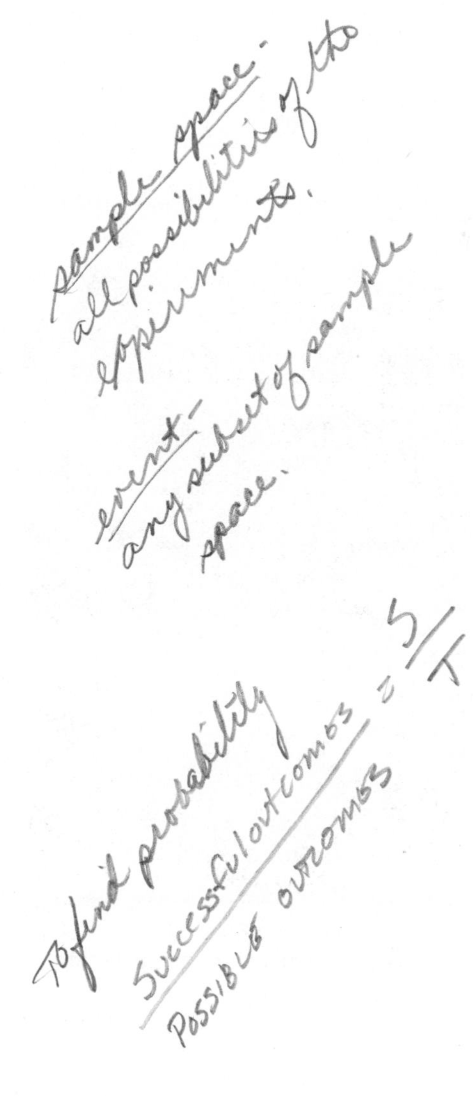

sample space. For example, when the experiment is tossing a die, the sample space is $\{1, 2, 3, 4, 5, 6\}$. An **event** is any subset of the sample space: $\{3\}$ and $\{1, 6\}$ are both events.

If we toss a die, what is the probability of obtaining a 3? If we are interested in the probability of obtaining a 3, then we are actually interested in the probability that the event $\{3\}$ will occur. Therefore before we proceed any further, let us define what we mean by the probability that an event A will occur.

If an experiment has a *total* of T equally likely possible outcomes, and if exactly S of them are considered **successful**—that is, they are the members of the event A—then the probability that event A will occur, denoted by $P(A)$, is

$$P(A) = \frac{\text{number of } \textit{successful} \text{ outcomes}}{\textit{total} \text{ number of possible outcomes}} = \frac{S}{T}$$

Using the idea of cardinality from chapter 1, we can also state this as

$$P(A) = \frac{\mathrm{n}(A)}{\mathrm{n}(T)}$$

where $\mathrm{n}(A)$ stands for the cardinality (number of members) of the set A.

A helpful way to remember this probability formula, $P(A) = \frac{S}{T}$, is to remember that to find the probability in favor of an event A, we find the number of *successful* outcomes and divide it by the *total* number of possible outcomes.

Hence the probability of obtaining a 3 in a toss of one die is $\frac{1}{6}$. We see that there is only one possible successful outcome, a 3. There are a total of six possible outcomes, namely, 1, 2, 3, 4, 5, and 6. To compute the probability of obtaining a 3 for this experiment, we have

$$P(3) = \frac{1}{6} \quad \left(\frac{\text{successful outcomes}}{\text{total number of possible outcomes}}\right)$$

EXAMPLE 1

Find the probability of getting a head when you flip a quarter.

Solution

There are 2 different possible outcomes (heads, tails), and only 1 of these outcomes (heads) is successful. Hence the probability in favor of obtaining heads is $\frac{1}{2}$.

$$P(\text{heads}) = \frac{1}{2}$$

EXAMPLE 2
Find the probability of obtaining an even number on a single toss of a die.

Solution
There are 6 different possible outcomes and 3 of these outcomes (2, 4, 6) are successful. Therefore the probability of obtaining an even number is $\frac{3}{6}$.

$$P(\text{even number}) = \frac{3}{6} = \frac{1}{2}$$

EXAMPLE 3
Find the probability of obtaining a number greater than 4 on a single toss of a die.

Solution
There are 6 different possible outcomes and 2 of these outcomes (5, 6) are successful. Hence the probability in favor of obtaining a number greater than 4 is $\frac{2}{6}$.

$$P(\text{number greater than 4}) = \frac{2}{6} = \frac{1}{3}$$

Ewing Galloway

If the compartments on this roulette wheel are numbered 1-36, what is the probability thât the winning number will be 7?

EXAMPLE 4
Find the probability of drawing a king (one pick) from a shuffled standard deck of 52 cards. (A standard deck of cards is the most common type of deck used in most card games. If you are not familiar with such a deck, see figure 2, which shows all 52 cards.)

Solution
There are 52 different possible outcomes. Four of these outcomes are successful: king of spades, king of hearts, king of clubs, king of diamonds. Therefore the probability in favor of obtaining a king is $\frac{4}{52}$.

$$P(K) = \frac{4}{52} = \frac{1}{13}$$

EXAMPLE 5
Find the probability of drawing a jack or a queen from a shuffled standard deck of 52 cards.

Solution
There are 52 different possible outcomes and 8 of these are considered successful outcomes, since if a jack *or* a queen is picked, we are

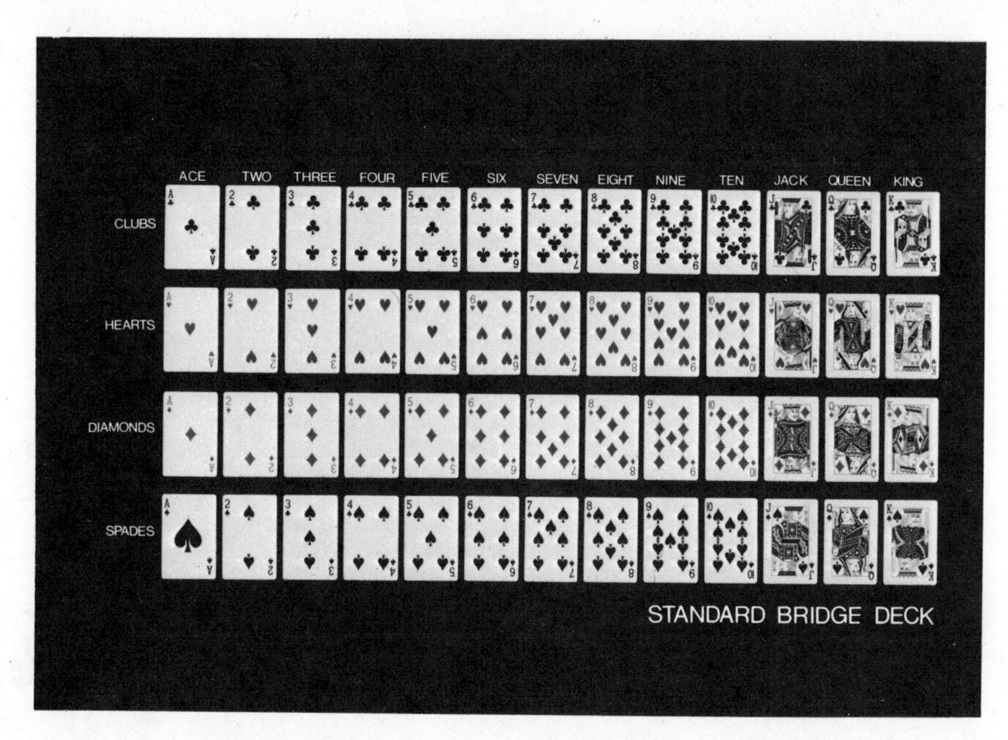

FIGURE 2 A standard deck of 52 cards

successful. Hence the probability in favor of obtaining a jack or a queen is $\frac{8}{52}$.

$$P(J \text{ or } Q) = \frac{8}{52} = \frac{2}{13}$$

EXAMPLE 6

Find the probability of obtaining a 7 on a single toss of one die.

Solution

There are 6 different possible outcomes and none of these outcomes would produce a 7. That is, zero of these outcomes would be successful. The probability in favor of obtaining a 7 on a single toss of one die is $\frac{0}{6}$, or 0.

$$P(7) = \frac{0}{6} = 0$$

When an event cannot possibly succeed, we say it is an **impossible event.** The probability of an impossible event is zero.

$$P(\text{impossible event}) = \frac{0}{T} = 0 \qquad (T \neq 0)$$

Ewing Galloway

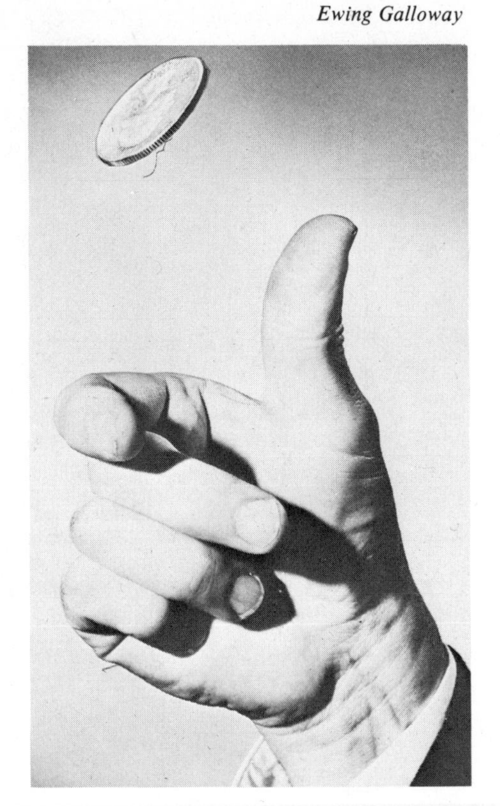

EXAMPLE 7
Find the probability of getting heads or tails when you flip a quarter.

Solution
You can't lose! There are 2 different possible outcomes (heads, tails) and both of these outcomes are successful. Hence the probability in favor of obtaining heads or tails is $\frac{2}{2}$, or 1.

$$P(\text{heads or tails}) = \frac{2}{2} = 1$$

When an event is sure to occur, then success is inevitable, and we say that the probability is 1. This is sometimes called certainty.

$$P(\text{certain event}) = \frac{T}{T} = 1 \qquad (T \neq 0)$$

For any event, $P(A) \geq 0$ and $P(A) \leq 1$, so we may write

$$0 \leq P(A) \leq 1$$

EXAMPLE 8
Find the probability of drawing an ace or a spade from a shuffled standard deck of 52 cards.

Solution
Your initial answer might be $\frac{17}{52}$, but this is not correct! There are 52 different possible outcomes, and there are 4 aces and 13 spades; hence, adding 4 and 13, we would have $\frac{17}{52}$. Why isn't this correct? If we examine the possible successful outcomes, we see there are indeed 13 spades, but one of them is an ace! We include the ace with our 13 spades, leaving 3 other aces to include in our successful number of outcomes. If you count the aces first, then you have included the ace of spades, and there are 12 spades left to include in the successful number of outcomes. Altogether, there are 16 successful outcomes, and the probability of drawing an ace or a spade is $\frac{16}{52}$, or $\frac{4}{13}$.

$$P(\text{ace or spade}) = \frac{16}{52} = \frac{4}{13}$$

Note that if you draw an ace *or* a spade, you are successful. You do not have to draw an ace *and* a spade, which would be the ace of spades: the probability of drawing the ace of spaces is only $\frac{1}{52}$.

EXAMPLE 9
One card is drawn at random from a shuffled standard deck of 52 cards. Find the probability that the card selected is *not* a king.

Solution
There are 52 different possible outcomes. There are 4 kings in a deck, so the other 48 cards are not kings, and these are the successful outcomes. Hence the probability that the card selected is not a king is $\frac{48}{52}$.

$$P(not\ K) = \frac{48}{52} = \frac{12}{13}$$

The sum of the probability that an event will occur and the probability that it will not occur is 1. Therefore, we can solve example 9 in another way.
Be sure to remember that if

$$P(A) = \frac{S}{T}$$

then

$$P(not\ A) = 1 - \frac{S}{T} \qquad (T \neq 0)$$

Alternate solution to example 9:
The probability of selecting a king is $\frac{4}{52}$ or $\frac{1}{13}$. Therefore the probability that the card selected is *not* a king is

$$P(not K) = 1 - \frac{1}{13} = \frac{13}{13} - \frac{1}{13} = \frac{12}{13}$$

EXERCISES FOR SECTION 3.2

1. On the single toss of one die, find the probability of obtaining—
   a. a 4
   b. an odd number
   c. an even number
   d. a number less than 4
   e. a number greater than 4
   f. an odd or an even number.

2. On the single toss of one die, find the probability of obtaining—
   a. a number divisible by 3 (Example: 6 is divisible by 3 because 3 divides 6 evenly; that is, the remainder is zero)
   b. a number divisible by 5
   c. a number divisible by 2
   d. a number divisible by 7 or 1

e. a number less than 1
f. a number less than 7.

3. On a single draw from a shuffled standard deck of 52 cards, find the probability of obtaining—
a. the ace of spades
b. the 2 of hearts
c. a deuce (2)
d. a red card
e. a diamond
f. a red jack.

4. On a single draw from a shuffled standard deck of 52 cards, find the probability of obtaining—
a. a spade
b. a club
c. a spade or a club
d. a spade and a club
e. a 3 or a heart
f. a 3 and a heart. (Be careful on this one!)

5. On a single draw from a shuffled standard deck of 52 cards, find the probability of obtaining—
a. a picture card (jack, queen, king)
b. a picture card or a heart
c. a jack or a heart
d. a jack and a heart (Be careful on this one.)
e. a one-eyed jack (jack of spades or jack of hearts)
f. a king with an axe (king of diamonds).

6. Harry Hose keeps all of his socks in the top drawer of his bureau. In the drawer there are 4 blue socks, 6 black socks, 7 brown socks (he lost one in the laundry), and 4 red socks. Harry reaches in and pulls a sock out at random. Find the probability that the sock chosen is—
a. brown
b. blue
c. red
d. black
e. not brown
f. neither brown nor blue.

7. Gloria Glove keeps all of her mittens on the top shelf of her hall closet. On the shelf are 4 blue mittens, 6 brown mittens, and 4 green mittens. Gloria reaches up and pulls a mitten out at random. Find the probability that the mitten chosen is—
a. blue or brown
b. blue or green
c. not red
d. green or red
e. neither blue nor green.

8. If it is dark in the hall and the light doesn't work, what is the greatest number of mittens Gloria (see exercise 7) will have to pull out to make sure that she has 2 mittens of the same color?

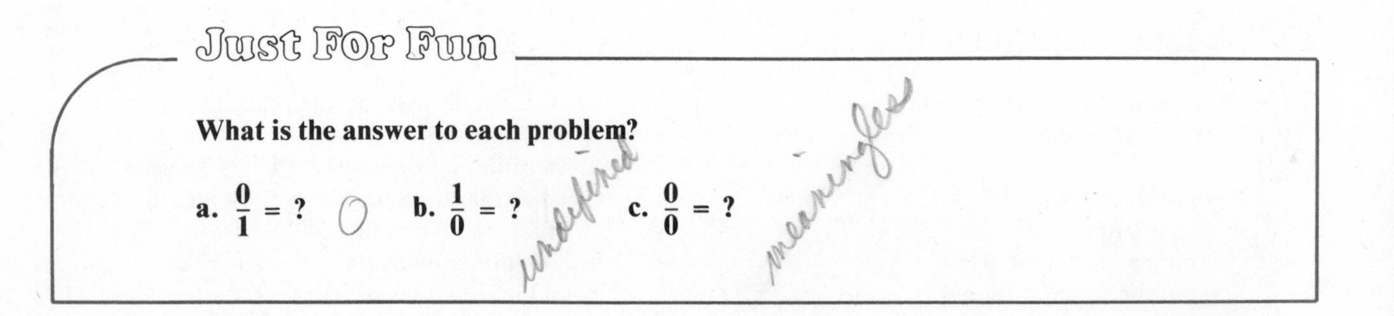

Just For Fun

What is the answer to each problem?

a. $\frac{0}{1} = ?$ **b.** $\frac{1}{0} = ?$ **c.** $\frac{0}{0} = ?$

3.3 SAMPLE SPACES

Many times when we want to compute the probability of some event, the total number of possible outcomes of the experiment is not easy to determine. In some instances it can be downright tricky. Consider the following examples.

EXAMPLE 1

A quarter is flipped and a die is tossed. What is the probability of obtaining heads *or* an odd number?

Solution

The gullible person might reason that there are 2 outcomes with the quarter and 6 outcomes with the die, so altogether there are 8 total possible outcomes. Also, this same person might reason that there is 1 successful outcome with the quarter (heads) and 3 successful outcomes with the die (1, 3, 5), so there are 4 successful outcomes altogether. Therefore the probability of getting heads or an odd number is $\frac{4}{8}$. This is WRONG!

In fact, there are 12 possible outcomes. The quarter may turn up 2 ways, heads or tails. If it lands heads up, then the heads can be matched with 6 different outcomes of the die. Namely, H—1, H—2, H—3, H—4, H—5, H—6. The same thing can happen with tails. If tails comes up, then we could have T—1, T—2, T—3, T—4, T—5, T—6. Altogether, we have 12 *different* outcomes. Making a list of these outcomes, we have

H—1	H—4	T—1	T—4
H—2	H—5	T—2	T—5
H—3	H—6	T—3	T—6

How many of these possible outcomes are successful? Remember, we wanted heads or an odd number. Six of the total outcomes contain a head, so they have to be considered successful. Three of the remaining possible outcomes are also successful: T—1, T—3, T—5. Hence, there are 9 successful outcomes, and the probability of obtaining heads or an odd number is $\frac{9}{12}$, or $\frac{3}{4}$.

Recall that in chapter 1 we studied the Cartesian product. We can use the Cartesian product to generate the outcomes in example 1 above. If we let $A = \{H, T\}$ and $B = \{1, 2, 3, 4, 5, 6\}$, then

$$A \times B = \{(H, 1), (H, 2), (H, 3), (H, 4), (H, 5), (H, 6), (T, 1), (T, 2), (T, 3), (T, 4), (T, 5), (T, 6)\}$$

Recall from section 3.2 that a list of all possible outcomes for an experiment is called a sample space. A handy method to determine the number of outcomes for two experiments together is to multiply the number of ways one can occur by the number of ways the other can occur. In example 1, a quarter can come up 2 ways (assuming it can't land on its edge), and a die can turn up 6 ways. Therefore, we have $2 \times 6 = 12$ outcomes for both experiments together.

In general, if one experiment has m different outcomes and a second experiment has n different outcomes, then the first and second experiments performed together have $m \times n$ different outcomes. This idea may be extended if there are other experiments to follow: we would have $m \times n \times r \times \cdots \times t$ outcomes. This is often called the **fundamental counting principle.**

EXAMPLE 2

List a sample space showing all possible outcomes when a pair of dice is tossed.

Solution

First of all, how many outcomes will we have in our sample space? There are two dice and each die has 6 possible outcomes. Using the counting principle, we have $6 \times 6 = 36$ total possible outcomes.

Let one die be red and the other white, so that we can distinguish the two dice. If a 1 comes up on the white die, it can be paired with each of the 6 numbers on the red die; that is, (1, 1), (1, 2), (1, 3), (1, 4), (1, 5), and (1, 6). We can do this in turn for each of the numbers that comes up on the white die, pairing it with all the numbers on the red die. This pairing is illustrated in figure 3.

Note that again we could use the Cartesian product to generate the above sample space. If $D = \{1, 2, 3, 4, 5, 6\}$, then $D \times D$ would yield the 36 ordered pairs that are listed in figure 3.

Harold M. Lambert

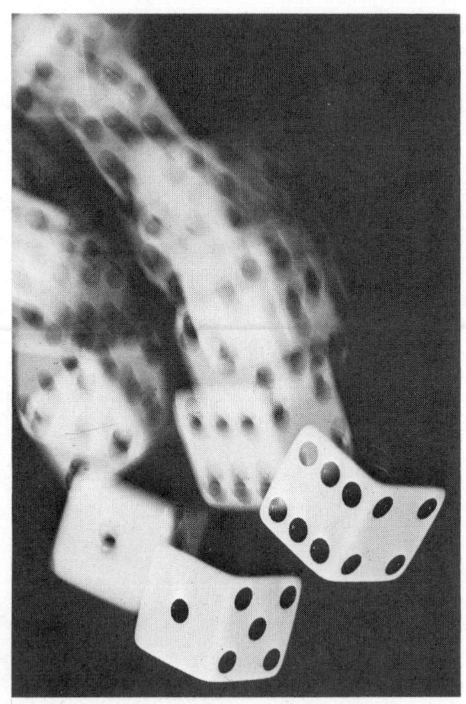

In games where a pair of dice is used, the sum of the two numbers is the primary concern. Note that there is only one outcome where the sum of the dice is 2 (the vernacular term is "snake eyes"); this is also the case for 12 ("box cars"). The sums that may be obtained by rolling a pair of dice are: 2, 3, 4, 5, 6, 7, 8, 9, 10, 11, 12. These are the only possible totals with two dice.

Examining the sample space illustrated in figure 3, we see that we have the following number of ways to obtain each of the sums:

sum:	2	3	4	5	6	7	8	9	10	11	12
number of ways:	1	2	3	4	5	6	5	4	3	2	1

FIGURE 3

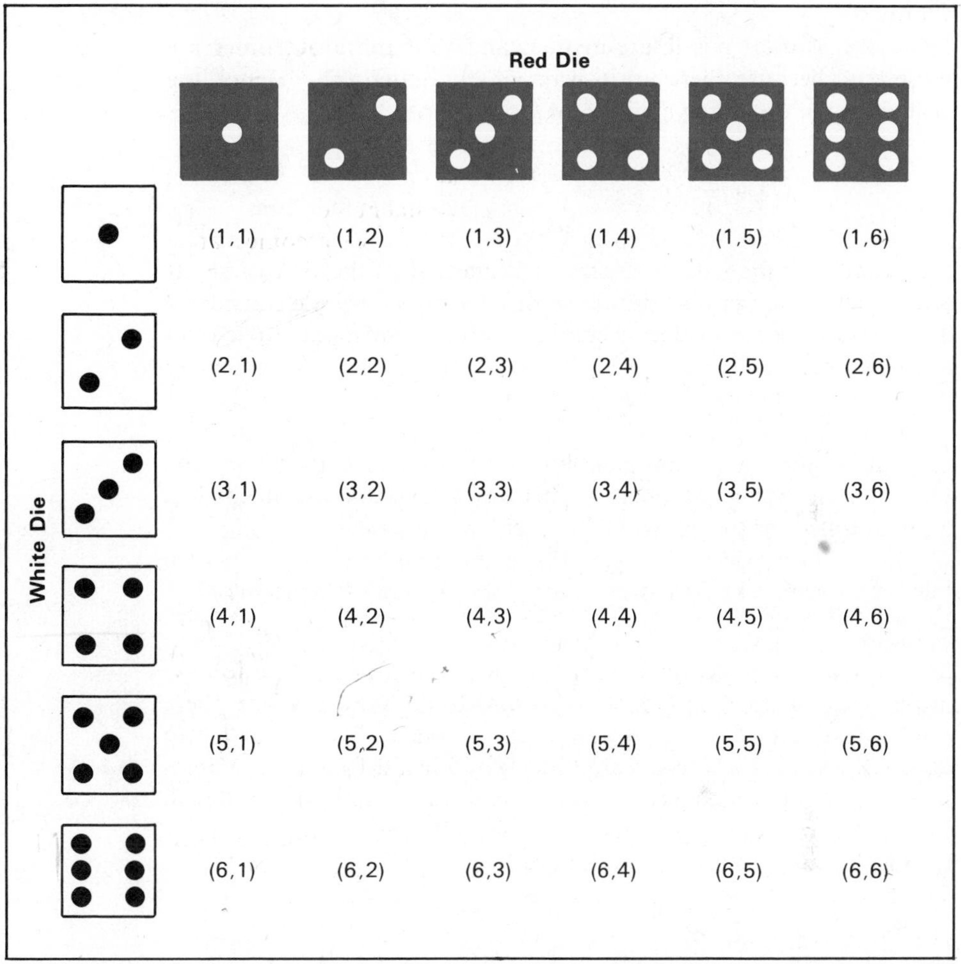

Red Die

White Die

(1,1)	(1,2)	(1,3)	(1,4)	(1,5)	(1,6)
(2,1)	(2,2)	(2,3)	(2,4)	(2,5)	(2,6)
(3,1)	(3,2)	(3,3)	(3,4)	(3,5)	(3,6)
(4,1)	(4,2)	(4,3)	(4,4)	(4,5)	(4,6)
(5,1)	(5,2)	(5,3)	(5,4)	(5,5)	(5,6)
(6,1)	(6,2)	(6,3)	(6,4)	(6,5)	(6,6)

Note that the sum 7 can occur in six ways and all of the other sums can occur in fewer ways. It is interesting (and helpful) to note that 6 and 8 can occur in five ways, the 5 and 9 can occur in four ways, and so on. This is a symmetric pattern, and it should help you to remember the total number of ways each sum can occur. First you must remember that a 7 can occur in six ways, and that this is the greatest number of ways that a sum may occur on a pair of dice.

EXAMPLE 3

Use the sample space for the total number of possible outcomes when a pair of dice is tossed to find the probability of obtaining a sum of 7 on a single toss of a pair of dice.

Solution

There are 36 total possible outcomes and 6 of these outcomes are successful, because there are 6 ways of obtaining a 7. Hence the probability of obtaining a 7 is $\frac{6}{36}$, or $\frac{1}{6}$.

$$P(7) = \frac{6}{36} = \frac{1}{6}$$

Note that a 7 is most likely to occur when a pair of dice is tossed. It is believed that this is one of the reasons why many people consider 7 their lucky number. It keeps coming up for them more often than other numbers.

EXAMPLE 4

A popular dice game in gambling establishments is "Over and Under." This is a game where the bettor may bet that the dice (when flipped, rolled, or tossed) will total over 7 or under 7.

Lucky Larry decides to play this game, and he wants to bet on "under," because he feels that "under" should win. Is he right?

Solution

Since Lucky Larry bet on "under," he will win if any of the following sums come up: 2, 3, 4, 5, 6. According to the sample space, there are 5 ways for a 6 to occur, 4 ways for a 5 to occur, 3 ways for a 4 to occur, 2 ways for a 3 to occur, and 1 way for a 2 to occur. Therefore there are 15 outcomes that can be considered successful. Recall that there are 36 total possible outcomes. Hence the probability of the dice turning up under 7 is $\frac{15}{36}$, or $\frac{5}{12}$.

$$P(\text{under } 7) = \frac{15}{36} = \frac{5}{12}$$

Is this a good bet? Not really; if Lucky Larry has a $\frac{5}{12}$ probability of winning, then he is more likely to lose. Out of a total of 36 possible outcomes, only 15 are considered successful. The other 21 outcomes can turn Lucky Larry into Larry the Loser. The probability that the dice will *not* be under 7 is $\frac{21}{36}$; since the sum of the probability in favor of an event and the probability against an event is 1, if $P(\text{under } 7)$ is $\frac{15}{36}$, then $P(\textit{not} \text{ under } 7)$ is $1 - \frac{15}{36}$, or $\frac{36}{36} - \frac{15}{36} = \frac{21}{36}$.

What happens if Lucky Larry changes his mind and decides to bet "over"? Has he got a better chance of winning? No! If we compute the probability of "over 7," we get $\frac{15}{36}$, and again Lucky Larry is more likely to lose because the probability of "*not* over 7" is $\frac{21}{36}$.

What happens if Lucky Larry decides to bet both "over" and "under"? The best he can do is get his own money back, because if he wins on one side, he loses on the other side. Note also that he could be a double loser if 7 comes up, because 7 is neither "over" nor "under"!

As we learned in chapter 1, one of the uses of Venn diagrams is to show and separate numerical data that has been compiled. Venn diagrams can also be used to solve probability problems. Consider the following example.

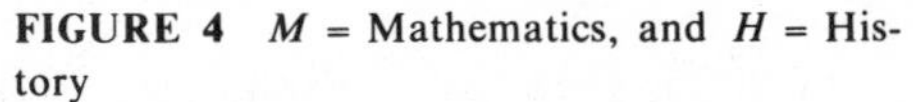

FIGURE 4 M = Mathematics, and H = History

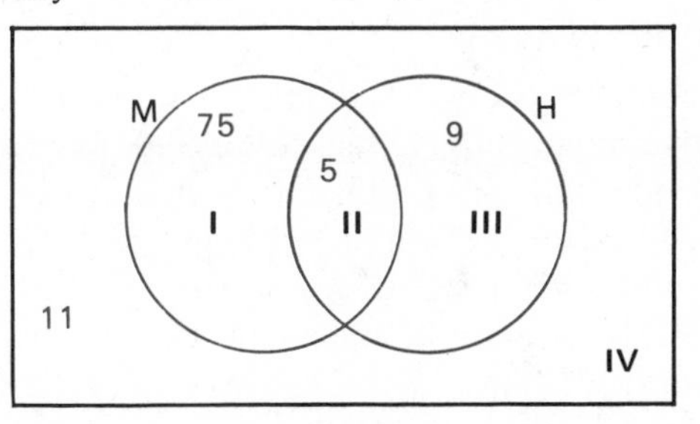

EXAMPLE 5

At a recent college registration, 100 students were interviewed. Eighty of the students stated that they had registered for a mathematics course, 14 of the students stated that they had registered for a history course, and 5 of the students stated that they had registered for a mathematics course and a history course. What is the probability that a student in this survey registered only for history?

Solution

A Venn diagram using the given information is shown in figure 4. With it, we can summarize the information that 11 of the students did not register for either a mathematics course or a history course, 9 students registered only for history, and 75 students registered only for mathematics.

This information can also be used to solve the probability problem. We note that the total number of students is 100, while 9 of them are registered only for history. Hence, the answer is $\frac{9}{100}$.

FIGURE 5 S = Statistics, C = Chemistry, and P = Psychology

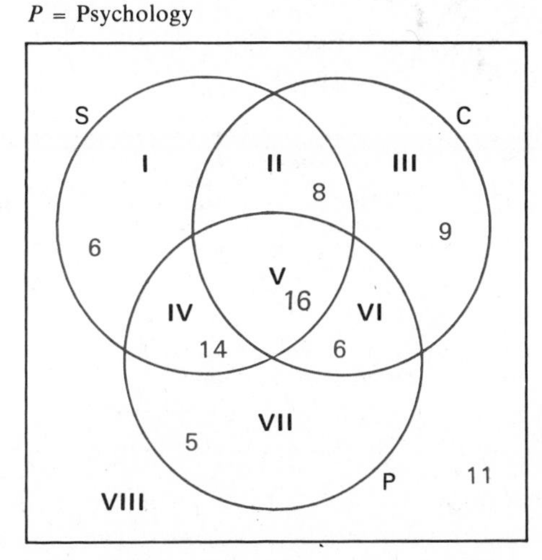

EXAMPLE 6

In a certain group of 75 students, it has been determined that 16 students are taking statistics, chemistry, and psychology; 24 students are taking statistics and chemistry; 30 students are taking statistics and psychology; 22 students are taking chemistry and psychology; 6 students are taking only statistics; 9 students are taking only chemistry; and 5 students are taking only psychology.

a. What is the probability that a student is not taking any of the three subjects?

b. What is the probability that a student is taking chemistry?

Solution

We first complete the necessary Venn diagram. See figure 5. Remember, it is best to start with the most definite piece of information. After completing the diagram, we can answer the questions.

a. The probability that a student is not taking any of the three subjects is $\frac{11}{75}$.

b. The probability that a student is taking chemistry is $\frac{39}{75}$.

The answer to question *b* is obtained by adding the number of students in each partition of the chemistry circle. Hence there are 39 students that are taking chemistry.

EXERCISES FOR SECTION 3.3

1. A quarter is flipped and a die is tossed. Use a sample space to find the probability of obtaining—
   a. heads and a 6
   b. heads and a 7
   c. heads or a 6
   d. heads or a 7
   e. tails and an even number
   f. tails or an even number.

2. A quarter is flipped and a die is tossed. Use a sample space to find the probability of obtaining—
   a. tails and a number less than 4
   b. tails and a number less than 1
   c. tails or a number less than 10
   d. heads and a number divisible by 2
   e. heads or a number divisible by 2
   f. tails or a number divisible by 7.

3. A bag contains 5 balls numbered 1 through 5. Two balls are chosen at random in succession. The first ball is replaced before the second is chosen.
   a. Construct a sample space for this experiment.
   b. What is the probability that the first ball has an odd number?
   c. What is the probability that the second ball has an odd number?
   d. What is the probability that both balls have odd numbers?
   e. What is the probability that the first or second ball has an odd number?
   f. What is the probability that neither ball has an odd number?

4. A box contains a one-dollar bill, a two-dollar bill, a five-dollar bill, a ten-dollar bill and a twenty-dollar bill. Two bills are chosen at random in succession. The first bill is not replaced before the second is drawn.
   a. Construct a sample space for this experiment.
   b. What is the probability that the first bill is even?
   c. What is the probability that the second bill is even?
   d. What is the probability that both bills are even?
   e. What is the probability that the first or second bill is even?
   f. What is the probability that neither bill is even?

5. Using the sample space for the possible outcomes when a pair of dice is tossed (see example 2 in this section), find the probability that—
   a. the sum of the numbers is 7
   b. the sum of the numbers is 11
   c. the sum of the numbers is 1
   d. the sum of the numbers is even
   e. the sum of the numbers is odd or even
   f. the sum of the numbers is divisible by 5.

6. Using the sample space for the possible outcomes when a pair of dice is tossed (see example 2 in this section), find the probability

that—

a. the same number appears on both dice
b. the number on one die is twice the number on the other die
c. the sum of the numbers is greater than 7
d. the sum of the numbers is at least 8
e. the sum of the numbers is greater than 1
f. the sum of the numbers is less than 1.

7. In a survey of 50 contestants at a track meet, the following information was obtained: 12 contestants were entered in both a running event and a field event, 32 contestants were entered in a running event, and 30 contestants were entered in a field event.
   a. Show the results of this survey in a Venn diagram.
   b. Find the probability that a contestant is entered in a running event only.
   c. Find the probability that a contestant is entered in a field event only.

8. At a meeting of 50 car salespeople, the following information was obtained: 12 salespeople sold Pintos, 15 salespeople sold Monzas, and 16 salespeople sold Darts. Four salespeople sold Pintos and Monzas, 6 salespeople sold Monzas and Darts, and 5 salespeople sold both Pintos and Darts. Two salespeople sold all three kinds of cars.

   Using a Venn diagram, find the probability that a salesperson at this meeting sold—
   a. Pintos only
   b. Monzas only
   c. Darts only
   d. Pintos and Monzas, but not Darts
   e. Monzas and Darts, but not Pintos
   f. Monzas or Pintos.

9. In a survey of 75 students who registered for courses, the following data were collected: 27 students were taking statistics, 26 were taking history, and 41 were taking English. Twelve students were taking statistics and history, 13 students were taking statistics and English, and 17 students were taking history and English. Four students were taking all three courses.

   Use a Venn diagram to find the probability that a student participating in this survey is taking—
   a. only statistics
   b. only English
   c. statistics and English, but not history
   d. statistics or English
   e. none of the three subjects
   f. exactly one of these three subjects.

Just For Fun

Given the digits 1, 2, 3, 4, 5, 6, and 7, construct an addition problem whose sum is 100. You may use each digit only once.

B.C. by permission of Johnny Hart and Field Enterprises, Inc., 1975

3.4 TREE DIAGRAMS

A sample space consists of all possible outcomes for a particular experiment. A technique that shows us the sample space for two or more experiments that are performed together is a **tree diagram.**

A tree diagram consists of a number of "branches" that illustrate the possible outcomes for the experiments. We may read the possibilities directly from the branches. The following examples illustrate the use of a tree diagram to obtain a sample space.

EXAMPLE 1

When a coin is flipped, it may turn up heads or tails. How many different outcomes are possible when 2 coins are tossed, and what are the possible outcomes?

Solution

There are 2 experiments and each experiment has 2 possible outcomes (heads or tails). Using the counting principle, we have $2 \times 2 = 4$ possible outcomes.

In constructing the tree diagram, remember that the first experiment has 2 possible outcomes and each of these may be matched with the 2 possible outcomes of the second experiment. Hence we have

	First Coin	Second Coin	Possible Outcomes: First Coin	Possible Outcomes: Second Coin
Start	*H*	*H*	*H*	*H*
		T	*H*	*T*
	T	*H*	*T*	*H*
		T	*T*	*T*

From the tree diagram we may obtain our sample space. The first branch gives us *H*—*H*, the second branch *H*—*T*, and so on, until we have the complete sample space: *H*—*H*, *H*—*T*, *T*—*H*, and *T*—*T*.

EXAMPLE 2

Mr. Testhappy is preparing a quickie quiz for his mathematics class to see if the students did their assignment. The quiz is to consist of 3 true-false questions. How many different arrangements of the answers are possible? What are the possible outcomes?

Solution

We have 3 questions and each question has 2 possible outcomes (true or false). Using the counting principle, we compute $2 \times 2 \times 2 = 8$ total possible outcomes.

We can determine the various outcomes by means of a tree diagram. Remember that the quickie quiz consists of 3 questions and the answer to each question is either true or false.

	First Question	Second Question	Third Question	Possible Outcomes: First Question	Second Question	Third Question
Start	*T*	*T*	*T*	*T*	*T*	*T*
			F	*T*	*T*	*F*
		F	*T*	*T*	*F*	*T*
			F	*T*	*F*	*F*
	F	*T*	*T*	*F*	*T*	*T*
			F	*F*	*T*	*F*
		F	*T*	*F*	*F*	*T*
			F	*F*	*F*	*F*

The sample space is listed beside the tree diagram.

EXAMPLE 3

Now that we have the sample space for the total number of possible outcomes for the three true-false questions, consider the following problem.

Benny Banana, a student in Mr. Testhappy's class, did not do the assignment. Benny took the quiz and guessed at the answers. Benny answered the questions *F—T—F*. What is the probability that he answered all three questions correctly?

Solution

The sample space in example 2 shows each of the 8 possible outcomes. Only 1 of these contains all three correct answers. Hence, the probability that *F—T—F* is the correct combination is $\frac{1}{8}$.

EXERCISES FOR SECTION 3.4

1. Use a tree diagram to find the sample space showing the possible arrangements of boys and girls in a family with exactly 2 children.
   a. What is the probability that both children are boys?
   b. What is the probability that both children are girls?
   c. What is the probability that the first child is a boy and the second child is a girl?
   d. What is the probability that at least 1 of the children is a girl?
   e. What is the probability that none of the children are girls?

2. A box contains a one-dollar bill, a two-dollar bill, a five-dollar bill, and a ten-dollar bill. Two bills are chosen in succession without replacement. Use a tree diagram to list the sample space for this experiment.
   a. What is the probability that both bills are even?
   b. What is the probability that neither bill is even?
   c. What is the probability that exactly 1 of the bills is even?
   d. What is the probability that at least 1 of the bills is even?
   e. What is the probability that the total value of the bills chosen is $15?

3. Use a tree diagram to list the sample space showing the possible arrangements of boys and girls in a family with exactly 3 children.
   a. What is the probability that all 3 children are boys?
   b. What is the probability that 2 children are boys and 1 is a girl?
   c. What is the probability that at least 1 of the children is a girl?
   d. What is the probability that none of the children are boys?
   e. What is the probability that all 3 children are of the same sex?
   f. If a family has 3 children, all boys, what is the probability that a fourth child would be a girl?

4. Use a tree diagram to list the sample space showing the possible arrangements of heads and tails when 4 coins are tossed. Then use the sample space to find the probability that—
   a. all 4 coins come up heads
   b. exactly 2 coins come up heads
   c. exactly 3 coins come up tails
   d. at most 2 coins come up heads
   e. at least 2 coins come up heads
   f. no more than 3 coins come up tails.

5. An urn contains 1 blue and 3 orange balls. An experiment consists of selecting 2 balls in succession, without replacement. Use a tree diagram to list the sample space for this experiment. (*Hint:* Balls of the same color are indistinguishable, so you may find it helpful to use the symbols O_1, O_2, and O_3 to distinguish the 3 orange balls.) Then use the sample space to find the probability that—
   a. both balls are orange
   b. at least 1 ball is orange
   c. the first ball is blue
   d. the second ball is blue
   e. at least 1 ball is blue
   f. 1 ball is blue and the other is orange.

*6. The Knicks won the first game against the Lakers in a basketball playoff. If the first team to win four games is the winner, construct a tree diagram for this playoff. (*Hint:* Some of the branches in the tree diagram will end sooner than others, as one of the teams will have won four games.)

Just For Fun

Hit songs are usually recorded on records that are approximately $6\frac{3}{4}$ inches in diameter and spin at 45 revolutions per minute. How many grooves are on such a record?

3.5 ODDS AND EXPECTATION

What odds will you give me?
The Browns are odds-on favorites to win the title.
The odds on Dead Last are 100 to 1.
The odds on getting a royal flush in poker are 649,739 to 1.

All of these expressions mention the word *odds*. We all have a nodding acquaintance with odds and probably have come in contact with them at some time or another. Most of us know that if the odds on a bet are 3 to 1 (3:1) and we bet a dollar and win, then we will win 3 dollars. But there is more to it than that.

Odds are usually expressed as a ratio, for example 3:2. But, they may also be expressed as 3 to 2, or even as a fraction, $\frac{3}{2}$.

Odds can occur in two ways. We may discuss the *odds in favor* of an event or the *odds against* an event.

Odds are computed by finding a ratio. The **odds in favor of an event A** are found by taking the probability that an event A will occur and dividing it by the probability that an event A will not occur. We may state this as

$$\text{odds in favor of } A = \frac{\text{probability in favor of } A}{\text{probability not in favor of } A}$$

EXAMPLE 1

What are the odds in favor of obtaining a sum of 7 when a pair of dice is tossed once?

Solution

Using our definition, we first find the probability of obtaining a sum of 7 when a pair of dice is tossed. It is $\frac{6}{36}$. Next, we find the probability of not getting a 7. Recall that if $P(A) = S/T$, then $P(not\ A) = 1 - S/T$. Hence, we have

$$P(7) = \frac{6}{36} \quad \text{and} \quad P(not\ 7) = 1 - \frac{6}{36} = \frac{36}{36} - \frac{6}{36} = \frac{30}{36}$$

Now we construct our ratio according to the definition.

$$\text{odds in favor of } A = \frac{\text{probability in favor of } A}{\text{probability not in favor of } A}$$

$$\text{odds in favor of a 7} = \frac{\frac{6}{36}}{\frac{30}{36}} = \frac{6}{36} \times \frac{36}{30} = \frac{6}{30} = \frac{1}{5}$$

The odds in favor of obtaining a 7 are $\frac{1}{5}$. But, remember that the odds may also be stated as 1 to 5 or 1:5.

What does this mean? If we roll a pair of dice, over the long run we should obtain a 7 once out of 6 times. The other 5 times we would lose. (*Note:* The odds in favor of obtaining a 7 are 1:5. Hence, the odds *against* obtaining a 7 are 5:1.)

EXAMPLE 2
Find the odds in favor of obtaining 2 heads in a single toss of 2 coins.

Solution
The probability in favor of obtaining 2 heads in a single toss of 2 coins is $\frac{1}{4}$. We can obtain this from the sample space ($H—H$, $H—T$, $T—H$, $T—T$). The probability against 2 heads is $1 - \frac{1}{4} = \frac{4}{4} - \frac{1}{4} = \frac{3}{4}$. Therefore we have

$$\frac{\frac{1}{4}}{\frac{3}{4}} = \frac{1}{4} \times \frac{4}{3} = \frac{1}{3}$$

The odds in favor of 2 heads are 1 to 3, or 1:3.

If we change the ratio in example 2 around (3 to 1, or 3:1), we will have the odds *against* obtaining 2 heads in a single toss of 2 coins. We must always make certain that we find the odds in favor of an event, if that is what we are asked to find. If we are asked to find the **odds against an event** ***A***, then we should use the following ratio:

$$\text{odds against } A = \frac{\text{probability not in favor of } A}{\text{probability in favor of } A}$$

We can also find the odds in favor of A and then reverse the ratio.

EXAMPLE 3
What are the odds against obtaining an 11 when a pair of dice is tossed once?

Solution
Using the definition, we still must find the probability in favor of getting an 11 and the probability against getting an 11 when a pair of dice is tossed. The probability of getting an 11 is $\frac{2}{36}$, and the probability of not getting an 11 is $1 - \frac{2}{36} = \frac{36}{36} - \frac{2}{36} = \frac{34}{36}$. Now we construct our ratio according to the definition.

$$\text{odds against an 11} = \frac{\frac{34}{36}}{\frac{2}{36}} = \frac{34}{36} \times \frac{36}{2} = \frac{34}{2} = \frac{17}{1}$$

The odds against obtaining an 11 are $\frac{17}{1}$, which we would usually write as 17 to 1, or 17:1. Therefore, from the odds (17:1), we can see that we probably are not going to get an 11 when we roll the dice. Over the long run, we should get an 11 once out of every 18 tries. This is not a very good percentage of successful outcomes.

Note that when we compute odds we must compute two probabilities: the probability in favor of an event, and the probability not in favor of (against) an event. It does not matter which probability we find first, but usually the probability in favor of an event is the easier one to find. What is important to remember is that we must be sure to find the odds that are asked for. Do we want the "odds in favor," or "the odds against"? Check the wording of the problem to make sure which is wanted.

Odds enable us to play and bet fairly on games. We have just computed the odds against obtaining an 11 on the toss of a pair of dice, and it was 17 to 1. Suppose that Nevada Nellie is rolling a pair of dice and she bets that she will roll an 11. If she bets a dollar, then, according to the odds, she should receive $17 if she does toss an 11. This would make the game a "fair" one. Nevada Nellie should get an 11 once every 18 tries, over the long run. Hence, in 18 tries she should lose $17, $1 on each of 17 failures, and she should win once, and that one time should pay $17.

Two things are usually against a shooter like Nevada Nellie. One, the owner of the game (the house) normally cuts down the odds a little when they pay a winner, to 15 to 1 or 14 to 1. Two, Nevada Nellie has to play over the long run in order to regain her losses, which means she has to play for a long time. People like Nevada Nellie usually lose all their money first, or just get tired and bored, and then quit. The house can keep its game going because the casinos in Las Vegas are open twenty-four hours a day.

Mathematical probability and odds lead to another related topic: **mathematical expectation.** We can describe expectation using an example, as follows: Suppose that Lucky Louie is betting on a certain dice game, and his probability of winning is P, and if Louie wins then he receives M dollars. We would say that Lucky Louie's *mathematical expectation* is $P \times M$. You may not understand mathematical expectation yet, but at least we have a method for finding it. First, let's describe the way we find mathematical expectation a little more formally. Let

M = the amount that will be won if an event occurs
P = the probability that the event will occur

Then

$$expectation = P \times M$$

Mathematical expectation tells us the "fair" price to pay to play a game, if the game is a fair one and can be described by the probability of equally likely outcomes. It also gives us the *average* amount of winnings we can expect for each game if we play a great many games.

Suppose Lucky Louie is playing a dice game at Brutus's Palace and he is betting on 7 coming up. If a 7 comes up, then Louie receives a payoff of \$18. According to the formula, Lucky Louie's expectation is

$$E = P(7) \times \$18$$

$$E = \frac{6}{36} \times \$18 = \$3$$

That is, Lucky Louie should be willing to bet \$3 for the privilege of betting on 7. But, unfortunately, the smallest denomination chip used at Brutus's Palace is \$5. Hence, Lucky Louie must bet at least \$5 each time he plays the game. We have already computed his expectation, and it is \$3. Now, if he has to bet \$5 each time he plays, then Lucky Louie can expect to lose an average of \$2 on each \$5 bet, over the long run.

EXAMPLE 4

Bobby and Jim are rolling one die. Bobby wins if an even number (2, 4, 6) comes up, and Jim wins if an odd number (1, 3, 5) comes up. Both Bobby and Jim put up a dollar for each throw of the die. What is the expectation for one of the players?

Solution

Let's compute Jim's expectation. The probability of getting an odd number with one die is $\frac{3}{6}$, and if he wins he gets \$2. Hence, we have

$$E = \frac{3}{6} \times \$2 = \frac{6}{6} = \$1$$

One dollar is Jim's expectation (it would also be the same for Bobby), and it is therefore a fair price to pay to play the game. Over the long run, he should win 3 times out of every 6. Each time he wins, he will win \$2, for a total of \$6 every 6 times he plays. Every 6 times he will have bet \$6 and also won \$6 and, therefore, he will be even—neither ahead nor behind. (*Note:* Even in fair games, a player may become exceptionally lucky or "hot" and defy the laws of probability for a while, but over the long run, he will not.)

EXAMPLE 5

The Brighton Fire Department is running a raffle in which the prize for the lucky ticket is \$1000. If 5000 tickets are sold at \$2 each, what is the expectation of Hugh, who buys 1 ticket?

Solution

The probability of having the winning ticket for this person is $\frac{1}{5000}$ and Hugh's expectation is

$$E = \frac{1}{5000} \times \$1000.00 = \frac{\$1.00}{5} = \$.20$$

Twenty cents represents a fair price to pay for a ticket. It appears that \$2.00 is too much to pay for a ticket, but this is how organizations make money.

EXAMPLE 6

Suppose Hugh changes his mind and decides to buy 5 tickets in the raffle for the \$1000 prize. Does this change his expectation?

Solution

Hugh has increased only his chances of winning. Now the probability of winning is $\frac{5}{5000}$, and his expectation is

$$E = \frac{5}{5000} \times \$1000.00 = \frac{5000}{5000} = \$1.00$$

Hugh's mathematical expectation is \$1.00 for five tickets; it is still \$.20 per ticket.

A bag contains 2 red, 5 white, and 3 green balls. A prize of \$10 is given if a red ball is drawn and a prize of \$1 is given if a green ball is drawn. What is the expectation?

First we must compute the probability of winning. The probability of drawing a red ball is $\frac{2}{10}$, and the probability of drawing a green ball is $\frac{3}{10}$. The expectation for the red ball is

$$E(\text{red}) = \frac{2}{10} \times \$10.00 = \frac{20}{10} = \$2.00$$

The expectation for the green ball is

$$E(\text{green}) = \frac{3}{10} \times \$1.00 = \frac{3}{10} = \$0.30$$

The total expectation is the sum of the two expectations above. Therefore, we would have

$$E = E(\text{red}) + E(\text{green}) = \$2.00 + \$0.30 = \$2.30$$

Remember that the \$2.30 represents a "fair" price to pay for the privilege of playing the game. It also represents the average amount you should expect to win when you play a great many games—that is, if play is over the long run—since these are equally likely events. Suppose you played the game 10 times. You should expect to win 2 times on red, which equals \$20.00 in prizes; you should win 3 times on green, which equals \$3.00 in prizes; and the other 5 times you should receive nothing. So, in 10 games you should receive \$23.00. In a fair game, you should also bet \$23.00, which is \$2.30 per try.

EXERCISES FOR SECTION 3.5

1. In a single toss of a pair of dice, find the odds (the number referred to is the sum of the numbers on the dice)—
   a. in favor of obtaining a 7
   b. in favor of obtaining an 11
   c. in favor of obtaining a 12
   d. against obtaining a 12
   e. against obtaining a 6
   f. against obtaining a 10.

2. In a single toss of a pair of dice, find the odds (the number referred to is the sum of the numbers on the dice)—
   a. in favor of obtaining a 3
   b. in favor of obtaining a 5
   c. against obtaining a 5
   d. against obtaining a double
   e. in favor of obtaining a 7 or 11
   f. against obtaining a number other than 7.

3. On a single draw from a shuffled standard deck of 52 bridge cards, find the odds—
   a. in favor of drawing an ace
   b. in favor of drawing a club
   c. in favor of drawing a red card
   d. against drawing a deuce (2)
   e. in favor of drawing a picture card (jack, queen, or king)
   f. against drawing a picture card or a diamond.

4. In a single toss of 3 coins, make a sample space and find the odds—
   a. in favor of obtaining 3 heads
   b. in favor of obtaining exactly 1 head
   c. in favor of obtaining at least 1 tail
   d. against obtaining 3 tails
   e. against obtaining no tails
   f. against obtaining at most 2 heads.

5. The odds against the Yankees winning the pennant are 7:2. What is the probability that the Yankees will win the pennant?

6. The odds against Sure Thing winning the featured race at Finger Lakes Race Track are 11:2. What is the probability that Sure Thing will win the race?

7. Find the probability that event A will happen if the odds are—
   a. 7:5 in favor of A
   b. 1:1 in favor of A
   c. 3:2 against A
   d. 5:9 against A.

8. In order to win a game, Nelson must throw a 7 in a single toss of a pair of dice. Larry bets that Nelson will, Benny bets that Nelson won't. What are the odds for and against this event? Do the odds favor Larry or Benny?

9. A player will win \$18 if he or she throws a double on the first toss of a pair of dice. What are the odds in favor of this player winning? What is a fair price to pay to play this game?

10. A box contains 3 one-dollar bills, 2 two-dollar bills, 1 five-dollar bill, and 1 ten-dollar bill. If you reach in and select a bill at random, what is your mathematical expectation?

11. The Association for Conservation is awarding a \$1000 cash prize to the winner of a raffle. A total of 5000 tickets are sold for \$2 each. What is the mathematical expectation for a person who buys 5 tickets?

12. The New York State Lottery offered a weekly prize of \$50,000. A person bought one ticket which cost \$.50. The winning ticket can be any 6-digit number from 000,000 to 999,999. What were the person's chances of winning? What was a fair price to pay for the ticket?

13. A special New York State Lottery offers a grand prize of \$100,000, 3 second prizes of \$10,000, and 10 third prizes of \$1000. The winning ticket can be any six-digit number from 000,000 to 999,999. If a person buys one ticket, what is a fair price to pay for the ticket?

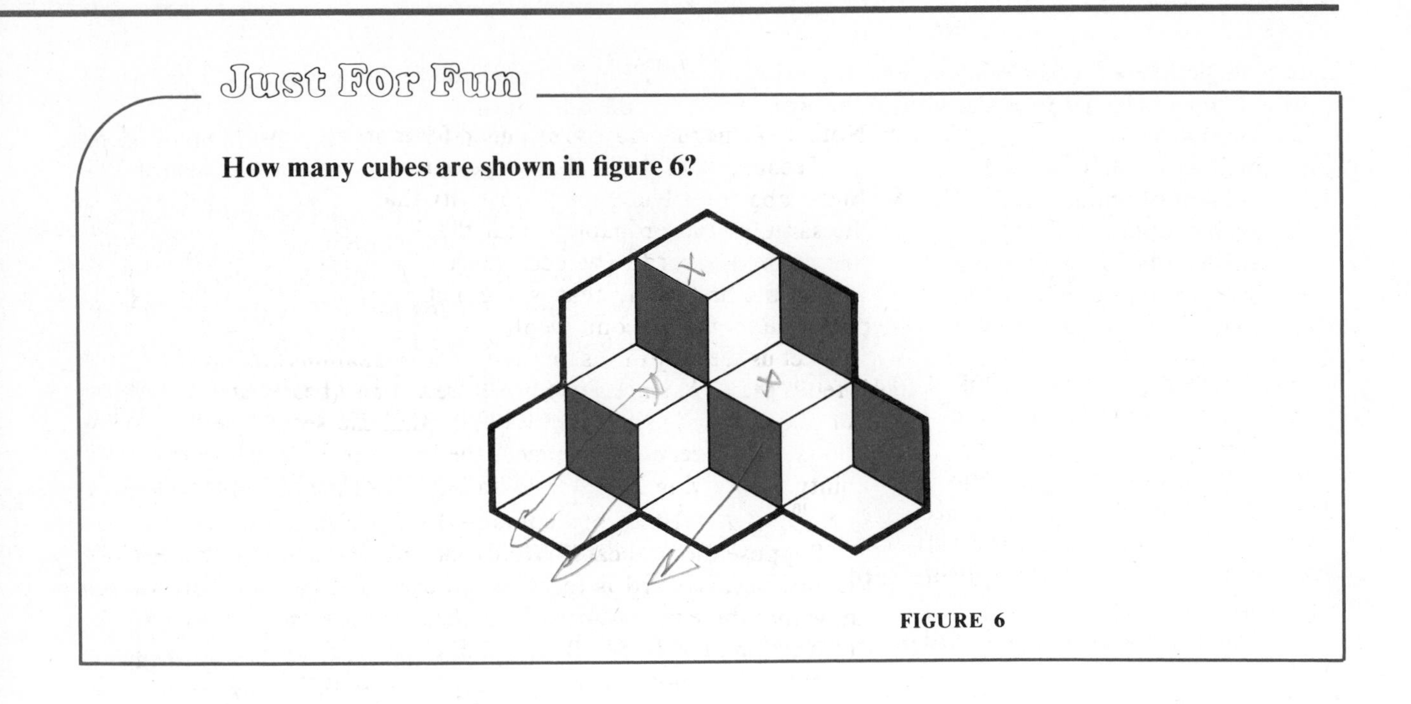

Just For Fun
How many cubes are shown in figure 6?

FIGURE 6

3.6 COMPOUND PROBABILITY

Many events are **compound events,** made up of two or more simpler events. If we draw a card from a shuffled deck of cards, look at it, replace it, shuffle the cards again, and then draw another card, we have made two single simple drawings; but together they constitute a compound event. The problem in this example could be stated as, "What is the probability of drawing 2 red cards from a shuffled deck of 52 cards, if the first card is replaced?"

This probability problem leads us to two important questions:

1. How do we compute compound probabilities?
2. How does replacement influence the probability?

First, let us figure out how to compute compound probabilities. Recall the *counting principle;* it said that if one experiment has m different outcomes and a second experiment has n different outcomes, then the first and second experiments performed together have $m \times n$ different outcomes. Thus, if the outcome of the first experiment does not influence the outcome of the second, we can use this idea to compute the probability of events occurring in succession: that is, we find the product of the probabilities of the two events. In general, we have

$$P(A \text{ and } B) = P(A) \times P(B)$$

Note that this rule may be extended for more than two events.

Second, let us decide how replacement influences the probability. In the above problem, the probability that the second card is red is the same as the probability that the first card is red, since we replaced the first card. The occurrence of the second event is **independent** of the first event; that is, the first event does not influence the probability of the second event.

Let us see why this is the case. The probability that the first card is red is $\frac{26}{52}$, or $\frac{1}{2}$, since there are 26 red cards (hearts and diamonds) out of a total of 52. The probability that the second card is red is also $\frac{26}{52}$, or $\frac{1}{2}$, because we replaced the first card. Therefore the probability of drawing 2 red cards in succession, with replacement, is $\frac{26}{52} \times \frac{26}{52}$, or $\frac{1}{2} \times \frac{1}{2} = \frac{1}{4}$.

Suppose the first card had not been replaced. The probability that the second card is red then becomes $\frac{25}{51}$. Why? Because we know that there is 1 less card in the deck; hence, the denominator of our fraction must be 51. But the 25 in the numerator confuses some students. Why not 26? How do we know that the first card is red? We don't necessarily know that it is red, but we must assume that it is, because the probability of success (that is, of drawing 2 red cards in succession) depends upon the first card being red. The probability of the second card being red is **dependent** on the probability of the first card being red. The first event *did* influence the second event. Two events are **dependent** if the occurrence of one affects the occurrence of the other.

EXAMPLE 1

What is the probability of being dealt 2 red cards in succession, without replacement, from a shuffled bridge deck?

Solution

These are dependent events. The probability that the first card is red is $\frac{26}{52}$, but the probability that the second card is red is $\frac{25}{51}$. Hence the probability that both cards are red is

$$\frac{26}{52} \times \frac{25}{51} = \frac{650}{2652} = \frac{25}{102}$$

EXAMPLE 2

A quarter is flipped 3 times. What is the probability of obtaining 3 tails in succession?

Solution
The flipping of a coin is an independent event. What happens the first time does not affect what will happen the next time. The probability that a flip will yield a tail is $\frac{1}{2}$. To compute the probability of obtaining three tails in a row we have

$$P(T \text{ and } T \text{ and } T) = P(T) \times P(T) \times P(T)$$

$$P(3 \text{ tails}) = \frac{1}{2} \times \frac{1}{2} \times \frac{1}{2} = \frac{1}{8}$$

EXAMPLE 3
On a certain Sunday in October, the probability that the Rams will win their football game is 0.6 and the probability that the Browns will win their football game is 0.4. Assuming that they are not playing each other, what is the probability that both teams will win their games on this given Sunday?

Solution
The probability that the Rams will win does not affect the probability that the Browns will win. We may compute the probability that the Rams and Browns will both win by multiplying their respective probabilities. Therefore we have

$$P(\text{Rams and Browns will win}) = 0.6 \times 0.4 = 0.24$$

The next question is how to compute the probability that *A or B* will occur.

Before computing the probability of *A or B*, we must first familiarize ourselves with a different kind of set of events. These are *mutually exclusive events*. Events that are **mutually exclusive** are events that *cannot* happen at the same time: only one of the events can occur at any one time. For example, when we flip a coin, we can get heads or tails. Only one of these—not both—can occur for any one flip. If we roll a die, we can get an odd or an even number with one roll, but not both. Two horses cannot win the same race at the same time: one or the other may win, but not both. These are mutually exclusive events. If two or more events cannot happen at the same time, they are mutually exclusive.

How do we compute the probability of *A or B*? Recall our discussion of the union of two sets *A* and *B* in chapter 1: The *union* of sets *A* and *B* is the set of all elements that are elements of *A or B*, or elements of both. When we count the elements of the union of two sets, we count the elements of the two sets, but if an element is in both sets, we only count it once.

If two events are mutually exclusive, then they are disjoint (have

FIGURE 7

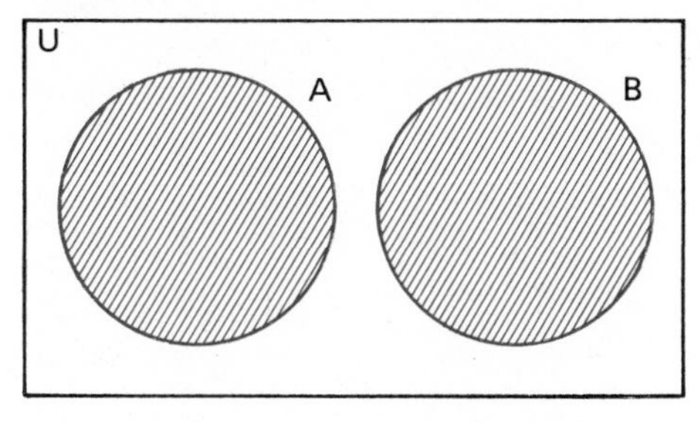

FIGURE 8

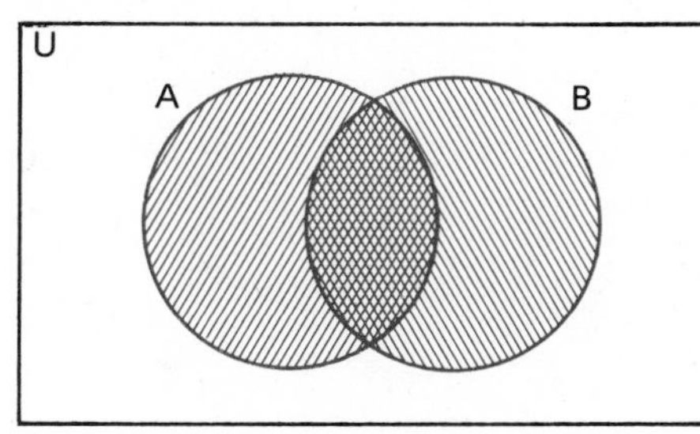

no common elements). The Venn diagram for this situation is shown in figure 7.

The probability of A *or* B, which we will denote by $P(A \text{ or } B)$, is the same as the probability of A union B, denoted by $P(A \cup B)$. From our diagram and the previous discussion, we have

$$P(A \cup B) = P(A) + P(B) = P(A \text{ or } B)$$

That is, in order to compute $P(A \text{ or } B)$, where A and B are mutually exclusive events, we add $P(A)$ and $P(B)$.

Suppose A and B are not mutually exclusive events; then what is $P(A \text{ or } B)$? In this case, A and B are not mutually exclusive events and therefore have a nonempty intersection. The Venn diagram for this situation is shown in figure 8.

Here again, $P(A \text{ or } B)$ is the same as $P(A \cup B)$, but remember that this time the events are not mutually exclusive. Hence we would have

$$P(A \cup B) = P(A) + P(B) - P(A \cap B)$$

We subtract $P(A \cap B)$, the probability of the intersection of A and B, from $P(A) + P(B)$ because $P(A \cap B)$ is included twice in $P(A) + P(B)$. Recall that the *intersection* of a set A and a set B is the set of elements common to both A *and* B. So $P(A \cap B)$ is the same as $P(A \text{ and } B)$, which is the probability of the outcomes which are common to both events A and B. Therefore, if A and B are not mutually exclusive events,

$$P(A \text{ or } B) = P(A) + P(B) - P(A \text{ and } B)$$

It should be noted that this equation is true even when A and B are mutually exclusive events; then $P(A \text{ and } B) = 0$, because A and B are disjoint, and so the intersection is empty. Then we have $P(A \text{ or } B) = P(A) + P(B) - 0$.

EXAMPLE 4

In a single toss of a pair of dice, find the probability of obtaining a 7 or 11.

Solution

A 7 and an 11 are mutually exclusive events. Hence, we have

$$P(\{7\} \cup \{11\}) = P(7 \text{ or } 11) = P(7) + P(11)$$

$$= \frac{6}{36} + \frac{2}{36} = \frac{8}{36} = \frac{2}{9}$$

EXAMPLE 5
One card is drawn from a shuffled deck of 52 cards. Find the probability that the card selected is a heart or a jack.

Solution
We must first determine if the events are mutually exclusive or not. Since a card may be selected that is both a heart and a jack, namely, the jack of hearts, the events are not mutually exclusive. Therefore we should use the formula

$$P(A \text{ or } B) = P(A) + P(B) - P(A \text{ and } B)$$

and we have

$$P(\text{heart } or \text{ jack}) = P(H) + P(J) - P(H \text{ and } J)$$

$$= \frac{13}{52} + \frac{4}{52} - \frac{1}{52} = \frac{16}{52} = \frac{4}{13}$$

EXAMPLE 6
Suppose that the probability that Fisher will win any chess tournament he enters is $\frac{3}{5}$ and the probability that Spasky will win any chess tournament he enters is $\frac{1}{3}$. What is the probability that Fisher or Spasky will win if both are in the same tournament?

Solution
If Fisher and Spasky are in the same tournament, then only one can win the tournament and these events are mutually exclusive. Hence the probability that Fisher or Spasky will win is

$$P(F \text{ or } S) = \frac{3}{5} + \frac{1}{3} = \frac{9}{15} + \frac{5}{15} = \frac{14}{15}$$

EXAMPLE 7
A quarter is flipped and a die is tossed. What is the probability of obtaining heads or a 6?

Solution
These events are not mutually exclusive. We can get heads on the quarter and a 6 on the die at the same time.

We can compute probability problems of this type in a variety of ways. One technique is to rely on a sample space and read our answer directly from it. For this problem we would have

H—1	H—4	T—1	T—4
H—2	H—5	T—2	T—5
H—3	H—6	T—3	T—6

The outcomes that are circled in the sample space are the successful outcomes, so $P(H \text{ or } 6) = \frac{7}{12}$.

Another way to solve this problem is to use the formula for events that are not mutually exclusive, namely, $P(A \text{ or } B) = P(A) + P(B) - P(A \text{ and } B)$. Then

$$P(H \text{ or } 6) = P(H) + P(6) - P(H \text{ and } 6)$$

$$P(H \text{ or } 6) = \frac{1}{2} + \frac{1}{6} - \frac{1}{12}$$

$$= \frac{6}{12} + \frac{2}{12} - \frac{1}{12} = \frac{7}{12}$$

EXERCISES FOR SECTION 3.6

1. One card is selected at random from a shuffled standard deck of 52 cards. Find the probability that the card chosen is—
   a. a picture card (jack, queen, king)
   b. a red picture card
   c. an ace or a king
   d. an ace or a heart
   e. a king and a diamond
   f. an ace and a heart.

2. A bag contains 25 colored balls, of which 10 are red, 8 are green, 4 are blue, and 3 are orange If one ball is selected at random from the bag, find the probability that the ball chosen is—
   a. red or green b. red or blue
   c. red or orange d. not red
   e. not blue f. neither red nor blue.

3. Two cards are selected at random from a shuffled standard deck of 52 cards, without replacement. Find the probability that—
   a. both cards are picture cards
   b. the first card is an ace and the second card is a picture card
   c. the first card is the ace of spades and the second card is a picture card
   d. both cards are kings
   e. both cards are diamonds
   f. both cards are of the same denomination.

4. A bag contains 25 marbles, of which 8 are red marbles, 8 are green, 4 are blue, and 5 are clear (colorless).
   a. What is the probability that a marble drawn at random is green?
   b. What is the probability of drawing 7 blue marbles in succession if there is no replacement after each draw?
   c. What is the probability of drawing 4 blue marbles in succession if there is replacement after each draw?
   d. If 2 marbles are drawn in succession (without replacement), what is the probability that both marbles are red?
   e. If 2 marbles are drawn in succession (without replacement), what is the probability that the first marble is clear and the second marble is blue?

5. An ice chest contains 5 cans of ginger ale, 3 cans of orange soda, and 4 cans of root beer. If 3 cans are drawn at random from the chest without replacement, find the probability of getting—
   a. 3 cans of root beer
   b. 3 cans of orange
   c. 3 cans of ginger ale
   d. a can of orange soda, then a can of root beer, then a can of ginger ale

e. a can of orange soda, then a can of ginger ale, then a can of root beer
f. no root beer.

6. One die is rolled, and then another die is rolled. Find the probability of obtaining—
   a. a 6 on the first die and a 5 on the second die
   b. a 6 on the first die or a 5 on the second die
   c. the same number on both dice
   d. an even number on the first die and an odd number on the second die
   e. an even number on the first die or an odd number on the second die
   f. an even number on the first die and 6 on the second die.

7. A cube from a certain game has an A on each of three faces, a B on each of two faces, and a C on the remaining face. If the cube is rolled 3 times in succession, what is the probability that the letters appearing will read C, A, B, in that order?

8. A track coach must decide who is going to run on the school's mile relay team. He must choose four runners from a list of 6 equally fast runners: Nenno, Nelson, Gilligan, Neanderthal, Clar, and Connelly. He decides that the easiest way to make the decision is draw names from a hat in succession. (Note that there is no replacement since the coach needs 4 different runners.) Find the probability that—
   a. he chooses Neanderthal first
   b. he chooses Neanderthal first and Nenno second
   c. he chooses Nelson or Gilligan first
   d. his third choice is a person whose name begins with N, given that Connelly and Clar were already chosen
   e. Nenno is chosen first, Neanderthal is chosen second, Nelson is chosen third, and Gilligan is chosen to run the anchor leg (fourth).

9. Each of the numbers 0 through 9 is painted on a separate ball. The 10 balls are put in a can and the can is shaken to mix up the balls. Find the probability of—
   a. drawing an even-numbered ball
   b. drawing a ball numbered 4 and then another numbered 4 on two successive draws with replacement
   c. drawing a ball numbered 4 and then another numbered 4 on two successive draws without replacement
   d. drawing a ball numbered 4 on the first draw, but not on the second draw, with replacement
   e. drawing an even number on the first draw and an odd number on the second draw, without replacement.

10. A bingo caller has a machine which contains 75 balls. The balls are marked in the following manner:

$$\begin{array}{l} B1,\ B2,\ B3,\ \ldots, B15 \\ I16,\ I17,\ I18,\ \ldots, I30 \\ N31, N32, N33, \ldots, N45 \\ G46, G47, G48, \ldots, G60 \\ O61, O62, O63, \ldots, O75 \end{array}$$

The balls are mixed and each ball is drawn at random.
   a. What is the probability of drawing a ball that has a B on it?
   b. What is the probability of drawing three B balls in succession, if there is no replacement?
   c. What is the probability of drawing a ball that has a double number (11, 22, . . .) on it?
   d. What is the probability of drawing a ball that has a B or a G on it?
   e. What is the probability of drawing 2 balls, without replacement, so that the first one has a B on it and the second one has an I or N on it?
   f. What is the probability of drawing a ball that has a number on it that is not divisible by 5?

Just For Fun

Take a standard deck of 52 cards and shuffle them as much as you want. Start with the top card and turn each card over 1 at a time. Is it a good bet that you will get 2 cards in a row that are of the same denomination and the same color (for example, 3 of clubs and 3 of spades, 7 of hearts and 7 of diamonds)?

3.7 COUNTING, ORDERED ARRANGEMENTS, AND PERMUTATIONS (Optional)

What is counting? When we count something, we want to know "how many." Often we have to count all of the possible outcomes for probability problems. The basic definition of the probability that an event A will occur is the *number* of successful outcomes of the experiment divided by the total *number* of possible outcomes. In section 3.3, we introduced the *counting principle* to enable us to efficiently count the possible outcomes for an experiment.

> **Counting Principle: If one experiment has m different outcomes, and a second experiment has n different outcomes, then the first and second experiments performed together have $m \times n$ different outcomes. This may be extended if there are other experiments to follow to $m \times n \times r \times \cdots \times t$.**

The counting principle helped us to determine the total number of outcomes when a pair of dice is tossed: there are $6 \times 6 = 36$ different outcomes. Let us apply the counting principle to a more complicated example: telephone numbers.

How many different telephone numbers are possible in your local calling area? Every phone number consists of 7 digits (10, if you also count the area code). As a counting example, we shall examine just the last 4 digits. There are 4 numbers of outcomes to multiply, 1 for each digit:

$$m \times n \times r \times s$$

In how many ways can each digit be chosen? Each digit may be any one of the numerals 0, 1, 2, 3, 4, 5, 6, 7, 8, 9, and, therefore, there are 10 possibilities for each digit. (We are assuming that each numeral may be repeated in more than one digit.) Hence each digit may be chosen in any one of 10 different ways, assuming no restrictions on the use of 0 or 1, so we have

$$10 \times 10 \times 10 \times 10 = 10{,}000$$

different telephone numbers in a local calling area or exchange. If we want to find out the number of telephone numbers in one general area which share the same area code, we can expand this problem to 7 digits.

We have already used the counting principle in some probability problems. Let's examine some more examples.

EXAMPLE 1

How many different license plates can be made if each license plate is to consist of 3 letters followed by 3 digits? (Assume replacement.)

Solution

Here order is important. There are 6 slots to fill. The first 3 slots are to be filled by letters of the alphabet; hence there are 26 choices for each of the first 3 slots. Each of the last 3 slots may be filled by any one of the 10 numerals 0, 1, 2, 3, 4, 5, 6, 7, 8, 9. Therefore the solution to the problem would be

$$26 \times 26 \times 26 \times 10 \times 10 \times 10 = 17{,}576{,}000$$

EXAMPLE 2

How many different license plates can be made if each license plate is to consist of 3 letters followed by 3 digits, and no letters or digits may be repeated?

Solution

This problem is similar to example 1, but now we have some restrictions. If no repetition is allowed, then we still have 26 choices for the first slot in the license plate, but there are only 25 choices for the second slot, because we have already used one letter in the first slot. Similarly, there are 24 choices left for the third slot, because we used one letter of the 25 in the second slot. The same thing happens when we fill in the slots for the digits. Hence the solution is

$$26 \times 25 \times 24 \times 10 \times 9 \times 8 = 11{,}232{,}000$$

EXAMPLE 3

Marie is planning her schedule for next semester. She must take the following 5 courses: English, history, geology, psychology, and mathematics.

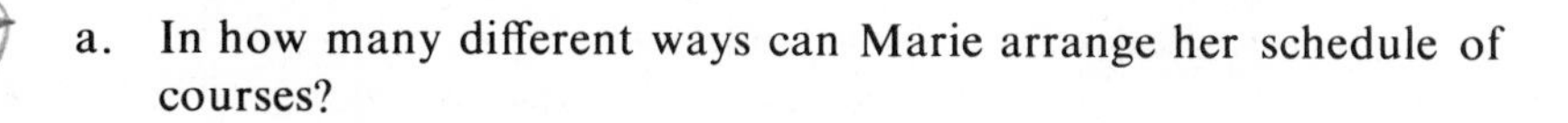

a. In how many different ways can Marie arrange her schedule of courses?

b. How many of these schedules have mathematics listed first?

Solution

a. Since Marie has to take 5 different courses, there are 5 time slots to consider:

$$m \times n \times r \times s \times t$$

There are 5 choices for the first time slot, 4 for the second, and so on. The solution is

$$5 \times 4 \times 3 \times 2 \times 1 = 120$$

b. If mathematics is to be listed first, then this is a restriction, and there is only 1 choice for the first time slot. Mathematics is used to fill the first slot, so we have 4 choices for the second slot, 3 for the third slot, etc. The solution is

$$1 \times 4 \times 3 \times 2 \times 1 = 24$$

EXAMPLE 4

Given the set of digits $\{1, 3, 4, 5, 6\}$.

a. How many three-digit numbers can be formed?

b. How many three-digit numbers can be formed if the number must be even?

c. How many three-digit numbers can be formed if the number must be even and no repetition of digits is allowed?

Solution

a. There are three places, or slots, to consider. Each slot can be filled by any one of the 5 digits. Therefore, the solution is

$$5 \times 5 \times 5 = 125$$

b. The three-digit number must be even, so it must end either in a 4 or 6, the even digits in our given set. This is a restriction, so we should attend to it first: we have 2 choices, 4 or 6, for the last place. The counting principle gives us

$$\_\_ \times \_\_ \times 2$$

three-digit numbers. We have 5 choices for the first place and 5 choices for the second. Hence, we have

$$5 \times 5 \times 2 = 50$$

three-digit numbers.

c. There are two restrictions here: The number must be even, and no digit may be repeated. We still fill in the last place with 2 choices.

$$\_ \times \_ \times 2$$

But now there are only 4 choices for the first place, because one of the digits has been used in the last place. There are then 3 choices for the second place, and we have

$$4 \times 3 \times 2 = 24$$

three-digit numbers.

EXAMPLE 5

Dan is a salesman at an auto agency. At the present time, the agency has the following cars in stock: 6 Dodges, 5 Plymouths, 4 Fords, and 2 Volkswagens. Dan must sell 2 of these cars today!

a. How many different possible ways can he sell 2 cars?

b. How many different ways can he sell 2 cars if at least 1 of them must be a Ford?

Solution

a. Dan must sell 2 cars and altogether there are 17 cars to choose from. So he has 17 choices for the first car and 16 for the second car; that is,

$$17 \times 16 = 272$$

b. If at least 1 car sold must be a Ford, we first figure out how many different sales are possible without Fords. We would have

$$13 \times 12 = 156$$

possible sales without Fords. There are 272 possible sales altogether. Hence, the number of sales having at least one Ford is

$$272 - 156 = 116$$

If a woman has individual photos of each of her three children, Mary, Scott, and Joe, how many ways can she arrange these photos in a row on her desk? From our previous discussion, we know that there are $3 \times 2 \times 1 = 6$ different possible outcomes, and they are:

Mary—Scott—Joe
Mary—Joe—Scott
Scott—Joe—Mary
Scott—Mary—Joe
Joe—Scott—Mary
Joe—Mary—Scott

Each of these arrangements is different. The pictures are in a definite order, and the order in which the pictures are arranged is important. When we have a group of things arranged in a definite order, we have a *permutation*. **A permutation** is a particular ordering of the elements of a given set.

Permutations may use all the elements from a given set (as we did in our example), or only a certain number of them. If we are given the digits 1, 2, 3, 4, and 5, how many two-digit numbers can we form? This is a permutation (arrangement) of 5 things taken 2 at a time. In this case, we would have $5 \times 4 = 20$ different arrangements. If we have n items and want to count the possible arrangements using r of them at a time, we say that we want the number of **permutations of n things taken r at a time.** The notation for this is

${}_nP_r$ the number of permutations of n things taken r at a time

The notation ${}_nP_r$ says that we want to fill r places or slots from a total group of n things. Recall that ${}_5P_2$ is a permutation of 5 things taken 2 at a time, which is $5 \times 4 = 20$. Similarly, ${}_6P_4$ is a permutation of 6 things taken 4 at a time, and ${}_6P_4 = 6 \times 5 \times 4 \times 3 = 360$. If we have ${}_nP_r$, a permutation of n things taken r at a time, then we have r slots and must fill them from n things. The first slot may be filled in n different ways, the second slot may be filled in $n - 1$ different ways, the third slot may be filled in $n - 2$ different ways, and so on. The computation of the number of permutations of n things taken r at a time is indicated below, along with two examples that we discussed previously.

$${}_5P_2 = 5 \times 4 = 20$$
$${}_6P_4 = 6 \times 5 \times 4 \times 3 = 360$$

$${}_nP_r = \underbrace{\overset{\textit{1st slot}}{n} \times \overset{\textit{2nd slot}}{(n-1)} \times \overset{\textit{3rd slot}}{(n-2)} \times \overset{\textit{4th slot}}{(n-3)} \times \cdots \times \overset{\textit{rth (last) slot}}{(n-r+1)}}_{\text{number of ways each slot may be filled}}$$

This leads to the formula

$$_nP_r = n \times (n - 1) \times (n - 2) \times (n - 3) \times \cdots \times (n - r + 1)$$

or

$$_nP_r = n(n - 1)(n - 2)(n - 3)(n - 4)\cdots(n - r + 1)$$

From this formula we now know that

$$_8P_3 = 8 \times (8 - 1) \times (8 - 3 + 1) = 8 \times 7 \times 6$$

n things *r* slots n $(n - 1)$ $(n - r + 1)$

What happens when $n = r$? If we examine $_nP_r$ when $n = r$, we have $_nP_r = n(n - 1)(n - 2)\cdots(n - r + 1)$ and the factor $(n - r + 1)$ becomes $(n - n + 1)$, or just 1. Thus when $n = r$, we have $_nP_n = n(n - 1)(n - 2)(n - 3)\cdots(3)(2)(1)$. The product of all the numbers from n down to and including 1 is called ***n* factorial.** The symbol for n factorial is $n!$. A table of factorials (table I) appears in the appendix. Some examples are:

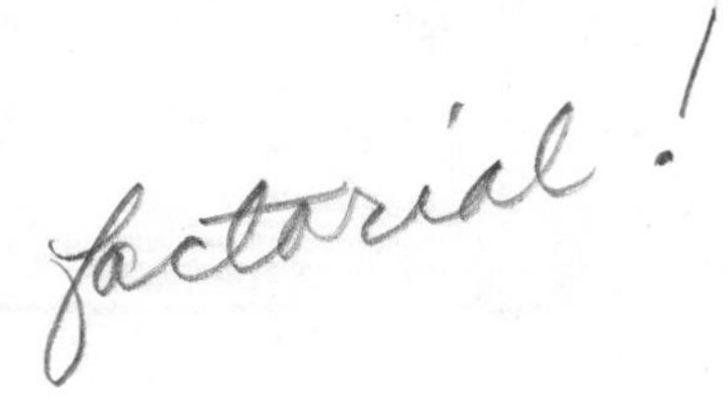

(n factorial)	$n! = n(n - 1)(n - 2)(n - 3)\cdots(3)(2)(1)$
(3 factorial)	$3! = 3 \times 2 \times 1$ or $(3)(2)(1)$ or $3 \cdot 2 \cdot 1 = 6$
(5 factorial)	$5! = 5 \times 4 \times 3 \times 2 \times 1$ or $(5)(4)(3)(2)(1)$ or $5 \cdot 4 \cdot 3 \cdot 2 \cdot 1 = 120$

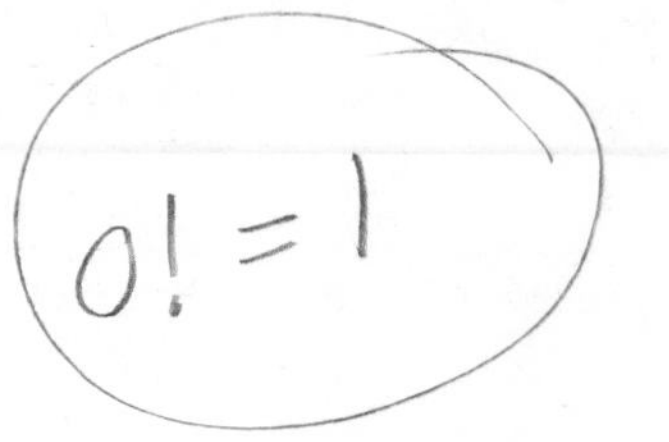

Zero factorial, 0!, cannot be defined by the rule used in the preceding examples. We agree to let 0! equal 1 as a matter of convenience. This special definition enables us to produce a simpler formula for $_nP_r$.

Examine the following examples:

$$_6P_2 = 6 \cdot 5 = 30 \quad \text{or} \quad _6P_2 = \frac{6!}{4!} = \frac{6 \cdot 5 \cdot 4 \cdot 3 \cdot 2 \cdot 1}{4 \cdot 3 \cdot 2 \cdot 1} = \frac{720}{24} = 30$$

$$_5P_3 = 5 \cdot 4 \cdot 3 = 60 \quad \text{or} \quad _5P_3 = \frac{5!}{2!} = \frac{5 \cdot 4 \cdot 3 \cdot 2 \cdot 1}{2 \cdot 1} = \frac{120}{2} = 60$$

In general, we have

$$_nP_r = \frac{n!}{(n - r)!}, \qquad r \leq n$$

For example,

$$_5P_3 = \frac{5!}{(5-3)!} = \frac{5!}{2!} = \frac{5\cdot4\cdot3\cdot2\cdot1}{2\cdot1} = 60$$

$$_6P_2 = \frac{6!}{(6-2)!} = \frac{6!}{4!} = \frac{6\cdot5\cdot4\cdot3\cdot2\cdot1}{4\cdot3\cdot2\cdot1} = 30$$

$$_4P_4 = \frac{4!}{(4-4)!} = \frac{4!}{0!} = \frac{4\cdot3\cdot2\cdot1}{1} = 24$$

Note that 0! appears in the denominator of the last example, $_4P_4$. In order for our formula to work for this case, we must define $0! = 1$.

EXAMPLE 6

Evaluate each of the following:

a. 4! b. 8! c. $_7P_4$ d. $_8P_3$ e. $_{10}P_{10}$ f. $_5P_0$

Solution

a. $4! = 4\cdot3\cdot2\cdot1 = 24$

b. $8! = 8\cdot7\cdot6\cdot5\cdot4\cdot3\cdot2\cdot1 = 40{,}320$

c. $_7P_4 = \frac{7!}{(7-4)!} = \frac{7!}{3!} = \frac{7\cdot6\cdot5\cdot4\cdot\cancel{3}\cdot\cancel{2}\cdot1}{\cancel{3}\cdot\cancel{2}\cdot1} = 840$

d. $_8P_3 = \frac{8!}{(8-3)!} = \frac{8!}{5!} = \frac{8\cdot7\cdot6\cdot\cancel{5}\cdot\cancel{4}\cdot\cancel{3}\cdot\cancel{2}\cdot1}{\cancel{5}\cdot\cancel{4}\cdot\cancel{3}\cdot\cancel{2}\cdot1} = 336$

e. $_{10}P_{10} = \frac{10!}{(10-10)!} = \frac{10\cdot9\cdot8\cdot7\cdot6\cdot5\cdot4\cdot3\cdot2\cdot1}{1} = 3{,}628{,}800$

f. $_5P_0 = \frac{5!}{(5-0)!} = \frac{5!}{5!} = \frac{\cancel{5}\cdot\cancel{4}\cdot\cancel{3}\cdot\cancel{2}\cdot1}{\cancel{5}\cdot\cancel{4}\cdot\cancel{3}\cdot\cancel{2}\cdot1} = 1$

EXAMPLE 7

If 7 people board an airplane, and there are 9 aisle seats, in how many ways can they be seated if they all choose aisle seats?

Solution

This is a permutation of 9 things taken 7 at a time. So we have

$$_9P_7 = \frac{9!}{(9-7)!} = \frac{9!}{2!} = \frac{9\cdot8\cdot7\cdot6\cdot5\cdot4\cdot3\cdot\cancel{2}\cdot1}{\cancel{2}\cdot1} = 181{,}440$$

EXAMPLE 8

A disc jockey can play 8 records in a thirty-minute segment of her show. For a particular thirty-minute segment, she has 12 records to select from. In how many ways can she arrange her program for the particular segment?

Solution

This is a permutation of 12 things taken 8 at a time. Therefore we have

$$_{12}P_8 = \frac{12!}{(12-8)!} = \frac{12!}{4!} = \frac{12 \cdot 11 \cdot 10 \cdot 9 \cdot 8 \cdot 7 \cdot 6 \cdot 5 \cdot \cancel{4} \cdot \cancel{3} \cdot \cancel{2} \cdot 1}{\cancel{4} \cdot \cancel{3} \cdot \cancel{2} \cdot 1}$$

$$= 19{,}958{,}400$$

We now know that $_3P_3$ is equal to 6. This means that there are 6 distinct arrangements of 3 things taken 3 at a time. Let's now look at a slightly different problem. Consider the word *ALL* and the number of arrangements that can be made from the letters in it. There are 3 letters, and therefore the number of arrangements is 3! Let's call the first *L* in the word L_1 ("*L* sub one") and the second *L* in the word L_2 ("*L* sub two"). Now we can distinguish between the two *L*'s. The arrangements of the 3 letters are

1. $A\,L_1\,L_2$
2. $A\,L_2\,L_1$
3. $L_1\,A\,L_2$
4. $L_2\,A\,L_1$
5. $L_1\,L_2\,A$
6. $L_2\,L_1\,A$

If we use the subscript notation as above, we can see the different arrangements. When we interchange the *L*'s, we technically have a different arrangement. But without the subscripts, it is impossible to tell if the *L*'s have been moved, and we have

1. *A L L*
2. *A L L*
3. *L A L*
4. *L A L*
5. *L L A*
6. *L L A*

Since we cannot tell if the *L*'s have been interchanged, the arrangements are not all distinct. Arrangements 1 and 2 are the same, as are 3 and 4, and 5 and 6. Hence, there are only three distinct arrangements, namely, *ALL, LAL,* and *LLA*. The two *L*'s can be arranged in 2! ways. In order to obtain the number of distinct arrangements, we must divide $_3P_3 = 3!$ by 2! That is, there are

$$\frac{3!}{2!} = \frac{3 \cdot 2 \cdot 1}{2 \cdot 1} = 3$$ distinct arrangements from the letters in *ALL*

If we want to find the number of **distinct arrangements** of n things when p of the things are alike, then we divide $n!$ by $p!$.

EXAMPLE 9

How many distinct arrangements can be formed from all the letters of *GENESEE*?

Solution

There are 7 letters. If they were all different, we would have 7! arrangements, but there are 4 E's, which can be arranged in $4! = 24$ ways. So we divide 7! by 4! and the answer is

$$\frac{7!}{4!} = \frac{7 \cdot 6 \cdot 5 \cdot \not{4} \cdot \not{3} \cdot \not{2} \cdot 1}{\not{4} \cdot \not{3} \cdot \not{2} \cdot 1} = 210$$

Suppose, in addition to p things of the same kind, we also have q things alike, and r things alike, and so on. If this occurs, then we just extend our computation to

$\dfrac{n!}{p!q!\ldots}$	the number of distinct arrangements of n things when p things are alike, q things are alike, etc.

EXAMPLE 10

In how many distinct ways can the letters of *OSMOSIS* be arranged?

Solution

There are 7 letters in the word *OSMOSIS*, and if they were all different we would have 7! arrangements. Since there are 2 O's, we must divide by 2!, and since there are 3 S's, we must divide by 3!. Hence we have

$$\frac{7!}{2!3!} = \frac{7 \cdot 6 \cdot 5 \cdot 4 \cdot \not{3} \cdot \not{2} \cdot 1}{2 \cdot 1 \cdot \not{3} \cdot \not{2} \cdot 1} = 420$$

EXAMPLE 11

In how many distinct ways can the letters of *INFATUATION* be arranged?

Solution

If all of the letters in the word *INFATUATION* were different, there would be 11! arrangements. But there are 2 I's, so we must divide by 2!; 2 N's, so we must again divide by 2!; 2 A's, so we again divide by 2!; and 2 T's, so we must divide by 2! once again. Therefore, we have

$$\frac{11!}{2!2!2!2!} = \frac{11 \cdot 10 \cdot 9 \cdot \cancel{8} \cdot 7 \cdot 6 \cdot 5 \cdot 4 \cdot 3 \cdot \cancel{2} \cdot 1}{\cancel{2} \cdot 1 \cdot \cancel{2} \cdot 1 \cdot \cancel{2} \cdot 1 \cdot \cancel{2} \cdot 1} = 2{,}494{,}800$$

EXERCISES FOR SECTION 3.7

1. Given the set of digits $\{1, 2, 3, 4, 5, 6\}$, how many three-digit numbers can be formed if no digit can be repeated? How many three-digit numbers can be formed if repetition is allowed?

2. If there are 50 contestants in a beauty pageant, in how many ways can the judges award first and second prizes?

3. The Kingston Karate Club has 30 members. A slate of officers consists of a president, a vice president, a secretary, and a treasurer. If a person can hold only one office, in how many ways can a set of officers be formed?

4. A baseball manager has 8 pitchers and 3 catchers on his squad. In how many ways can the manager select a starting battery (pitcher and catcher) for a game?

5. A conference room has 4 doors. In how many ways can a person enter and leave the conference room by a different door?

6. Given the set of digits $\{4, 5, 6, 7, 8\}$, how many four-digit numbers can be formed if no digit can be repeated? How many of these will be odd? How many of these will be divisible by 5? How many of these will be over 6000? How many will be over 4000?

7. How many license plates can be made if each plate must consist of 2 letters followed by 4 digits, if we assume no repetition? How many are possible if the first digit cannot be zero? How many are possible if the first letter cannot be *O* and the first digit cannot be zero?

8. How many four-letter words can be formed from the set of letters $\{m, o, n, e, y\}$? Assume that any arrangement of letters is a word. How many four-letter words can be formed if the first letter must be y and the last letter must be m? (Assume no repetition.)

9. The Rochester Tennis Club is having a mixed-doubles tournament. If 8 women and their husbands sign up for the tournament, how many mixed-doubles teams are possible? How many can be formed if no woman is paired with her husband?

10. At Finger Lakes Race Track, there are 8 horses in each race. The daily double consists of picking the winning horses in the first and second races. If a bettor wanted to purchase all possible daily double tickets, how many would he have to purchase?

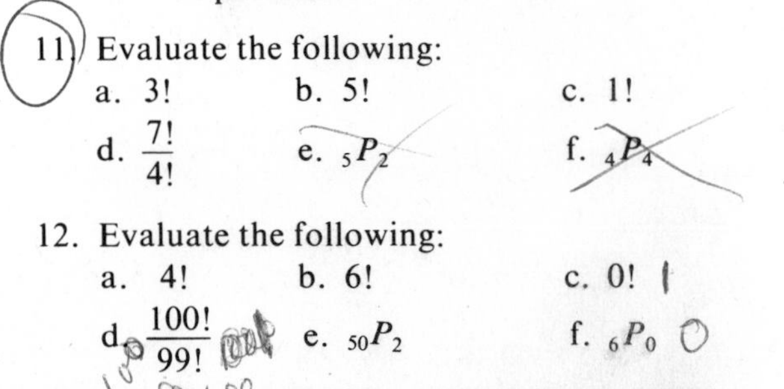

11. Evaluate the following:
    a. $3!$ b. $5!$ c. $1!$
    d. $\dfrac{7!}{4!}$ e. ${}_5P_2$ f. ${}_4P_4$

12. Evaluate the following:
    a. $4!$ b. $6!$ c. $0!$
    d. $\dfrac{100!}{99!}$ e. ${}_{50}P_2$ f. ${}_6P_0$

13. On a naval vessel, a signal can be formed by running up 2 flags on a flag pole, one above the other. If there are 10 different flags to choose from, how many different signals can be formed?

14. How many different ways can 7 students be seated in 7 seats on a subway car?

15. A baseball manager must hand in the lineup card before the game begins. The manager has decided which 9 players will play, but not the batting order. In how many ways can a batting order be chosen?

16. Given the set of digits $\{3, 4, 5, 6, 7, 8, 9\}$, how many different numbers between 3000 and 6000 can be written using these digits if repetition of digits is allowed?

17. In how many ways can a basketball coach select a guard and then a center from a squad of 12 players?

18. A dictionary, an almanac, a catalog, and a diary are to be placed on a shelf. In how many ways can they be arranged?

19. In how many distinct ways can the letters of each word be arranged?
    a. *ALGEBRA* b. *STATISTICS*
    c. *CALCULUS* d. *SCIENCE*

20. In how many distinct ways can the letters of each word be arranged?
    a. *OHIO* b. *ALABAMA*
    c. *MISSISSIPPI* d. *CONNECTICUT*

21. Given the set of digits $\{1, 2, 3, 4, 5\}$, how many different numbers consisting of 4 digits can be formed when repetition is allowed?

22. A traveling book salesperson has 5 copies of a certain statistics book, 4 copies of a certain geometry book, and 3 copies of a certain calculus book. If these books are to be stored on a shelf in the salesperson's van, how many distinct arrangements are possible?

23. Almost all students have a Social Security number. In many schools, a student's Social Security number is also his or her ID number. How many possible Social Security numbers are there? (Assume repetition of digits. A Social Security number contains 9 digits.)

24. Telephone numbers consist of 7 digits: 3 digits for the exchange, followed by 4 more digits. In order to call long distance, you must also use an area code which consists of 3 more digits. How many long-distance telephone numbers are there if the first digit cannot be zero or 1, and the fourth digit cannot be zero or 1? (Assume repetition of digits.)

Just For Fun

Name the southernmost state in the United States.

3.8 COMBINATIONS (Optional)

In section 3.7, we were concerned with the counting of ordered arrangements, or permutations. The techniques developed can be applied to situations like the following: If a tennis squad has 5 players on it, the coach can select the first singles player in 5 ways and the second singles player in 4 ways. Hence, there are $4 \times 5 = 20$ different ways that the coach can select the first and second singles players. This type of choice is a permutation of 5 things taken 2 at a time.

Now, let's consider a slightly different type of choice. Suppose the 5 members of the squad are Arthur, Stan, Billie Jean, Rosemary, and Pancho, and we want to know in how many ways the coach can select a doubles team from this group of 5 players. This is different because when the coach selects a doubles team, it does not matter what person is chosen first and what person is chosen second. That is, the order of choice is not important. The twenty different ordered arrangements are

Arthur, Stan	Stan, Arthur	Billie Jean, Arthur
Arthur, Billie Jean	Stan, Billie Jean	Billie Jean, Stan
Arthur, Rosemary	Stan, Rosemary	Billie Jean, Rosemary
Arthur, Pancho	Stan, Pancho	Billie Jean, Pancho

Rosemary, Arthur	Pancho, Arthur
Rosemary, Stan	Pancho, Stan
Rosemary, Billie Jean	Pancho, Billie Jean
Rosemary, Pancho	Pancho, Rosemary

We can see that the doubles team of Arthur and Stan is the same doubles team as that of Stan and Arthur. This is also the case for Rosemary and Billie Jean and Billie Jean and Rosemary. If we eliminate the duplicate doubles teams, we have

Arthur, Stan	Stan, Billie Jean
Arthur, Billie Jean	Stan, Rosemary
Arthur, Rosemary	Stan, Pancho
Arthur, Pancho	Billie Jean, Rosemary
Rosemary, Pancho	Billie Jean, Pancho

There are 10 different doubles teams that the coach can select. The 20 distinct ordered arrangements of 2 people are reduced to 10 groups of 2 people when we disregard order.

The problem we have just discussed can be classified as a *combination* problem. A **combination** is a distinct group of objects without regard to their arrangement. Committees are good examples of combinations. We are only concerned with who is on a committee, not with who is first, who is second, and so on.

Consider the set of letters $\{x, y, z\}$. How many different three-letter arrangements can be formed from this set? The answer is 3!, or 6, and they are

xyz	yzx	zxy
xzy	yxz	zyx

How many combinations (distinct groups) can be formed from this set? We see that all 6 arrangements constitute the same group of elements. Therefore there is only one combination. We can say that the number of combinations of 3 things taken 3 at a time is 1. Symbolically, we can write

$$_3C_3 = 1$$

In the doubles problem, we had combinations of 5 things taken 2 at a time. The number of these combinations was equal to 10; hence

$$_5C_2 = 10$$

Note that

$$_5C_2 = \frac{_5P_2}{2!} = \frac{5!}{(5-2)!2!} = \frac{5!}{3!2!} = \frac{5 \cdot 4 \cdot \not{3} \cdot \not{2} \cdot 1}{\not{3} \cdot 2 \cdot 1 \cdot \not{2} \cdot 1} = 10$$

and

$$_3C_3 = \frac{_3P_3}{3!} = \frac{3!}{(3-3)!3!} = \frac{3!}{0!3!} = 1$$

In order to obtain the number of distinct combinations, we must eliminate the different ordered arrangements within each group, and we do this by division. From the set of 5 tennis players, we have 5 choices for the first selection and 4 for the second, but those two selections can order themselves in 2! ways, so we must divide by 2!

The general formula for the number of combinations of n things taken r at a time is

$$_nC_r = \frac{n!}{(n-r)!r!}$$

Recall the formula for the number of permutations of n things taken r at a time:

$$_nP_r = \frac{n!}{(n-r)!}$$

If you know this formula, then the formula for $_nC_r$ is easy to remember because it is similar to $_nP_r$. If:

$$_nP_r = \frac{n!}{(n-r)!} \quad \text{and} \quad _nC_r = \frac{n!}{(n-r)!r!}$$

then

$$_nP_r = {_nC_r} \cdot r! \quad \text{or} \quad _nC_r = \frac{_nP_r}{r!} = \frac{n!}{(n-r)!r!}$$

You should be thoroughly familiar with both formulas.

$$_nP_r = \frac{n!}{(n-r)!}$$ the number of *permutations* of n things taken r at a time.

$$_nC_r = \frac{n!}{(n-r)!r!}$$ the number of *combinations* of n things taken r at a time.

(*Note:* Some books use the notation $\binom{n}{r}$ to represent the number of combinations of n things taken r at a time. We shall not use this notation, but be aware that $\binom{n}{r} = \frac{n!}{(n-r)!r!}$.)

EXAMPLE 1

Two co-captains are to be selected from the starting 5 for a basketball team. In how many ways can this be done?

Solution

This is a combination problem since order is not important. We have a combination of 5 things taken 2 at a time. Therefore, we have

$$_5C_2 = \frac{5!}{(5-2)!2!} = \frac{5!}{3!2!} = \frac{5 \cdot 4 \cdot \cancel{3} \cdot \cancel{2} \cdot 1}{\cancel{3} \cdot 2 \cdot 1 \cdot \cancel{2} \cdot 1} = 10$$

Harlem Globetrotters

EXAMPLE 2

The student association each year elects a council consisting of 7 members. If there are 10 candidates for the 7-member council, how many different councils may be elected?

Solution

Since order is not important, we treat this as a combination of 10 things taken 7 at a time. Therefore, we have

$$_{10}C_7 = \frac{10!}{(10-7)!7!} = \frac{10!}{3!7!} = \frac{10 \cdot 9 \cdot 8 \cdot \cancel{7} \cdot \cancel{6} \cdot \cancel{5} \cdot \cancel{4} \cdot \cancel{3} \cdot \cancel{2} \cdot 1}{3 \cdot 2 \cdot 1 \cdot \cancel{7} \cdot \cancel{6} \cdot \cancel{5} \cdot \cancel{4} \cdot \cancel{3} \cdot \cancel{2} \cdot 1} = 120$$

EXAMPLE 3

There are 100 senators in the United States Senate. If the Senate Foreign Relations Committee is composed of 5 senators, in how many different ways can the Foreign Relations Committee be chosen?

Solution
This is a combination of 100 things taken 5 at a time.

$$_{100}C_5 = \frac{100!}{(100 - 5)!5!} = \frac{100!}{95!5!}$$

The calculation is quite time-consuming if we have to write out 100! and 95!, but by using the idea that $n! = n(n - 1)!$, we can write

$$_{100}C_5 = \frac{100!}{95!5!} = \frac{100!}{5!95!} = \frac{100 \cdot 99 \cdot 98 \cdot 97 \cdot 96 \cdot \cancel{95!}}{5 \cdot 4 \cdot 3 \cdot 2 \cdot 1 \cdot \cancel{95!}} = 75{,}287{,}520$$

The calculation is still time-consuming, but it is shorter since cancellation can be used.

EXAMPLE 4
How many different poker hands can be dealt from a standard deck of 52 cards? (Here we assume a poker hand consists of 5 cards.)

Order is not important, since a poker hand consisting of king of hearts, king of clubs, queen of clubs, jack of hearts, and king of spades is the same as one consisting of king of hearts, queen of clubs, king of spades, jack of hearts, and king of clubs. We have a combination of 52 things taken 5 a time; that is,

$$_{52}C_5 = \frac{52!}{(52 - 5)!5!} = \frac{52!}{47!5!} = \frac{52!}{5!47!}$$

$$= \frac{52 \cdot 51 \cdot 50 \cdot 49 \cdot 48 \cdot \cancel{47!}}{5 \cdot 4 \cdot 3 \cdot 2 \cdot 1 \cdot \cancel{47!}} = 2{,}598{,}960$$

EXAMPLE 5
How many committees can be selected from 4 teachers and 100 students if each committee must have 2 teachers and 3 students?

This combination problem is somewhat different from the others we have considered. Here we must choose 2 teachers from 4 and 3 students from 100 to form our committee. Hence we must use the counting principle. If we can choose the teachers in t ways and the students in s ways, then together the choosing can be done in $t \times s$ ways. Therefore the number of committees consisting of 2 teachers and 3 students may be found by multiplying ${}_4C_2$ (4 teachers taken 2 at a time) by ${}_{100}C_3$ (100 students taken 3 at a time).

$$ {}_4C_2 = \frac{4!}{(4-2)!2!} = \frac{4!}{2!2!} = \frac{4\cdot 3\cdot \not{2}\cdot 1}{2\cdot 1\cdot \not{2}\cdot 1} = 6 $$

$$ {}_{100}C_3 = \frac{100!}{(100-3)!3!} = \frac{100!}{97!3!} = \frac{100\cdot 99\cdot 98\cdot \not{97!}}{3\cdot 2\cdot 1\cdot \not{97!}} = 161{,}700 $$

Therefore

$$ {}_4C_2 \cdot {}_{100}C_3 = 6\cdot 161{,}700 = 970{,}200 $$

In doing problems of this nature, we must carefully examine the problem to determine if we are doing a permutation problem or a combination problem. If we want distinct arrangements, then we are doing permutations. If the distinct arrangements are not to be counted, only the different groups, then we are doing combinations. Always be careful of special conditions in the problem. Try to take care of these first; then proceed with the rest of the problem.

EXERCISES FOR SECTION 3.8

1. Evaluate the following:
   a. ${}_5C_3$ b. ${}_5C_2$ c. ${}_7C_4$
   d. ${}_7C_3$ e. ${}_{10}C_{10}$ f. ${}_{10}C_0$
2. Evaluate the following:
   a. ${}_8C_5$ b. ${}_8C_3$ c. ${}_{52}C_2$
   d. ${}_{52}C_{50}$ e. ${}_{52}C_5$ f. ${}_{52}C_{47}$
3. If the Xerox Corporation has to transfer 4 of its 10 junior executives to a new location, in how many ways can the 4 executives be chosen?
4. A newspaper boy discovers in delivering his papers that he is 3 papers short. He has 8 houses left to deliver to, but only 5 papers left. In how many ways can he deliver the remaining newspapers?
5. Alice has a penny, a nickel, a dime, a quarter, and a half dollar. She may spend any 3 coins. In how many ways can Alice do this? What is the most money she can spend using just 3 coins?
6. Joe has to take a math exam that consists of 10 questions. He must answer only 7 of the 10 questions. In how many ways can Joe choose the 7 questions? If he must answer the first and last questions and still only answer a total of 7, in how many ways can he do this?
7. In a mathematics class of 15 students, 10 students must do problems at the board on a given day. In how many ways can the 10 students be chosen?
8. A football coach has 40 candidates out for the squad. In how many ways can a starting eleven be selected without regard to the position that a candidate will play? (Indicate your answer; do not evaluate.)
9. At registration, a student needs 2 more courses to complete her schedule. If there are 7 possible courses left to pick from, in how many ways can she choose the 2 courses?

10. A committee of 11 people, 6 women and 5 men, is forming a subcommittee that is to be made up of 2 women and 3 men. In how many ways can the subcommittee be formed?

11. A baseball squad consists of 8 outfielders and 7 infielders. If the baseball coach must choose 3 outfielders and 4 infielders, in how many ways can this be done?

12. An urn contains 6 blue balls and 4 orange balls.
    a. In how many ways can we select a group of 3 balls?
    b. In how many ways can we select 2 blue balls and 1 orange ball?
    c. In how many ways can we select 2 orange balls and 1 blue ball?

13. From a group of 12 sprinters and 10 distance runners, a medley relay team is to be formed. The relay team must consist of 2 sprinters and 2 distance runners. How many possible medley relay teams are there?

14. Don has to take a history exam that consists of 15 multiple-choice questions and 5 essay questions. If Don has to answer 10 multiple-choice questions and 2 essay questions, in how many ways can he choose them?

15. A student belongs to a record club. This month she has to purchase 2 records and 3 tapes. If there are 10 records and 10 tapes to choose from, in how many ways can she choose her purchases?

16. If 6 points are drawn on a plane, no 3 of which are on the same straight line, how many straight lines can be formed? (Two points determine a line.) $_6C_2 = 15$

17. How many different committees, each composed of 2 Democrats, 2 Republicans, 2 Liberals, and 1 Conservative, can be formed from 12 Democrats, 11 Republicans, 5 Liberals, and 3 Conservatives?

18. The Speaker of the House wants to appoint a committee consisting of 3 representatives from New York, 3 representatives from California, 2 representatives from Ohio, and 3 representatives from Illinois. How many different committees can be formed if 8 representatives from New York, 10 representatives from California, 5 representatives from Ohio, and 6 representatives from Illinois are eligible?

Just For Fun

Is a "combination" lock really a combination lock? Why or why not?

3.9 MORE PROBABILITY (Optional)

We can utilize the counting principle, permutations, and combinations in solving many probability problems. Some of them are similar to those that we discussed previously, while others are a little more involved.

Consider the example solved in example 1 of section 3.6. We want to find the probability of being dealt 2 red cards in succession, without replacement, from a shuffled deck of 52 cards. We found that the probability was

$$\frac{26}{52} \cdot \frac{25}{51} = \frac{650}{2652} = \frac{25}{102}$$

There are 26 red cards out of 52 for the first card, and then 25 red cards out of 51 remaining cards for the second card. We multiplied the two probabilities together to find the probability of getting a red card *and* a red card.

Another way to solve this problem is by combinations. We are not concerned with the order in which the cards appear, just as long as they are red. In how many ways can 2 cards be chosen from a deck of 52? We have

$$_{52}C_2 = \frac{52!}{(52-2)!2!} = \frac{52!}{50!2!} = \frac{52 \cdot 51 \cdot 50!}{50! \cdot 2 \cdot 1} = \frac{52 \cdot 51}{2 \cdot 1} = 1326$$

In how many ways can 2 red cards be chosen? This is the number of successful outcomes. There are 26 red cards in the deck, and a successful outcome is any combination of 2 of these cards. Therefore, we compute

$$_{26}C_2 = \frac{26!}{(26-2)!2!} = \frac{26!}{24!2!} = \frac{26 \cdot 25 \cdot 24!}{24! \cdot 2 \cdot 1} = \frac{26 \cdot 25}{2 \cdot 1} = 325$$

Hence the probability of being dealt 2 red cards in succession is

$$\frac{_{26}C_2}{_{52}C_2} = \frac{325}{1326} = \frac{25}{102}$$

Let us look at some other examples.

EXAMPLE 1

A student has to complete registration for the next semester by choosing 3 more courses. The courses left to choose from are 5 humanities courses and 4 science courses. If the 3 courses are chosen at random, what is the probability that they will all be humanities courses?

Solution

There are 9 courses to choose from, and the student must choose 3 of them. There are ${}_9C_3$ ways of choosing 3 courses. The 3 humanities courses may be chosen from the 5 offered in ${}_5C_3$ ways. Hence we have

$$ {}_5C_3 = \frac{5!}{(5-3)!3!} = \frac{5!}{2!3!} = \frac{5\cdot 4\cdot 3!}{2\cdot 3!} = 10 $$

$$ {}_9C_3 = \frac{9!}{(9-3)!3!} = \frac{9!}{6!3!} = \frac{9\cdot 8\cdot 7\cdot 6!}{6!\cdot 3\cdot 2\cdot 1} = 84 $$

$$ P(3 \text{ humanities}) = \frac{{}_5C_3}{{}_9C_3} = \frac{10}{84} $$

EXAMPLE 2

Find the probability of being dealt a hand in five-card poker which is all spades. (A *flush* is a hand where all the cards are of the same suit: all hearts, all spades, all diamonds, or all clubs.)

Solution

There are 52 cards in a deck, and there are ${}_{52}C_5$ possible ways of being dealt 5 cards. There are 13 spades in a deck, and a successful outcome is being dealt any 5 of these spades. There are ${}_{13}C_5$ ways of being dealt 5 spades. Therefore

$$ {}_{13}C_5 = \frac{13!}{(13-5)!5!} = \frac{13!}{8!5!} = \frac{13\cdot 12\cdot 11\cdot 10\cdot 9\cdot \cancel{8!}}{\cancel{8!}\cdot 5\cdot 4\cdot 3\cdot 2\cdot 1} = 1287 $$

$$ {}_{52}C_5 = \frac{52!}{(52-5)!5!} = \frac{52!}{47!5!} = \frac{52\cdot 51\cdot 50\cdot 49\cdot 48\cdot \cancel{47!}}{\cancel{47!}\cdot 5\cdot 4\cdot 3\cdot 2\cdot 1} = 2{,}598{,}960 $$

$$ P(5 \text{ spades}) = \frac{{}_{13}C_5}{{}_{52}C_5} = \frac{1287}{2{,}598{,}960} = \frac{33}{66{,}640} $$

EXAMPLE 3

Find the probability of being dealt a *full house* (3 cards of one denomination and 2 of another) consisting of kings over deuces (3 kings and 2 twos).

Solution

There are 52 cards in a deck and there are ${}_{52}C_5$ possible ways of being dealt 5 cards. A successful outcome consists of getting 3 kings and 2 twos. There are 4 of each kind of card in a deck and so the number of ways of getting 3 kings is ${}_4C_3$. The number of ways of getting 2 twos is ${}_4C_2$. Therefore the total number of ways of obtaining 3 kings and 2 twos is ${}_4C_3 \cdot {}_4C_2$. So we have

$${}_4C_2 = \frac{4!}{(4-2)!2!} = \frac{4!}{2!2!} = \frac{4 \cdot 3 \cdot \not{2} \cdot 1}{2 \cdot 1 \cdot \not{2} \cdot 1} = 6$$

$${}_4C_3 = \frac{4!}{(4-3)!3!} = \frac{4!}{1!3!} = \frac{4 \cdot \not{3} \cdot \not{2} \cdot 1}{1 \cdot \not{3} \cdot \not{2} \cdot 1} = 4$$

$${}_{52}C_5 = 2{,}598{,}960 \qquad \text{(see example 2)}$$

and the probability of obtaining a full house consisting of 3 kings and 2 twos is

$$P(\text{3 kings and 2 twos}) = \frac{{}_4C_3 \cdot {}_4C_2}{{}_{52}C_5} = \frac{24}{2{,}598{,}960} = \frac{1}{108{,}290}$$

Table 1 provides some interesting information about 5-card poker. It shows the number of various kinds of hands possible when a 52 card deck is used and nothing is wild. Note that the total number of different hands is 2,598,960, which is ${}_{52}C_5$.

TABLE 1

straight flush	40
four of a kind	624
full house	3,744
flush	5,108
straight	10,200
three of a kind	54,912
two pairs	123,552
one pair	1,098,240
no pair	1,302,540
Total	2,598,960

Camerique

A poker hand consisting of the ten, jack, queen, king, and ace of any one suit is called a *royal flush*. The probability of being dealt a royal flush is $\frac{4}{{}_{52}C_5}$, or $\frac{4}{2{,}598{,}960}$.

EXERCISES FOR SECTION 3.9

1. Find the probability of being dealt 2 queens when you are dealt 2 cards from a shuffled deck of 52 cards.

2. Find the probability of being dealt 3 kings when you are dealt 3 cards from a shuffled deck of 52 cards.

3. You are dealt 3 cards from a shuffled deck of 52 cards. Find the probability that all 3 cards are hearts.

4. You are dealt 5 cards from a shuffled deck of 52 cards. Find the probability that all 5 cards are picture cards (king, queen, or jack). What is the probability that none of the cards are picture cards? (Indicate your answer; do not evaluate.)

5. On the track team of the York Athletic Club, there are 8 sprinters and 10 distance runners. A relay team consisting of 4 people must be chosen at random (without regard to who runs first, second, etc.). Find the probability that the relay team will—
   a. consist of sprinters only
   b. consist of distance runners only
   c. consist of 2 sprinters and 2 distance runners
   d. consist of 3 sprinters and 1 distance runner
   e. contain at least 1 sprinter.

6. You are dealt 5 cards from a shuffled deck of 52 cards. Find the probability that—
   a. all 5 cards are aces
   b. 3 cards are aces and 2 cards are picture cards
   c. 4 cards are aces and 1 card is a picture card.

7. A five-man committee is to be formed at random from 7 Democrats and 9 Republicans. Find the probability that the committee will consist of—
   a. all Democrats
   b. all Republicans
   c. 2 Democrats and 3 Republicans
   d. 2 Republicans and 3 Democrats
   e. 4 Democrats and 1 Republican
   f. 4 Republicans and 1 Democrat.

8. A football coach has 5 guards and 7 tackles trying out for the squad. As a final cut, five of these players will be cut at random. Find the probability that the group of players cut will—
   a. consist of guards only
   b. consist of tackles only
   c. consist of 3 guards and 2 tackles
   d. consist of 4 tackles and 1 guard
   e. consist of 3 tackles and 2 guards
   f. consist of 4 guards and 1 tackle.

9. In how many ways can 13 cards (a bridge hand) be selected from a deck of 52 cards? Find the probability of being dealt a bridge hand that consists of all spades. (Indicate your answer; do not evaluate.)

10. A case of soda pop contains 6 bottles of root beer, 6 bottles of orange soda, 7 bottles of cola, and 5 bottles of ginger ale. If you select 3 bottles at random, find the probability that—
    a. all 3 bottles are cola
    b. all 3 bottles are root beer
    c. 2 bottles are ginger ale and 1 bottle is cola
    d. 1 bottle is orange soda and 2 bottles are root beer
    e. 1 bottle is cola, 1 bottle is orange soda, and 1 bottle is ginger ale
    f. 1 bottle is cola, 1 bottle is orange soda, and 1 bottle is root beer.

11. A committee consisting of 4 people is to be selected at random from a group of 7 people which includes Bob and Carol and Ted and Alice. Find the probability that the committee will consist of Bob, Carol, Ted, and Alice.

12. A jury (12 people) is to be randomly selected from a group consisting of 6 men and 8 women. Find the probability that the jury will—
    a. consist of women only
    b. consist of men only
    c. consist of 6 women and 6 men
    d. consist of 7 women and 5 men
    e. consist of 7 men and 5 women
    f. contain at least 1 man.

13. An urn contains 6 orange balls, 4 blue balls, and 3 red balls. If 3 balls are drawn at random, indicate (but do not evaluate) the probability of selecting—
    a. 3 orange balls
    b. 3 blue balls
    c. 3 red balls
    d. 2 orange balls and 1 blue ball
    e. 2 blue balls and 1 red ball
    f. 2 red balls and 1 orange ball.

14. Pocket billiards (pool) is played with 15 balls numbered 1 through 15. The balls with numbers greater than 8 are striped, while the rest are solid colors. Fast Eddie, a pool hustler, sinks two balls with a single shot. Find the probability that the balls made by Fast Eddie are—
    a. both striped
    b. both solids
    c. 1 striped and 1 solid
    d. the number 1 ball and the number 15 ball.

15. Find the probability of being dealt a flush (5 cards of the same suit) in hearts in five-card poker.

16. What is the probability of being dealt a flush in hearts or spades in five-card poker?

17. Find the probability of being dealt 2 queens and 3 jacks in five-card poker.

18. Find the probability of being dealt 2 queens, 2 kings, and 1 ace in five-card poker.

19. Find the probability of being dealt a full house (3 of a kind, together with 2 of a kind) in five-card poker.

Just For Fun

Can you make $1.00 using 21 coins? (There is more than one solution.)

3.10 SUMMARY

Probability is a topic that we encounter in everyday living. We all encounter it in the weather forecast with the probability of precipitation. The manufacturer of a television set uses it when he decides to offer a one-year guarantee on the picture tube. Life insurance companies use laws of probability to calculate the premiums on the policies they issue to people.

A *sample space* is the set of all possible outcomes to a given experiment. An *event* is any subset of a sample space. If an experiment has a total of T *equally likely* outcomes, and if exactly S of these outcomes are considered successful—that is, they are the members of the event A—then the probability that event A will occur is

$$P(A) = \frac{S}{T} = \frac{\text{number of } \textit{successful} \text{ outcomes}}{\textit{total} \text{ number of possible outcomes}}$$

When two or more experiments are performed together, the *counting principle* can sometimes be used to determine the total number of possible outcomes in the sample space. If one experiment has m different outcomes, and a second experiment has n different outcomes, then the first and second experiments performed together have $m \times n$ different outcomes. A *tree diagram* may be used to determine a sample space for a particular problem, because it illustrates the possible outcomes for an experiment.

The *odds in favor of an event A* are found by taking the probability that the event A will occur and dividing it by the probability that the event A will not occur:

$$\text{odds in favor of } A = \frac{\text{probability in favor of } A}{\text{probability not in favor of } A}$$

Odds are usually expressed as a ratio: if the odds in favor of A are $\frac{3}{2}$, then we may express the odds as 3 to 2, or 3:2. If the odds in favor of A are 3:2, then the *odds against A* are 2:3.

Mathematical expectation is found by multiplying P, the probability that an event will occur, times M, the amount that will be won if the event occurs.

$$E = P \times M$$

Mathematical expectation tells us what is a fair price to pay to play a game. Expectation can also be thought of as the average amount of winnings we can expect for each game when we play a great number of games.

Independent events are events where the occurrence of one event does not affect the occurrence of a second event. *Mutually exclusive events* are events that cannot happen at the same time. To compute the probability of two or more events occurring together, we multiply the various probabilities together. That is,

$$P(A \text{ and } B) = P(A) \times P(B)$$

When we have mutually exclusive events and we want to compute the probability of A *or* B, we have

$$P(A \text{ or } B) = P(A) + P(B) \quad \text{(where } A \text{ and } B \text{ are mutually exclusive)}$$

If A and B are not mutually exclusive events and we want to compute the probability of A *or* B, we have

$$P(A \text{ or } B) = P(A) + P(B) - P(A \text{ and } B)$$

(where A and B are *not* mutually exclusive)

Permutations are ordered arrangements of things. The number of permutations of n things taken r at a time is denoted by ${}_nP_r$, and

$$ {}_nP_r = \frac{n!}{(n - r)!} $$

Combinations are distinct groups of things without regard to their arrangement. The number of combinations of n things taken r at a time is denoted by ${}_nC_r$, and

$$ {}_nC_r = \frac{n!}{(n - r)!r!} $$

Permutations and combinations are useful in solving many probability problems, particularly those that involve a great many possible outcomes, such as the various probabilities for different poker hands. (For more examples, see section 3.9.)

Review Exercises for Chapter 3

1. On a single draw from a bag containing 4 red, 6 blue, and 3 green balls, find the probability of obtaining—
   a. a red ball
   b. a blue ball
   c. a red or a green ball
   d. a ball that is not red
   e. a ball that is not green
   f. a ball that is neither red nor blue.

2. A pair of dice is tossed. Find the probability that—
   a. the sum of the numbers is 7
   b. the sum of the numbers is not 7
   c. the same number appears on both dice
   d. the sum of the numbers is 7 or 11
   e. the sum of the numbers is greater than 7
   f. the sum of the numbers is not even.

3. A box contains a one-dollar bill, a two-dollar bill, a five-dollar bill, and a ten-dollar bill. An experiment is conducted which consists of randomly selecting two bills in succession without replacement.
   a. In how many ways can this be done?
   b. Construct a sample space showing the possible outcomes for this event.
   c. Find the probability that the sum of money for the 2 bills selected is $11.

4. By means of a tree diagram, list the sample space showing the possible arrangements when 3 coins are tossed.
   a. Find the probability that all 3 coins are heads.
   b. Find the probability that at least 1 coin is heads.
   c. Find the probability that no coins are heads.

5. On a single draw from a shuffled deck of 52 cards, find the odds—
   a. in favor of drawing a king
   b. against drawing a king
   c. in favor of drawing a picture card (jack, queen, king)
   d. in favor of drawing a club or jack
   e. in favor of drawing a club and a jack
   f. against drawing the ace of spades.

6. If the odds are 8 to 1 against the Giants winning the Super Bowl, what is the probability that the Giants will win the Super Bowl?

7. The probability that the Bruins will win the Stanley Cup is $\frac{3}{11}$. Find the odds in favor of the Bruins winning the Stanley Cup.

8. Five thousand tickets are sold for a drawing on a yacht valued at \$10,000. If a woman buys 1 ticket, what is her expectation?

9. A fraternity sold 500 raffle tickets at \$2 each on a color television set valued at \$400. If Joe Kool buys 5 tickets, what is his mathematical expectation?

10. A bag contains 5 balls numbered 1 through 5. Two balls are chosen in succession. The first ball is replaced before the second is drawn. Are these events *dependent* or *independent*? Why?

11. The object of a game is to obtain a 7 or 11 with a single toss of a pair of dice. Are these events *mutually exclusive*? Why?

12. Two cards are randomly selected in succession from a shuffled deck of 52 cards, without replacement. Find the probability that—
    a. both cards are red
    b. both cards are 3s
    c. the first card is red and the second card is black
    d. the first card is a heart and the second card is a club
    e. the first card is a picture card and the second card is not a picture card
    f. the first card is an ace and the second card is a king.

13. One card is randomly selected from a shuffled deck of 52 cards and then a die is rolled. Find the probability of obtaining—
    a. a king and a 1
    b. a king or a 1
    c. a red card and a 1
    d. a red card or a 1
    e. a red ace and a 1

14. There is a game called *poker dice*, which is based on rolling 5 dice. The resulting outcome is then treated as a poker hand.
    a. How many different outcomes are possible?
    b. What is the probability of getting five of a kind?

(*Note:* Exercises 15–25 are based on the optional topics in this chapter.)

*15. How many license plates can be made if each plate must consist of 2 letters followed by 3 digits? (Assume no repetition.) How many are possible if the first letter cannot be *O* and the first digit cannot be zero? How many are possible if the letter *Q* cannot be used and zero cannot be used?

*16. How many different four-letter "words" (that is, arrangements of 4 letters) can be formed from the letters of the alphabet if each letter can only be used once and none of the vowels *a, e, i, o, u* may be used?

*17. In a certain collegiate basketball conference, there are 10 teams and each team plays every other team in the conference twice. How many league games are played in a season?

*18. How many distinct arrangements are possible in using all the letters of the word *ECOLOGY*?

*19. Evaluate the following:

a. $4!$ b. $0!$ c. $\frac{6!}{3!2!}$

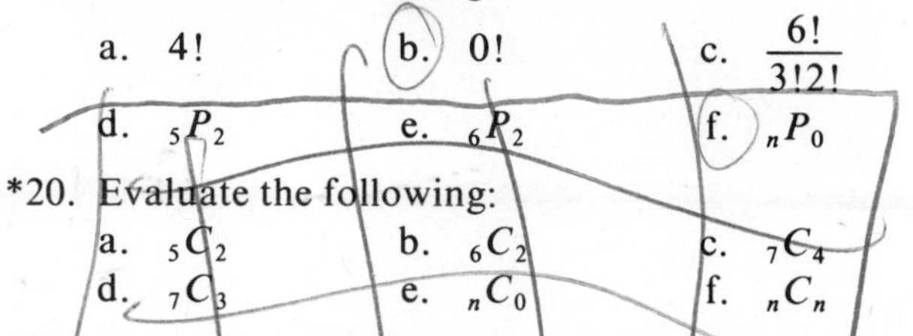

d. ${}_5P_2$ e. ${}_6P_2$ f. ${}_nP_0$

*20. Evaluate the following:

a. ${}_5C_2$ b. ${}_6C_2$ c. ${}_7C_4$

d. ${}_7C_3$ e. ${}_nC_0$ f. ${}_nC_n$

*21. From 5 teachers and 50 students, how many committees can be selected if each committee is to have 2 teachers and 3 students?

*22. From a group of 5 freshmen, 6 sophomores, 4 juniors, and 3 seniors, a staff of 3 freshmen, 3 sophomores, 2 juniors, and 2 seniors is to be chosen for the school's radio station. In how many ways can this be done?

*23. Three balls are drawn simultaneously at random from a bag containing 4 red, 4 blue, and 2 yellow balls. Find the probability that—
   a. all 3 balls are blue
   b. all 3 balls are red
   c. 2 balls are yellow and 1 is red
   d. 2 balls are red and 1 blue
   e. 2 balls are red and 1 is yellow.

*24. You are dealt 5 cards from a shuffled deck of 52 cards. Indicate (but do not evaluate) the probability that—
   a. all 5 cards are red
   b. none of the 5 cards are red
   c. all 5 cards are picture cards
   d. none of the 5 cards are picture cards
   e. 3 of the cards are picture cards and 2 are not.

*25. In the game of five-card poker, find the probability of being dealt a hand that is—
   a. 4 of a kind
   b. a full house
   c. a flush.

Just For Fun

Do you think that 2 people in your class have the same birthday (month and day)? Try it; you might be surprised. If there are 25 people in your class, the probability that 2 people have the same birthday is greater than 0.5. If there are 50 people in your class, the probability that 2 people have the same birthday is very close to 1.0 (certainty).

19

4 STATISTICS

After studying this chapter, you will be able to do the following:

1. State three measures of **central tendency,** and distinguish among them
2. Compute the **mean, median,** and **mode** for a given set of data
3. Compute the **range** and **midrange** for a given set of data, and use the range to estimate the central tendency of the data
4. Compute the **standard deviation** for a given set of data
5. Find the **percentile** or **quartile** of a single datum in a given set of data
6. Construct **circle graphs**
7. Construct a **frequency distribution table** and **histogram** or **frequency polygon** from a given set of data
8. Determine what percentage of normally distributed data is within a given number of **standard deviations** from the mean.

Symbols frequently used in this chapter

$\bar{x}$ =	*mean*	(read "x-bar")
a^2	a squared	(for example, $3^2 = 3 \times 3 = 9$)
$\sqrt{a}$	the positive square root of a	(for example, $\sqrt{9} = 3$)
σ	standard deviation	(lowercase Greek letter sigma)

4.1 INTRODUCTION

Whenever we watch television, listen to the radio, or read newspapers, magazines, or books, we encounter statistics. We can find statistics in articles on business, the state of the economy, politics, science, education, sports, and many other subjects. The headline in the newspaper article below uses the word *median,* a statistical concept. In order to understand the article, we must possess some understanding of statistics.

Most people first encounter statistics in elementary school when they take standardized achievement tests: schools want to know what the "average" student has learned, or how a student compares to other students that have taken the standardized tests.

Harold M. Lambert

Median Price Of 1986 Home: $90,000?

WASHINGTON (AP) — The median price for a new home will reach $90,000 by 1986 with the average down payment averaging close to $23,000, a Princeton University economist told a Senate committee today.

Professor Kenneth Rosen said, "The United States is today in the midst of a housing crisis of unprecedented proportions. Most new entrants to the housing market cannot afford to purchase any home at all."

He predicted the prices for new homes will continue to increase faster than the over-all inflation rate. Young families "will be confronted with a major explosion in housing prices and down payment requirements. The situation will become much worse," Rosen told the Senate Banking Committee.

The median price of a new home will approximately double from the present $45,000 by 1986, he predicted. He said that the inflation rate for new housing is likely to be 10 per cent per year over the next decade, or at least 3 per cent higher than the over-all inflation rate.

Rosen testified in favor of a bill by Sen. Edward Brooke, R-Mass., designed to make the purchase of an initial home easier for young families.

Brooke's bill would provide for lower payments in the early years of a mortgage. The bill assumes that families' incomes will rise, making possible higher payments in the later years of a mortgage.

The Brooke bill also would allow creation of tax-exempt savings accounts for home buyers to accumulate funds for down payments.

Reprinted by permission of The Associated Press

A **statistic** is a number derived from a set of data which (in some way) is used to represent or describe the data. Unfortunately, statistics are often misused or abused. Some people maintain that a statistician can prove whatever he wants to. A more familiar quote is "Figures don't lie, but liars figure." Adults and children alike are inundated by statistics. All too often they have no real understanding of the facts presented.

Advertising is one area where statistics are often abused. Sometimes the statistics are distorted by the manner in which the facts are gathered, or by the manner in which the facts are presented. Statistics can be distorted in diagrams by emphasizing the wrong facts, exaggerating comparisons, or simply not showing all of the data.

In this chapter, we shall discuss the basic concepts of statistics in order to prepare you better to understand and interpret statistics when you encounter them. It has often been said that if a person is going to be an intelligent member of today's society, he or she must possess some understanding of statistics.

4.2 AVERAGE—THE MEASURE OF CENTRAL TENDENCY

One of the first concepts of statistics that people encounter, and one that is familiar to everyone, is the concept of an "average." Whether a person can calculate an average or not, he has an intuitive idea of what it is. We are all familiar with phrases like "average miles per gallon," "average precipitation for the month," the "batting average" of a baseball player, or the "earned run average" (E.R.A.) of a pitcher.

The term *average* is used in different ways by different people. An average is a *measure of central tendency.* A **measure of central tendency** describes a set of data by locating the middle region of the set. The three most common measures of central tendency are the *arithmetic mean,* the *median,* and the *mode.* The arithmetic mean is usually referred to simply as the *mean.* Each of the three measures of central tendency, mean, median, or mode, is an average, since each describes the middle region of the data. Each one has its advantages and disadvantages; in a given situation, it may be more desirable to use one as opposed to the other two.

The mean (arithmetic mean) is the most familiar type of average that we encounter. You have probably found the mean for a set of data before, but instead of calling it a mean, you probably called it an average.

The mean for a set of data, scores, or facts is found by determining the sum of the data and dividing this sum by the total number of elements in the set.

The scores for 5 students on a quiz are 40, 20, 30, 25, and 15. To find the mean score for this group of students, we first find the sum of the scores:

$$40 + 20 + 30 + 25 + 15 = 130$$

We then divide the sum by 5, the number of scores:

$$\frac{130}{5} = 26$$

The mean for a set of data is often represented by the notation $\overline{x}$ (read "x-bar"). Therefore, we can write $\overline{x} = 26$ for our given set of data. Even though the mean score is 26, note that none of the scores is 26. There are two scores, 40 and 30, above the mean and three scores, 25, 20, and 15, below the mean. This single measure of central tendency is used to represent the whole group of 5 students. It gives us information as to how the group of students performed, but it does not tell us anything about a particular student in the group.

EXAMPLE 1

Find the mean for the set of test scores

71, 75, 60, 84, 71, 63, 66

Solution

The sum of the scores is $71 + 75 + 60 + 84 + 71 + 63 + 66 = 490$, and we have 7 scores; hence we divide 490 by 7.

$$\bar{x} = \frac{490}{7} = 70$$

EXAMPLE 2

At the Surf and Sand Restaurant, a waitress earns \$10 a night in tips for five nights during the week. But on weekends (Saturday and Sunday) she earns \$20 a night in tips. What is her average daily earnings in tips?

Solution

We must find the total income and then divide this sum by the number of elements in our set of data. We have $5 \times \$10 = \$50$, $2 \times \$20 = \$40$, and total tips are $\$40 + \$50 = \$90$. Now we divide \$90 by 7:

$$\bar{x} = \frac{\$90}{7} \approx \$12.86 \qquad \text{daily average earnings in tips}$$

In example 1, the mean score of 70 tells us that the average score for the set was 70, but it does not necessarily represent any particular score. Note that the mean of 70 is somewhat centrally located in the set of data. However, a mean can be misleading in describing data, since it can be affected by extreme values in the data. Example 3 illustrates this.

EXAMPLE 3

Find the mean for the set of scores

82, 81, 80, 87, 20

Solution
The sum of the scores is $82 + 81 + 80 + 87 + 20 = 350$; hence the mean is

$$\bar{x} = \frac{350}{5} = 70$$

The mean is only 70, although most of the scores are in the 80's.

Although the average, or mean, of the scores in example 3 was 70, it is obvious that this number does not give a very accurate idea of the typical score. Suppose, however, that we arrange the scores in example 3 in order, from lowest to highest:

20, 80, 81, 82, 87

The middle number in this arrangement is 81. This number, called the *median* of the data, gives a more accurate idea of the typical score in cases where the data include extreme values.

Licht/Stockmarket

Based on this photograph, do you think the median of the heights of these three fifteen-year-old boys would be equal to the mean of their heights?

The median for a set of data is found by arranging the data (numbers) in sequential order, from smallest to largest, and finding the middle number.

The scores for 5 students on a quiz are 40, 20, 30, 24, and 15. To find the median, we must first arrange the scores in sequential order, i.e., 15, 20, 24, 30, 40. The median (middle number) for these 5 scores is 24.

Suppose we had 6 scores instead of 5? Consider the scores 40, 20, 30, 24, 28, and 15. What is the median? We have a problem: the median is the middle number, but we have no single middle number because we have an even number of test scores. When we have an even number of scores, it is customary to use the number halfway between the two middle scores (that is, the mean of the two middle numbers) as the median. The two middle numbers in our given set of data are 24 and 28. Therefore the median is

$$\frac{24 + 28}{2} = \frac{52}{2} = 26$$

The median is the measure of central tendency that determines the middle of a given set of data, that is, the number, value, or score such that the number of scores below the median is the same as the number of scores above the median.

EXAMPLE 4
Find the median for the set of test scores:

71, 75, 60, 84, 63, 66, 74

Solution
We must first arrange the test scores in sequential order; we have

60, 63, 66, 71, 74, 75, 84

and the middle number is 71. (The number of scores is odd.) Since the median is 71, we know that 3 of the scores are greater than 71, and 3 of the scores are less than 71.

EXAMPLE 5
Find the median for the given set of data:

13, 16, 14, 12, 20, 19, 10, 18

Solution
Arranging the numbers in sequential order, we have

10, 12, 13, 14, 16, 18, 19, 20

Note that there are 8 data (an even number), so we must find the mean of the two middle numbers, 14 and 16:

$$\frac{14 + 16}{2} = \frac{30}{2} = 15$$

Therefore the median is 15. Four of the pieces of data are less than 15, and four of the pieces of data are greater than 15.

EXAMPLE 6
Find the median for the given set of data:

26, 25, 24, 26, 25, 24, 25, 25, 24, 26, 26

Solution
When we arrange the numbers in sequential order, we have

24, 24, 24, 25, 25, 25, 25, 26, 26, 26, 26

There are 11 numbers, so the median is the sixth one, 25. Five numbers are before it, and five numbers are after it.

EXAMPLE 7
A group of students, Tom, Carlos, Janie, Irene, and Frank, reported that they had earned the following amounts during summer vacation:

Tom	\$800	Irene	\$900
Carlos	\$900	Frank ...	\$600
Janie	\$300		

Find the mean and median summer income for this group of students.

a. To find the mean income, we must find the total income and then divide this sum by 5 (the number of students).

$$\$800 + \$900 + \$300 + \$900 + \$600 = \$3500$$

$$\frac{\$3500}{5} = \$700 \qquad \text{(mean income)}$$

b. To find the median income, we must arrange the amounts in sequential order and find the middle amount. (The number of students is odd.)

\$300, \$600, \$800, \$900, \$900

The middle amount, \$800, is the median income.

In example 7, we calculated both the mean and the median for a given set of data. Notice, however, that although the mean was \$700 and the median was \$800, more people in the group earned \$900 than earned any other amount. This figure, \$900, is called the *mode* of the data in example 7.

The mode for a given set of data is that number, item, or value that occurs most frequently.

The scores for 5 students on a quiz are 40, 24, 30, 24, and 15. The mode for this set of scores is 24 since it occurs twice, while each of the other scores occurs only once.

There are times when the mode is the most meaningful of the three measures of central tendency. Consider the case of Harry, who operates Harry's Hamburger Haven. During a typical eight-hour shift, Harry sells approximately 2400 hamburgers. This is a mean of 300 hamburgers per hour. But Harry's customers do not come into his Hamburger Haven at the same rate each hour. He is much busier some hours than others. In planning how many hamburgers to have ready each hour, Harry is interested in what hour(s) he sells the most hamburgers. Likewise, a shoe manufacturer is more interested in the most frequently sold shoe size than in the mean shoe size or median shoe size.

You should be aware that a set of data always has a mean and a median, but does not necessarily have a mode. Consider the set of data 1, 2, 3, 4, 5. The mean is 3 and the median is 3, but there is no mode, since there is no number that occurs most frequently.

It is also the case that a set of data can have more than one mode. Consider the set of data 1, 2, 3, 3, 4, 5, 5, 6. We see that the number 3 occurs twice and the number 5 occurs twice. Hence, there are two modes, 3 and 5. Since this given set of data has two modes, we refer to it as **bimodal.**

EXAMPLE 8
Find the mode for the following data:

12, 10, 11, 13, 11, 14, 13, 11, 17

Solution
The mode is 11. It occurs 3 times in the set of data.

EXAMPLE 9
Find the mode for the following data:

18, 16, 13, 14, 12, 10, 11, 15, 17, 19

Solution
There is no mode for this set of data because each value occurs only once.

EXAMPLE 10
Find (*a*) the mean, (*b*) the median, (*c*) the mode for the following data:

72, 77, 74, 82, 74, 84, 83

Solution

a. To find the mean, we first find the total of the set of values and then divide this sum by 7 (the number of values).

$$72 + 74 + 74 + 77 + 82 + 83 + 84 = 546$$

$$\frac{546}{7} = 78$$

b. To find the median, we must arrange the values in sequential order and find the middle value (we have an odd number of values). Arranging the data in order, we have:

72, 74, 74, 77, 82, 83, 84

The middle value is 77; therefore the median is 77.

c. The mode is that number which occurs most frequently. The mode for this set of data is 74, since it occurs twice and the other values occur only once.

EXAMPLE 11

Find (*a*) the mean, (*b*) the median, (*c*) the mode for the following data:

52, 74, 74, 77, 82, 83, 104

Solution

a. Since $52 + 74 + 74 + 77 + 82 + 83 + 104 = 546$, the mean is

$$\frac{546}{7} = 78$$

b. The median is 77.

c. The mode is 74.

Note that in examples 10 and 11 we had the same mean, the same median, and the same mode, but the two sets of data were not exactly the same. In the first set of data, we have a low value of 72 and a high value of 84, or a spread of only 12 between the two extremes. But in example 11, the low value is 52, while the high value is 104—a spread of 52. This spread of values is called the **range.** The range for a set of data is found by subtracting the smallest value in the given set of data from the largest value. We shall discuss the range in the next section.

You should be aware that each of the three measures of central tendency is a useful measuring device. There are instances when one of the three (mean, median, mode) is more representative than the other two. The mean is not a true indication of the "average" if there are values in the given set of data that are extreme values at one end or the other. A typical example that illustrates this phenomenon is one that involves wages. Consider the annual salaries of the employees of the Custom Moving Company. The manager earns $28,000 per year, one driver earns $11,000 per year, and another driver earns $10,000 per year. One helper earns $6,000 per year, while three other helpers each earn $5,000 per year. The mean salary for these 7 people is computed as follows:

$$\$28{,}000 + \$11{,}000 + \$10{,}000 + \$6{,}000 + \$5{,}000 + \$5{,}000 + \$5{,}000 = \$70{,}000$$

$$\frac{\$70{,}000}{7} = \$10{,}000$$

Carrying the discussion a little further, suppose the owner earns $50,000 a year from the Custom Moving Company. The mean salary for all eight people would be

$$\$50{,}000 + \$28{,}000 + \$11{,}000 + \$10{,}000 + \$6{,}000 + \$5{,}000 + \$5{,}000 + \$5{,}000 = \$120{,}000$$

$$\frac{\$120{,}000}{8} = \$15{,}000$$

The mean salary is $15,000. This certainly is not a sensible representation of the average salary. This is a case where the mean should not be used to describe the situation for the given set of data. In contract bargaining, the management would probably like to use the mean ($15,000) as a basis for negotiations, while the union would probably like to use the mode ($5,000). Remember that the mean is not a true indication of the average if the given set of data contains extreme values at one end.

The median is the middle value (number). The number of data below the median is the same as the number of data above the median. The median is not affected by extreme values in the set of data. But it may not be a true representation of the average if the data occurs in distinct, separate groups.

The mode is that item that occurs most frequently in a set of data. There can be times when a set of data has no mode, that is, when each value occurs an equal number of times. Other times a set of data can have more than one mode, that is, when two or more different values occur the same number of times. The mode can be misleading at times: it does not take into account the other numbers in the set of data, as do the mean and median.

EXERCISES FOR SECTION 4.2

In exercises 1-10, find the mean, median, and mode for each set of data. (Round off any decimal answer to the nearest tenth.)

1. 1, 2, 3, 4, 4, 5, 9, 12
2. 10, 16, 14, 10, 17
3. 1, 2, 3, 4, 5, 6, 7, 8, 9, 10
4. 1, 1, 5, 5, 7, 10, 12, 11, 10
5. 2, 4, 8, 10, 6, 12
6. 5, 7, 13, 1, 3, 9, 11, 15
7. 11, 99, 77, 88, 66, 44, 55, 22, 33
8. 99, 19, 89, 29, 79, 39, 69, 49, 59
9. 1492, 1776, 1941, 1812
10. 1984, 2004, 1974, 1954, 2004
11. An art teacher grades all student projects 1, 2, or 3, as follows:

    1—excellent
    2—acceptable
    3—unacceptable (must be done over)

    Given a set of data consisting of all the grades assigned to student projects during the term, which measure of central tendency would you use to determine the grade received by the greatest number of projects?

12. Women employees of a certain firm have complained of sex discrimination in the company's pay scale. If the women win their case, the firm will have to raise the salaries of all underpaid employees. The employees and their salaries are listed below:

Ms. O'Brien	$25,000	Mr. Ponti	$15,000
Ms. Jones	$10,000	Mr. Hansen	$15,000
Ms. Chung	$10,000	Mr. Steinberg	$15,000

    a. What is the mean salary for all employees?
    b. What is the mean salary for women? For men?
    c. What is the median salary for women? For men?
    d. If you were a lawyer acting for the firm, which measure of central tendency would you use to describe the average salaries of male and female employees? Which measure of central tendency would you use if you were a lawyer arguing on behalf of Ms. Jones and Ms. Chung?

13. In Joe Kool's physics course, a mean score of 80 on 10 quizzes is necessary for a grade of B. Joe's mean score for the 10 quizzes was 79 and his instructor, Ms. Molecule, gave him a grade of C. Joe protested that since he was so close, his instructor should give him the "one lousy point" and hence a B. Did Joe need only one point for a B? Explain your answer.
14. In a certain week, the New York Mets played 8 baseball games. The number of runs scored in the respective games were 3, 2, 0, 6, 9, 4, 5, and 3. Find the mean, median, and mode for this set of data. Did the Mets win all 8 games?
15. The mean score on a set of 10 scores is 71. What is the sum of the 10 test scores?
16. The mean score on a set of 13 scores is 77. What is the sum of the 13 test scores?
17. The mean score on 4 of a set of 5 scores is 75. The fifth score is 90. What is the sum of the 5 scores? What is the mean of the 5 scores?
18. Two sets of data are given: the first set of data has 10 scores with a mean of 70, and the second set of data has 20 scores with a mean of 80. What is the mean for both sets of data combined?

19. Janet and Larry took the same courses last semester: calculus (4 credits), geology (4 credits), English (3 credits), history (3 credits), and physical education (1 credit). Janet received A, A, B, C, C, respectively, and Larry received C, C, B, A, A, respectively. Janet and Larry bet a dinner as to who would have the higher average. Janet maintains she won, while Larry maintains that they are even since they both got two A's, one B, and two C's. Who is right? (*Hint:* A grade point average is found as follows: Allow 4 points for an A, 3 points for a B, and 2 points for a C. Multiply the number of points equivalent to the letter grade received in each course by the number of credits for the course to arrive at the total points earned in each course. Divide the sum of the points by the total number of credit hours. The answer is the grade point average.)

20. Below is the score card for Scott and Joe, who played nine holes of miniature golf. Find the mean, median, and mode of the scores for Scott and Joe. In your opinion, who is the better player?

HOLE	1	2	3	4	5	6	7	8	9	TOTAL
Scott	2	4	1	5	3	3	2	2	2	24
Joe	3	2	2	2	5	2	5	1	2	24

Just For Fun

If you are given 10 pennies and 3 coffee cups, can you place all of the pennies in the 3 coffee cups so that there is an odd number of coins in each coffee cup? Each cup has to contain at least 1 penny.

4.3 MEASURES OF DISPERSION

In the previous section, we discussed measures of central tendency, mean, median, and mode. They are called measures of central tendency because each of them tells us something about the average of the data, that is, where the data tends to center or cluster. Measures of dispersion tell us how much the data tends to disperse or scatter, that is, the spread of the data.

When we discussed examples 10 and 11 in section 4.2, we mentioned one measure of dispersion, the *range.*

The range for a set of data is found by subtracting the smallest value from the largest value in the given set of data.

Consider the respective quiz scores in a history class for Cathy and Juanita:

Juanita:	72,	74,	74,	77,	80,	83,	86
Cathy:	58,	74,	74,	77,	81,	82,	100

Computing the mean score for Juanita, we have

$$\frac{72 + 74 + 74 + 77 + 80 + 83 + 86}{7} = \frac{546}{7} = 78$$

The mean score for Cathy is

$$\frac{58 + 74 + 74 + 77 + 81 + 82 + 100}{7} = \frac{546}{7} = 78$$

The mean score for each is 78. Comparing further, we see that the median for each is 77, and the mode for each is 74. Cathy and Juanita have the same measures of central tendency for their quiz scores. The one significantly different thing about their scores is the range: the range for Juanita's scores is $86 - 72 = 14$, while the range for Cathy's scores is $100 - 58 = 42$.

Cathy's scores had the greater range. Since Cathy and Juanita cannot be compared by means of the measures of central tendency, we might use the range. But we must be careful. We might argue that Juanita is the more consistent of the two since her score range is only 14, but we might also argue that Cathy showed the most improvement since her score range is 42.

EXAMPLE 1

Find the mean and the range for the two sets of test scores.

Test A:	75,	75,	70,	70,	70,	65,	65,	65,	55,	40
Test B:	70,	70,	67,	66,	65,	65,	65,	62,	60,	60

Solution

Test A: mean =

$$\frac{75 + 75 + 70 + 70 + 70 + 65 + 65 + 65 + 55 + 40}{10}$$

$$= \frac{650}{10} = 65$$

range $= 75 - 40 = 35$

Test B: mean =

$$\frac{70 + 70 + 67 + 66 + 65 + 65 + 65 + 62 + 60 + 60}{10}$$

$$= \frac{650}{10} = 65$$

range = 70 − 60 = 10

Note that both test A and test B had a mean score of 65, but the range of test A was 35, while the range for test B was only 10. There were both higher and lower grades on test A. There is less dispersion for the scores on test B.

An interesting property of the range is that, besides being a measure of dispersion, it can be used to find an approximation for a measure of central tendency. This approximation for a measure of central tendency is called the *midrange*. We can think of the midrange as the number in the middle of the range, or as the number midway between the end points of the range.

The midrange is found by adding the least value in the given set of data to the greatest value and dividing the sum by 2.

$$\textbf{midrange} = \frac{L + G}{2}$$

In example 1, the least value that occurs for test B is 60 and the greatest value is 70. Therefore the midrange for test B is

$$\frac{60 + 70}{2} = \frac{130}{2} = 65$$

This is also the mean and median for test B.

The midrange should only be used for a quick estimate of the central tendency of the data. It is a coincidence that the midrange for test B in example 1 is exactly the same as the mean and median for test B. For test A, the least value is 40 and the greatest value is 75; hence the midrange is

$$\frac{40 + 75}{2} = \frac{115}{2} = 57.5$$

Recall that the mean of test A is 65; thus, the midrange is 7.5 points from the mean.

EXAMPLE 2
Find the range and midrange for the following data:

72, 74, 74, 77, 82, 83, 84

Solution

$$\text{range} = 84 - 72 = 12$$

$$\text{midrange} = \frac{72 + 84}{2} = \frac{156}{2} = 78$$

Note that the range (a measure of dispersion) is 12, which means that the scores do not differ much. The midrange is 78, which tells us that 78 is a reasonable estimate for the mean.

EXAMPLE 3
Find the range, midrange, and mean for the following data:

56, 74, 74, 77, 82, 83, 100

Solution

$$\text{range} = 100 - 56 = 44$$

$$\text{midrange} = \frac{56 + 100}{2} = \frac{156}{2} = 78$$

$$\text{mean} = \frac{56 + 74 + 74 + 77 + 82 + 83 + 100}{7} = \frac{546}{7} = 78$$

The range, though easily computed, is not considered to be the best measure of dispersion since it only involves the use of two extreme values. It does not tell us anything about the remaining values in our set of data.

One measure of dispersion (variation) that considers all scores in a given set of data is the **standard deviation.** You will see later that a large standard deviation indicates that the data is scattered widely about the mean, while a small standard deviation indicates that the data is closely grouped about the mean.

In order to find the standard deviation for a given set of data, we must perform a number of tasks. First we must find the **deviation from the mean** for each value in the given set of data. As an example of this process, consider the set of scores {60, 65, 70, 70, 70}. The mean for this set of data is 67. The deviation from the mean is found by subtracting the mean from each value in the set of data.

The deviation for 70 is 70 − 67 = 3.
The deviation for 65 is 65 − 67 = −2.
The deviation for 60 is 60 − 67 = −7.

This is summarized in table 1.

TABLE 1

	Value	Deviation from the mean
	70	70 − 67 = 3
	70	70 − 67 = 3
	70	70 − 67 = 3
	65	65 − 67 = −2
	60	60 − 67 = −7
Sum	335	0

$$\text{mean} = \frac{335}{5} = 67$$

After we have found the deviation from the mean for each score, we square each of the deviations (see table 2). We need to do this in order to find a value that is useful in measuring dispersion or deviation about the mean. Note that the sum of the deviations is zero. This is true for any set of data. Therefore you can use this information to check the accuracy of your work, particularly the accuracy of your mean.

TABLE 2

	Value	Deviation	(Deviation)^2
	70	3	9
	70	3	9
	70	3	9
	65	−2	4
	60	−7	49
Sum	335	0	80

Now that we have squared each deviation, we have a set of positive values. The sum of these values in table 2 is 80. We next find the mean of these squared deviations, called the **variance.** The mean of the squared values is 80 ÷ 5 = 16. The variance also indicates dispersion, but we only mention it in passing; we need the variance in order to find the standard deviation. We find the standard deviation by finding the principal square root of the variance. Hence for our example,

$$\text{standard deviation} = \sqrt{16} = 4$$

The standard deviation for a set of data is usually represented by the lowercase Greek letter σ, called *sigma*. Therefore we can write $\sigma = 4$ for our given set of data.

Admittedly, there is a lot of work involved in finding the standard deviation, but, if it is done in an orderly and precise manner, it is not difficult.

To find the standard deviation for a given set of data, we must follow these steps:

1. **Find the mean for the given set of data.**
2. **Find the deviation from the mean for each value in the set of data.**
3. **Square each deviation.**
4. **Find the mean of the squared deviations.**
5. **Find the principal square root of this number.**

EXAMPLE 4
Find the standard deviation for the following data:

$$\{5, 7, 9, 13, 16\}$$

Solution
We follow the steps outlined previously:

1. The mean is

$$\frac{5 + 7 + 9 + 13 + 16}{5} = \frac{50}{5} = 10$$

2. The deviation from the mean for each value is:

$$5 - 10 = -5, \quad 7 - 10 = -3, \quad 9 - 10 = -1$$
$$13 - 10 = 3, \quad 16 - 10 = 6$$

3. Now we square each deviation:

$$(-5)^2 = 25, \quad (-3)^2 = 9, \quad (-1)^2 = 1, \quad (3)^2 = 9, \quad (6)^2 = 36$$

4. To find the mean of the squared deviations, we sum the squares and divide by 5 (the number of values):

$$\frac{25 + 9 + 1 + 9 + 36}{5} = \frac{80}{5} = 16$$

5. We now find the principal square root of this number: $\sqrt{16} = 4$

6. Hence

$$\text{standard deviation} = \sigma = 4$$

The fact that we obtained a standard deviation of 4, as we did in the discussion prior to this example, is pure coincidence.

EXAMPLE 5

Find the standard deviation for the given set of data:

$$\{20, 22, 26, 26, 28, 34\}$$

Solution

We combine the necessary operations in the table below.

	Value	Deviation from the mean	$(\text{Deviation})^2$
	34	$34 - 26 = 8$	64
	28	$28 - 26 = 2$	4
	26	$26 - 26 = 0$	0
	26	$26 - 26 = 0$	0
	22	$22 - 26 = -4$	16
	20	$20 - 26 = -6$	36
Sum	156	0	120

$$\text{mean} = \frac{156}{6} = 26$$

$$\text{mean of squared deviations} = \frac{120}{6} = 20 \qquad \text{(variance)}$$

$$\text{standard deviation} = \sigma = \sqrt{20} \approx 4.5$$

The standard deviation for the set of data given in example 5 is approximately 4.5. We approximated $\sqrt{20}$ by means of table II in the appendix.

EXAMPLE 6
Find the standard deviation for the following test scores:

$$\{72, 74, 74, 77, 82, 83, 84\}$$

Solution
We combine the necessary operations in the table below.

	Score	Deviation	(Deviation)^2
	84	6	36
	83	5	25
	82	4	16
	77	−1	1
	74	−4	16
	74	−4	16
	72	−6	36
Sum	546	0	146

$$\text{mean} = \frac{546}{7} = 78$$

$$\text{mean of squared deviations} = \frac{146}{7} = 20.86 \qquad \text{(variance)}$$

$$\text{standard deviation} = \sigma = \sqrt{20.86} \approx 4.6$$

Note that in example 6 the mean of the squared deviations did not come out evenly; that is, $\frac{146}{7}$ was rounded off to 20.86, and then we approximated the square root of 20.86 by 4.6, using table II in the Appendix.

A large standard deviation indicates that the data is scattered widely about the mean, while a small standard deviation indicates that the data is closely grouped about the mean. As an example, consider the following two sets of data:

$A = \{4, 5, 6, 7, 8\}$	$B = \{2, 4, 6, 8, 10\}$
The mean for set A is 6 and the standard deviation is $\sqrt{2} \approx 1.41$	The mean for set B is 6 and the standard deviation is $\sqrt{8} \approx 2.82$

Now that we are able to calculate the standard deviation, what can we do with it? Suppose that Tony scored 78 on a mathematics exam. The mean score for this exam was 73 and the standard deviation was 5. What does Tony know? First, he scored above the mean, that is, he obtained a better than average score on his exam. Secondly, he scored one standard deviation above the mean. On the next exam in his mathematics class, Tony scored 47. Did he do better or worse on the second test? In order to answer this, we must know the mean and the standard deviation. Suppose the second exam had a mean of 41 and a standard deviation of 3. Therefore, Tony again scored above the mean and, in fact, he scored 2 standard deviations above the mean $(41 + (2 \times 3) = 47)$, which indicates that he performed better on the second exam than he did on the first.

EXAMPLE 7

Sue and Jim took their midterm exams in their respective statistics courses. Sue is in Mr. Data's class, where she scored 79. In Mr. Data's class the mean was 75 and the standard deviation was 4. Jim is in Ms. Mode's class, where he also scored 79. But in Ms. Mode's class the mean was 73 and the standard deviation was 3. Who performed better in their respective class?

Solution

Sue's score was 1 standard deviation above the mean $(75 + 4 = 79)$, while Jim's score was 2 standard deviations above the mean for his class $(73 + (2 \times 3) = 79)$. Jim performed better compared to his classmates in Ms. Mode's class than did Sue compared to her classmates in Mr. Data's class.

EXAMPLE 8

Frank, Louie, Janie, and Irene are all in the same history class. Their scores for the first two exams in their history class are listed below.

	Exam 1	Exam 2
Frank	78	66
Louie	66	66
Janie	90	70
Irene	72	70

The first exam had a mean of 78 and a standard deviation of 6, while the second exam had a mean of 66 and a standard deviation of 4. Which of the 4 people improved on the second exam, which did worse, and which performed the same, in relation to the rest of the class?

Solution

On each of the exams, Frank scored the mean score (78 on exam 1 and 66 on exam 2), so he performed the same on both exams. Louie scored 66 on the first exam, two standard deviations below the mean ($78 - (2 \times 6) = 66$). On the second exam, he also scored 66, but that was the mean for exam 2; hence Louie improved. Janie scored two standard deviations above the mean on the first exam ($78 + (2 \times 6) = 90$), while on the second exam she only scored one standard deviation above the mean ($66 + (1 \times 4) = 70$). Hence Janie did poorer on the second exam compared to the first. On the first exam, Irene scored one standard deviation below the mean ($78 - (1 \times 6) = 72$), while on the second exam, she scored one standard deviation above the mean ($66 + (1 \times 4) = 70$). Therefore Irene's performance improved on the second exam.

EXERCISES FOR SECTION 4.3

In exercises 1–10, find the standard deviation for each set of data. (Round off any decimal answer to the nearest tenth.)

1. {8, 10, 10, 16}
2. {2, 4, 6, 8}
3. {16, 13, 13, 12, 9, 9}
4. {13, 11, 11, 9, 9, 7}
5. {38, 44, 46, 48, 48, 48, 50}
6. {20, 22, 23, 23, 24, 24, 25}
7. {1, 2, 7, 10, 15}
8. {11, 13, 17, 19}
9. {68, 68, 70, 72, 66, 73, 78, 65, 80, 70}
10. {82, 80, 81, 87, 86, 86, 88, 84, 82, 84}
11. A sample of 10 bowlers was taken in a tournament. The number of strikes recorded for each bowler in his or her first game was as follows: {2, 3, 4, 5, 5, 6, 3, 2, 7, 3}. For this set of data, find the following:
    a. mean
    b. median
    c. mode
    d. range
    e. midrange
    f. standard deviation
12. On a physics lab quiz, the following scores were made in a class of 10 students: 80, 75, 88, 95, 90, 72, 67, 73, 78, 82. For this set of scores, find the following:
    a. mean
    b. median
    c. mode
    d. range
    e. midrange
    f. standard deviation
13. The back nine holes of the Southampton Golf Club have the following lengths in yards:

150	370	310
340	200	450
490	420	375

    For this set of measurements, find the following:
    a. mean
    b. median
    c. mode
    d. range
    e. midrange
    f. standard deviation
14. Hugh, Norma, Ruth, Joe, and Doris are all in the same statistics class. Their scores for the first two exams in their statistics class are listed

below. The first exam had a mean of 84 and a standard deviation of 6, while the second exam had a mean of 78 and a standard deviation of 4.

	Exam 1	Exam 2
Hugh	84	78
Norma	90	74
Ruth	66	78
Joe	78	70
Doris	84	78

a. Who improved on the second exam?
b. Who did the poorest on the second exam?
c. Who performed the same on both exams?

15. Two history exams were given. On the first exam, the mean was 74 and the standard deviation was 8. On the second exam, the mean was 48 and the standard deviation was 10. The scores of 6 students who took the exams are listed below.

	Exam 1	Exam 2
Eric	82	48
Maria	78	53
Rudy	70	58
Maureen	58	58
Jeff	74	53
Mark	90	78

a. Who improved on the second exam?
b. Who improved the most on the second exam?
c. Who did not improve on the second exam?
d. Considering both exams, which student did the poorest?
e. Who performed the same on both exams?
f. Considering both exams, which student has the best grade so far?

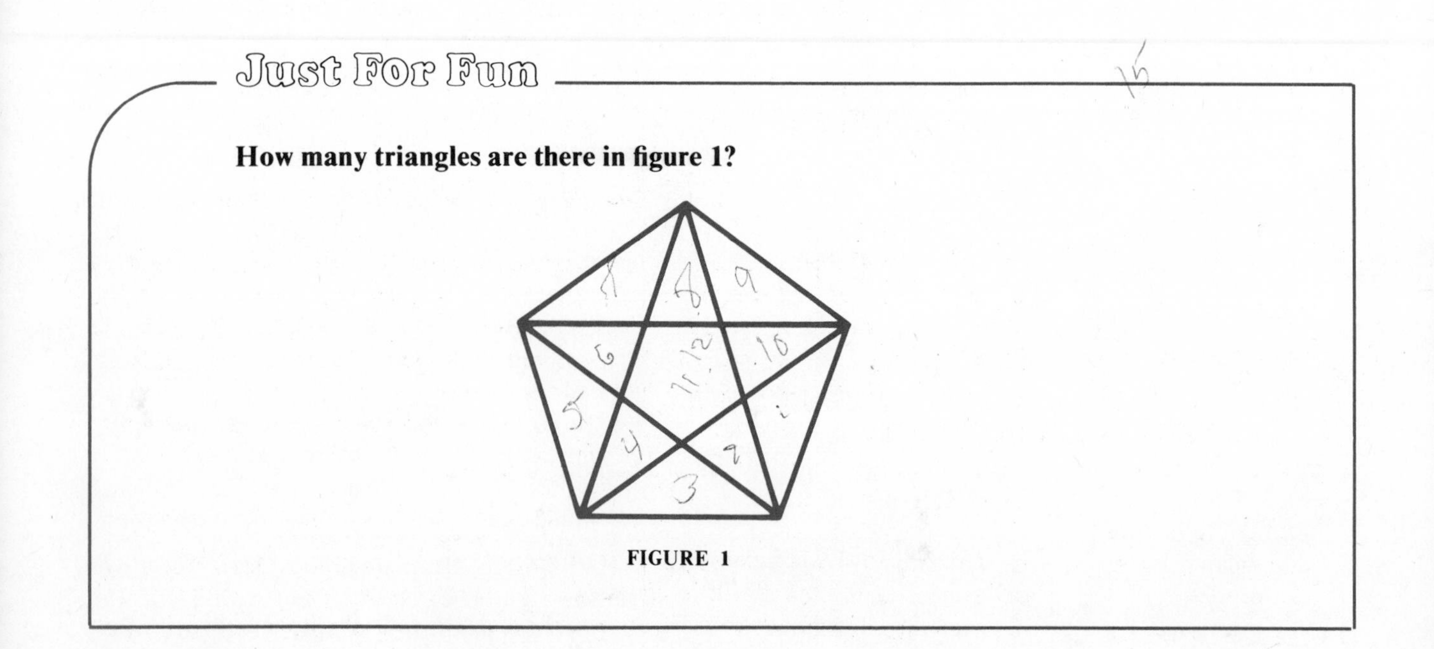

Just For Fun

How many triangles are there in figure 1?

FIGURE 1

4.4 MEASURES OF POSITION (Percentiles)

We have examined measures of central tendency—those values that indicate where the data tends to center or cluster. We have also examined measures of dispersion—those values that tell us about the spread of the data. We can also determine where a datum is located relative to the mean in terms of the standard deviation.

But suppose we want to locate a particular datum in relation to the whole set of data or to another datum? In order to do this, we must find the position of the datum relative to the other data, comparing it with the other data to obtain a clearer picture of its value.

Consider the case of Sue and Dale, who are applying for a job: both people have to fill out an application form. On the form, there is a question which asks for the applicant's high school class rank. Sue's rank in her class was 30th, while Dale's class rank was 10th. That is, Sue was 30th in her class, while Dale was 10th in his class. At first glance it seems that Dale was the better student and probably should get the job. But this is not necessarily the case, since Sue and Dale went to different high schools. We do not have enough information for a fair comparison.

Suppose Sue's rank was 30th in a class of 300, while Dale's rank was 10th in a class of 50. This points out that rank is useless as a measure of position unless we know how many are in the total group. It would appear that Sue's rank is similar to Dale's. In order better to describe their relative positions, we can find the **percentile rank,** or **percentile,** for Sue and Dale.

In Sue's class, there were $300 - 30 = 270$ people ranked below her. Therefore

$$\frac{270}{300} = \frac{27}{30} = \frac{9}{10} = 0.90 = 90\%$$

of Sue's classmates were ranked lower than she was. We can say that Sue was in the 90th percentile. Again, this means that 90% of her classmates were ranked below her.

Computing Dale's percentile rank, we find that $50 - 10 = 40$ people rank below him. Hence

$$\frac{40}{50} = \frac{4}{5} = 0.80 = 80\%$$

of Dale's classmates were ranked lower than he was. Dale was in the 80th percentile in his class.

Percentiles divide sets of data into 100 equal parts; hence 100%

is the basis of measure. Practically every student has encountered a percentile rank at one time or another. Every graduating senior has a percentile rank in his or her graduating class. Standardized achievement tests, intelligence tests, perception tests, etc., give results in terms of percentiles. When a student is told that he scored at the 77th percentile on an exam, he then knows that he scored better than 77% of those that took the exam, or that 77% of those taking the exam scored lower than he did.

EXAMPLE 1
In a class of 100 students, Bob has a rank of 12th. What is Bob's percentile rank in the class?

Solution
There are 100 − 12 = 88 students ranked below Bob. Hence, we compare 88 to 100: $\frac{88}{100} = 0.88 = 88\%$. Bob's percentile rank is 88; he is in the 88th percentile.

EXAMPLE 2
In a class of 120 students, Pat has a rank of 30th. What is her percentile rank in the class?

Solution
There are 120 − 30 = 90 students ranked below Pat. Therefore, we compare 90 to 120: $\frac{90}{120} = \frac{9}{12} = \frac{3}{4} = 0.75 = 75\%$. Pat is in the 75th percentile.

EXAMPLE 3
In a class of 120 students, Sam has a rank of 32nd. What is Sam's percentile rank?

Solution

$$120 - 32 = 88, \qquad \frac{88}{120} = \frac{22}{30} = \frac{11}{15} = 0.73 = 73\%$$

Sam's percentile rank is 73.

EXAMPLE 4
In a statistics class of 40 students, an exam was given and Pam scored at the 75th percentile. How many students scored lower than Pam?

Solution
Since Pam scored at the 75th percentile, 75% of the students scored less than she did.

$$75\% = 0.75, \qquad 0.75 \times 40 = 30$$

Therefore 30 students scored lower than Pam on the exam.

EXAMPLE 5
In a class of 30 students, Larry has a percentile rank of 70. What is Larry's rank in the class?

Solution
Larry's percentile rank indicates that 70% of the students have scores less than Larry's on the tests given to the class.

$$70\% \text{ of } 30 = 0.70 \times 30 = 21$$

Twenty-one students have scored lower than Larry. Hence, Larry's rank in the class is 9th, because $30 - 21 = 9$.

In order to determine at what percentile a student scored on an exam, we need to find the number of students that scored below the individual and divide that number by the total number of students in the class.

EXAMPLE 6
The scores of 10 students on a math quiz were

4, 10, 12, 14, 16, 18, 20, 20, 22, 22

Tracy's score on the quiz was 16. What is her percentile rank?

Solution
Since Tracy's score was 16, there were 4 scores lower than hers. We divide 4 by 10, the total number of students that took the quiz: $4 \div 10 = 0.40 = 40\%$. Therefore, Tracy's percentile rank is 40; she is in the 40th percentile.

EXAMPLE 7
The scores of 10 students on a history quiz were

8, 11, 13, 15, 17, 19, 20, 21, 24, 25

Mike's score on the quiz was 21.

a. What is the rank (from the top) for Mike's score?

b. What is Mike's percentile rank?

70

Solution

a. There were 2 scores higher than Mike's, 24 and 25. His score of 21 was third.

b. Since Mike's score was third, there were 7 scores lower than his score of 21. We divide 7 by 10: $7 \div 10 = 0.70 = 70\%$. Hence Mike is in the 70th percentile.

The 25th percentile, 50th percentile, and 75th percentile are probably the most commonly used percentiles in educational testing. They are unique in that they divide sets of data into fourths, or quarters; hence they are referred to as **quartiles.** The **first quartile** for a set of data is that value which has 25%, or one-fourth, of the data (scores) below it. The **second quartile** is that value which has 50%, or one-half, of the data (scores) below it. The **third quartile** is that value which has 75%, or three-fourths, of the data (scores) below it.

EXAMPLE 8

Given the following set of scores: {60, 70, 72, 73, 73, 80, 82, 84, 84, 85, 87, 88}. Find—

a. the first quartile

b. the median

c. the third quartile.

Solution

a. The first quartile is that value which has 25% of the scores below it. There are 12 scores; 25% (one-fourth) of 12 is 3. So the value for the first quartile is the fourth score (3 scores will lie below it), which is 73.

b. The median is that measure that determines the middle of a given set of data. There are 12 scores (an even number), so we must find the mean of the sixth and seventh scores.

$$\frac{80 + 82}{2} = 81$$

The median is 81.

c. The third quartile is that value which has 75% of the scores below it; 75% (three-fourths) of 12 is 9. The value for the third quartile is the tenth score (9 scores lie below it), which is 85.

EXAMPLE 9

Given the following set of data: {30, 35, 36, 37, 38, 39, 40, 41}. Find—

a. the first quartile

b. the third quartile.

Solution

a. There are 8 data; $\frac{1}{4}$ of 8 = 2. Hence, the third piece of data from the bottom, 36, is the first quartile, or 25th percentile.

b. Three-fourths of 8 is 6. Therefore, the seventh piece of data from the bottom, 40, is the third quartile, or 75th percentile.

You should be aware that a datum or score has no significance or meaning by itself. If Helen reported that she scored 92 on her last exam in history, many people would jump to the conclusion that she did well on the exam. But that might not be the case. If Helen got 92 points out of a possible 200, then the score has more meaning, and we would conclude that Helen did not do well on the exam. On the other hand, if Helen got 92 points out of a possible 100, then we could conclude that she did do well on the exam. We could also better judge Helen's performance on the exam if we knew her score's location relative to the other students' scores on the exam. If, for instance, Helen reported that her score was at the 90th percentile, then we know that 90% of the scores are below Helen's.

EXERCISES FOR SECTION 4.4

1. In a class of 200 students, Julia has a rank of 12th. What is Julia's percentile rank in the class?

2. In a senior class of 300 students, Erin has a rank of 30th. What is Erin's percentile rank in the class?

3. In a statistics class of 30 students, an exam was given and Eddie scored at the 80th percentile. How many students scored lower than Eddie?

4. In Hugh's mathematics class, there are 32 students including Hugh. On the last exam, Hugh's score was at the third quartile. How many students scored lower than Hugh?

5. In a class of 40 students, Don has a percentile rank of 80. What is Don's rank in the class?

6. Doris is ranked at the first quartile in her senior class of 300 students. What is her rank in the class?

7. Jessie is ranked at the third quartile in her economics class of 60 students. What is her rank in the class?

8. The following data represents the heights in inches of the starting five for the Dunkem Basketball team: 73, 78, 80, 82, 85.
   a. What is the rank (from the top) of the height of 82 inches?
   b. What is the percentile rank of the height of 82 inches?

Just For Fun

Almost every word that begins with a *q* has a *u* as the second letter. Can you name two *q* words that do not have *u* as the second letter? Can you name one?

By Charles M. Schulz

© 1974 United Feature Syndicate, Inc.

9. Given the 12 scores 62, 72, 74, 75, 75, 82, 85, 86, 86, 87, 89, and 90.
   a. What is the rank (from the top) of a score of 87?
   b. What is the percentile rank of a score of 87?
   c. What is the percentile rank of a score of 89?
   d. What score is at the first quartile?

10. Given the 20 scores 62, 64, 66, 69, 70, 72, 74, 75, 80, 82, 83, 86, 87, 88, 90, 91, 92, 94, 97, and 98.
    a. What is the rank (from the top) of 87?
    b. What is the percentile rank of a score of 87?
    c. What score is at the first quartile?
    d. What score is at the third quartile?
    e. What score is at the 90th percentile?
    f. What score is at the 80th percentile?

11. Roberto is ranked 75th in his senior class of 200 students. Larry is in the same senior class and Larry has a percentile rank of 75. Of the two seniors, who has the higher standing in the class?

12. Sarah is ranked 8th in her mathematics class of 35 students. Helen is in the same mathematics class, and Helen has a percentile rank of 85. Who ranks higher, Sarah or Helen?

13. Daniel is ranked at the third quartile in his English class of 60 students. Peter is in the same English class, and he is ranked 13th. Who ranks higher, Daniel or Peter?

14. In a class of 50 students, Jay has a rank of 50th. What is his percentile rank?

15. The following is a list (distribution) of first round scores in a golf tournament for 40 golfers. In golf the lowest score wins; therefore, we list the lowest score first and the highest score last.

Score	Frequency (number of golfers making that score)
65	2
66	1
67	2
68	5
69	2
70	4
71	6
72	8
73	3
74	7

a. What is the rank (from the top) of a score of 66?

b. What is the percentile rank of a score of 66?

c. What score is at the third quartile?

d. What score is the 95th percentile?

e. What is the percentile rank of Mike Mulligan, if he shot a 72?

f. If even par is 72, what percentage of the golfers scored below par?

4.5 PICTURES OF DATA

Pictures are often used to transmit or summarize information. It has been said that a picture is worth a thousand words. It is true that most of us can gain information more easily from pictures than from written material. Architects and contractors use drawings (blueprints) to exchange information. Large corporations use various types of graphs in their reports to stockholders to show how the company has performed. Newspapers and magazines use graphs to show the reader some types of information, usually the presentation of data.

Graphs are used in mathematics to show relationships between sets of numbers. Graphs are useful in the field of statistics because they can show the relationships in a set of data. In this section we shall examine the most common types of graphs used in statistics: the *bar graph,* the *divided bar graph,* the *circle* or *pie graph,* the *histogram,* and the *frequency polygon* or *line graph.*

A **bar graph** consists of a series of bars of uniform width, with some form of measure on a vertical or horizontal axis. Figure 2 is an example of a typical bar graph.

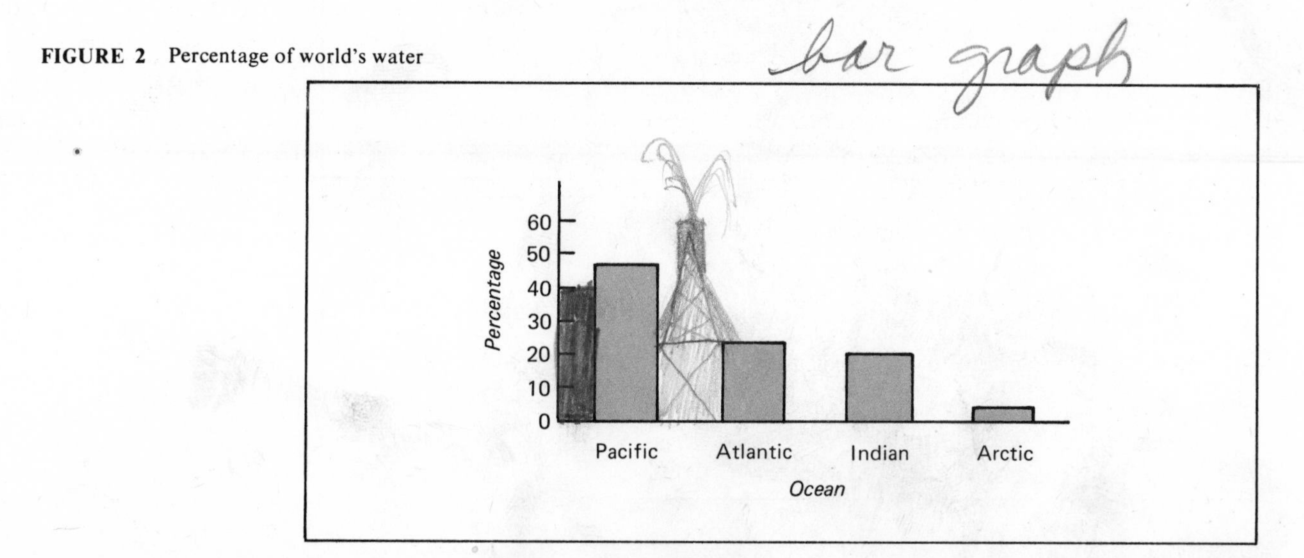

FIGURE 2 Percentage of world's water

EXAMPLE 1
Using figure 2, answer the following:

a. What percentage of the world's water does the Indian Ocean contain?

b. Approximately what percentage of the world's water does the Pacific Ocean contain?

Solution

a. We locate the top of the bar representing the Indian Ocean and then read across to the corresponding percentage. The Indian Ocean contains 20% of the world's water.

b. Approximately 46%.

The **circle** or **pie graph** is another type of graph that is used quite often. It is particularly useful in illustrating how a whole quantity is divided into parts. Budgets are often illustrated in this manner. Circle graphs used by government agencies often show how the tax dollar is spent or how the tax dollar is collected. Figure 3 is an example of a typical circle graph.

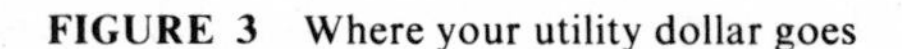

FIGURE 3 Where your utility dollar goes

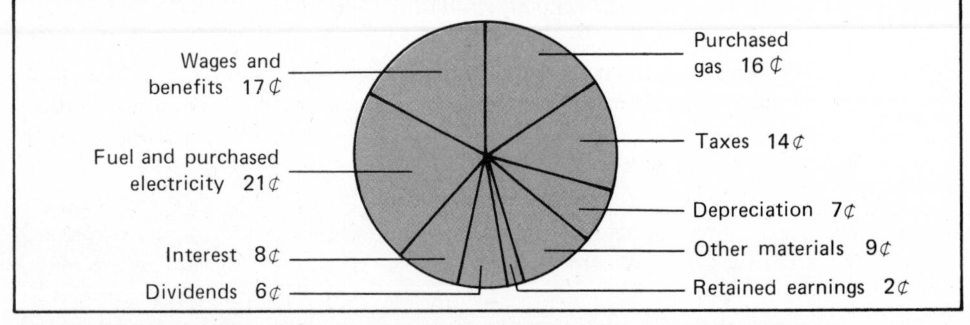

EXAMPLE 2
Using figure 3, answer the following:

a. How much of every utility dollar is used for wages and benefits?

b. How much of every utility dollar is used for taxes?

Solution

a. 17¢ b. 14¢

When we want to show how a whole quantity is divided into parts, we can use a circle graph. Consider the case of Sam Student. On any given weekday, Sam spends 8 hours sleeping, 6 hours in school, 4 hours doing homework, and 3 hours eating and doing odd jobs. He uses the remaining 3 hours for recreation, leisure, and miscellaneous. We can illustrate how Sam spends his time for a 24-hour period by means of a circle graph.

First recall that a circle contains 360 degrees. In order to illustrate the given data, we must find the percentage of a 24-hour day spent in each activity and then multiply the percentages by 360 degrees, so that we can divide the circle into proportional parts.

Sleeping:	8 hours	$\frac{8}{24} = \frac{1}{3} = 33\frac{1}{3}\%$;	$33\frac{1}{3}\%$ of $360° = 120°$
School:	6 hours	$\frac{6}{24} = \frac{1}{4} = 25\%$;	25% of $360° = 90°$
Homework:	4 hours	$\frac{4}{24} = \frac{1}{6} = 16\frac{2}{3}\%$;	$16\frac{2}{3}\%$ of $360° = 60°$
Meals, chores:	3 hours	$\frac{3}{24} = \frac{1}{8} = 12\frac{1}{2}\%$;	$12\frac{1}{2}\%$ of $360° = 45°$
Miscellaneous:	3 hours	$\frac{3}{24} = \frac{1}{8} = 12\frac{1}{2}\%$;	$12\frac{1}{2}\%$ of $360° = 45°$
	24 hours	100%	$360°$

We then use a protractor to construct the corresponding angles in a circle. Figure 4 shows the completed circle graph. It does not matter how large or how small the circle is, since every circle represents 100% of the data and can be divided into proportional parts.

Another way to illustrate how Sam Student spends his time during a 24-hour period is to use a bar graph sometimes called a **divided bar graph** or a **distribution bar graph.** In order to show Sam's day, we choose an arbitrary length to represent 100 percent of the day

FIGURE 4

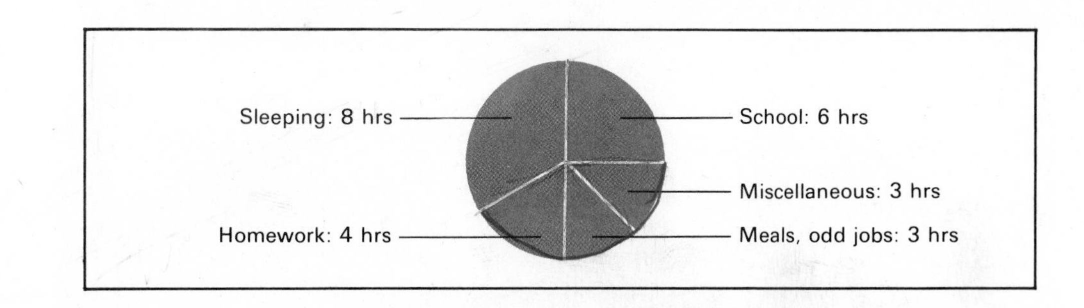

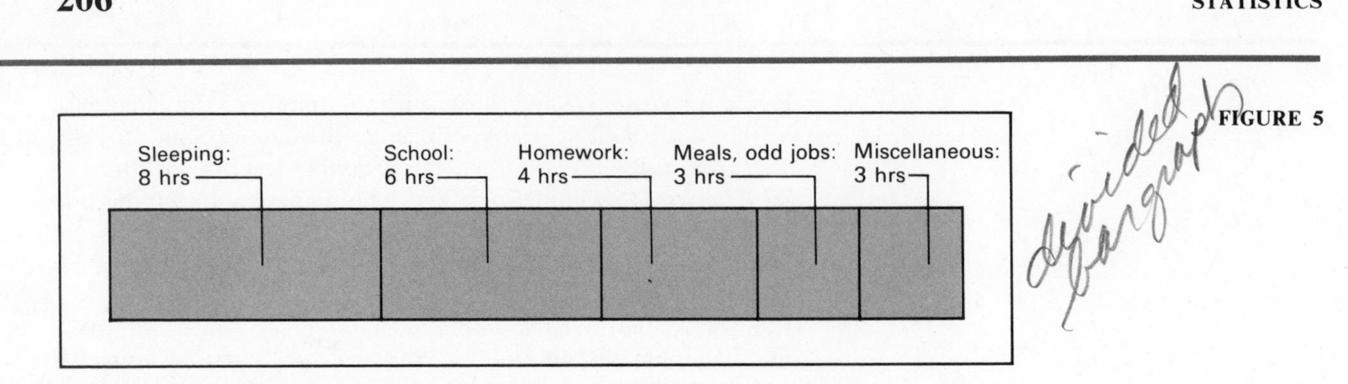

FIGURE 5

(24 hours), and then divide the length into parts proportional to the given percentages. See figure 5.

Circle graphs and divided bar graphs are used in showing how a whole quantity is divided into parts, but they are not so useful when we wish to picture other types of data. Consider the following set of 35 scores for an exam in a statistics class:

42, 48, 43, 47, 46, 45, 42
49, 50, 42, 44, 48, 47, 42
42, 48, 50, 45, 46, 42, 49
45, 44, 48, 49, 45, 47, 46
46, 46, 44, 46, 49, 50, 45

Examining these scores does not give us much information because the scores have been presented in the order of their occurrence. We probably could find the highest and lowest score without too much trouble, but we can get a better understanding of the scores if we do the following: first, list the scores in order from highest to lowest, and then count how many of each score we have. The number of times a score occurs is the **frequency** of the score, and table 3 is called a **frequency distribution.**

TABLE 3

Score	Tally	Frequency
50	///	3
49	////	4
48	////	4
47	///	3
46	~~////~~/	6
45	~~////~~	5
44	///	3
43	/	1
42	~~////~~/	6

TABLE 4

Score	Tally	Frequency
48–50	~~||||~~ ~~||||~~ /	11
45–47	~~||||~~ ~~||||~~ ////	14
42–44	~~||||~~ ~~||||~~	10

The frequency distribution or frequency table in table 3 lists all of the different scores and the frequency with which each score occurs. This frequency distribution can be illustrated by means of a **histogram.** A histogram consists of series of bars that are drawn all with the same size widths on the horizontal axis, and uniform units on the vertical axis. The frequencies are shown on the vertical axis. See figure 6.

It is important that each bar in a histogram should be drawn in proportion to the frequency of values that occur. Note that both the horizontal and vertical scales have been labeled completely. If a frequency distribution contains too many different scores, so that the resulting histogram is unwieldy, then we can reduce the number of categories by creating classes, that is, intervals that contain more than one score. Table 3 contains nine categories. We can reduce this number by creating categories that contain more than one score. For example, table 4 contains three categories for the same data we used in table 3. Note that each class interval is the same width, and each score belongs in exactly one interval.

A **frequency polygon,** sometimes called a **line graph,** can also be used to graph a frequency distribution. It is constructed in much the same manner as a histogram, using the same kind of vertical and

FIGURE 6

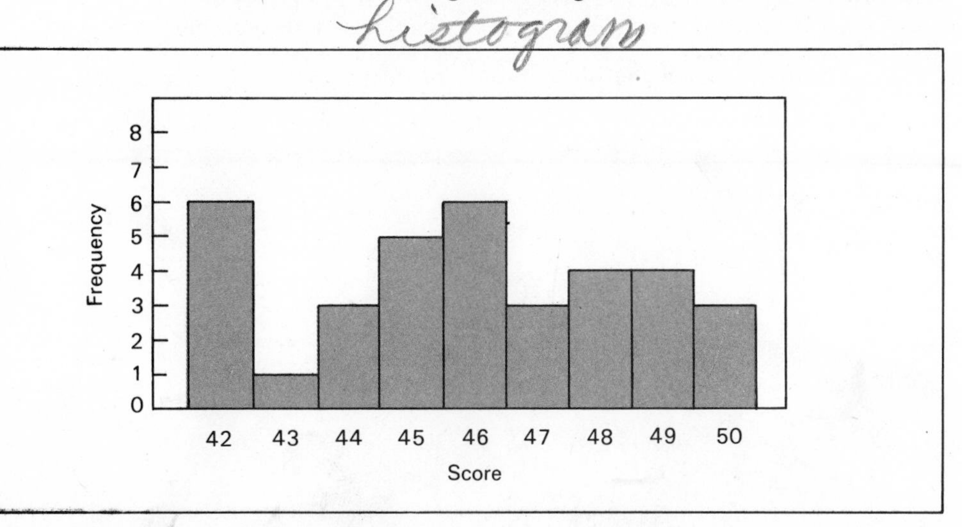

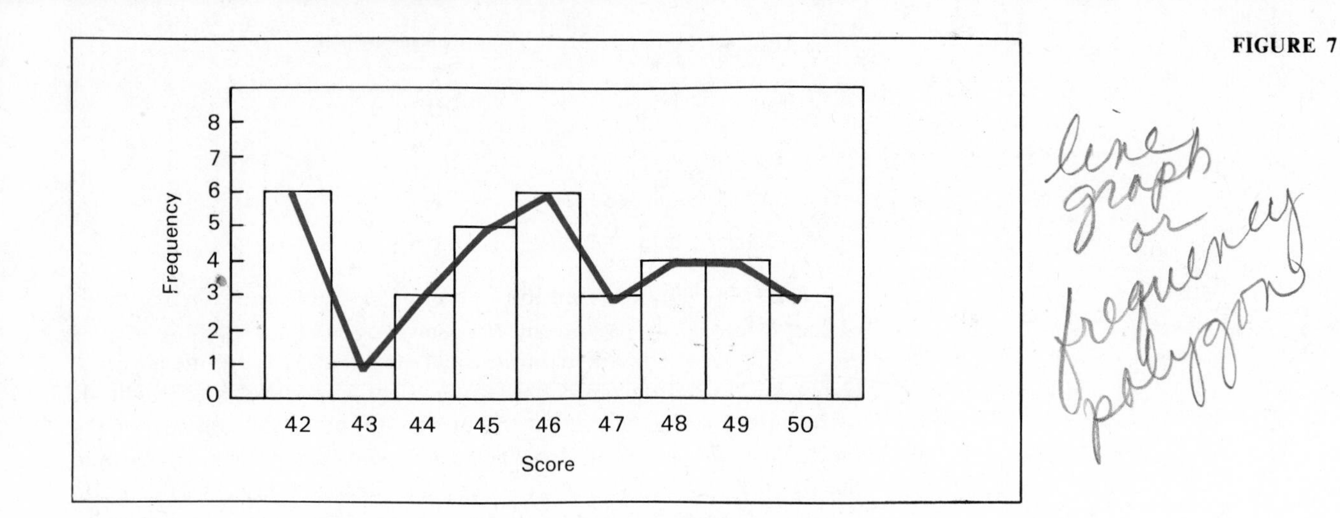

FIGURE 7

horizontal scales. If we connect the midpoints of the top of each bar in a histogram, then the resulting line graph, shown in figure 7, is a frequency polygon. Figure 7 represents the same data as figure 6.

EXAMPLE 3

A survey of 30 customers was made at Dan's Donut Shop. Below is the set of 30 breakfast checks, in cents, for each customer. Construct a histogram for this data.

92, 93, 90, 91, 84, 85, 86, 87, 89, 88
93, 93, 84, 85, 85, 92, 92, 91, 88, 87
86, 92, 92, 86, 90, 86, 88, 88, 87, 86

Solution

First we construct a frequency distribution for the given data.

Amount of check	Tally	Frequency
93	///	3
92	~~////~~	5
91	//	2
90	//	2
89	/	1
88	////	4
87	///	3
86	~~////~~	5
85	///	3
84	//	2

FIGURE 8

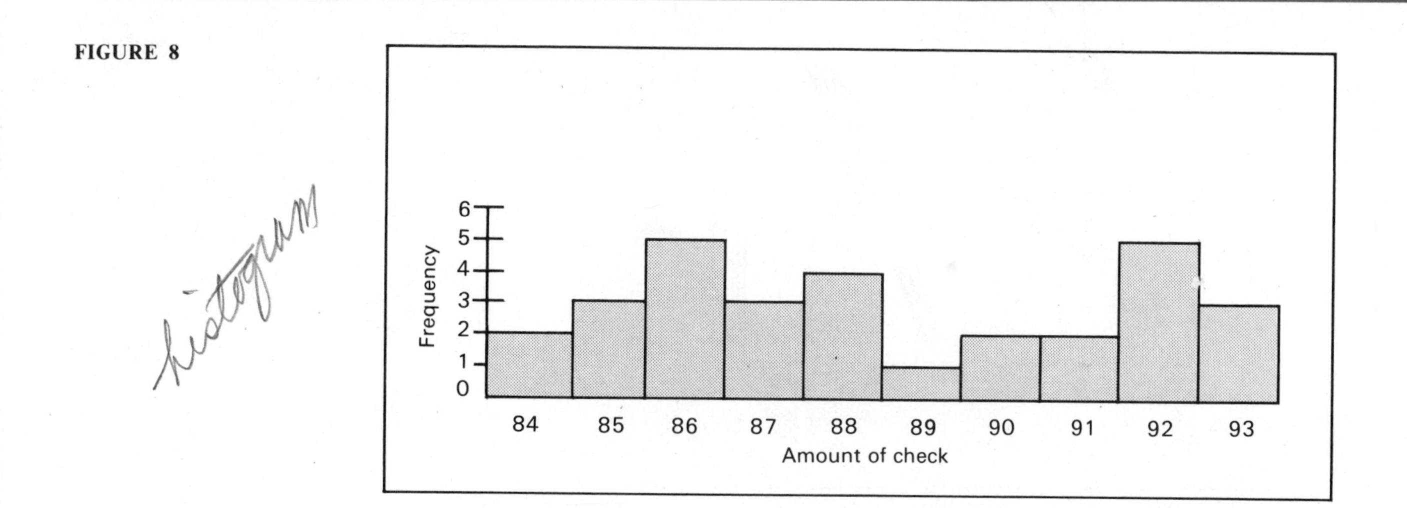

Next, we construct a histogram, making sure that the frequencies are shown on the vertical axis in uniform intervals, and that all of the bars have the same width. The result is shown in figure 8.

EXAMPLE 4

A student rolled a die 30 times and obtained the results shown below. Construct a frequency polygon for this data.

6, 5, 4, 4, 5, 6, 1, 2, 1, 6, 4, 3, 3, 3, 4,
2, 2, 5, 6, 4, 1, 2, 4, 3, 5, 5, 3, 3, 4, 2

Solution

First we construct a frequency distribution for the given data.

Number	Tally	Frequency
6	////	4
5	卌	5
4	卌 //	7
3	卌 /	6
2	卌	5
1	///	3

To aid us in drawing the frequency polygon, we construct a histogram for this data and then connect the midpoints of the tops of each bar in the histogram. The resulting line graph is the frequency polygon shown in figure 9.

FIGURE 9

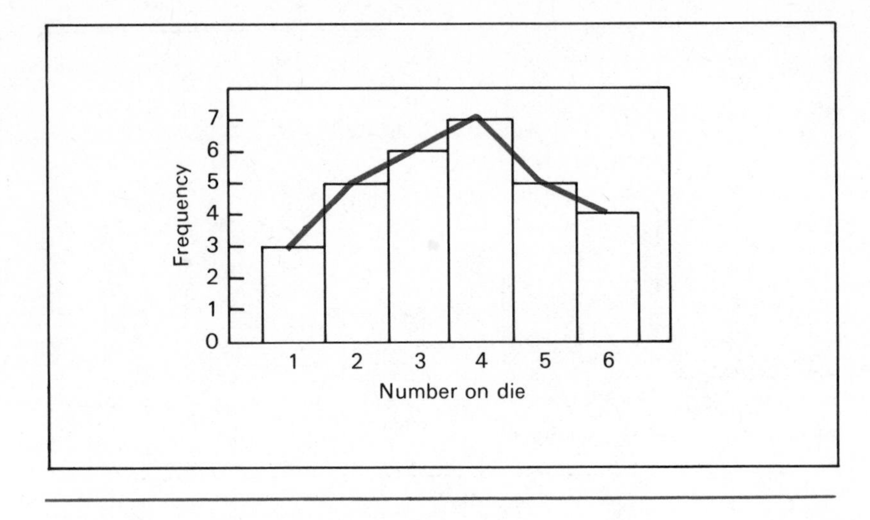

EXERCISES FOR SECTION 4.5

1. In a survey of 180 students, it was determined that 90 students came to school by bus, 60 students came to school by car, and 30 students walked to school. Construct a circle graph representing this information.
2. Construct a divided bar graph representing the information given in exercise 1.
3. The Garcia family has a budget. Each month they use their income in the following manner: 30% for food, 25% for household expenses, 20% for transportation, 10% for savings, 5% for entertainment, and 10% for unexpected expenses. Construct a circle graph representing this information.
4. Construct a divided bar graph representing the information in exercise 3.
5. The Land of Taxes obtains its money from taxes. Each dollar that the Land of Taxes collects is obtained in the following manner: 25¢ comes from personal income taxes, 25¢ comes from corporate income taxes, 15¢ comes from excise taxes, 20¢ comes from sales taxes, 10¢ comes from highway taxes, and 5¢ comes from miscellaneous taxes. Construct a circle graph representing this information.
6. The following table indicates the approximate land area of the 7 continents. Construct a circle graph (to the nearest degree) representing this information.

Continent	Area in square miles	Percentage of world's land
Asia	16,988,000	29.5
Africa	11,506,000	20.0
North America	9,390,000	16.3
South America	6,795,000	11.8
Europe	3,745,000	6.5
Australia	2,975,000	5.2
Antarctica	5,500,000	9.6

7. A student rolled a die 40 times; the results are shown below. Construct a frequency distribution and a histogram to represent this data.

   6, 4, 3, 5, 6, 2, 4, 4, 5, 2, 1, 5, 2, 4, 1, 6, 6, 6, 1, 5
   2, 6, 3, 1, 6, 1, 6, 2, 1, 2, 2, 4, 4, 6, 4, 1, 3, 4, 5, 5

8. The heights of 28 students in a mathematics class were recorded, as shown below. Construct a frequency distribution and a histogram to represent this data.

65, 69, 64, 65, 70, 71, 75, 60, 60, 61, 68, 70, 67, 69
67, 70, 67, 66, 67, 71, 70, 72, 70, 66, 68, 70, 72, 64

9. The following test scores were received by 33 students in a statistics class.

56, 91, 85, 66, 72, 81, 60, 90, 70, 71, 77
84, 75, 58, 89, 67, 98, 96, 70, 87, 74, 64
64, 59, 87, 73, 91, 63, 86, 81, 72, 72, 73

a. Construct a frequency distribution for these scores using the intervals 95–99, 90–94, 85–89, etc.
b. Use the frequency distribution from part *a* to construct a histogram that represents this data.

10. A student tossed 3 coins together 32 times and after each toss recorded the number of heads. The table below represents the results. Construct a frequency polygon that represents this data.

Number of heads	0	1	2	3
Frequency	4	12	12	4

11. A survey of 40 students was made in the cafeteria to determine the cost (in cents) of each student's lunch, as shown below. Construct a frequency distribution and a frequency polygon for this data.

92, 93, 86, 93, 93, 92, 90, 84, 92, 91
85, 86, 84, 85, 90, 86, 92, 86, 87, 92
87, 92, 88, 89, 91, 88, 87, 88, 88, 86
93, 84, 86, 84, 85, 86, 85, 92, 92, 88

12. The weights (in pounds) of 50 elementary school students are given below.

62, 43, 62, 90, 84, 78, 46, 53, 44, 92
65, 73, 61, 66, 76, 53, 58, 87, 83, 71
94, 87, 83, 71, 96, 64, 58, 77, 76, 58
85, 74, 68, 63, 47, 68, 86, 75, 77, 71
90, 42, 84, 84, 53, 58, 84, 62, 68, 74

a. Construct a frequency distribution for these weights using the intervals 95–99, 90–94, 85–89, etc.
b. Using the frequency distribution from part *a*, construct a frequency polygon that represents this data.

13. Given the following distribution:

Scores	56–60	61–65	66–70	71–75	76–80	81–85	86–90	91–95	96–100
Frequency	2	4	6	7	6	10	6	5	4

a. Construct a histogram and a frequency polygon for this data.
b. What do you think a "good guess" would be for the mode?
c. What do you think a "good guess" would be for the median?

14. The circle graph in figure 10 is titled "Where Your Tax Dollar Goes." Does the graph give enough information? Why or why not?

FIGURE 10

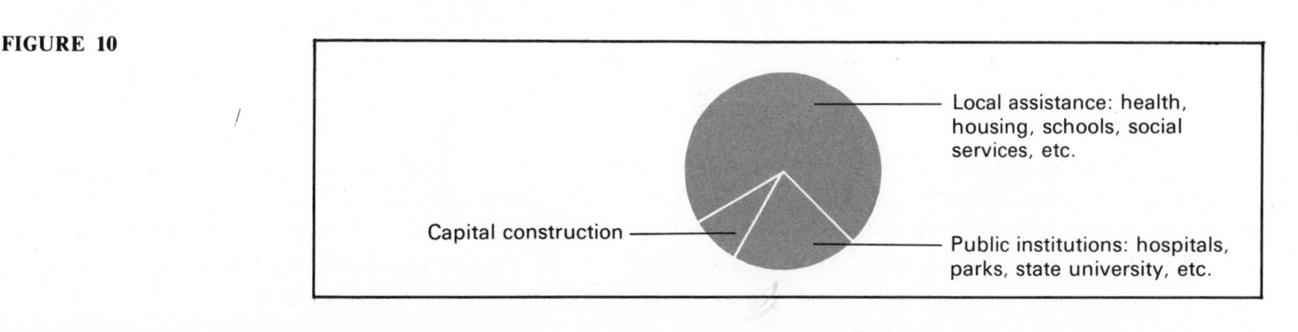

15. What can you deduce from the statement "Statistics show that more men than women are involved in auto accidents"?

16. Toss a die 30 times and tabulate the results.
    a. Construct a frequency distribution showing the results.
    b. Construct a histogram that represents the results.
    c. Construct a frequency polygon that represents the results.

17. Toss 3 coins together 32 times and tabulate the number of heads appearing on each toss, i.e., 0, 1, 2, or 3 heads.
    a. Construct a frequency distribution showing the results.
    b. Construct a frequency polygon that represents the results.
    c. Construct a percentage distribution for part *a*.
    d. Compare the percentages in *c* with the theoretical percentages: 0 heads—12.5 percent, 1 head—37.5 percent, 2 heads—37.5 percent, 3 heads—12.5 percent.
    e. Toss the three coins 32 more times and tabulate the results with those of part *a*. Construct a frequency polygon for the 64 tosses.

Just For Fun

Can you make 10 by using just 9 matches?

4.6 THE NORMAL CURVE

Two students, Scott and Joe, decided to toss 4 coins 100 times, record the number of heads appearing for each toss, and then construct a frequency polygon representing the outcomes. Their results are shown below.

1, 2, 2, 4, 2, 2, 1, 1, 0, 3, 1, 0, 2, 2, 3, 1, 4, 2, 2, 3
2, 1, 1, 4, 2, 2, 2, 1, 2, 4, 2, 3, 2, 2, 2, 1, 2, 2, 3, 2
2, 0, 3, 3, 2, 1, 1, 3, 2, 1, 4, 0, 2, 3, 2, 1, 2, 1, 3, 1
1, 3, 3, 3, 3, 3, 2, 1, 2, 2, 1, 1, 2, 1, 0, 3, 2, 3, 2, 3
3, 3, 2, 2, 1, 0, 2, 1, 2, 1, 4, 1, 4, 0, 0, 2, 1, 3, 2, 2

FIGURE 11

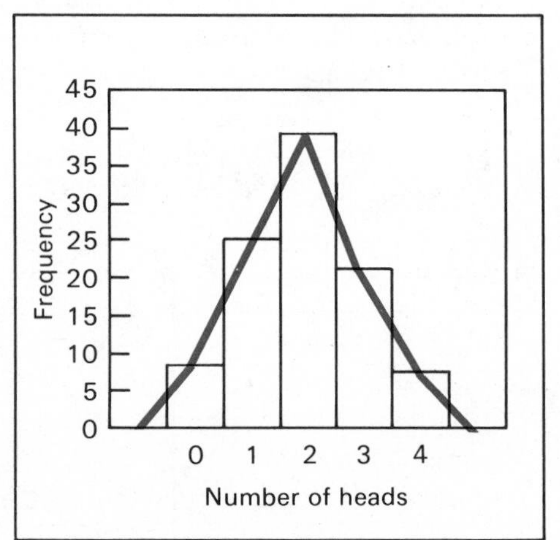

After recording the number of heads appearing on each toss, Scott and Joe next constructed the frequency distribution in table 5. After constructing the frequency distribution, they constructed the frequency polygon in figure 11.

FIGURE 12

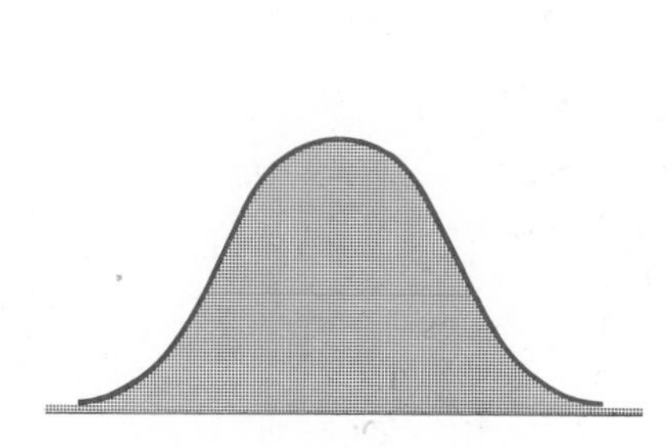

TABLE 5

Number of heads	Tally	Frequency
0	~~||||~~ |||	8
1	~~||||~~ ~~||||~~ ~~||||~~ ~~||||~~ ~~||||~~	25
2	~~||||~~ ~~||||~~ ~~||||~~ ~~||||~~ ~~||||~~ ~~||||~~ ~~||||~~ ||||	39
3	~~||||~~ ~~||||~~ ~~||||~~ ~~||||~~ |	21
4	~~||||~~ ||	7

Computing the mean for this experiment, we find it is 1.94. The mode is 2, as is the median. If Scott and Joe had increased the number of tosses, their outcomes would have more closely approached the outcomes for tossing 4 coins according to the laws of probability. For example, they tabulated 39 outcomes out of 100 where 2 heads appeared. Recall from exercise 4b, section 3.4 that if 4 coins are tossed, the probability of obtaining 2 heads is $\frac{6}{16}$, or 0.375, which is the same as 375 out of 1000. This ratio is quite close to the results that Scott and Joe got $(\frac{39}{100})$. Note that the frequency polygon in figure 11 is similar in shape to the bell-shaped curve in figure 12.

There is one particular bell-shaped curve that has been studied extensively by statisticians and named the **normal curve.** The normal curve has some unique properties. If we have a **normal distribution,** then the mean, median, and mode all have the same value, and all occur exactly at the center of the distribution. For normally distributed data, it can also be shown that approximately 68% of the data will be included within an interval of 1 standard deviation about the mean—that is, from 1 standard deviation above the mean to 1 standard deviation below the mean. For 2 standard deviations about the mean, that is, from 2 standard deviations above the mean to 2 standard deviations below the mean, approximately 95% of the data will be included. Practically all of the data, about 99.7%, will be included in the interval from 3 standard deviations above the mean to 3 standard deviations below the mean. The given percentages for a normal curve are illustrated in figure 13. The area under the curve represents all of the frequencies for some normal distribution.

Note that the tails of the curve do not touch the horizontal axis and they will not, no matter how far they may be extended. For all practical purposes, we need only consider 3 standard deviations above or below the mean, since this includes 99.7% of the data. Data that lie more than 3 standard deviations from the mean are rare.

An interesting occurrence in natural phenomena is that the data in large samples is often distributed so that the frequency polygon

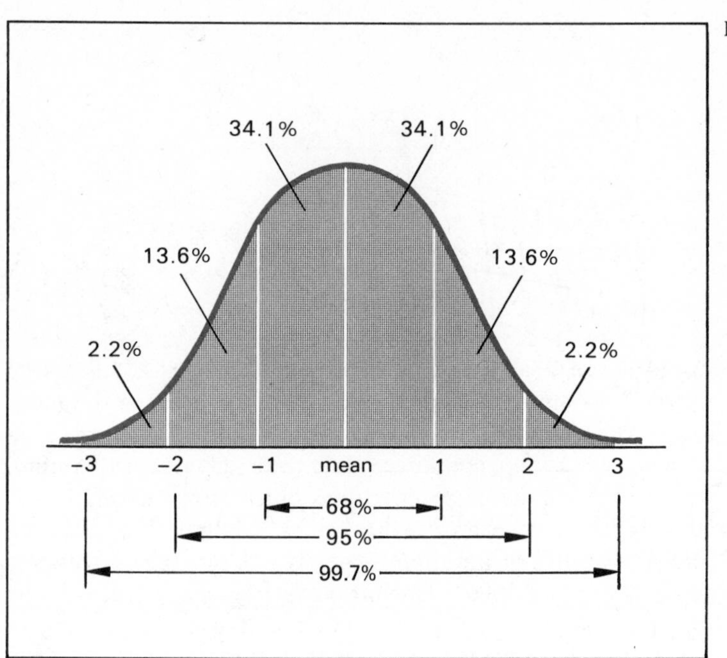

FIGURE 13 The normal curve

approximates the normal curve. A classic example is the distribution of intelligence quotient (IQ) scores for the entire population of a city, state, or country.

Other sets of data that are usually approximately normally distributed include the heights of males (or females) of a particular country, the weights of males (or females) of a particular country, the useful lifetimes of manufactured objects (washing machines, lightbulbs), the times necessary for monkeys to learn certain tasks, and the amounts of fluid dispensed by many dispensing machines.

EXAMPLE 1

The weights of the first graders in an elementary school are found to be approximately normally distributed with a mean of 60 pounds and a standard deviation of 5 pounds.

a. What percentage of the students in this group weigh between 55 and 65 pounds?

b. What percentage of these students weigh between 55 and 70 pounds?

81.8

Solution

Given that the weights are approximately normally distributed with a mean of 60 and a standard deviation of 5, we can draw a normal curve for this group of first graders, as shown in figure 14.

FIGURE 14

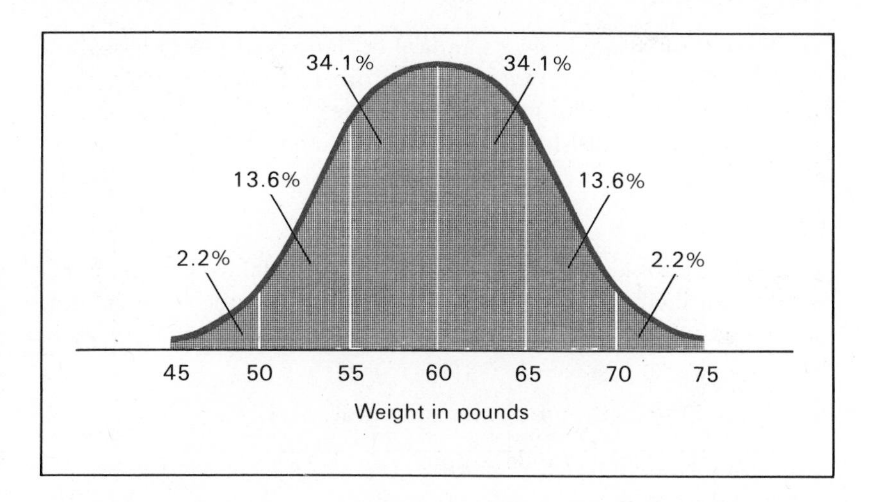

a. Using figure 14, we see that 34.1 + 34.1 = 68.2% of the students weigh between 55 and 65 pounds.

b. 34.1 + 34.1 + 13.6, or 81.8%, of the students weigh between 55 and 70 pounds.

EXAMPLE 2

The results of an exam in a statistics class are approximately normally distributed, with a mean of 70 and a standard deviation of 10. The instructor decides that a student will receive an A on the exam if the student scores more than 2 standard deviations above the mean. What is the lowest grade a student can get and still receive an A?

Solution

Since the scores are approximately normally distributed with a mean of 70 and a standard deviation of 10, we can draw a normal curve for this class. See figure 15.

FIGURE 15

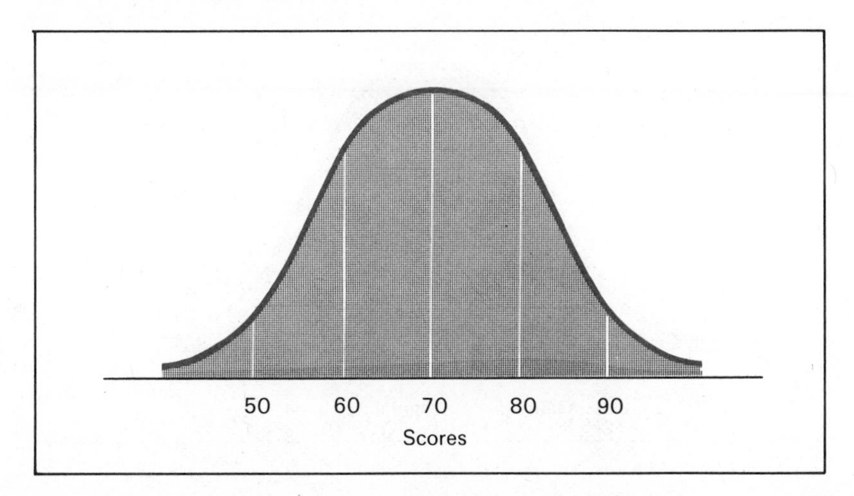

A score of 90 is exactly 2 standard deviations above the mean of 70. But, the instructor said that only those students scoring *more* than 2 standard deviations above the mean would receive an A. Hence, a student must have a score greater than 90 in order to receive an A.

EXAMPLE 3

A statistician took a survey of the mileage on 1000 cars in the parking lot of a factory. The mileage on each car was recorded. A frequency polygon of the mileages approximated the normal curve in figure 16.

Standard Oil Co. (Indiana)

a. What is the mean mileage for this set of cars?

b. What is the standard deviation?

c. What percentage of the cars had less than 20,000 miles on the odometer?

d. How many cars had more than 50,000 miles on the odometer?

Solution

a. Since the data are approximated by the normal curve in figure 16, we see that the mean is 30,000 miles.

b. Vertical lines have been drawn on the graph to indicate the standard deviations from the mean. The standard deviation is 10,000 miles.

c. Since the mean is 30,000 miles and the standard deviation is 10,000 miles, those cars with less than 20,000 miles on the odom-

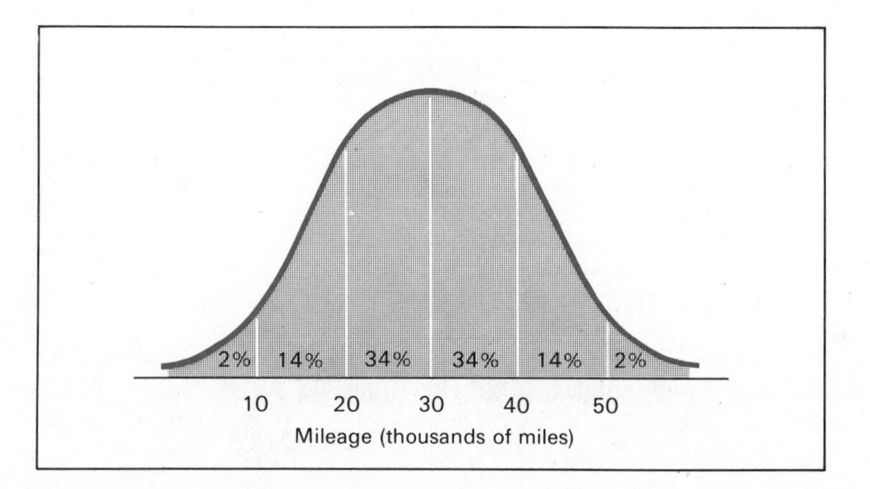

FIGURE 16

eter must be more than 1 standard deviation below the mean. Therefore, 14 + 2 = 16% of the cars have been driven less than 20,000 miles.

d. Cars with more than 50,000 miles on the odometer are more than 2 standard deviations above the mean. From the curve, we see that this is 2% of the data; 2% of 1000 = 0.02 × 1000 = 20 cars.

EXAMPLE 4

Testing indicates that the lifetimes of 1000 Dimlite light bulbs are approximately normally distributed with a mean of 100 hours and a standard deviation of 8. How many of these 1000 Dimlite bulbs will last—

a. more than 116 hours?

b. less than 92 hours?

Solution

Since the lives of the light bulbs are approximately normally distributed with a mean of 100 and a standard deviation of 8, the normal curve shown in figure 17 reflects the lifetimes of this set of bulbs.

FIGURE 17

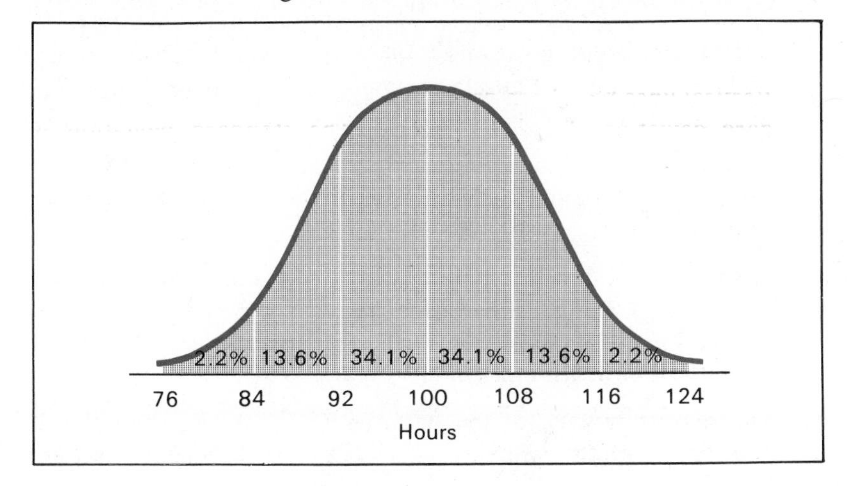

a. The light bulbs lasting more than 116 hours are those that lie more than 2 standard deviations above the mean. That is, 2.2% of the 1000 Dimlite bulbs will last more than 116 hours; 2.2% of 1000 = 0.022 × 1000 = 22 bulbs.

b. The light bulbs lasting less than 92 hours are those that lie more than 1 standard deviation below the mean. That is, 13.6 + 2.2, or about 15.8% of the 1000 bulbs will last less than 92 hours; 15.8% of 1000 = 0.158 × 1000 = 158 bulbs.

FIGURE 18

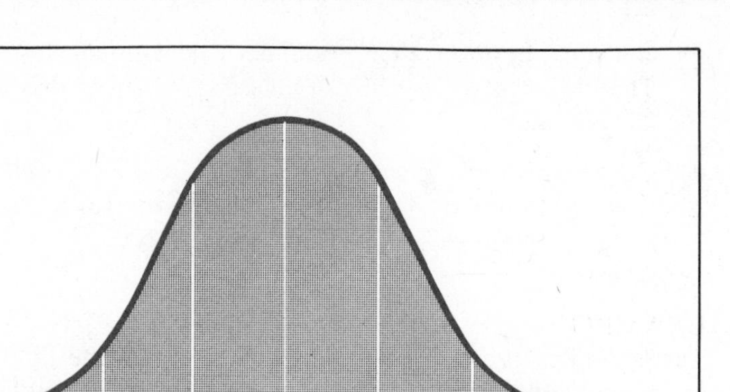

EXAMPLE 5
Tests show that the lifetimes of a shipment of Failsafe automobile batteries are approximately normally distributed with a mean of 3 years and a standard deviation of 6 months ($\frac{1}{2}$ year).

a. What percentage of these batteries will last more than $3\frac{1}{2}$ years?

b. If the batteries are guaranteed for 2 years, what percentage of the batteries will the Failsafe company have to replace?

Solution
Given that the lives of the batteries are normally distributed with a mean of 3.0 and a standard deviation of 0.5, we can use the normal curve in figure 18 for this shipment of batteries.

a. The batteries that will last more than $3\frac{1}{2}$ years are those that are more than 1 standard deviation above the mean, or $13.6 + 2.2 = 15.8\%$ of the batteries. That is, approximately 15.8% of the batteries will last more than $3\frac{1}{2}$ years.

b. Approximately 2.2% of the batteries will last less than the 2 years for which the company will have to honor its guarantee.

In an earlier discussion, it was pointed out that approximately 68% of normally distributed data is included within an interval of plus or minus 1 standard deviation about the mean, or 34.1% of the data is between the mean and 1 standard deviation above the mean. Similarly, approximately 95% of normally distributed data is included within an interval of plus or minus 2 standard deviations about the mean, and 99.7% of the data is included within 3 standard deviations about the mean. Sometimes we may wish to consider a portion of the data that is a fractional part of a standard deviation from the mean. Table 6 indicates what percentage of the data is contained between various tenths of a standard deviation from the mean.

TABLE 6

Amount of standard deviation from mean (in one direction only)	Percentage of data	Amount of standard deviation from mean (in one direction only)	Percentage of data
0.1	4.0	1.6	44.5
0.2	7.9	1.7	45.5
0.3	11.8	1.8	46.4
0.4	15.5	1.9	47.1
0.5	19.2	2.0	47.7
0.6	22.6	2.1	48.2
0.7	25.8	2.2	48.6
0.8	28.8	2.3	48.9
0.9	31.6	2.4	49.2
1.0	34.1	2.5	49.4
1.1	36.4	2.6	49.5
1.2	38.5	2.7	49.65
1.3	40.3	2.8	49.7
1.4	41.9	2.9	49.8
1.5	43.3	3.0	49.87

Many problems require fractional parts of standard deviations. Consider the following example. In a statistics class, the scores on the last exam were approximately normally distributed. The instructor decided to assign a grade of C for any score within 1.2 standard deviations of the mean. What percentage of the class will receive a grade of C? Table 6 indicates that 38.5% of the data is within 1.2 standard deviations of the mean on only one side of the mean. Since the normal curve is symmetric, we have the same percentage on the other side, and we therefore double the percentage: $2 \times 38.5 = 77\%$ of the class will receive a grade of C.

EXAMPLE 6

Testing indicates that the lifetimes of a new shipment of 1000 Dimlite light bulbs are approximately normally distributed with a mean of 100 hours and a standard deviation of 10. How many of the light bulbs will last between 83 and 117 hours?

Solution

This problem is similar to example 4, but the boundaries involve fractional parts of standard deviations from the mean. The mean is 100 and the standard deviation is 10.

$$100 - 83 = 17, \quad \frac{17}{10} = 1.7$$

and

$$117 - 100 = 17, \quad \frac{17}{10} = 1.7$$

Therefore, both boundaries are 1.7 standard deviations from the mean. Table 6 indicates that 45.5% of the data lies within 1.7 standard deviations of the mean. We have 1.7 standard deviations on either side of the mean; hence $2 \times 45.5 = 91\%$, or 910 light bulbs, will last between 83 and 117 hours.

EXAMPLE 7

The final exam results in a statistics class were approximately normally distributed. Grades were assigned in the following manner:

A for a score more than 1.6 standard deviations above the mean

B for a score between 1.1 and 1.6 standard deviations above the mean

C for a score between 1.1 standard deviations above the mean and 1.1 standard deviations below the mean

D for a score between 1.1 and 1.6 standard deviations below the mean

F for a score more than 1.6 standard deviations below the mean

What percentage of the class received each grade?

Solution

a. Table 6 indicates that 44.5% of the population lies within 1.6 standard deviations above the mean. Hence, $50 - 44.5 = 5.5\%$ of the class was above 1.6 standard deviations from the mean and received an A.

b. Since 44.5% of the population lies within 1.6 standard deviations and 36.4% lies within 1.1 standard deviations, we subtract $44.5 - 36.4$ to determine that 8.1% of the class received a grade of B.

c. Since 36.4% of the class lies within 1.1 standard deviations from the mean, and we want those both above and below the mean, we multiply 36.4 by 2 to determine that 72.8% of the class received a grade of C.

d. This is similar to part *b*, but we are concerned with standard deviations below the mean. Hence, we have 44.5 − 36.4 = 8.1% of the class received a grade of D.

e. Calculations like those in part *a* show that 5.5% of the class received a grade of F.

EXERCISES FOR SECTION 4.6

1. The IQ scores for a certain group of elementary school students are approximately normally distributed with a mean of 100 and a standard deviation of 10.
   a. What percentage of the students have IQ scores between 90 and 110?
   b. What percentage of the students have IQ scores between 80 and 120?
   c. What percentage of the students have IQ scores between 70 and 130?

2. An examination is given to all entering students at a certain college. The scores are approximately normally distributed with a mean of 100 and a standard deviation of 15.
   a. What percentage of the students have scores between 85 and 115?
   b. What percentage of the students have scores between 70 and 130?
   c. What percentage of the students have scores between 55 and 145?
   d. What percentage of the students have scores above 115?

3. A group of students in a statistics class recorded the mileages of 2000 cars in the student parking lot at their college. A frequency polygon of the mileages approximated the normal curve in figure 19.

FIGURE 19

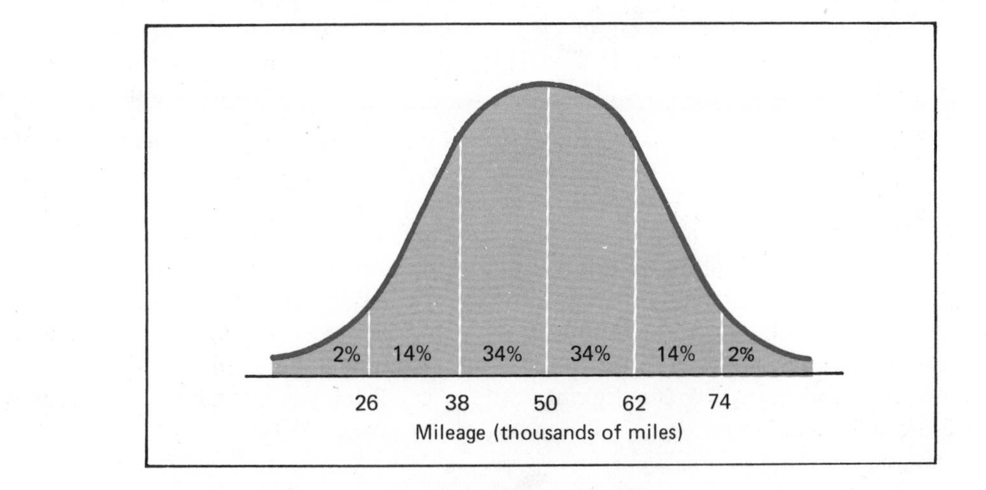

a. What is the mean mileage for this set of cars?
b. What is the standard deviation?
c. What percentage of the cars had less than 62,000 miles on the odometer?
d. How many had more than 62,000 miles on the odometer?

4. A statistician recorded the heights of 1000 students at a particular college. A frequency polygon of the heights approximated the normal curve in figure 20.

c. Approximately how many of the lighters will light less than 700 times?
d. Approximately how many of the lighters will light between 800 and 1100 times?

6. The scores on a statistics exam for a class of 50 students were approximately normally distributed with a mean of 30 and a standard deviation of 5.
a. Approximately how many of the students scored above 35?

FIGURE 20

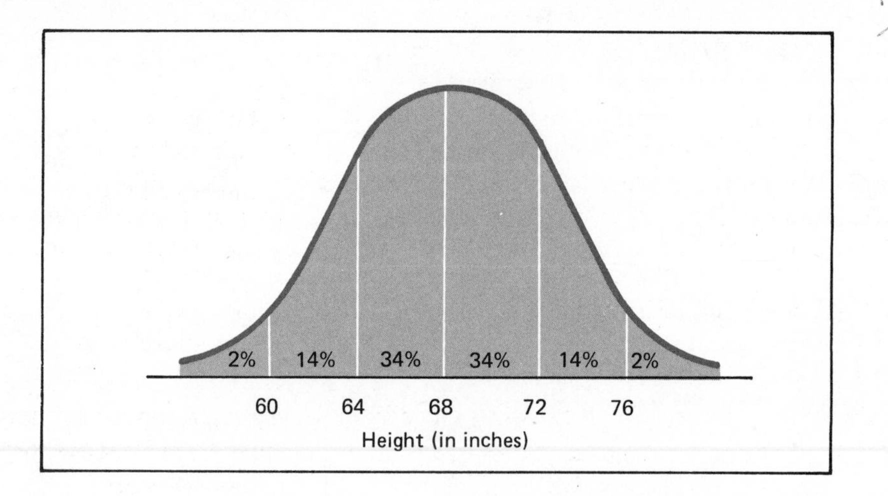

a. What is the mean height for this set of students?
b. What is the standard deviation?
c. What percentage of the students were more than 5 feet tall and less than 6 feet tall?
d. How many of the students were less than 5 feet tall?

5. Testing indicates that the lifetimes of a shipment of disposable butane lighters are approximately normally distributed with a mean of 1000 lights and a standard deviation of 100. A shipment contains 5000 of these lighters.
a. Approximately how many of the lighters will light more than 1100 times?
b. Approximately how many of the lighters will light more than 1200 times?

b. Approximately how many of the students scored above 40?
c. Approximately how many of the students scored below 25?
d. Approximately how many of the students scored between 20 and 35?

7. It has been shown that the lifetimes for a certain type of belted radial tire are approximately normally distributed with a mean of 40,000 miles and a standard deviation of 5000 miles.
a. What percentage of these tires will last more than 50,000 miles?
b. What percentage of these tires will last less than 35,000 miles?
c. What percentage of these tires will last between 35,000 and 50,000 miles?

d. What percentage of these tires will last between 30,000 and 55,000 miles?

8. The manufacturer of Z A P television sets guarantees its picture tube for 2 years. Testing indicates that the lifetimes of Z A P picture tubes are approximately normally distributed with a mean of 3 years and a standard deviation of 6 months ($\frac{1}{2}$ year). What percentage of Z A P television sets will have to be replaced under the guarantee?

9. The scores on a mathematics exam were approximately normally distributed, and the instructor assigned grades in the following manner:
   A for a score more than 1.7 standard deviations above the mean
   B for a score between 1.1 and 1.7 standard deviations above the mean
   C for a score between 1.1 standard deviations above the mean and 1.2 standard deviations below the mean
   D for a score between 1.2 and 1.8 standard deviations below the mean
   F for a score more than 1.8 standard deviations below the mean

   What percentage of the class received each grade?

10. Given that the scores on a certain exam are approximately normally distributed with a mean of 50 and a standard deviation of 10, use table 6 in this section to find—
    a. the percentage of the scores greater than 55
    b. the percentage of the scores greater than 65
    c. the percentage of the scores greater than 71
    d. the percentage of the scores between 45 and 55
    e. the percentage of the scores between 33 and 44
    f. the percentage of the scores less than 44.

11. The heights of the students at a local school are approximately normally distributed with a mean height of 64 inches and a standard deviation of 4 inches. Use Table 6 in this section to find—
    a. the percentage of the students between 62 and 66 inches tall
    b. the percentage of the students between 58 and 66 inches tall
    c. the percentage of the students over 6 feet tall
    d. the percentage of the students shorter than 5 feet 6 inches
    e. the percentage of the students between 5 feet 6 inches and 6 feet tall.

12. The coffee machines in a certain snack bar are supposed to dispense 5 ounces of coffee in each cup. The amount of coffee in each cup is approximately normally distributed with a mean of 5.0 ounces and a standard deviation of 0.5 ounce.
    a. What percentage of the cups contain between 4 and 5.5 ounces of coffee?
    b. What percentage of the cups contain between 5 and 5.8 ounces of coffee?
    c. What percentage of the cups contain between 4.8 and 5.8 ounces of coffee?
    d. What percentage of the cups will overflow if 6-ounce cups are used in the machine?

13. A survey indicates that a postmarking machine in a certain post office cancels an average (mean) of 65 letters per minute, with a standard deviation of 10 letters per minute. Assume that the number of letters cancelled by the machine is approximately normally distributed.
    a. What percentage of the time does the machine cancel more than 90 letters per minute?
    b. What percentage of the time does the machine cancel less than 50 letters per minute?

14. The average (mean) length of a phone call from a certain pay phone is 5 minutes, with a standard deviation of 1 minute. Assume that the lengths of the phone calls are approximately normally distributed.

a. What percentage of the calls last more than $6\frac{1}{2}$ minutes?
b. What percentage of the phone calls last between $5\frac{1}{2}$ and $6\frac{1}{2}$ minutes?
c. What percentage of the calls cost extra, if a caller has to pay extra for any call over 3 minutes in length?

15. A manufacturer has been awarded a government contract to manufacture bleeps, which must be between 1.45 and 1.55 inches in length. Testing indicates that the manufacturer's machines produce bleeps whose lengths are normally distributed with a mean length of 1.5 inches and a standard deviation of 0.1 inch.
    a. What percentage of the bleeps meet the government specifications?
    b. What percentage of the bleeps do not meet the government specifications?

16. The results of an exam in a statistics class are approximately normally distributed with a mean of 40 and a standard deviation of 8.
    a. What percentage of the scores are above 52?
    b. What percentage of the scores are above 60?
    c. If 100 students took the exam and all students with grades from 52 to 60 received a B, how many students received a B?

*17. Testing indicates that the lifetimes of a certain lot of light bulbs are approximately normally distributed with a mean of 100 hours and a standard deviation of 10.
    a. What is the probability that a bulb selected at random will last more than 100 hours?
    b. What is the probability that a bulb selected at random will last more than 105 hours?
    c. What is the probability that a bulb selected at random will last between 95 and 108 hours?

*18. The results of an exam in a history class are normally distributed with a mean of 60 and a standard deviation of 6.
    a. What is the probability that a student who took the exam will have a score between 57 and 63?
    b. What is the probability that a student who took the exam will have a score between 45 and 75?

Just For Fun

Can you connect all nine dots in figure 21? Use only four connecting line segments and do not raise your pencil from the paper.

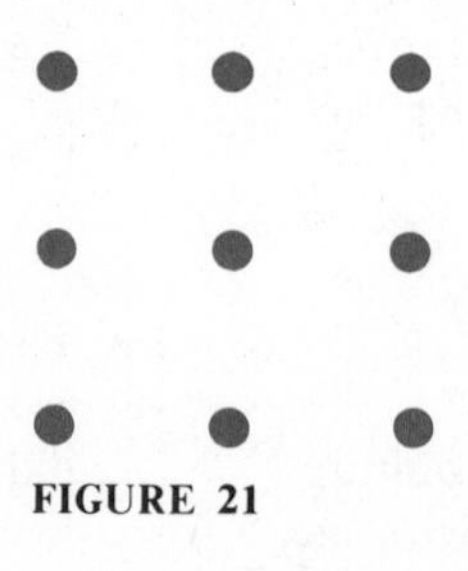

FIGURE 21

4.7 SUMMARY

The introduction to statistics presented here should enable you to understand some of its basic aspects. You should be aware that this chapter has merely been an introduction to some of the topics in statistics. If time permits, you should read *How To Lie With Statistics* by Darrell Huff (New York: W. W. Norton and Co., Inc., 1954) for an enjoyable discussion of statistics.

As we saw in this chapter, a *statistic* is a number derived from a set of data which is used in some way to represent or describe the data. The *measures of central tendency* and *measures of dispersion* are devices used to acquire knowledge from the data.

Measures of central tendency tell us where the data tends to center or cluster. The *mean* for a set of data is found by determining the sum of the data, and dividing this sum by the total number of elements in the set. The *median* for a set of data is found by arranging the data in sequential order and finding the middle number. When we have an even number of data, it is customary to use the number halfway between the two middle numbers, that is, the mean of the two middle numbers. The median is the measure of central tendency that determines the middle of a given set of data. The *mode* for a given set of data is that item or value that occurs most frequently. There can be times when a set of data has no mode—that is, when each value occurs an equal number of times. A set of data can also have more than one mode.

Measures of dispersion tell us about the spread of the data. The *range* for a set of data is found by subtracting the smallest value in the set from the largest value in the set. The range only involves the use of two extreme values; it does not tell us anything about the remaining values in the set of data. The *midrange* is found by adding the least value in the given set of data to the greatest value, and dividing the sum by 2. The midrange is used as a quick approximation to the measures of central tendency. Thus we use the range (a measure of dispersion) to help us get an approximation for the measures of central tendency.

A measure of dispersion that depends upon all of the values in a given set of data is the *standard deviation.* To find the standard deviation for a given set of data, we must follow these steps:

1. Find the *mean* of the given set of data.
2. Find the *deviation from the mean* for each value in the set of data.
3. Square each deviation.

4. Find the mean of the squared deviations.
5. Find the principal square root of this number.

After determining where data tends to center or cluster and then where data tends to disperse or scatter, we next examined the location of data in relation to the whole set or to other data. *Percentiles* and *quartiles* are *measures of location* of data. *Percentiles* divide sets of data into 100 equal parts; hence 100% is the basis of the measure. In order to determine the percentile at which a datum lies, we need to find the number of data that are ranked below the given datum and divide that number by the total number of data in the given set. The 25th percentile, 50th percentile, and 75th percentile are unique since they divide sets of data into quarters; hence, they are referred to as *quartiles*. The *first quartile* for a set of data is the value which has 25% of the data below it. The *second quartile* is the value which has 50% of the data below it, and the *third quartile* is the value which has 75% of the data below it.

Pictures of data are often used to transmit or summarize statistical information. In this chapter we studied the *divided bar graph*, the *circle* or *pie graph*, the *histogram*, and the *frequency polygon* (*line graph*). A *circle graph* or a *divided bar graph* is used in showing how a whole quantity is divided into parts. A *histogram* consists of a series of bars that are all the same width. Uniform units are used on the vertical axis to show the frequency of the data. A *frequency distribution table* is used to tabulate the given data and to determine the number of bars in the histogram. A *frequency polygon* (*line graph*) can also be used to graph the information in a frequency distribution table. It is constructed by connecting the midpoints of the tops of each bar in a histogram.

When we have a *normal distribution*, the area under the curve represents the entire set of data, or population. The *normal curve* has some unique properties. If we have data that is approximately normally distributed, then the mean, median, and mode all have approximately the same value. It can also be shown that approximately 68% of the data will be included within an interval of 1 standard deviation about the mean, and approximately 95% of the data will be included in the interval of 2 standard deviations about the mean. Practically all of the data (99.7%) will be included in the interval from 3 standard deviations below the mean to 3 standard deviations above the mean. These proportions will remain the same, regardless of the mean and standard deviation for a specific experiment, so long as the data is normally distributed. All normal curves do not appear the same, but the percentage of the data between any two corresponding standard deviations is the same.

Review Exercises for Chapter 4

1. a. Name three measures of central tendency.
   b. Which measure of central tendency gives the value that occurs most frequently?
   c. Which measure of central tendency gives the middle value of the given data?
   d. Which measure of central tendency gives the average value of the given data?

In exercises 2–5, find the mean, median, and mode for each set of data. (Round off any decimal answer to the nearest tenth.)

2. 12, 18, 16, 12, 19
3. 2, 3, 4, 5, 5, 6, 10, 13
4. 2, 3, 5, 7, 11, 13, 17, 19
5. 99, 11, 88, 22, 33, 77
6. In order to receive a grade of C in his statistics class, Scott needs a mean score of 72 on five tests. If Scott had scores of 62, 78, 80, and 68 on his first 4 tests, what is the lowest score that Scott can get on his last test and still receive a grade C in the course?
7. The mean of 4 of a set of 5 scores is 74. The fifth score is 88. What is the sum of the 5 scores? What is the mean of the 5 scores?
8. Find the range, midrange, and mean for the following set of test scores: {70, 70, 67, 66, 65, 65, 65, 62, 60, 60}.
9. Find the range, midrange, and mean for the following set of data: {58, 76, 76, 79, 84, 85, 100}.
10. Find the standard deviation for the following set of data: {10, 11, 13, 13, 14, 17}. (Round your answer to the nearest tenth.)
11. Find the standard deviation for the following set of data: {10, 14, 16, 26, 14, 16}. (Round your answer to the nearest tenth.)
12. In a senior class of 300 students, Joe has a rank of 45th. What is Joe's percentile rank in the class?
13. Andy is ranked at the third quartile in his statistics class of 32 students. Julie is in the same statistics class, and she is ranked 6th. Who ranks higher, Andy or Julie?
14. A student tossed a die 50 times. His results are shown below. Construct (*a*) a frequency distribution, (*b*) a histogram, and (*c*) a frequency polygon that represent this data.

    2, 5, 4, 4, 1, 5, 3, 5, 6, 6
    2, 5, 2, 5, 5, 1, 1, 5, 4, 6
    2, 4, 4, 6, 6, 3, 2, 6, 1, 1
    6, 1, 6, 4, 5, 4, 1, 6, 3, 2
    3, 4, 4, 5, 6, 4, 6, 1, 2, 2

15. For the fiscal year 1978, the United States government received the indicated part of each tax dollar from the following sources:

Individual income taxes:	38¢
Social insurance receipts:	28¢
Corporation income taxes:	13¢
Borrowing:	13¢
Excise taxes:	4¢
Other:	4¢

    Construct a circle graph representing this information. (Find degree measurements to the nearest degree.)

16. The heights (in inches) of 30 students in a physical education class were recorded as shown below.

    65, 69, 64, 65, 70, 71, 75, 69, 70, 64
    72, 70, 66, 68, 70, 72, 64, 68, 70, 69
    67, 70, 67, 66, 67, 71, 70, 75, 69, 70

a. Construct a frequency distribution for this data.
b. Construct a histogram for this data.
c. Construct a frequency polygon for this data.
d. Does the frequency polygon obtained in part c approximate a normal curve? Why?

17. One thousand test scores are approximately normally distributed with a mean of 60 and a standard deviation of 8.
    a. What percentage of the scores are above 68?
    b. What percentage of the scores are above 76?
    c. What percentage of the scores are below 52?
    d. What percentage of the scores are between 52 and 76?
    e. If a grade of C is assigned to the scores between 52 and 68, how many scores will receive a grade of C?

18. Two exams were given to a certain mathematics class. On the first exam, the mean was 64 and the standard deviation was 8. On the second exam, the mean was 48 and the standard deviation was 12. The scores of 4 students who took both exams are given below.

	Exam 1	Exam 2
Bill	64	63
Louie	70	60
Jack	60	54
Steve	80	66

    a. Who improved on the second exam?
    b. Who did not do as well on the second exam as on the first?
    c. Who performed the same on both exams?
    d. Who has the best grade for the two exams combined?

19. In 1973, O. J. Simpson set a rushing record for professional football by gaining 2003 yards in fourteen games. The following is a list of the number of times he carried the ball in each game: 29, 22, 24, 27, 22, 14, 39, 20, 20, 20, 15, 24, 22, 34. For this set of data, find the following:

    a. mean b. median c. mode
    d. range e. midrange

20. Use the set of data {20, 22, 26, 26, 28, 34} to find the following:

    a. mean b. median c. mode
    d. range e. midrange
    f. standard deviation (to the nearest tenth)

21. Given the following set of 12 scores: {29, 22, 24, 27, 22, 20, 20, 20, 15, 24, 22, 34}.

    a. What is the rank (from the top) of 27?
    b. What is the percentile rank of 27?
    c. What score is at the third quartile?
    d. What is the median?
    e. What is the mode?

Just For Fun

Can you number the rest of the squares in figure 22 so that each row, column, and diagonal totals 33? No number may be used more than once. (*Hint:* The 9 numbers used are consecutive.)

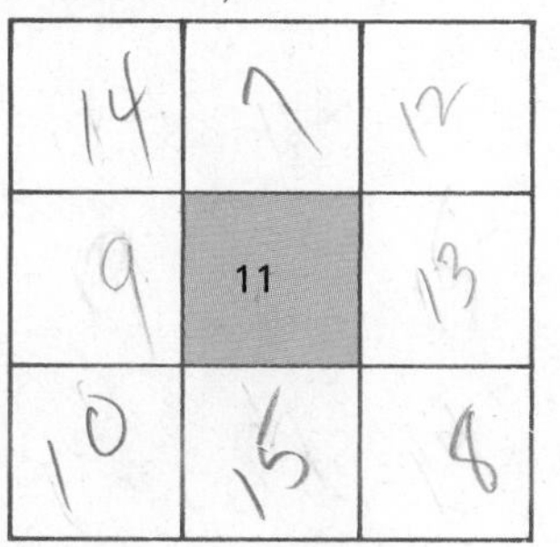

FIGURE 22

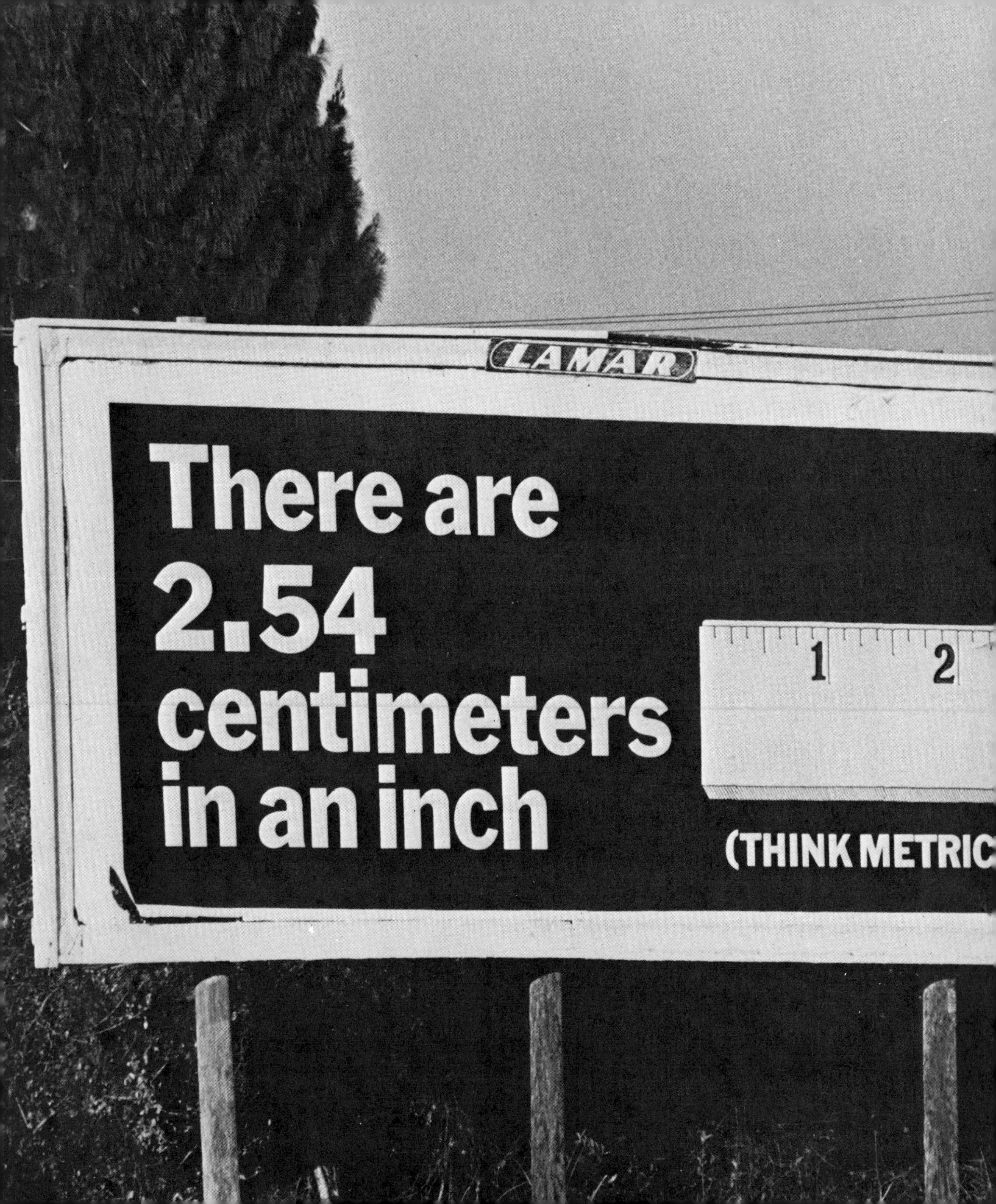

LAMAR
There are
2.54
centimeters
in an inch
1
2
(THINK METRIC

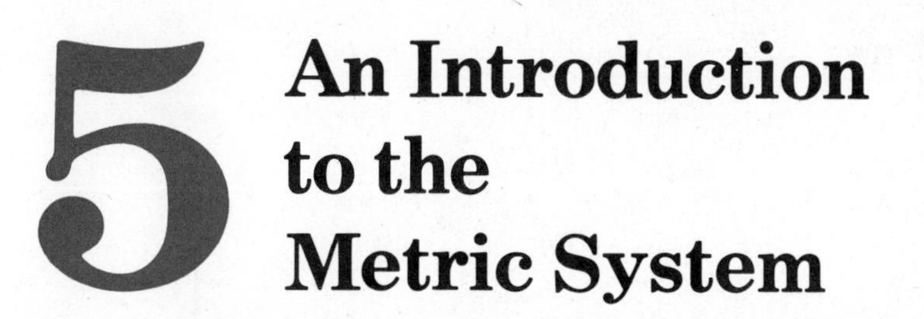

5 An Introduction to the Metric System

After studying this chapter, you will be able to do the following:

1. Describe the basic characteristics and advantages of the **metric system of measurement**
2. Identify and use the following prefixes in the metric system: **kilo, hecto, deka, deci, centi,** and **milli**
3. Name the basic unit of length in the metric system and describe its origin
4. Convert between the various metric units for measuring length, and use conversion tables to convert our customary units of length to metric units and vice versa
5. Name the basic unit of volume in the metric system, and describe the relationship between this unit and the basic unit of length
6. Convert between the various metric units for measuring volume, and use conversion tables to convert our customary units of volume to metric units and vice versa
7. Name the basic unit of weight in the metric system, and describe the relationship between this unit and the basic unit of volume
8. Convert between the various metric units for weight, and use conversion tables to convert our customary units of weight to metric units and vice versa
9. Describe the **Celsius** thermometer, and convert Fahrenheit degree readings to Celsius and vice versa
10. Make approximate conversions from metric measurements to customary measurements and vice versa, without computation.

Symbols frequently used in this chapter

Basic metric unit	Symbol
metre	m
litre	ℓ
gram	g
Celsius degree	°C

Gross/Stock, Boston

Prefix	Symbol	Meaning
kilo	k	1000
hecto	h	100
deka	da	10
deci	d	0.1
centi	c	0.01
milli	m	0.001

5.1 INTRODUCTION

The **metric system of weights and measures** is an international system derived from the International System of Units (SI). In this chapter, we shall use the internationally agreed upon spellings for *metre, litre,* and *tonne*.* Many books and pamphlets dealing with the metric system approach the topic from strictly a metric viewpoint. We have not used this approach here. The purpose of this chapter is to introduce you to the metric system and its advantages, and to help you understand how the system works.

Our present system of measures is a little of this and a little of that, a hodgepodge. It may not seem like a bad system, or one that is awkward to work with, but that is because you have used it all your life. However, it can be a confusing system. If you buy a quart of strawberries, you are buying a different amount than if you buy a quart of milk. The reason is that one is a dry quart and the other is a liquid quart. A dry quart is 16% greater in volume than a liquid quart. If you buy a pound of hamburger and a pound of gold, which weighs more? This question is not as silly as it seems. A pound of hamburger is weighed by avoirdupois weight and contains 7000 grains, but a pound of gold is weighed by troy weight and contains 5760 grains. Therefore, a pound of hamburger is heavier. These inconsistencies do not occur in the metric system.

The metric system is a system of measurement that has basic units of measure for length, width, area, volume, and weight which are in a decimal relationship to each other. For example, 1 metre = 10 decimetres = 100 centimetres. On the other hand, the United States uses a system of measurement that does not relate units in a decimal manner. For example, 12 inches = 1 foot, 3 feet = 1 yard,

*As recommended in the *Interstate Consortium on Metric Education: Final Report* (Sacramento: California State Department of Education, 1975).

FIGURE 1

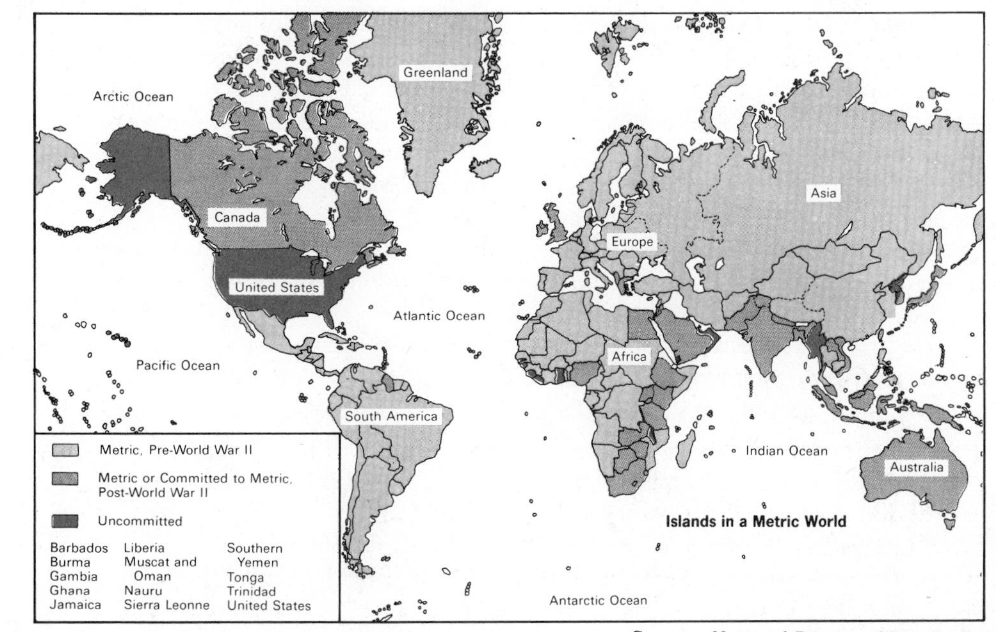

Source: National Bureau of Standards.

16 ounces = 1 pound, and so on. You will find that it is much simpler to multiply or divide by 10 than it is to use 12 for one calculation, 3 for another, 16 for another, and 5280 for yet another (5280 feet = 1 mile).

Practically every country in the world either uses the metric system now or is in the process of converting to the metric system. Figure 1 is a map of the world which shows the present status of the metric system.

As you can see from the map, only a few countries, such as Liberia, Burma, South Yemen, and the United States, are not committed to the metric system. Even our neighbors to the north and south, Canada and Mexico, are metric countries. In 1970, Canada announced a commitment to the metric system, and on April 1, 1975, started its conversion to the metric system by giving weather reports with Celsius temperature readings.

Originally, the metre was not immediately welcomed by the general public for some of the same reasons that apply today. It was inconvenient for people to learn a new system. But it was gradually accepted, and by 1875 all of Europe (except England) had adopted the metric system. Throughout the history of the United States many people, including Benjamin Franklin, have recommended that we adopt the metric system. In 1790, Thomas Jefferson recommended that the United States adopt a uniform system of weights and measures that used a decimal system. Alexander Graham Bell spoke before a Congressional committee in 1906 and stated that the metric system should be provided for the whole of the United States. On December 23, 1975, President Gerald Ford signed *The Metric Conversion Act* into law. This act established a national policy in support of metric measurement in the United States. Shortly after this, the Bureau of Alcohol, Tobacco, and Firearms of the Treasury Department established a policy that all distilled spirits will be sold in six standard metric sizes. After January 1, 1980, the labels on wine and liquor bottles will give metric measures only.

We are already using the metric system in many ways. Movie cameras use 8 or 16 millimetre film, and 35 millimetres is a popular size for photographic film. Skis are sold in centimetre lengths. Many of the items that we purchase at the supermarket are now labeled in both customary and metric units. For example, a can of pepper whose net weight is 4 ounces is also labeled with 113 grams, and a jar of mayonnaise contains 32 fluid ounces, 1 quart, or 0.95 litres. In many places, it is already possible to purchase a soft drink in one-litre bottles.

Stock, Boston

Road signs like this one, showing both miles and kilometres per hour, will soon be a common sight.

Most foreign cars are constructed using the metric system, and some American-made autos have many parts that are built using the metric system. Over one-fourth of the cars being used in this country require some metric tools. Engine displacement in many cars is now being described in metric terms; for example, a car might have a 1.6 litre engine. Speedometers on new cars contain guages that give readings in miles per hour and kilometres per hour. Road signs give both metric and English information. Temperatures will soon be reported in terms of degrees Celsius, precipitation figures will be given in centimetres (or millimetres), and wind speeds will be given in kilometres per hour instead of miles per hour. International track events such as the Olympic Games are now conducted using the metric system. The 100-yard dash is changed to a slightly longer race, the 100-metre dash. Similarly, the 220-yard dash becomes the 200-metre dash, and the mile run becomes the 1500-metre run.

Obviously, we will not stop using all of our customary units of measure. Even when the metric system is adopted, the height of horses will probably still be measured in terms of *hands* and depths at sea will still be measured in *fathoms.* Wood may still be sold by the *cord.* In a game like football, distances will probably always be given in yards; it will never be "first down, 9.14 metres to go." It is also doubtful that the Indianapolis 500-mile race will become Indianapolis 804.7-kilometre race.

Conversion to the metric system will be made in those areas where it is advantageous to do so. Conversion to the metric system will make many computations easier; the metric system is also easier to learn. Conversion will help our economic situation in world markets, because countries that already use the metric system will find American products more acceptable; this will benefit the entire population of the United States.

Just For Fun

$

What is the origin of the dollar sign?

5.2 HISTORY OF SYSTEMS OF MEASUREMENT

Regardless of what period of time we examine, man has always developed some method for weighing and measuring things. History tells us that early man used a measuring system for making weapons, building places to live in, and even making clothing. The **cubit** is one of the first recorded units of measurement. Noah supposedly built his ark 300 cubits long, 50 cubits wide, and 30 cubits high. According to Egyptian records (4000 B.C.), a cubit was the distance from the tip of the middle finger of the outstretched hand to the point of the elbow (approximately 19 inches). It is interesting to note that the side of one of the pyramids is 500 cubits long. Volume was measured by filling a container with seeds and then counting the number of seeds in the container. Stones were commonly used for determining the weight of heavy objects.

Man first used parts of his body or things that he could easily obtain, such as seeds and stones, as measuring instruments. A common brick is supposed to be a span long and one-half span wide. A **span** is the distance from the tip of the little finger to the tip of the thumb of an outstretched hand. A **palm** is the distance across the base of the four fingers that form your palm. A **digit** is the thickness or width of the middle of the middle finger, approximately three-fourths inch. The Romans used the idea of the digit to invent the inch. According to the Romans, an **inch** was the thickness or width of a thumb. There are no reliable facts on how the **foot** was invented.

According to most accounts, King Henry I decreed that a **yard** was the distance from the tip of his nose to the end of the thumb on his outstretched hand. It was by means of such royal commands that many standards of measurement were determined. For example, Queen Elizabeth I changed the measure of the mile from 5000 feet to 5280 feet. She did this because 1 furlong equaled 220 yards (660 feet), and if 1 mile equaled 5280 feet, then a mile would equal 8 furlongs. Thus, a partial list of English measures about 1500 A.D. was:

12 inches	=	1 foot
3 feet	=	1 yard
5 feet	=	1 pace
125 paces	=	1 furlong
8 furlongs	=	1 mile
12 furlongs	=	1 league

England became a world power and, by means of trade and colonization, the English system became established in many parts of the world. But other systems were also being developed in other countries. Bariel Mouton, vicar of St. Paul's Church in Lyons, France, proposed in 1670 that a standard unit of length be one minute of arc of a great circle of the earth. (A great circle is any circumference of the earth; the meridians that pass through the poles of the earth are examples of great circles.) Another proposal was made by Jean Picard, a French astronomer, who proposed a unit of length that was the length of a pendulum that takes one second to swing back and forth. But since a pendulum swings faster at the north and south poles than it does at the equator, nothing ever became of Picard's proposal.

After Mouton's proposal, not much was done toward developing a standard unit of measurement for over a hundred years. In 1790, at the request of the French government, the French Academy of Sciences devised a new system of measurement. The new basic unit of length was a portion of a meridian of the earth, similar to the unit proposed by Mouton. The new unit was called a *metre,* which was taken from the Greek word *metron,* "to measure."

Since the scientists wanted a unit similar in length to a yard, they chose a portion of the meridian that was approximately the same length. But they also wanted the unit to be part of a base ten, or decimal, system. Therefore they calculated the distance from the north pole to the equator along the meridian that runs through Dunkirk and Paris, and then took one ten-millionth $(\frac{1}{10{,}000{,}000})$ of that distance as the standard unit of measure, the **metre.** The French Academy of Sciences recommended this unit because all future calculations could be done using the decimal system. There would be no need to divide by 5280, multiply by 16, and so on; all quantities that were larger or smaller than the metre could be converted by multiplying or dividing by 10 or powers of 10. Recall that it is quite easy to multiply or divide a number by 10: we simply move the decimal point to the right or left.

A metre is about 39.37 inches long—a little longer than a yard. By keeping this in mind, you will be able to visualize how long a metre is. It will also help to give you an idea of the size of other metric measurements. The metric unit used for determining mass (weight) is called a *gram,* and the metric unit used for determining volume is called a *litre.* We shall examine grams and litres in greater detail later in this chapter.

Before we proceed with a study of the metric system, we must first develop a familiarity with a set of prefixes used throughout the system. Listed below are the most common prefixes and their meanings.

Prefix	Symbol	Meaning
kilo	k	1000
hecto	h	100
deka	da	10
deci	d	$\frac{1}{10}$ or 0.1
centi	c	$\frac{1}{100}$ or 0.01
milli	m	$\frac{1}{1000}$ or 0.001

H. Armstrong Roberts

Skis are sold in centimetre lengths.

For example, a **kilometre** is 1000 metres and a **centimetre** is $\frac{1}{100}$ of a metre.

To help you remember these prefixes, observe that prefixes containing an *i* (*deci, centi, milli*) are all fractional parts of one unit. The prefix *deci* should remind you of a word involving 10, such as *decade* or *decimal. Centi* should remind you of a *century,* which is 100 years. The prefix *milli* should remind you of *millennium,* a period of 1000 years.

The prefixes *deka, hecto,* and *kilo* are prefixes that indicate multiplication by 10, 100, and 1000, respectively. A *kilowatt* is a unit of measure used in electricity and is equivalent to 1000 watts. An easy way to remember the correct multiple for *hecto* is to notice that *hecto* and *hundred* both begin with the same letter, *h.* The prefix *deka* is similar to *deci* and should also remind you of words involving 10.

We did not list two prefixes, *micro* and *mega,* as they are not frequently used. These prefixes refer to one million. A *megaton* is one million tons. A *micrometre* is one-millionth of a metre. The prefix *micro* may remind you of *microfilm,* the very small film that libraries use to keep copies of printed matter.

Regardless of the fact that we do not yet know the meaning of *gram* or *litre,* we should be able to give an interpretation to the following terms: *kilogram, decigram, hectolitre,* and *millilitre.* A *kilogram* is 1000 grams, while a *decigram* is $\frac{1}{10}$ of a gram. A *hectolitre* is 100 litres, and a *millilitre* is $\frac{1}{1000}$ of a litre.

EXAMPLE 1

Using the prefixes as a guide, find equivalent expressions in grams, metres, or litres for each of the following:

a. centigram
b. kilometre
c. decilitre
d. hectometre

Solution

a. The prefix *centi* tells us that 1 centigram = 0.01 grams.

b. The prefix *kilo* indicates that 1 kilometre = 1000 metres.

c. The prefix *deci* indicates *ten* and the *i* tells us that it is a fractional part; that is, 1 decilitre = 0.1 litres.

d. The prefix *hecto* indicates that 1 hectometre = 100 metres.

EXAMPLE 2

What prefix can be used to indicate each number?

a. 0.001
b. 10
c. 0.1
d. 0.01

Solution

a. 0.001 is the same as $\frac{1}{1000}$ and is indicated by *milli.*

b. 10 is indicated by *deka.*

c. 0.1 is the same as $\frac{1}{10}$ and is indicated by *deci.*

d. 0.01 is the same as $\frac{1}{100}$ and is indicated by *centi.*

EXERCISES FOR SECTION 5.2

1. List two reasons why the United States should convert to the metric system.

2. a. Where did the metric system originate?

   b. What is the basic unit of length of the metric system? meter

   c. How was this basic unit of length obtained (what was it taken from)?

3. Name four units of measurement used by ancient civilizations.

4. Name the three basic units of measure in the metric system.

5. What prefix can be used to indicate each number?
   a. 10 b. 0.1 c. 1000
   d. 0.01 e. 100 f. 0.001

6. Name the prefix that means—
   a. one thousand times b. one-tenth of
   c. ten times d. one-hundredth of
   e. a hundred times f. one-thousandth of

7. Given the fact that one metre is approximately equal to 39.37 inches, how many inches are contained in each of the following?
   a. 1 decimetre b. 1 dekametre
   c. 1 centimetre d. 1 hectometre
   e. 1 millimetre f. 1 kilometre

8. Complete each of the following:
   a. 1 metre = ____ centimetres
   b. 1 metre = ____ decimetres
   c. 1 metre = ____ millimetres
   d. 1 kilometre = ____ metres
   e. 1 kilometre = ____ dekametres
   f. 1 kilometre = ____ decimetres

9. Complete each of the following:
   a. 1 dekametre = ____ metres
   b. 1 hectometre = ____ dekametres
   c. 1 kilometre = ____ hectometres
   d. 1 metre = ____ kilometres
   e. 1 decimetre = ____ metres
   f. 1 centimetre = ____ millimetres

10. Complete each of the following:
   a. 1 decigram = ____ grams
   b. 1 hectogram = ____ grams
   c. 1 dekalitre = ____ litres
   d. 1 decilitre = ____ centilitres
   e. 1 kilogram = ____ hectograms
   f. 1 hectolitre = ____ dekalitres

11. Complete each of the following:
   a. 10 hectometres = ____ kilometres
   b. 1000 metres = ____ hectometres
   c. 10 dekagrams = ____ grams
   d. 100 grams = ____ hectograms
   e. 10 millilitres = ____ centilitres
   f. 10 decilitres = ____ litres

12. Complete each of the following:
   a. 17 metres = ____ centimetres
   b. 2400 metres = ____ kilometres
   c. 32 kilograms = ____ grams
   d. 80 milligrams = ____ grams
   e. 44 hectolitres = ____ litres
   f. 37 centilitres = ____ millilitres

Just For Fun

The United States system of measurement uses two different miles, the statute mile and the nautical mile. Which is longer and by how much?

5.3 LENGTH AND AREA

As we stated earlier, the basic unit of length in the metric system is the metre. A metre is slightly longer than a yard; 1 metre ≈ 39.37 inches. (Recall that the symbol ≈ means approximately equal.) The symbol for metre is a lowercase *m*. Most likely there is a *metre stick* in your classroom. If there is, examine it in order to get an idea of how long 1 metre is.

Figure 2 shows a ruler. Note that one edge is marked in inches and the other edge is marked in centimetres. A centimetre is 0.01 metres; that is, 1 metre = 100 centimetres. Each small division on the metric edge of the ruler is 1 millimetre; 10 millimetres = 1 centimetre. The symbol for centimetre is *cm* and the symbol for millimetre is *mm*.

Suppose we want to find the length of line segment *AB* in figure 3. Using inches, segment *AB* is approximately $3\frac{3}{16}$ inches long. But if you measure it using the metric edge, you will find that its length is 8 centimetres. This illustrates another reason why the metric system is favored over our customary units of measurement:

FIGURE 2

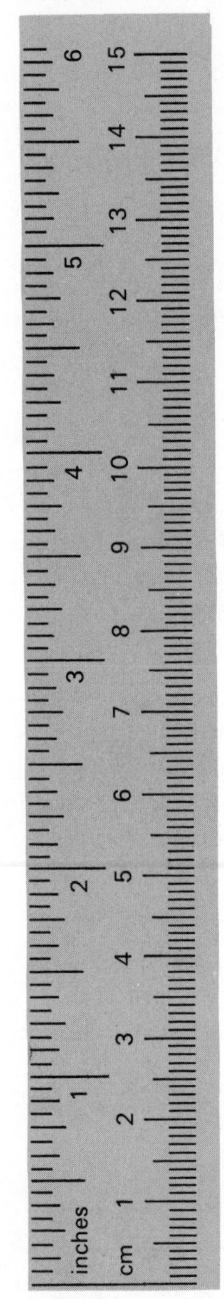

FIGURE 3

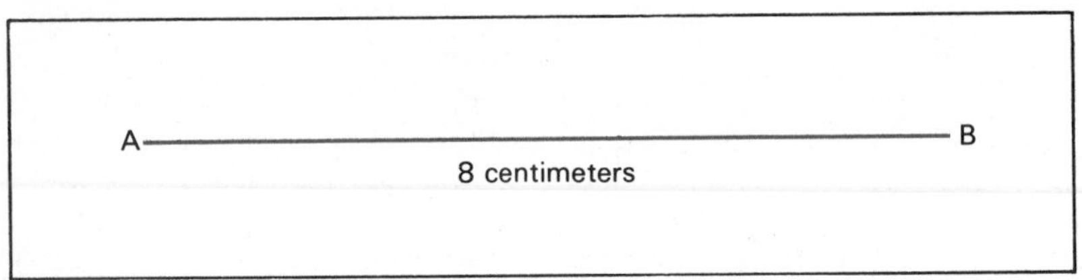

the metric system eliminates fractions such as $\frac{1}{4}$, $\frac{3}{8}$, $\frac{5}{16}$, and $\frac{1}{32}$. Granted, a line segment that is 2 inches long (see figure 4) is easy to measure in terms of inches, but it is also not difficult to measure using the metric system. A line segment 2 inches long is a little longer than 5 centimetres; in fact, it is 51 millimetres in length. Note that we do not have to use a fraction to express this length.

FIGURE 4

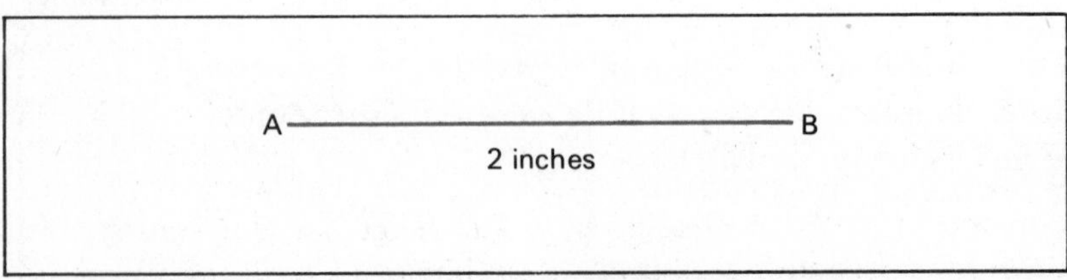

EXAMPLE 1

Measure each of the given line segments to the nearest millimetre (mm).

a. ______________________________

b. ________

c. ________________

d. ____________________________

Solution

a. 55 mm b. 15 mm c. 32 mm d. 50 mm

EXAMPLE 2

Complete each of the following:

a. 2 metres = ____ centimetres

b. 3 centimetres = ____ millimetres

c. 15 millimetres = ____ centimetres

d. 3 metres = ____ millimetres

e. 5 centimetres = 50 ____

f. 5 metres = 5000 ____

Solution

a. A centimetre is $\frac{1}{100}$ of a metre, so there are 100 centimetres (cm) in 1 metre (m) and 2 m = 200 cm.

b. There are 10 millimetres in 1 centimetre (see the metric scale on the ruler in figure 2). Therefore 3 cm = 30 mm.

c. 15 mm = 1.5 cm

d. A millimetre is $\frac{1}{1000}$ of a metre. Therefore there are 1000 millimetres in 1 metre and 3 m = 3000 mm.

e. 5 cm = 50 mm

f. There are 1000 millimetres in each metre, so 5 m = 5000 mm.

Although the metre is the basic unit of measurement in the metric system, many lengths (and thicknesses) are measured in terms of parts of a metre such as decimetre (dm), centimetre (cm), and millimetre (mm). In measuring greater lengths and distances, we can use the dekametre and hectometre, but the kilometre is the most commonly used unit for longer lengths. For example, all of the races in the Olympic Games are described in metres—100-metre dash, 200-metre dash, 400-metre dash, 800-metre run, 1500-metre run, 5000-metre run, and so on—but the distance between two cities is measured in kilometres. The symbol for kilometre is *km.* The distance from Los Angeles to New York is 4690 kilometres; from Dallas to Chicago, 1506 kilometres.

The table below summarizes the units related to the metre. Note that the first part of the symbol indicates the prefix and the second part (m) indicates *metre.*

Symbol	Word	Meaning
km	kilometre	1000 metres
hm	hectometre	100 metres
dam	dekametre	10 metres
m	metre	1 metre
dm	decimetre	0.1 metre
cm	centimetre	0.01 metre
mm	millimetre	0.001 metre

How long is a kilometre? From the prefix *kilo,* we know that 1 kilometre = 1000 metres. But does that give you any idea how long it is? Let's compare it with something that is familiar. How does it compare with a mile? A metre is 39.37 inches long; therefore, a kilometre is 39,370 inches long. A mile contains 5280 feet and 1 foot = 12 inches. Therefore, 1 mile = 5280 × 12 = 63,360 inches. A mile is longer than a kilometre. In fact, a mile is approximately 1.6 kilometres.

As another illustration, a football field is 120 yards long (including the end zones); therefore a football field is 120 × 36 = 4320 inches long. Nine football fields have a total length of 38,880 inches long. Since a kilometre is 39,370 inches in length, a kilometre is 490 inches longer than 9 football fields placed end to end. A kilometre is approximately 0.6 mile in length.

Keeping in mind that 1 mile ≈ 1.6 kilometres, we can also discuss speed in terms of kilometres per hour. For example, if a cyclist is pedaling his bike at the rate of 5 miles per hour (mi/hr), his rate of speed is 8 kilometres per hour (km/hr): since 1 mile = 1.6 kilometres, we have 5 × 1.6 = 8 kilometres per hour. If an automobile travels at the rate of 40 miles per hour, then it is also traveling at 64 (40 × 1.6) kilometres per hour.

If a person never drives faster than 90 kilometres per hour, how fast would this be in miles per hour? In order to convert kilometres to miles, we simply multiply the number of kilometres by 0.6 (1 kilometre $\approx$ 0.6 mile). Therefore, 90 kilometres per hour is approximately the same as $90 \times 0.6 = 54$ miles per hour. (*Note:* Although most of our conversion factors are approximate, we will use the equals sign for convenience.)

EXAMPLE 3

Convert each given measurement to the indicated measurement.

a. 10 mi = ____ km b. 25 mi = ____ km

c. 150 km = ____ mi d. 25 km = ____ mi

Solution

a. Since 1 mi = 1.6 km, 10 mi = $10 \times 1.6 = 16$ km. (*Note:* A better approximation of 1 mile is 1.61 kilometres, but we shall use 1 mi = 1.6 km.)

b. 25 mi = $25 \times 1.6 = 40$ km

c. Since 1 km = 0.6 mi, 150 km = $150 \times 0.6 = 90$ mi.

d. 25 km = $25 \times 0.6 = 15$ mi

EXAMPLE 4

Convert each speedometer reading to the indicated measurement.

a. 30 mi/hr = ____ km/hr b. 45 mi/hr = ____ km/hr

c. 100 km/hr = ____ mi/hr d. 120 km/hr = ____ mi/hr

Solution

This example is similar to example 3. To convert miles to kilometres, multiply the number of miles by 1.6. To convert kilometres to miles, multiply the number of kilometres by 0.6.

a. 30 mi/hr = $30 \times 1.6 = 48$ km/hr

b. 45 mi/hr = $45 \times 1.6 = 72$ km/hr

c. 100 km/hr = $100 \times 0.6 = 60$ mi/hr

d. 120 km/hr = $120 \times 0.6 = 72$ mi/hr

Area is measured in square units. The floor of a room that measures 8 feet by 10 feet has an area of 80 square feet. This means that there are 80 squares, each measuring 1 foot by 1 foot, that will cover the surface of the floor. Area is sometimes referred to as *surface area.*

A square whose measurements are 1 centimetre by 1 centimetre is said to have an area of 1 square centimetre; this is denoted by 1 cm^2. See figure 5. The square centimetre is used to find the area of relatively small regions, such as the area of this page. Larger regions are measured in terms of square metres (m^2). Since 1 metre = 100 centimetres, a square metre contains 10 000 square centimetres ($100 \times 100 = 10\,000$). (*Note:* We use a space instead of a comma in numbers of 10,000 or more when discussing metric measurements.)

FIGURE 5 Area = 1 cm^2

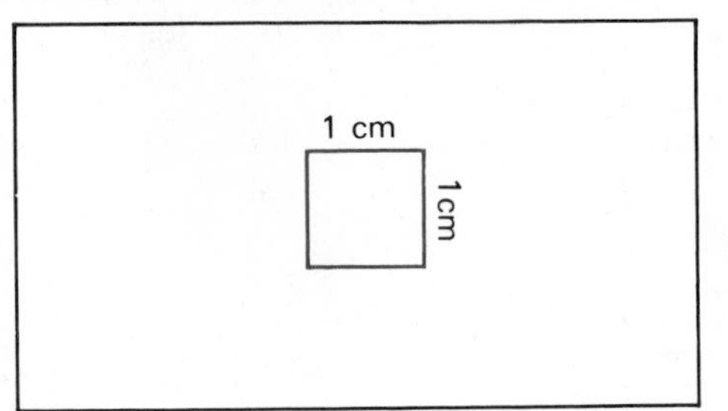

The area of very large regions is measured in *hectares* (10 000 square metres). Land that is measured in terms of acres can also be measured in hectares. Since 1 hectare = 10 000 square metres, a hectare is the area of a square that measures 100 metres on each side. It is highly unlikely that the hectare will be used in measuring the area of land when the United States converts to the metric system. Land will probably continue to be sold in terms of acres for two reasons: (1) land cannot be shipped overseas, that is, we will not export land to metric countries as we do machinery and other products; and (2) it would be impossible to change all of the property deeds in the United States so that the area would be in terms of hectares. However, the other metric measures of area, such as the square centimetre and the square metre, will be used. Another metric area measure is the *are* (pronounced *air*). It is a square measuring 10 metres on each side; therefore its area is 100 square metres.

You may find the metric conversion factors in table 1 helpful. *Remember, these are all approximate conversions.*

TABLE 1. Metric conversion factors

Length

To metric

1 inch (in.)	=	2.5 centimetres
1 foot (ft)	=	30 centimetres
1 yard (yd)	=	0.9 metres
1 mile (mi)	=	1.6 kilometres

From metric

1 millimetre (mm)	=	0.04 inch
1 centimetre (cm)	=	0.4 inch
1 metre (m)	=	3.3 feet
1 metre (m)	=	1.1 yards
1 kilometre (km)	=	0.6 mile

TABLE 1 (continued). Metric conversion factors

Area	
To metric	
1 square inch (in.^{2})	= 6.5 square centimetres
1 square foot (ft^2)	= 0.09 square metres
1 square yard (yd^2)	= 0.8 square metres
1 square mile (mi^2)	= 2.6 square kilometres
1 acre (A)	= 0.4 hectares
From metric	
1 square centimetre (cm^2)	= 0.16 square inch
1 square metre (m^2)	= 1.2 square yards
1 square kilometre (km^2)	= 0.4 square mile
1 hectare (ha)	= 2.5 acres

EXERCISES FOR SECTION 5.3

1. Measure each of the given line segments to the nearest millimetre.
   a. ________________
   b. ______________________________________
   c. ______________________
   d. ____
   e. ___________________________
   f. _____________________
   g. ____

2. Complete each of the following:
   a. 3 m = ____ mm b. 5 cm = ____ mm
   c. 10 mm = ____ cm d. 11 km = ____ m
   e. 40 cm = ____ m f. 36 m = ____ cm

3. Complete each of the following:
   a. 18 cm = ____ dm b. 35 mm = ____ cm
   c. 3 km = ____ cm d. 3500 m = ____ km
   e. 40 hm = ____ km f. 400 m = ____ dam

4. If the distance from Miami to Atlanta is 1070 kilometres, how many miles is it?

5. If the distance from Seattle to New Orleans is 2625 miles, how many kilometres is it?

6. If the distance from Mexico City to Chicago is 2082.5 miles, how many kilometres is it?

7. If the distance from Los Angeles to New York City is 4690 kilometres, how many miles is it?

8. The speed limit in New York State is 55 miles per hour. What is the speed limit in New York State in kilometres per hour?

9. The speed limit in a certain town is 75 kilometres per hour. If radar records Larry's speed as 50 miles per hour, should he get a ticket?

10. The distance from the earth to the sun is approximately 93 million miles. How many kilometres is it?

11. Mercury is the closest planet to the sun; its distance from the sun is 58 million kilometres. How many miles is it?

12. Find the perimeter of each polygon below. Your answer should be in terms of the indicated unit.

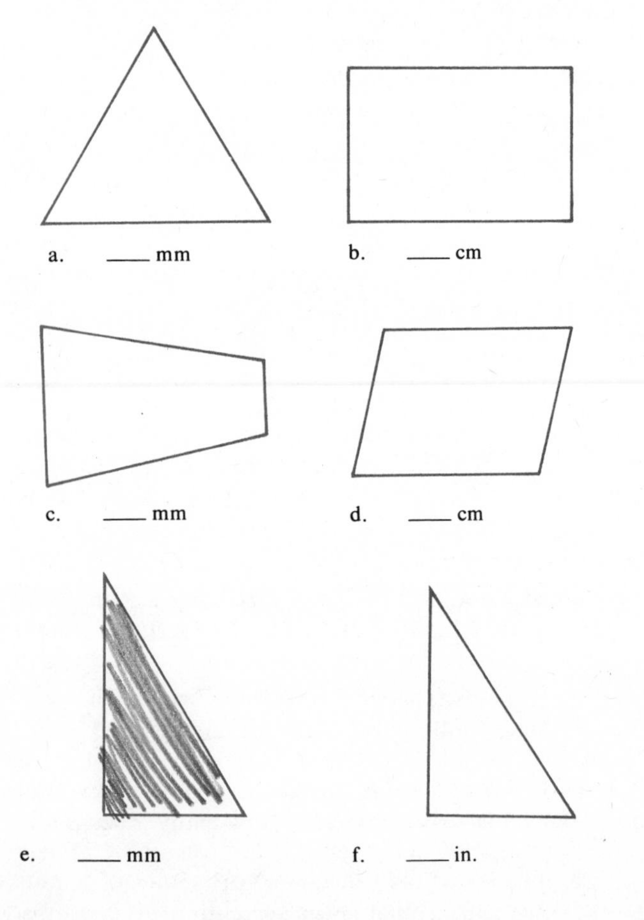

13. A square that measures 2 centimetres by 2 centimetres has what area?

14. A hectare is the area of a square that measures ___ metres on each side.

15. One mile is approximately 1.6 kilometres. Approximately how many square kilometres are in 1 square mile?

16. Which is the longer measurement in each case?
    a. 1 inch or 1 centimetre
    b. 1 yard or 1 metre
    c. 1 mile or 1 kilometre
    d. 6 inches or 13 centimetres
    e. 6 feet or 2 metres
    f. 1 foot or 35 centimetres
    g. 4 inches or 8 centimetres
    h. 2 inches or 55 millimetres
    i. 5 miles or 10 kilometres
    j. 12 miles or 19 kilometres

17. Which measurement of area is greater in each case?
    a. 1 square inch or 1 square centimetre
    b. 1 square foot or 1 square metre
    c. 1 square yard or 1 square metre
    d. 1 square mile or 1 square kilometre
    e. 1 square acre or 1 square hectare
    f. 1 square metre or 1 are

18. What are your metric measurements?
    a. height: feet ______ inches ______; metres ______ centimetres ______
    b. waist: inches ______; centimetres ______
    c. neck: inches ______; centimetres ______
    d. wrist: inches ______; centimetres ______
    e. biceps: inches ______; centimetres ______
    f. foot length: inches ______; centimetres ______

Just For Fun

Given the following equivalences, how many inches are contained in a distance that measures 2 miles 3 furlongs 4 rods 5 yards 2 feet?

12 inches	=	**1 foot**
3 feet	=	**1 yard**
$5\frac{1}{2}$ yards	=	**1 rod**
40 rods	=	**1 furlong**
8 furlongs	=	**1 statute mile**

5.4 VOLUME

FIGURE 6

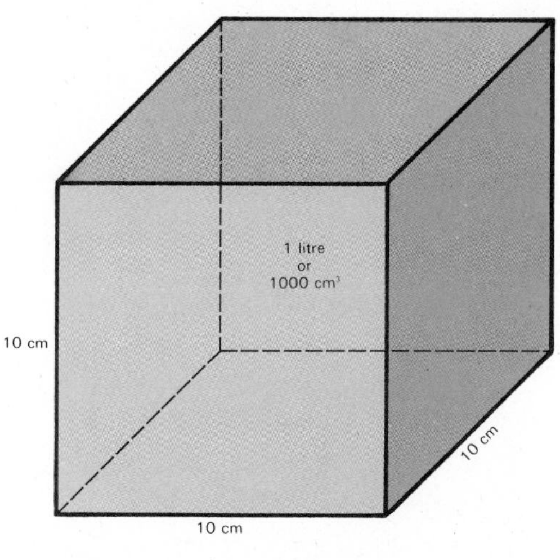

Volume is the measure of how much a container can hold, that is, its capacity. Unfortunately our system of measuring volumes and weights is quite confusing. For example, some soft-drink bottles contain 16 ounces, and some cans of coffee contain 16 ounces, but these are two different kinds of ounces. The soft-drink bottle contains 16 **fluid ounces,** which is equivalent to 1 pint. The 16 ounces in the can of coffee are units of weight, equivalent to 1 pound.

In the metric system, the *litre* is the basic unit used to measure capacity. A **litre** is defined as the volume of a cubic decimetre. In other words, a litre is the capacity of a cube (box) that is 1 decimetre long, 1 decimetre wide, and 1 decimetre high (see figure 6). The symbol for litre is the script letter ℓ. The volume of 1 litre is 1 cubic decimetre, or 1000 cubic centimetres. The symbol for cubic centimetres is cm^3.

One of the reasons that volume is easier to work with in the metric system than in our present system is that one set of units is used for all volume measure, whether liquid or dry. Using our present system, the volume of a pint of blueberries is different from the volume of a pint of cream. But in the metric system, cream is sold in litre containers and blueberries are sold in litre boxes.

FIGURE 7 One litre of milk contains slightly more than a quart of milk

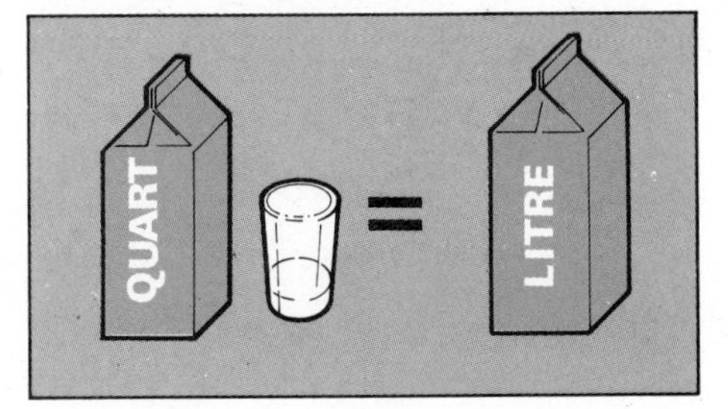

We shall compare a litre to a unit of volume with which you are already familiar, a quart. A litre contains a little more than a quart. One litre is approximately 1.06 liquid quarts, and one liquid quart is approximately 0.95 litres.

Two other common units of volume measure in the metric system are the **millilitre** and the **cubic metre.** Recall that the prefix *milli* means $\frac{1}{1000}$, or 0.001; hence, a millilitre is $\frac{1}{1000}$ of a litre. Since a litre contains 1000 cubic centimetres, a millilitre is 1 cubic centimetre. We can think of a millilitre as a cube whose length, width, and height each measure 1 centimetre. The millilitre and the litre are the two most commonly used units of volume.

FIGURE 8 One millilitre (1 cubic centimetre)

1 cm
1 cm
1 cm

The millilitre is a small unit of volume measure. For example, 5 millilitres = 1 teaspoon, and 15 millilitres = 1 tablespoon. Most liquid prescriptions obtained at pharmacies are sold in millilitres.

A cubic metre (m^3) is used to measure large volumes. One cubic metre is equivalent to 1.3 cubic yards. Items such as large amounts of sand and concrete are sold by the cubic yard at present. In the metric system, they would be sold by the cubic metre. Extremely large quantities of liquids would be measured in terms of the **kilolitre,** which is equivalent to 1000 litres. The capacity of fuel oil trucks and gasoline trucks, for example, would be expressed in kilolitres.

Remember that a litre is defined as the volume of a cube that is 10 centimetres long, 10 centimetres wide, and 10 centimetres high. The volume of 1 litre is 1000 cubic centimetres. The following table lists metric volume measures and their equivalent measure in litres.

Symbol	Word	Meaning
kl	kilolitre	1000 litres
hl	hectolitre	100 litres
dal	dekalitre	10 litres
ℓ	litre	1 litre
dl	decilitre	0.1 litre
cl	centilitre	0.01 litre
ml	millilitre	0.001 litre

EXAMPLE 1

a. 2 litres = ____ millilitres b. 3 litres = ____ centilitres

c. 5 kilolitres = ____ litres d. 10 litres = ____ kilolitres

e. 3000 millilitres = ____ litres f. 1000 litres = ____ hectolitres

Solution

a. A litre is 1000 millilitres, so 2 litres = 2000 ml.

b. A litre contains 100 centilitres. Therefore 3 litres = 300 cl.

c. A kilolitre is 1000 litres. Therefore 5 kl = 5000 litres.

d. A litre is $\frac{1}{1000}$ of a kilolitre, so 10 litres is $\frac{10}{1000}$ or $\frac{1}{100}$ of a kilolitre, that is, 10 litres = 0.01 kl.

e. Since 1000 millilitres = 1 litre, 3000 ml = 3 litres.

f. A hectolitre is 100 litres. Therefore 1000 litres = 10 × 100 litres = 10 hl.

You may find the metric conversion factors in table 2 helpful. *Remember, these are all approximate conversions.* It should be noted that some texts list a cup as equivalent to 250 millilitres, but according to the National Bureau of Standards, 1 cup = 0.24 litres, which is 240 millilitres.

TABLE 2. Metric conversion factors (continued)

Volume		
To metric		
1 teaspoon (tsp)	=	5 millilitres
1 tablespoon (Tbsp)	=	15 millilitres
1 fluid ounce (fl oz)	=	30 millilitres
1 cup (c)	=	0.24 litres
1 pint (pt)	=	0.47 litres
1 quart (qt)	=	0.95 litres
1 gallon (gal)	=	3.8 litres
1 cubic foot (ft^3)	=	0.03 cubic metres
1 cubic yard (yd^3)	=	0.76 cubic metres
From metric		
1 millilitre (ml)	=	0.03 fluid ounce
1 litre (ℓ)	=	2.1 pints
1 litre	=	1.06 quarts
1 litre	=	0.26 gallon
1 cubic metre (m^3)	=	35 cubic feet
1 cubic metre	=	1.3 cubic yards

One of the areas where the metric system can provide some help is in the kitchen. For example, a cake recipe that calls for

$2\frac{1}{4}$ cups of flour
$1\frac{1}{2}$ cups of sugar
3 teaspoons of baking powder
1 teaspoon of salt
$\frac{1}{3}$ cup of shortening
1 cup of milk
1 tablespoon of flavoring
2 eggs

would call for the following in the metric system:

540 millilitres of flour
360 millilitres of sugar
15 millilitres of baking powder
5 millilitres of salt
80 millilitres of shortening
240 millilitres of milk
15 millilitres of flavoring
2 eggs

Note that in the first recipe, we have to use fractions, as well as different units of measure. In the metric recipe, everything (except the eggs) is measured in the same unit, millilitres. If the metric recipe did call for fractional amounts, they would all be in decimal notation, because the metric system only uses powers of ten.

Using the metric conversions in table 2, we can convert any recipe to a metric recipe. Since 1 cup = 0.24 litres = 240 millilitres, $2\frac{1}{4}$ cups = $2\frac{1}{4} \times 240$ = 540 millilitres. The other conversions are done in a similar manner, using table 2.

EXAMPLE 2

Convert the given recipe to a metric recipe.

Blueberry Pie Filling

$\frac{1}{2}$ cup of sugar
1 tablespoon of lemon juice
$2\frac{1}{3}$ cups of drained blueberries
$\frac{1}{3}$ cup of blueberry juice
$2\frac{1}{2}$ tablespoons of flour
$\frac{1}{2}$ teaspoon of cinnamon
1 tablespoon of shortening

Solution

$\frac{1}{2}$ cup of sugar = $\frac{1}{2} \times 240$ = 120 millilitres of sugar
1 tablespoon of lemon juice = 15 millilitres of lemon juice
$2\frac{1}{3}$ cups of blueberries = $2\frac{1}{3} \times 240$ = 560 millilitres of blueberries
$\frac{1}{3}$ cup of blueberry juice = $\frac{1}{3} \times 240$ = 80 millilitres of blueberry juice
$2\frac{1}{2}$ tablespoons of flour = $2\frac{1}{2} \times 15$ = 37.5 millilitres of flour
$\frac{1}{2}$ teaspoon of cinnamon = $\frac{1}{2} \times 5$ = 2.5 millilitres of cinnamon
1 tablespoon of shortening = 15 millilitres of shortening

Remember that in the metric system one type of unit, the litre or some multiple of the litre, is used to measure both liquid and dry volume. The litre and the millilitre are the two most commonly used units of volume. One litre is defined to be the volume of a cube that is 10 centimetres long, 10 centimetres wide, and 10 centimetres high. A millilitre is the volume of a cube that is 1 centimetre long, 1 centimetre wide, and 1 centimetre high.

EXERCISES FOR SECTION 5.4

1. Complete each of the following:
   a. 3 litres = ____ ml
   b. 500 ml = ____ litres
   c. 1 kl = ____ litres
   d. 1 kl = ____ hl
   e. 20 dal = ____ litres
   f. 20 dl = ____ litres

2. Complete each of the following:
   a. 12 hl = ____ litres
   b. 2 dl = ____ ml
   c. 3000 litres = ____ kl
   d. 10 ml = ____ cl
   e. 60 cl = ____ litres
   f. 42 hl = ____ dal

3. Arrange the following measurements of volume in descending order beginning with the largest: dekalitre, litre, hectolitre, millilitre, kilolitre, decilitre, centilitre.

4. Which has the greatest volume in each pair?
   a. 1 quart or 1 litre
   b. 1 gallon or 3 litres
   c. 2 pints or 1 litre
   d. 1 teaspoon or 1 millilitre
   e. 1 cubic foot or 1 litre
   f. 4 pints or 2 litres

5. Find the volume of a box that is 1 metre long, 40 centimetres wide, and 50 centimetres high. (*Hint:* Convert all measurements to the same unit before finding the volume, where volume = length × width × height. Express your answer in cubic metres.)

6. Find the volume of a box that is 80 centimetres long, 0.5 metres wide, and 50 millimetres high. Express your answer in cubic metres.

7. How much water will an aquarium hold if it is 1 metre long, 60 centimetres wide, and 600 millimetres high? Express your answer in cubic metres.

8. What is the storage capacity of a food freezer whose inside measurements are 1.5 metres by 1 metre by 80 centimetres? Express your answer in cubic metres.

Just For Fun

In the late 1880s, liquid measure was also known as wine measure because it was used to measure liquors and wines. Given the following equivalences, how many gills are contained in 1 hogshead 1 barrel 20 gallons 3 quarts 1 pint?

4 gills	=	**1 pint**
2 pints	=	**1 quart**
4 quarts	=	**1 gallon**
$31\frac{1}{2}$ gallons	=	**1 barrel**
2 barrels	=	**1 hogshead**

9. Convert the given recipe to a metric recipe.

Clam Chowder

1 teaspoon salt	$\frac{1}{2}$ cup water
$\frac{1}{4}$ cup butter	1 pint minced clams
2 cups milk	2 cups diced potatoes
$\frac{1}{4}$ cup minced onions	

10. Convert the given recipe to a metric recipe.

Rice Pudding

$\frac{1}{2}$ cup uncooked rice	$\frac{1}{3}$ cup seedless raisins
$2\frac{1}{2}$ cups milk	$\frac{1}{2}$ tablespoon cinnamon
$\frac{1}{4}$ cup sugar	$\frac{1}{2}$ teaspoon salt

5.5 WEIGHT

Weight is a measure of the earth's gravitational pull. (Gravity is the force which holds you on earth.) **Mass** is the measure of the amount of matter—that is, atoms and molecules—that objects are made of. In space, the mass of an object does not change, but its weight does. *Weight* and *mass* are not the same thing, but on earth the mass of an object is always proportional to the weight of the object. Therefore, for this course, we shall assume that weight and mass mean the same thing.

In the metric system, the most common measures of weight (mass) are the kilogram (kg), the gram (g), and the milligram (mg). The basic unit of mass in the metric system is the gram. The weight of a paper clip is approximately one gram, while a nickel weighs five grams.

Imagine constructing a leakproof cubic centimetre out of weightless material (see figure 8, section 5.4) and filling it with very cold water. The weight of the water in such a container is 1 gram. Recall that one of the advantages of the metric system is that all of the measures (distance, volume, weight) are related. The weight of 1 millilitre of water is 1 gram, and a millilitre is the volume of a cube whose length, width, and height each measure one centimetre.

Now we can list some metric weight measures and their equivalent measure in grams.

Symbol	Word	Meaning
kg	kilogram	1000 grams
hg	hectogram	100 grams
dag	dekagram	10 grams
g	gram	1 gram
dg	decigram	0.1 grams
cg	centigram	0.01 grams
mg	milligram	0.001 grams

FIGURE 9 The weight of one nickel is 5 grams

Figure 9. A-165

IGA

How many kilograms does a 2-pound package of hamburger weigh?

A kilogram is equivalent to 1000 grams. Therefore 1000 millilitres—that is, 1 litre—filled with very cold water will approximate the weight of 1 kilogram. One pound is approximately 0.45 kilograms and one kilogram is approximately 2.2 pounds. Larger foodstuffs such as meats and fish are weighed in terms of kilograms. A piece of meat that weighs 1 kilogram is ample for 4 people

A milligram is equivalent to 0.001 of a gram. It is an extremely small unit of weight and is used only in medical prescriptions and some areas of science. For example, if you examine the label on a bottle of cold tablets, you will note that the amounts of different substances that each tablet contains are given in terms of milligrams. A common cold tablet contains, among other things, 225 milligrams of aspirin, 30 milligrams of caffeine, and 50 milligrams of ascorbic acid.

The weights of objects that are quite heavy, such as automobiles, are given in *metric tonnes.* One **metric tonne** is equivalent to 1000 kilograms. Since one kilogram is approximately 2.2 pounds, a metric tonne (1000 kilograms) is equivalent to 2200 pounds, or 1.1 tons (1 ton = 2000 pounds).

It is interesting to note that the weight of one gram is the weight of one millilitre of water, the weight of one kilogram is the weight of one litre of water, and the weight of the metric tonne is the weight of one cubic metre of water.

In order to develop a sense of the metric weights discussed, remember that a common paper clip weighs approximately 1 gram and a nickel weighs about 5 grams. A kilogram is equivalent to 2.2 pounds, so a man who weighs 220 pounds weighs 100 kilograms. A milligram ($\frac{1}{1000}$ of a gram) is a minute quantity and is difficult to approximate. A small grain of sand weighs about 1 milligram. The metric tonne is used in measuring the weight of very heavy objects. A metric tonne (1000 kilograms) is equivalent to 2200 pounds.

You may find the metric conversion factors in table 3 helpful. *These are all approximate conversions.*

TABLE 3. Metric conversion factors (continued)

Weight

To metric		From metric	
1 ounce (oz)	= 28 grams	1 gram (g)	= 0.035 ounces
1 pound (lb)	= 0.45 kilograms	1 kilogram (kg)	= 2.2 pounds
1 ton (T)	= 0.9 tonnes	1 tonne (t)	= 1.1 ton

EXAMPLE 1

Complete each of the following:

a. 3 kilograms = ____ grams

b. 2000 grams = ____ kilograms

c. 1 tonne = ____ kilograms

d. 2 grams = ____ milligrams

e. 8 g = ____ mg

f. 1000 g = ____ kg

g. 3000 mg = ____ g

h. 3000 kg = ____ t (tonnes)

Solution

a. A kilogram is equivalent to 1000 grams, so 3 kg = $3 \times 1000 = 3000$ g.

b. Since 1000 g = 1 kg, 2000 g = 2 kg.

c. 1 t = 1000 kg.

d. A milligram is $\frac{1}{1000}$ of a gram, so 1 g = 1000 mg. Therefore 2 g = 2000 mg.

e. 8 g = 8000 mg.

f. 1000 g = 1 kg

g. 3000 mg = 3 g

h. One metric tonne is equivalent to 1000 kilograms. Therefore 3000 kg = 3 t.

EXAMPLE 2

Recall that one gram is the weight of 1 millilitre of water whose temperature is approximately 39° Fahrenheit. Give the weight (in grams) of—

a. 1 litre of water

b. 1 cubic metre of water.

Solution

a. A litre is the capacity of a cube that is 10 centimetres long, 10 centimetres wide, and 10 centimetres high. The volume of one litre is 1000 cubic centimetres. A millilitre is the volume of one cubic centimetre. Therefore, 1 litre contains 1000 millilitres and 1 litre of water weighs 1000 grams.

As an alternate solution note that a millilitre is $\frac{1}{1000}$ of a litre (recall the meaning of the prefix *milli*). Therefore, 1 litre contains 1000 millilitres and 1 litre of water weighs 1000 grams.

b. A metre contains 100 centimetres, so a cubic metre contains 1 000 000 cubic centimetres. Thus a cubic metre of water contains 1 000 000 millilitres of water and weighs 1 000 000 grams. Note that 1 000 000 = 1000 × 1000; therefore, 1 000 000 grams is the same as 1000 kilograms. But 1000 kilograms is equivalent to 1 metric tonne. The weight of 1 cubic metre of very cold water is 1 metric tonne.

EXAMPLE 3

Convert each of the following to the indicated weight.

a. 16 oz = ____ g
b. 10 lb = ____ kg
c. 50 kg = ____ lb
d. 100 g = ____ oz
e. 4000 lb = ____ t
f. 180 lb = ____ kg

Solution

a. In order to convert ounces to grams, we must multiply the number of ounces by 28. Therefore 16 oz = 16 × 28 = 448 g.

b. One pound is equivalent to 0.45 kilograms. Therefore 10 lb = 10 × 0.45 = 4.5 kg.

c. One kilogram is equivalent of 2.2 pounds. Therefore 50 kg = 50 × 2.2 = 110 lb.

d. One gram is equivalent to 0.035 ounce. Therefore 100 g = 100 × 0.035 = 3.5 oz.

e. 2000 pounds = 1 ton, so 4000 pounds = 2 tons. Using table 3, 1 T = 0.9 t and 4000 lb = 2 T = 2 × 0.9 = 1.8 t.

f. One pound is equivalent to 0.45 kilograms. Therefore, 180 lb = 180 × 0.45 = 81 kg.

EXAMPLE 4

Convert the given recipe to a metric recipe.

Macaroni and Cheese

12 ounces macaroni
1 teaspoon salt
2 cups milk
8 ounces cheddar cheese
$\frac{1}{2}$ teaspoon pepper

Solution

To convert these measures to the metric system, remember that 1 ounce = 28 grams and 1 cup = 240 millilitres.

12 oz macaroni = 12 × 28 = 336 g macaroni
1 tsp salt = 1 × 5 = 5 ml salt
2 c milk = 2 × 240 = 480 ml milk
8 oz cheddar cheese = 8 × 28 = 224 g cheddar cheese
$\frac{1}{2}$ tsp pepper = $\frac{1}{2}$ × 5 = 2.5 ml pepper

The metric system is more consistent than our customary system of weights and measures. In the metric system, length, volume, and weight are directly related to each other. For example, the volume of a cube that has length, width, and height one centimetre (0.01 metre) is called a millilitre, and if we fill the millilitre with very cold water, it will have a weight of 1 gram. Similarly, a litre is the volume of a cube that has length, width, and height 10 centimetres. If we fill the litre with very cold water, it will have a weight of 1 kilogram.

Just For Fun

In the late 1800s, apothecaries (pharmacists) used the standard apothecaries' weight to make up different medicines. Given the following equivalences, how many grains are contained in 2 pounds 3 ounces 2 drams 2 scruples?

20 grains = 1 scruple
3 scruples = 1 dram
8 drams = 1 ounce
12 ounces = 1 pound

EXERCISES FOR SECTION 5.5

1. Complete each of the following:
   a. 5 g = ____ mg b. 3 kg = ____ g
   c. ____ g = 2 kg d. ____ kg = 5000 g
   e. 6 kg = ____ hg f. 10 g = ____ mg

2. Complete each of the following:
   a. 4000 mg = ____ g b. ____ g = 5 dag
   c. 60 cg = ____ g d. 10 mg = ____ cg
   e. ____ dag = 36 hg f. ____ mg = 7 dg

3. Convert each of the following to the indicated weight.
   a. 8 oz = ____ g b. 20 lb = ____ kg
   c. 32 oz = ____ g d. 140 lb = ____ kg
   e. 10 T = ____ t f. 6000 lb = ____ t

4. Convert each of the following to the indicated weight.
   a. 200 g = ____ oz b. 50 kg = ____ lb
   c. 100 kg = ____ lb d. 500 g = ____ oz
   e. 10 000 mg = ____ oz
   f. 10 kg = ____ oz

5. A certain vitamin tablet contains 150 milligrams of vitamin A, 225 milligrams of vitamin B, and 125 milligrams of vitamin C. What is the weight of the vitamin tablet in grams?

6. A box that is 60 centimetres long, 40 centimetres wide, and 50 centimetres high weighs 3 kilograms when it is empty. What will it weigh when it is filled with cold water? Express your answer in kilograms.

7. How much water (in litres) will an aquarium hold if it is 1 metre long, 60 centimetres wide, and 60 centimetres high? What is the weight of the water in kilograms if it is approximately 39° Fahrenheit?

8. Assuming that each of the following containers is filled with very cold water, give the weight (in grams) of—
   a. 1 gallon b. 1 quart
   c. 1 pint d. 1 cup
   e. 3 gallons f. 3 quarts

9. Which has the greater weight?
   a. 1 lb or 1 kg b. 2 oz or 20 g
   c. 10 lb or 5 kg d. 4000 lb or 3 t
   e. 40 oz or 2 kg f. 280 g or 9 oz

10. Convert the given recipe to a metric recipe.

    Swedish Meatballs

    1 pound ground beef
    $\frac{1}{2}$ cup minced onion
    1 tablespoon minced parsley
    $\frac{1}{2}$ teaspoon pepper
    1 egg
    $\frac{1}{2}$ pound ground pork
    $\frac{3}{4}$ cup fine dry bread crumbs
    $1\frac{1}{2}$ teaspoons salt
    1 teaspoon Worcestershire sauce
    $\frac{1}{2}$ cup milk

5.6 TEMPERATURE

One type of measure with which we are very familiar is temperature. All of us are concerned about weather on any given day, and one of the first questions we ask is "What's the temperature?" Wherever we go, there are time-temperature clocks that indicate how warm or how cold it is. It is even possible to dial a number and hear a recorded message stating the current time and temperature.

Gabriel Fahrenheit (1686–1736) used a brine solution (salt,

water, and ice) to devise a scale for the mercury thermometer such that the boiling point of water was 212° and the freezing point of water was 32°. Zero degrees was the lowest point on the thermometer, since that was the coldest temperature that he could get with the brine solution. But there are colder temperatures. A temperature of −81° was once recorded in Canada, −90° in Russia, and −127° in Antarctica.

If the temperature is 30 °C, it's hot!

Shortly after Fahrenheit developed his scale, Anders Celsius (1701–1744), a Swedish astronomer, developed another scale. Celsius developed his scale so that the boiling point of water was 100° and the freezing point of water was 0°. You may be familiar with the Celsius thermometer as the centigrade thermometer. Recall that *centi* means $\frac{1}{100}$, or 0.01; there are 100 intervals between 0° centigrade and 100° centigrade. Since the scale that Celsius developed was so convenient, the centigrade thermometer was adopted by scientists, and because 100 is a power of 10 ($100 = 10^2$), all of the countries using the metric system have also adopted the Celsius thermometer. This thermometer was formerly known as the centigrade thermometer, but, in honor of the man who developed the scale, it is now officially known as the Celsius thermometer, and each unit is called a degree Celsius (°C).

Pictured below are a Celsius thermometer and a Fahrenheit thermometer with indicated temperatures. See figure 10.

FIGURE 10

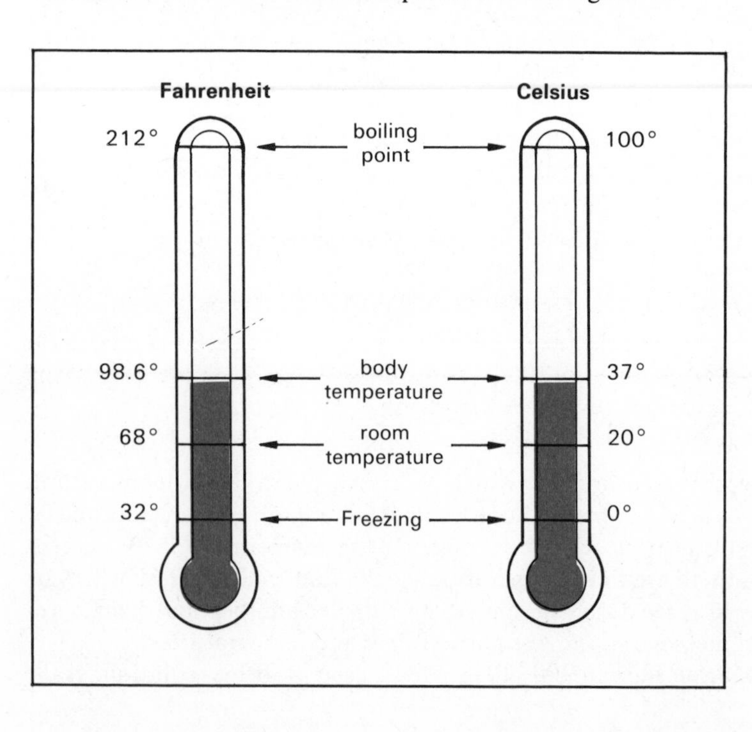

As shown in figure 10, the boiling point of water is 212 °F or 100 °C, and the freezing point of water is 32 °F or 0 °C. Most of us will be able to remember these comparative temperatures on the Celsius thermometer. But we would like to be able to interpret other Celsius thermometer readings as well. When someone tells us that their temperature is 37 °C, we should be aware that this is normal, or 98.6 °F. Note also that figure 10 indicates that a room temperature of 68 °F is the same as a room temperature of 20 °C. It is obviously important to identify a degree measurement as degrees Celsius or degrees Fahrenheit.

The following two formulas will enable you to convert from Fahrenheit to Celsius and vice versa. When you know the Fahrenheit temperature and wish to convert to Celsius, you should use the formula

$$C = \tfrac{5}{9}(F - 32)$$

When you know the Celsius temperature and wish to convert to Fahrenheit, you should use

$$F = \tfrac{9}{5}C + 32$$

Let's work out a problem using the first formula. Suppose we wish to find the temperature on the Celsius thermometer equivalent to 32 °F. From our previous discussion, you should know that the answer will be 0 °C. Using the formula,

$$\begin{aligned} C &= \tfrac{5}{9}(F - 32) && \\ C &= \tfrac{5}{9}(32 - 32) && \text{substituting 32 for } F \\ C &= \tfrac{5}{9}(0) && \text{subtracting 32 from 32} \\ C &= 0 && \text{multiplying by 0} \end{aligned}$$

Therefore, 32 °F = 0 °C.

Let's try another example using the same formula. Suppose the given temperature is 212°F and we wish to find the equivalent temperature on the Celsius thermometer.

$$\begin{aligned} C &= \tfrac{5}{9}(F - 32) && \\ C &= \tfrac{5}{9}(212 - 32) && \text{substituting 212 for } F \\ C &= \tfrac{5}{9}(180) && \text{subtracting 32 from 212} \\ C &= \tfrac{900}{9} && \text{multiplying 5 times 180} \\ C &= 100 && \text{dividing 900 by 9} \end{aligned}$$

Therefore, 212 °F = 100 °C.

Suppose the given Celsius temperature is 100 °C and we wish to find the equivalent temperature on the Fahrenheit thermometer. We use the formula $F = \frac{9}{5}C + 32$:

$F = \frac{9}{5}(100) + 32$	substituting 100 for C
$F = \frac{900}{5} + 32$	multiplying 9 times 100
$F = 180 + 32$	dividing 900 by 5
$F = 212$	adding 180 + 32

Therefore, 100 °C = 212 °F.

EXAMPLE 1

Most people set their thermostats at 68 °F. Find the equivalent temperature on the Celsius thermometer.

Solution

In order to convert Fahrenheit to Celsius, we use the formula $C = \frac{5}{9}(F - 32)$.

$C = \frac{5}{9}(68 - 32)$	substituting 68 for F
$C = \frac{5}{9}(36)$	subtracting 32 from 68
$C = \frac{180}{9}$	multiplying 5 times 36
$C = 20$	dividing 180 by 9

Therefore, 68 ° F = 20 °C.

EXAMPLE 2

Usually a person in good health has a body temperature of 98.6 °F. Find the equivalent temperature on the Celsius thermometer.

Solution

$C = \frac{5}{9}(F - 32)$	
$C = \frac{5}{9}(98.6 - 32)$	substituting 98.6 for F
$C = \frac{5}{9}(66.6)$	subtracting 32 from 98.6
$C = \frac{333.0}{9}$	multiplying 5 times 66.6
$C = 37$	dividing 333 by 9

Therefore, 98.6 °F = 37 °C.

EXAMPLE 3

A recipe in a metric cookbook calls for an oven setting of 175 °C. Find the equivalent temperature on the Fahrenheit thermometer.

Solution

In order to convert Celsius to Fahrenheit, we use the formula $F = \frac{9}{5}C + 32$.

$F = \frac{9}{5}(175) + 32$	substituting 175 for C
$F = \frac{1575}{5} + 32$	multiplying 9 times 175
$F = 315 + 32$	dividing 1575 by 5
$F = 347$	adding 315 and 32

Therefore, 175 °C = 347 °F. (*Note:* We could also divide 175 by 5, which equals 35, and then multiply 35 by 9, which also equals 315 and brings us to the last step.)

EXAMPLE 4
Convert 37 °C to Fahrenheit.

Solution

$F = \frac{9}{5}(37) + 32$	substituting 37 for C
$F = \frac{333}{5} + 32$	multiplying 9 times 37
$F = 66.6 + 32$	dividing 333 by 5
$F = 98.6$	adding 66.6 and 32

Therefore, 37 °C = 98.6 °F.

Besides being able to convert from Fahrenheit to Celsius and vice versa, you should develop a "feel" for temperatures that are given in terms of the Celsius scale. For example, a temperature of 30 °C is a hot day in most parts of the country and a temperature of 40 °C is a scorcher anywhere. Similarly, if someone has a body temperature of 40 °C then that person is very sick, and if his temperature rises to 41 °C, then he is near death.

EXERCISES FOR SECTION 5.6

1. Convert each Fahrenheit temperature to Celsius.
   a. 104 °F b. 50 °F c. 41 °F
   d. 95 °F e. 86 °F f. 23 °F

2. Convert each Fahrenheit temperature to Celsius.
   a. 59 °F b. 77 °F c. 5 °F
   d. 113 °F e. 32 °F f. 14 °F

3. Convert each Celsius temperature to Fahrenheit.
   a. 20 °C b. 50 °C c. 85 °C
   d. 10 °C e. 0 °C f. 65 °C

4. Convert each Celsius temperature to Fahrenheit.
   a. 95 °C b. 30 °C c. 5 °C
   d. 15 °C e. 25 °C f. −5 °C

5. Convert each temperature to the indicated scale.
   a. 14 °F = ____ °C b. 15 °C = ____ °F
   c. 25 °C = ____ °F d. 86 °F = ____ °C

6. Convert each temperature to the indicated scale.
   a. 95 °F = ____ °C b. 35 °C = ____ °F
   c. −10 °C = ____ °F d. −4 °F = ____ °C

7. A sick person has a high fever if his temperature is 40°C. Find the equivalent temperature on the Fahrenheit thermometer.

8. A recipe in a metric cookbook calls for an oven setting of 200 °C. Find the equivalent temperature on the Fahrenheit thermometer.

9. Iron will melt at 5432 °F. Find the equivalent temperature on the Celsius thermometer.

10. The highest temperature ever recorded in the state of Ohio is 113 °F. Find the equivalent temperature on the Celsius thermometer.

11. The lowest temperature ever recorded in the state of Alaska is −76 °F. Find the equivalent temperature on the Celsius thermometer.

12. In Montana, the temperature once dropped from 44°F to −56 °F in a period of 24 hours. This is a change in temperature of 100 °F. Approximately how much of a change is it on the Celsius scale? (Think carefully!)

13. The highest temperature ever recorded in the United States is 134 °F, recorded in Death Valley. Find the equivalent temperature (to the nearest whole degree) on the Celsius thermometer.

B.C. by permission of Johnny Hart and Field Enterprises, Inc., 1975.

Just For Fun

In the United States, dry measure is used in measuring such things as fruits and vegetables. Consider the following table:

2 pints = 1 quart
8 quarts = 1 peck
4 pecks = 1 bushel

Using the given table, how many pints are contained in 3 bushels 3 pecks 3 quarts 1 pint?

5.7 SUMMARY

The question is not *if* the United States will convert to the metric system, but *when.* Practically every country in the world uses the metric system or is presently converting to the metric system. The United States has not yet made an official act to convert to the metric system, but it is slowly inching its way in that direction.

The advantages of the metric system compared to our customary system of weights and measures are many. The metric system is a well-planned, logical system, based on powers of ten. Once we have learned the meaning of the various prefixes, the system becomes quite simple. The common prefixes in the metric system are:

Prefix	Symbol	Meaning
kilo	k	1000
hecto	h	100
deka	da	10
deci	d	$\frac{1}{10}$ or 0.1
centi	c	$\frac{1}{100}$ or 0.01
milli	m	$\frac{1}{1000}$ or 0.001

There are other reasons for converting to the metric system. The United States will probably be able to export more manufactured goods if the products are made according to metric specifications. Countries that are already metric (most of them are) give preference to metric products. Also, the metric system would eliminate many of the different kinds of measurements that we now use. For example, consider shoe sizes: In our current system there are babies' shoe sizes, children's shoe sizes, men's shoe sizes, and women's shoe sizes. In the metric system, people's feet are measured in centimetres.

You should know that a kilogram is 1000 grams, and a millilitre is 0.001 litres. For reference, remember that a metre is a little longer than 1 yard, a litre is a little more than 1 quart, and a gram is approximately the weight of a paper clip; also, a nickel weighs about 5 grams.

A room temperature of 20 °C is the same as 68 °F. Other important Celsius temperatures to keep in mind are: 0 °C = 32 °F (freezing) and 100 °C = 212 °F (boiling).

Table 4 is a table of metric conversion factors. Remember, these are all approximate conversions.

TABLE 4. Metric conversion factors

To metric	From metric
Length	
1 inch (in.) = 2.5 centimetres (cm)	1 millimetre (mm) = 0.04 inch (in.)
1 foot (ft) = 30 centimetres (cm)	1 centimetre (cm) = 0.4 inch (in.)
1 yard (yd) = 0.9 metres (m)	1 metre (m) = 3.3 feet (ft)
1 mile (mi) = 1.6 kilometres (km)	1 metre (m) = 1.1 yards (yd)
	1 kilometre (km) = 0.6 mile (mi)
Area	
1 square inch (in.^{2}) = 6.5 square centimetres (cm^2)	1 square centimetre (cm^2) = 0.16 square inch (in.^{2})
1 square foot (ft^2) = 0.09 square metres (m^2)	1 square metre (m^2) = 1.2 square yards (yd^2)
1 square yard (yd^2) = 0.8 square metres (m^2)	1 square kilometre (km^2) = 0.4 square mile (mi^2)
1 square mile (mi^2) = 2.6 square kilometres (km^2)	1 hectare (ha) = 2.5 acres
Volume	
1 teaspoon (tsp) = 5 millilitres (ml)	1 millilitre (ml) = 0.03 fluid ounce (fl oz)
1 tablespoon (Tbsp) = 15 millilitres (ml)	1 litre (ℓ) = 2.1 pints (pt)
1 fluid ounce (fl oz) = 30 millilitres (ml)	1 litre (ℓ) = 1.06 quarts (qt)
1 cup (c) = 0.24 litres (ℓ)	1 litre (ℓ) = 0.26 gallon (gal)
1 pint (pt) = 0.47 litres (ℓ)	1 cubic metre (m^3) = 35 cubic feet (ft^3)
1 quart (qt) = 0.95 litres (ℓ)	1 cubic metre (m^3) = 1.3 cubic yards (yd^3)
1 gallon (gal) = 3.8 litres (ℓ)	
1 cubic foot (ft^3) = 0.03 cubic metres (m^3)	
1 cubic yard (yd^3) = 0.76 cubic metres (m^3)	
Weight (mass)	
1 ounce (oz) = 28 grams (g)	1 gram (g) = 0.035 ounce (oz)
1 pound (lb) = 0.45 kilograms (kg)	1 kilogram (kg) = 2.2 pounds (lb)
1 ton (T) = 0.9 tonnes (t)	1 tonne (t) = 1.1 ton (T)
Temperature	
$C = \frac{5}{9}(F - 32)$	$F = \frac{9}{5}C + 32$

For more information (free) on the metric system, write to:
United States Department of Commerce, National Bureau of Standards,
Metric Information Office, Washington, D.C. 20234

Review Exercises for Chapter 5

1. Describe the basic characteristics of the metric system and list some of its advantages.

2. Name the prefix that indicates each of the following:
   a. 1000 b. 10 c. 0.1
   d. 100 e. 0.01 f. 0.001

3. How was the length of a metre determined?

4. Complete each of the following:
   a. 10 dam = ____ m b. 20 hm = ____ km
   c. 40 mm = ____ cm d. 35 m = ____ dm
   e. 32 km = ____ m f. 15 m = ____ dam

5. If 1 metre = 39.37 inches, then how many inches are contained in each of the following?
   a. 1 kilometre b. 1 millimetre
   c. 1 hectometre d. 1 centimetre
   e. 1 dekametre f. 1 decimetre

6. If the distance from New York City to Seattle is 4672 kilometres, approximately how many miles is it?

7. If the distance from New Orleans to Montreal is 1583 miles, approximately how many kilometres is it?

8. Convert each of the following to the indicated unit.
   a. 150 centimetres = ____ inches
   b. 300 feet = ____ metres
   c. 5 miles = ____ kilometres
   d. 10 metres = ____ yards
   e. 5 square metres = ____ square yards
   f. 10 square inches = ____ square centimetres

9. How is the volume of a litre determined?

10. Fill in the blank for each of the following:
    a. 2 litres = ____ millilitres
    b. 3 hectolitres = ____ dekalitres
    c. 5 kilolitres = ____ hectolitres
    d. 100 centilitres = ____ litres
    e. 20 centilitres = ____ millilitres
    f. 200 dekalitres = ____ litres

11. Find the volume of a box that is 1 metre long, 30 centimetres wide, and 30 centimetres high.

12. How much water will an aquarium hold if it is 80 centimetres long, 0.5 metres wide, and 600 millimetres high?

13. How is the weight of a gram determined?

14. Fill in the blank for each of the following:
    a. 4 kilograms = ____ grams
    b. 8 grams = ____ milligrams
    c. 30 dekagrams = ____ hectograms
    d. 20 centigrams = ____ milligrams
    e. 40 hectograms = ____ kilograms
    f. 18 grams = ____ centigrams

15. The lowest temperature ever recorded was −127 °F, recorded in Antarctica. Find the equivalent temperature (to the nearest degree) on the Celsius thermometer.

16. The highest temperature ever recorded was 58 °C, recorded in Africa. Find the equivalent temperature (to the nearest degree) on the Fahrenheit thermometer.

For questions 17–24, choose the best answer.

17. The height of a basketball player 6 feet 6 inches tall is approximately—
    a. 1 kilometre b. 100 centimetres
    c. 2.5 metres d. 3000 millimetres

18. The weight of a football player weighing 200 pounds is approximately—
    a. 1 tonne b. 1000 grams
    c. 90 kilograms d. 100 kilograms

19. Five gallons of apple cider is approximately equal to—
    a. 30 litres b. 19 litres
    c. 100 centilitres d. 5 litres

20. On a very hot day in August in Phoenix, Arizona, the temperature is likely to be—
    a. 60 °C b. 20 °C c. 25 °C d. 40 °C

Just For Fun

Can you find at least 15 metric terms hidden in this puzzle?

```
A K E I B C U B I C M E T R E
M I L L I G R A M A B C D E R
I L O U L A E I O U H F M G T
L O C E L S I U S H E I E J E
L L E K L M N O P Q C R T R M
I I N J S T U S C O T T R V A
M T T O N N E W X E A Y I Z K
E R I E A G R A M M R M C C E
T E M B C D E O A F E G H I D
R J E K L M L R D S H T A M E
E O T P Q I G I R N E T R S R
T U R V K A W X T Y A T A E A
Z T E E K L F O U R B L E M U
O H T E N O A D D I E E S K Q
T A D C U B I C M E T R E I S
```

21. If the temperature outside is 35 °C, what should you wear?
    a. a heavy sweater
    b. a ski parka
    c. a T-shirt

22. If you purchased 10 kilograms of groceries at the supermarket, in which of the following should you carry the groceries?
    a. a shopping cart
    b. a small bag
    c. a trailer

23. What would you be most likely to do if you purchased 400 millilitres of soda pop at the store?
    a. Call your friends to have a party.
    b. Drink it and satisfy your thirst.
    c. Store all the cases in the garage.

24. If your home was located 1 kilometre from your college, how would you get home?
    a. take a bus b. take a plane c. walk

25. Assuming that each of the following containers is filled with very cold water, give the weight (in

grams) of—

a. 1 litre
b. 1 cubic metre
c. 1 centilitre
d. 1 gallon
e. 1 quart
f. 3 pints

26. Convert each of the following to the indicated unit.
a. 2 tsp = ____ ml
b. 1 qt = ____ litres
c. 100 litres = ____ qt
d. 2 oz = ____ g
e. 200 lb = ____ kg
f. 95 °F = ____ °C

27. Convert the given recipe to a metric recipe.

Sour Cream Cookies

$\frac{1}{2}$ cup sour cream	$\frac{1}{2}$ pound butter
1 teaspoon vanilla	1 teaspoon baking soda
1 cup brown sugar	$2\frac{1}{2}$ cups flour
2 eggs	Bake at 350 °F

MUSIC ALARM CLOCK
AUTOMATIC
110
90
150
170
AUTOMATIC
240
AUTOMATIC
AUTOMATIC
225
220
AUTOMATIC

6 MATHEMATICAL SYSTEMS

After studying this chapter, you will be able to do the following:

1. Add, subtract, and multiply in the 12-hour clock system
2. Identify the basic parts of a **mathematical system**
3. Determine whether a set is **closed** with respect to a given operation
4. Determine whether an **identity element** exists for a given operation, and whether each element in the set has an **inverse** for a given operation
5. Determine whether a set is **associative** with respect to a given operation, and whether the set is **commutative** with respect to a given operation
6. Determine whether a mathematical system is a **group**
7. Add, subtract, and multiply in other **modular systems**
8. Evaluate problems in an abstract system, given a table that defines an operation for the elements in the system, and determine whether the properties of a group are satisfied by the system
9. Identify the basic parts of an **axiomatic system**
10. Construct a **diagram** (model) for which all of the axioms of a system are satisfied, and prove a **theorem** given the undefined terms, defined terms, and axioms for the system.

6.1 INTRODUCTION

In this chapter, we shall examine mathematical systems and their properties. That is, we shall study the nature and structure of mathematical systems. Regardless of what area of mathematics is examined (sets, logic, etc.), there are certain basic common characteristics that these topics possess. For the present time, we shall think of a **mathematical system** as a set of elements together with one or more operations (rules) for combining elements of the set. We shall expand upon this idea later in the chapter, but for now this concept of a system is sufficient.

Wolinsky/Stock, Boston

One of the first mathematical systems to which we are exposed in school is the set of counting numbers $\{1, 2, 3, \ldots\}$ together with the operation of addition. This system is considered to be an infinite system, since there are an infinite number of elements in the set. We shall begin our study of mathematical systems by examining a finite system, one that has some unusual properties.

FIGURE 1

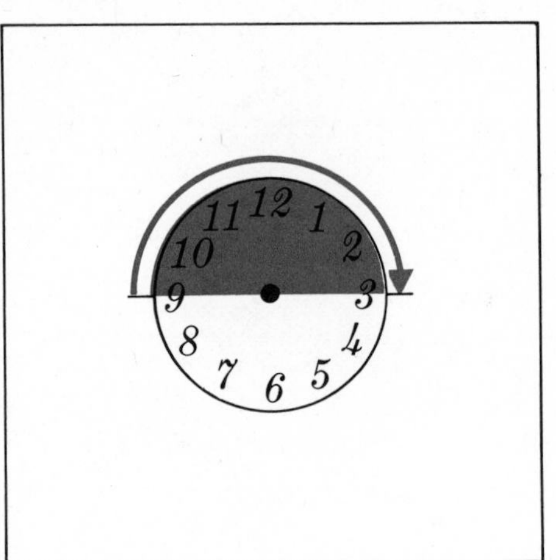

6.2 CLOCK ARITHMETIC

Clock arithmetic is an example of a finite mathematical system that will enable us to understand the nature and structure of mathematical systems.

Consider the following addition problems:

$$1 + 2 = 3 \qquad 5 + 7 = 12 \qquad 9 + 10 = 7$$
$$3 + 4 = 7 \qquad 5 + 8 = 1 \qquad 11 + 11 = 10$$
$$5 + 6 = 11 \qquad 6 + 12 = 6 \qquad 9 + 9 = 6$$

Each addition problem listed is correct; there are no mistakes. The reason all of these examples are correct is that they come from a system called clock arithmetic. The first four examples in the list look exactly like examples from ordinary arithmetic, but $5 + 8 = 1$ and $9 + 9 = 6$ do not. In clock arithmetic, $5 + 8 = 1$ because 1:00 comes 8 hours after 5:00. Similarly, 6:00 comes 9 hours after 9:00, so $9 + 9 = 6$. It also follows that, if it is 6:00 now, then 12 hours from now it will be 6:00, so $6 + 12 = 6$. Hence, we see that our mathematical system, clock arithmetic, has a set of elements, $\{1, 2, 3, 4, 5, 6, 7, 8, 9, 10, 11, 12\}$, the numerals 1 through 12 on the

TABLE 1

+	1	2	3	4	5	6	7	8	9	10	11	12
1	2	3	4	5	6	7	8	9	10	11	12	1
2	3	4	5	6	7	8	9	10	11	12	1	2
3	4	5	6	7	8	9	10	11	12	1	2	3
4	5	6	7	8	9	10	11	12	1	2	3	4
5	6	7	8	9	10	11	12	1	2	3	4	5
6	7	8	9	10	11	12	1	2	3	4	5	6
7	8	9	10	11	12	1	2	3	4	5	6	7
8	9	10	11	12	1	2	3	4	5	6	7	8
9	10	11	12	1	2	3	4	5	6	7	8	9
10	11	12	1	2	3	4	5	6	7	8	9	10
11	12	1	2	3	4	5	6	7	8	9	10	11
12	1	2	3	4	5	6	7	8	9	10	11	12

face of a clock, and it also has an operation (addition) which consists of counting hours in a clockwise direction.

Using the clock face in figure 1, we can see that $9 + 6 = 3$. We start at 9 and count 6 units (hours) in a clockwise direction. We complete the counting at 3. Therefore, $9 + 6 = 3$ in clock arithmetic. Any of the examples listed earlier can be figured out in this manner.

Using this technique, we can also verify that $6 + 12 = 6$, $9 + 10 = 7$, $11 + 11 = 10$, and $9 + 9 = 6$. In fact, we can construct a table of addition facts for a 12-hour clock using the set of elements $\{1, 2, 3, 4, 5, 6, 7, 8, 9, 10, 11, 12\}$ and the operation of addition. See table 1.

Table 1 gives us the answer when we add any two numbers on a 12-hour clock. All answers are included, since we have combined each element in the set with every other element in the set. This underscores the fact that we are working with a finite mathematical system. Every answer in the table is a member of the original set $\{1, 2, 3, 4, 5, 6, 7, 8, 9, 10, 11, 12\}$, so there are no new elements in the set of answers. This is a unique characteristic for some systems and is called **closure.**

A system is said to be *closed* with respect to an operation (in this case addition) if, when we operate on any two elements in the system, the result is also an element in the system. In the case of addition of clock numbers, when we add any two clock numbers, the sum is also a clock number.

More formally, we can say that:

> **A system consisting of a set of elements $\{a, b, c, \ldots\}$ and an operation $*$ is closed if for any two elements a and b in the set, $a * b$ (read "a operation b") is also a member of the set.**

EXAMPLE 1

Using 12-hour clock addition, evaluate each of the following:

a. $4 + 7$ b. $7 + 7$ c. $10 + 9$

d. $5 + 8$ e. $9 + 7$ f. $8 + 12$

Solution

We use the table of addition facts for a 12-hour clock (table 1) to find the answer to each problem.

a. $4 + 7 = 11$ b. $7 + 7 = 2$ c. $10 + 9 = 7$

d. $5 + 8 = 1$ e. $9 + 7 = 4$ f. $8 + 12 = 8$

It may have occurred to you that the technique of using a table to solve the problems in example 1 is not the only way that these problems could be done. There is a more efficient way that we shall now explore.

The armed forces and many factories operate on a 24-hour clock. These clocks begin the same as the 12-hour clock, but once it becomes noon, the 12-hour system starts over, while the 24-hour system continues. For example, 1:00 P.M. becomes 1300 hours, or 13:00. Similarly, 2:00 P.M. is the same as 14:00, 3:00 P.M. is the same as 15:00, and so on. This idea can be used to express any number as one of the numbers on a 12-hour clock, that is, as one of the numbers in the set $\{1, 2, 3, 4, 5, 6, 7, 8, 9, 10, 11, 12\}$. For example, 13 can be expressed as 1, since 13 hours is the same as 1 rotation around the clock (12 hours) plus 1 additional hour, that is, $13 = 12 + 1$. The number 15 can be expressed as 3, since $15 = 12 + 3$.

The number 12 has a special property in the 12-hour clock system: whenever we add 12 to a number, we obtain that number as a solution ($8 + 12 = 8$, $2 + 12 = 2$, and so on). The number 12 is the *identity element* in this system.

U.S. Navy

This clock can be used to tell time in either the 12-hour system or the 24-hour system.

A system consisting of a set of elements $\{a, b, c, \ldots\}$ and an operation * has an identity element (we will call it e) if for every element a in the system,

$$a * e = a \quad \text{and} \quad e * a = a$$

The identity element does not change any element when it is operated on together with that element. In ordinary arithmetic, the identity element is 0 (zero) for the operation of addition, since $4 + 0 = 0 + 4$, $99 + 0 = 0 + 99$, etc.

Since 12 is the identity element in the 12-hour clock system, we can use it to express any number as one of the numbers in that system. We have already changed some numbers to one of the numbers in the system, but suppose we want to change a number such as 55. It can be expressed as 7 in the 12-hour clock system as follows: starting at 12, we complete 4 rotations around the clock (48 hours), plus 7 more hours, to get 55 hours. That is, $55 = 12 + 12 + 12 + 12 + 7$. Hence, the number 55 can be expressed as 7 in our system, because 12 is the identity element and does not change the identity of the number 7. Therefore 55 and 7 are in the same position on a 12-hour clock.

Another way to show that 55 can be expressed as 7 is by means of division. If we divide 55 by 12, the remainder is 7:

$$\begin{array}{r} 4 \\ 12\overline{)55} \\ \underline{48} \\ 7 \end{array}$$

This division indicates that there are four 12s in 55. The 12s do not affect the value of the number in the 12-hour clock system; therefore the remainder, 7, is our answer.

In order to convert any number into a number in the 12-hour clock system, we divide it by 12 and record the remainder. The number 116, for example, can be expressed as 8:

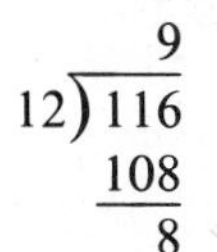

The nine 12s contained in 116 do not affect the value of the number in the 12-hour system, so the remainder, 8, is our answer.

EXAMPLE 2

Find the equivalent of each of the following on a 12-hour clock.

a. 124 b. 258 c. 2000 d. 300

Solution

a. We divide 124 by 12 and record the remainder.

$$\begin{array}{r} 10 \\ 12\overline{)124} \\ \underline{12} \\ 04 \\ \underline{0} \\ 4 \end{array}$$

Hence 124 is equivalent to 4 on a 12-hour clock.

b. Using the same technique we used in part *a*, we divide 258 by 12. The remainder is 6.

c. The number 2000 is equivalent to 8 on a 12-hour clock, since the remainder when 2000 is divided by 12 is 8.

d. The number 300 is equivalent to 12 on a 12-hour clock. Recall that 12 is the identity element in our system under the operation of addition; it has the same property as zero for the operation of addition in ordinary arithmetic. Thus 300 hours would take 25 complete rotations around the clock and would stop at the same place it began, at 12. Hence, after 300 hours, the clock is again at the beginning position.

EXAMPLE 3

Evaluate each of the following on a 12-hour clock.

a. $8 + 4$ b. $9 + 3$ c. $6 + 6$

Solution

a. $8 + 4 = 12$ b. $9 + 3 = 12$ c. $6 + 6 = 12$

The problems presented in example 3 all have the answer 12. Recall that 12 is the identity element in our system, and that anytime 12 is added to a number the identity or position of the number will not be changed. That is, $2 + 12 = 2$, $3 + 12 = 3$, and so on. In the problems in example 3, we have added one number to another number and obtained the identity element as the result. These problems illustrate another property found in mathematical systems. In the problem $8 + 4 = 12$, 4 is called the *inverse* or *additive inverse* of 8 because when we add 4 to 8 we obtain the identity element 12. Given any clock number in the 12-hour clock system $\{1, 2, 3, 4, 5, 6, 7, 8, 9, 10, 11, 12\}$, we can find another clock number such that the sum of the two numbers is the identity element.

Each element in a system consisting of a set of elements $\{a, b, c, \ldots\}$ and an operation $*$ has an inverse if for every element a in the system there exists an element b (also in the system) such that

$$a * b = e \quad \text{and} \quad b * a = e$$

where e is the identity element of the system.

Note that if a system has no identity element, then the inverses of elements cannot occur. In the 12-hour clock system, every element has an inverse. For example, 11 is the inverse of 1, because $1 + 11 = 12$ and $11 + 1 = 12$.

EXAMPLE 4

Find the additive inverse of each number in the 12-hour clock system.

a. 7 b. 2 c. 12

FIGURE 2

Solution

a. The additive inverse of 7 is 5. We have $5 + 7 = 12$ and $7 + 5 = 12$.

b. The additive inverse of 2 is 10: $2 + 10 = 12$ and $10 + 2 = 12$.

c. The additive inverse of 12 is 12, because in a 12-hour clock system $12 + 12 = 12$. Thus 12 is its own inverse.

Let us next consider subtracting numbers in our system. What is the answer to the problem $2 - 3$ on a 12-hour clock? Using the clock face in figure 2, we can see that $2 - 3 = 11$: we start at 2 and count 3 units (hours) in a counterclockwise direction; we complete the counting at 11. Therefore $2 - 3 = 11$ in clock arithmetic.

This problem may also be solved in another manner. Recall that 12 is the identity element in our system, so if we add 12 to a number we will not change its value in the system. Therefore when we add 12 to a number we will not change its position on the clock face. Hence 2:00 is the same as 14:00, and we can think of the problem $2 - 3$ as $14 - 3$. Therefore $2 - 3 = 14 - 3 = 11$ in the 12-hour clock system.

EXAMPLE 5

Evaluate each of the following on a 12-hour clock.

a. $4 - 7$ b. $3 - 8$ c. $4 - 12$

Solution

a. In the 12-hour clock system, we can add 12 to a number and not change its identity. Hence $4 - 7 = 16 - 7 = 9$.

b. $3 - 8 = 15 - 8 = 7$

c. $4 - 12 = 16 - 12 = 4$ is one way of solving this problem. An alternate method is to recall that 12 may also be thought of as zero in our system. Therefore we have $4 - 12 = 4 - 0 = 4$.

Thus far in our discussion of clock arithmetic, we have encountered three properties of a mathematical system: the clock system has the *closure* property with respect to the operation of addition, it has an *identity element* for addition, and each element has an *inverse* with respect to addition. This system also has a property that has not been mentioned yet: the *associative property for addition.*

An operation is associative if the location of parentheses in a problem does not affect the answer. To be more specific, consider the following addition problem in ordinary arithmetic:

$$4 + 6 + 9$$

We can find the sum by adding 4 and 6 first, obtaining 10, and then adding 9 to get an answer of 19; that is, $(4 + 6) + 9 = 10 + 9 = 19$. Or we might add 6 and 9 first, obtaining 15, and then add 4 to get an answer of 19; that is, $4 + (6 + 9) = 4 + 15 = 19$. Regardless of which two numbers are added first, the answer is the same. Hence we can say that

$$(4 + 6) + 9 = 4 + (6 + 9)$$

If a system consists of a set of elements $\{a, b, c, \ldots\}$ and an operation $*$, we say that the operation is associative (or has the associative property) if, for all of the elements in the system,

$$(a * b) * c = a * (b * c)$$

The associative property does not hold for all operations. We have seen that $(4 + 6) + 9 = 4 + (6 + 9)$ in ordinary arithmetic. But, consider the operation of subtraction in ordinary arithmetic. Let us see if

$$(7 - 4) - 3 = 7 - (4 - 3)$$

is true. In computing the answer, we always operate inside the parentheses first. Therefore $(7 - 4) - 3 = 3 - 3 = 0$, while $7 - (4 - 3) = 7 - 1 = 6$, and we have shown that

$$(7 - 4) - 3 \neq 7 - (4 - 3)$$

Hence the associative property does not hold for the operation of subtraction in ordinary arithmetic, because one example showing that a property does not hold is sufficient to illustrate that a property does not work for all elements in the system. Such an example is sometimes called a **counterexample.**

On a 12-hour clock, we have the set of elements $\{1, 2, 3, 4, 5, 6, 7, 8, 9, 10, 11, 12\}$. We can illustrate the associative property for addition by considering the following example:

$$(7 + 8) + 10 \stackrel{?}{=} 7 + (8 + 10)$$

Working with the left side of the equation first, we have $(7 + 8) + 10 = 15 + 10$. But on a 12-hour clock, 15 is the same as 3; hence,

$15 + 10 = 3 + 10 = 13$, which is the same as 1 on a 12-hour clock. Hence, $(7 + 8) + 10 = 1$ on a 12-hour clock. Now working with the right side, we have $7 + (8 + 10) = 7 + 18$. But 18 on a 12-hour clock is the same as 6, so $7 + 18 = 7 + 6 = 13$, which is the same as 1 on a 12-hour clock. Hence $7 + (8 + 10) = 1$ on a 12-hour clock. We have verified that the associative property holds for the example $(7 + 8) + 10 = 7 + (8 + 10)$.

Space does not permit the verification of the associative property for addition for all of the elements in the 12-hour clock system, but the associative property for addition does hold for this system.

Thus far, the 12-hour clock system, consisting of the set of elements $\{1, 2, 3, 4, 5, 6, 7, 8, 9, 10, 11, 12\}$ and the operation of addition—

1. is *closed* with respect to addition
2. contains an *identity element* with respect to addition
3. contains an *inverse* for each of its elements with respect to addition
4. is *associative* with respect to addition.

When a set of elements and an operation satisfy these properties, we say that the elements form a **group** under the operation (in this case addition). The group operation must be a **binary operation,** that is, one which combines two elements to produce a third element. Addition is a binary operation since it acts on two numbers to produce a third. Squaring a number is not a binary operation, since it acts on only one number.

EXAMPLE 6

Consider the set of counting numbers, $\{1, 2, 3, 4, \ldots\}$, and the operation of addition. Does this system form a group? Why or why not?

Solution

No; in order to form a group, a system must satisfy the closure property, identity property, inverse property, and associative property. The set of counting numbers $\{1, 2, 3, 4, \ldots\}$ does not have an identity element under the operation of addition, since zero is not included in the given set of elements. In addition, the elements have no additive inverses.

EXAMPLE 7

Consider the set of counting numbers, $\{1, 2, 3, 4, \ldots\}$, and the operation of multiplication. Does this system form a group? Why or why not?

Solution

In order to form a group under the operation of multiplication, the system must satisfy the four properties: closure, identity, inverse, and associative. We shall check to see if these properties hold under the operation of multiplication.

a. The system is closed under multiplication. Whenever we multiply two counting numbers, the product is a counting number.

b. The system does contain an identity element under multiplication, the number 1. The number 1 does not change the identity of a number when the two are multiplied together. That is, $2 \times 1 = 1 \times 2 = 2$, $3 \times 1 = 1 \times 3 = 3$, and so on.

c. The system does *not* contain an inverse element under multiplication for each element. The number 1 is the only element in the set that has an inverse, and it is its own inverse: $1 \times 1 = 1$. Two does not have an inverse in the set of elements: there is no number b in the set such that $2 \times b = 1$. This counterexample shows that the set of counting numbers $\{1, 2, 3, 4, \ldots\}$ under the operation of multiplication *does not* form a group.

When you find the sum of 4 and 5, whether you add them as $4 + 5$, or as $5 + 4$, the answer is 9. That is, $4 + 5 = 5 + 4$. Similarly, if you multiply 4 and 5 together, you can multiply 4×5 or 5×4, and the answer is 20. So $4 \times 5 = 5 \times 4$. No matter what order you do the operation in, the answer is the same. In other words, we can switch the elements around; we can *commute* them. When we can do this for all elements in a system using a given operation, we say that the operation is *commutative.*

Given a system consisting of a set of elements $\{a, b, c, \ldots\}$ and an operation $*$, we say that the operation is commutative if for all elements a and b in the system,

$$a * b = b * a$$

Not all operations are commutative. We have seen that addition and multiplication of counting numbers are commutative operations. But consider the operation of subtraction: does $7 - 6 = 6 - 7$? The answer is no, since $7 - 6 = 1$, while $6 - 7 = -1$. This verifies that $7 - 6 \neq 6 - 7$, and that subtraction is not commutative. A group with operation $*$ is called a **commutative group** if $a * b = b * a$ for all elements a and b in the group.

Thus far in this section, we have performed the operations of addition and subtraction on the numbers in a 12-hour clock system.

Now let's examine the operation of multiplication. Consider the problem 3×9 on a 12-hour clock. We can consider this as an addition problem, because $3 \times 9 = 9 + 9 + 9$. On a 12-hour clock, $9 + 9 = 18 = 6$, so $(9 + 9) + 9$ becomes $6 + 9$, and $6 + 9 = 3$. Therefore, on a 12-hour clock, $3 \times 9 = 3$. This process is quite tedious. We can multiply two numbers on a 12-hour clock in a more efficient manner: first multiply the numbers as you would in ordinary arithmetic, and then convert your answer to its equivalent on the 12-hour clock. Using this technique for 3×9, we have $3 \times 9 = 27$, which is equivalent to 3 on the 12-hour clock. Hence $3 \times 9 = 3$ on a 12-hour clock.

EXAMPLE 8
Evaluate each of the following on a 12-hour clock.

a. 4×5 b. 6×7 c. 8×10

d. 11×11 e. 10×12

Solution

a. $4 \times 5 = 20$ and 20 is equivalent to 8; hence $4 \times 5 = 8$

b. $6 \times 7 = 42$ and 42 is equivalent to 6; hence $6 \times 7 = 6$

c. $8 \times 10 = 80$ and 80 is equivalent to 8; hence $8 \times 10 = 8$

d. $11 \times 11 = 121$ and 121 is equivalent to 1; hence $11 \times 11 = 1$

e. $10 \times 12 = 120$ and 120 is equivalent to 12; hence $10 \times 12 = 12$

We shall not explore in detail the operation of division in a 12-hour clock system because of the problems that arise in doing division problems in this system. Consider the problem $8 \div 4$. In ordinary arithmetic, the answer is 2; we can verify this because $2 \times 4 = 8$. But in a 12-hour clock system, $8 \div 4 = 2$, $8 \div 4 = 5$, $8 \div 4 = 8$, and $8 \div 4 = 11$, because $4 \times 2 = 8$, $4 \times 5 = 8$, $4 \times 8 = 8$, and $4 \times 11 = 8$ on a 12-hour clock. Thus some problems have more than one answer, while other problems, such as $8 \div 3$, have no answer. In ordinary arithmetic, the answer to this is $3\frac{2}{3}$, but in a 12-hour clock system there is no answer. In order for 3 to divide 8 in a 12-hour clock system, the answer must be one of the elements in the set $\{1, 2, 3, 4, 5, 6, 7, 8, 9, 10, 11, 12\}$. But not one of these numbers will satisfy the statement $3 \times a = 8$. Hence there is no number in this system that when multiplied by 3 will yield 8 as an answer. In other words, $8 \div 3$ has no answer in a 12-hour clock system.

EXERCISES FOR SECTION 6.2

1. Evaluate each sum on a 12-hour clock.
   a. $2 + 4$ b. $7 + 6$ c. $9 + 2$
   d. $11 + 11$ e. $10 + 11$ f. $9 + 10$
2. Evaluate each sum on a 12-hour clock.
   a. $8 + 8$ b. $8 + 9$
   c. $8 + 10$ d. $8 + (9 + 11)$
   e. $(9 + 7) + 6$ f. $9 + (8 + 7)$
3. Find the equivalent of each number on a 12-hour clock.
   a. 33 b. 44 c. 55
   d. 66 e. 277 f. 188
4. Find the equivalent of each number on a 12-hour clock.
   a. 342 b. 201 c. 400
   d. 1984 e. 2001 f. -17
5. Evaluate each difference on a 12-hour clock.
   a. $5 - 7$ b. $6 - 8$ c. $7 - 11$
   d. $9 - 10$ e. $2 - 5$ f. $3 - 7$
6. Evaluate each difference on a 12-hour clock.
   a. $2 - 6$ b. $4 - 9$
   c. $3 - 11$ d. $2 - 12$
   e. $2 - (3 - 8)$ f. $4 - (7 - 9)$
7. Evaluate each product on a 12-hour clock.
   a. 6×8 b. 4×6 c. 3×7
   d. 4×10 e. 5×11 f. 9×9
8. Evaluate each product on a 12-hour clock.
   a. 7×11 b. 8×11
   c. 10×10 d. $3 \times (4 + 7)$
   e. $2 \times (9 + 10)$ f. $5 \times (6 - 10)$
9. State the property of the 12-hour clock system that is illustrated by each of the following problems:
   a. $(4 + 3) + 5 = 4 + (3 + 5)$
   b. $7 + 6 = 1$, a number in the 12-hour clock system
   c. $9 + 12 = 9$
   d. $8 + 9 = 9 + 8$
   e. $7 \times 7 = 1$
   f. $4 \times 5 = 8$, a number in the 12-hour clock system
10. Construct a complete table of multiplication facts for the 12-hour clock system. Use your table to answer each question.
    a. What is the identity element for multiplication in this system?
    b. Does the closure property hold for this system?
    c. Does the commutative property hold for this system?
    d. Verify that the associative property holds for one specific instance in this system.
    e. What elements in this system have an inverse?
11. Determine whether each statement is true or false. Given the set of counting numbers $\{1, 2, 3, 4, \ldots\}$, it—
    a. is closed with respect to addition
    b. is closed with respect to subtraction
    c. is associative with respect to addition
    d. is commutative with respect to division
    e. contains an identity element for addition
    f. contains an identity element for multiplication.
12. Determine whether each statement is true or false for the set of all integers, $\{\ldots, -2, -1, 0, 1, 2, \ldots\}$.
    a. The set is closed with respect to addition
    b. The set is closed with respect to subtraction
    c. The set is closed with respect to division
    d. The set is a system that contains an identity element for addition
    e. The set contains an identity element for multiplication
    f. The set contains an additive inverse element for each element in the set.
13. Replace each question mark with a number from the 12-hour clock system to give a true statement.
    a. $6 + ? = 1$ b. $? - 4 = 9$
    c. $5 \times ? = 1$ d. $5 \times (8 + 7) = ?$
    e. $? \times 7 = 9$ f. $? \times (2 + 11) = 2$

14. Replace each question mark with a number from the 12-hour clock system to give a true statement.

    a. $3 + ? = 1$
    b. $4 - ? = 9$
    c. $4 - 7 = ?$
    d. $3 \times (4 + ?) = 9$
    e. $8 + ? = 2$
    f. $4 + ? = 8 - ?$

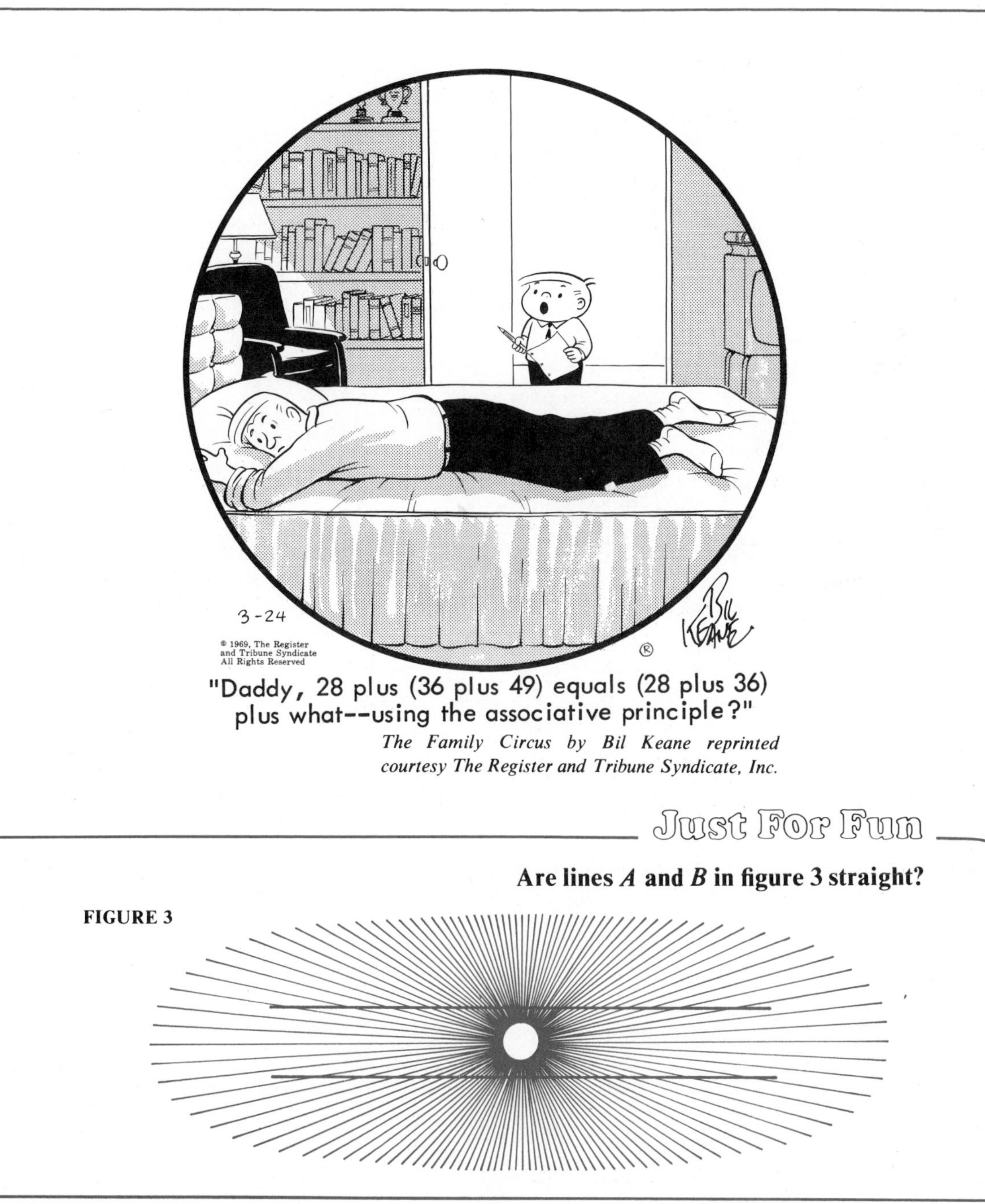

"Daddy, 28 plus (36 plus 49) equals (28 plus 36) plus what--using the associative principle?"

The Family Circus by Bil Keane reprinted courtesy The Register and Tribune Syndicate, Inc.

Just For Fun

Are lines *A* and *B* in figure 3 straight?

FIGURE 3

6.3 MORE NEW SYSTEMS

In the preceding section, we examined the nature of clock arithmetic. Using the set of numbers $\{1, 2, 3, 4, 5, 6, 7, 8, 9, 10, 11, 12\}$ and the properties of a 12-hour clock, we constructed a new system of arithmetic.

There are many other systems of arithmetic that can be created. For another example, let

1 = spring	3 = fall
2 = summer	4 = winter

If it is now spring, then 3 seasons from now it will be winter. That is, $1 + 3 = 4$, or spring + fall = winter. Suppose it is now winter. What season will it be 5 seasons from now? Starting with winter and counting off 5 seasons, we wind up at spring. This suggests that $4 + 5 = 1$ in our new system. That is, we have a cycle of 4 seasons, and so once we reach 4 in our counting, we start over again. For the previous problem we have $4 + 5 = 9$ in ordinary arithmetic, but $9 = 4 + 4 + 1$, so in our new system 9 is equivalent to 1. To help us in our calculations, we could think of this new system in terms of a 4-hour clock. Tables 2 and 3 show the addition and multiplication facts for the seasons system.

TABLE 2

+	1	2	3	4
1	2	3	4	1
2	3	4	1	2
3	4	1	2	3
4	1	2	3	4

TABLE 3

×	1	2	3	4
1	1	2	3	4
2	2	4	2	4
3	3	2	1	4
4	4	4	4	4

EXAMPLE 1

Let spring = 1, summer = 2, fall = 3, and winter = 4. Evaluate the following:

a. summer + summer

b. fall + summer

c. summer × fall

d. fall × fall

Solution

a. Since summer = 2, we have $2 + 2 = 4$, and 4 = winter, so summer + summer = winter.

b. Fall = 3 and summer = 2; therefore, we have $3 + 2 = 1$, according to table 2. Since 1 = spring, fall + summer = spring.

As an alternate solution, $3 + 2 = 5$, which is equivalent to 1 in this system, and $1 =$ spring.

c. Summer $= 2$ and fall $= 3$; therefore, we have $2 \times 3 = 2$, according to table 3. Since $2 =$ summer, summer $\times$ fall $=$ summer.

Alternately, we have $2 \times 3 = 6$, which is equivalent to 2 in this system, and $2 =$ summer.

d. Fall $= 3$; therefore, we have $3 \times 3 = 1$, according to table 3. Since $1 =$ spring, fall $\times$ fall $=$ spring.

As an alternate solution, we have $3 \times 3 = 9$, which is equivalent to 1 in this system, and $1 =$ spring.

Recall that table 2 illustrates the addition facts for the seasons system. If we substitute the names of the seasons for the numbers in table 2, we would have table 4, which may appear strange at first glance. But, remember, the only thing that we have done is substitute the names of the seasons in place of the numbers in table 2. This system may seem more abstract, but it is still a system.

TABLE 4

+	spring	summer	fall	winter
spring	summer	fall	winter	spring
summer	fall	winter	spring	summer
fall	winter	spring	summer	fall
winter	spring	summer	fall	winter

Note that we have closure in this system, because whenever we combine any two seasons the result is a member of the set {spring, summer, fall, winter}. What is the identity element in this system? Recall that the identity element does not change any element when it is operated on together with that element. Upon examination of table 4, we see that winter is the identity element in our seasons system for the operation of addition. We can verify this by checking each season's addition with winter: for example, spring + winter = spring, and winter + spring = spring.

EXAMPLE 2

Using table 4, find the additive inverse of each of the following:

a. spring b. summer c. fall d. winter

Solution

Recall that each element a in a system has an *inverse* element if there exists an element b such that $a * b = e$ and $b * a = e$, where e is the identity element of the system and $*$ is the operation of the system. Since winter is the identity element in the seasons system, we examine the table to see what season added to a given season yields winter as the answer.

a. The additive inverse of spring is fall.

b. The additive inverse of summer is summer.

c. The additive inverse of fall is spring.

d. The additive inverse of winter is winter.

Suppose that today is Wednesday, the fourth day of the month. What day of the week is the twenty-seventh? One way we could determine this is to examine a calendar. Another interesting technique is the following: Since Wednesday corresponds to 4 (it is the fourth day of the month), then Thursday corresponds to 5, Friday corresponds to 6, and so on. Twenty-one days from now it will be Wednesday again (the days of the week are in a cycle of 7), and the date will be the twenty-fifth; hence, the twenty-seventh will fall on a Friday. We can also solve this problem in another way: Since every 7 days brings us back to the same day of the week, 27 is equivalent to 6 in a week system, because $27 = 7 + 7 + 7 + 6$. Recall that Wednesday corresponded to 4; hence Friday must correspond to 6, and the twenty-seventh must be a Friday. This discussion suggests that we can use the days of the week to create another new system of arithmetic.

H. Armstrong Roberts

Let

1 = Sunday	5 = Thursday
2 = Monday	6 = Friday
3 = Tuesday	7 = Saturday
4 = Wednesday	

If today is Sunday, then 3 days from now it will be Wednesday; that is, $1 + 3 = 4$. If today is Tuesday, then 4 days from now it will be Saturday; that is, $3 + 4 = 7$. Suppose that today is Monday; what day will it be 8 days from now? We know that 8 days from now it will be Tuesday, because 7 days from now it will be Monday again, and 1 more day will bring us to Tuesday. But we could also say Monday = 2, and $2 + 8 = 10$. Ten is equivalent to 3 in the week system, since 7 acts as an identity element in this system for the operation of addition and 10 = 3 + 7.

Table 5 shows the addition facts for the week system. Note that whenever 7 is added to another number in the system, the result is always the original number. We also have closure in this new system, because the results of adding elements in the system are always in the system. There is an identity element (7), so we can determine the additive inverse for an element in this system. For example, the additive inverse of 2 is 5, since 2 + 5 = 7 and 5 + 2 = 7.

TABLE 5

+	1	2	3	4	5	6	7
1	2	3	4	5	6	7	1
2	3	4	5	6	7	1	2
3	4	5	6	7	1	2	3
4	5	6	7	1	2	3	4
5	6	7	1	2	3	4	5
6	7	1	2	3	4	5	6
7	1	2	3	4	5	6	7

EXAMPLE 3

Given that Sunday = 1, Monday = 2, Tuesday = 3, Wednesday = 4, Thursday = 5, Friday = 6, and Saturday = 7, evaluate the following:

a. Sunday + Tuesday

b. Monday + Wednesday

c. Friday + Tuesday

d. Wednesday + Friday

Solution

a. Since Sunday = 1 and Tuesday = 3, we have 1 + 3 = 4, and 4 = Wednesday; Sunday + Tuesday = Wednesday.

b. Monday = 2 and Wednesday = 4; therefore we have 2 + 4 = 6, and 6 = Friday; Monday + Wednesday = Friday.

c. Friday = 6 and Tuesday = 3; therefore, we have 6 + 3 = 2 according to the table, and 2 = Monday. Hence, Friday + Tuesday = Monday.

   As an alternate solution, 6 + 3 = 9, which is equivalent to 2 in this system, and 2 = Monday.

d. Wednesday = 4 and Friday = 6; hence, 4 + 6 = 3 according to the table, and 3 = Tuesday. Therefore, Wednesday + Friday = Tuesday.

We could also note that $4 + 6 = 10$, which is equivalent to 3 in this system, and $3 =$ Tuesday.

EXAMPLE 4

Using the week system, evaluate the following:

a. Wednesday − Thursday
b. Monday × Tuesday
c. Wednesday × Thursday
d. (Friday × Monday) − Wednesday

Solution

a. Wednesday = 4 and Thursday = 5; therefore we have $4 - 5$. In order to subtract 5 from 4, we must change 4 to an equivalent number: $4 + 7 = 11$, so 4 is equivalent to 11. Hence, in our system, $4 - 5 = 11 - 5 = 6$, and 6 = Friday. Hence Wednesday − Thursday = Friday.

b. Monday = 2 and Tuesday = 3; $2 \times 3 = 6$, and 6 = Friday. Hence Monday × Tuesday = Friday.

c. Wednesday = 4 and Thursday = 5; $4 \times 5 = 20$, which is equivalent to 6, since, in the week system, $20 = 7 + 7 + 6$, and 6 = Friday. Hence Wednesday × Thursday = Friday.

d. Friday = 6, Monday = 2, and Wednesday = 4. Working inside the parentheses first, we have (6×2), which equals 12, and $12 - 4 = 8$, which is equivalent to 1 in the week system. Hence (Friday × Monday) − Wednesday = Sunday.

EXERCISES FOR SECTION 6.3

In exercises 1–12, assume that 1 = spring, 2 = summer, 3 = fall, and 4 = winter. Evaluate each of the following, giving your answer in terms of a season:

1. summer + fall
2. fall + winter
3. winter + summer
4. winter − spring
5. spring − summer
6. fall − winter
7. summer × summer
8. spring × summer
9. fall × winter
10. fall × (winter + spring)
11. fall × (winter − spring)
12. winter × winter
13. What is the identity element in the seasons system for the operation of multiplication?
14. What is the multiplicative inverse of winter?
15. What is the multiplicative inverse of fall?

In exercises 16–32, assume that 1 = Sunday, 2 = Monday, 3 = Tuesday, 4 = Wednesday, 5 = Thursday, 6 = Friday, and 7 = Saturday. Evaluate each of the following, giving your answer in terms of a day of the week:

16. Sunday + Friday
17. Monday + Saturday
18. Tuesday + Monday
19. Thursday + Friday
20. Friday + Friday
21. Sunday + Sunday
22. Friday − Wednesday
23. Thursday − Friday
24. Friday − Saturday
25. Thursday − Saturday
26. Wednesday − Saturday
27. Tuesday − Thursday
28. Tuesday × Friday
29. Wednesday × Saturday
30. Saturday × Saturday
31. Tuesday × (Monday + Friday)
32. Friday × (Friday − Saturday)
33. What is the identity element in the week system for the operation of addition?
34. What is the identity element in the week system for the operation of multiplication?
35. What is the additive inverse of—
    a. Monday b. Tuesday c. Wednesday
36. What is the multiplicative inverse of—
    a. Monday b. Friday
37. Verify that Sunday × (Tuesday × Friday) = (Sunday × Tuesday) × Friday.
38. Does the week system form a group under the operation of multiplication?
39. Does the week system form a group under the operation of addition?

Just For Fun

Take any size piece of paper and fold it in half, then fold it in half again, and keep doing this as many times as possible. You probably cannot do it 8 times.

FIGURE 4

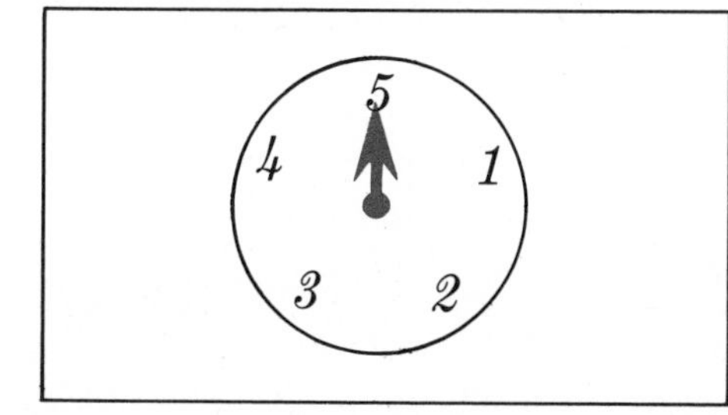

6.4 MODULAR SYSTEMS

Thus far in this chapter, we have examined various new systems of arithmetic. One of the systems discussed was the 12-hour clock system, since everyone is familiar with the 12-hour clock. We will next consider a mathematical system that is also based on a clock, but this time we will consider a 5-hour clock.

A 5-hour clock might be one like the one shown in figure 4, but to make our new clock a little easier to understand, we will make it

look like a timer. You have probably seen some sort of a timer before—for example, an egg timer or a stop watch. These timers have zero at the top of the face, rather than some other number. Figure 5 shows our new 5-hour clock. This 5-hour clock system contains the elements 0, 1, 2, 3, 4. If we begin to count in this system (clockwise), we have 1, 2, 3, 4, 0, and then the system starts to repeat itself; that is, 1, 2, 3, 4, 0, 1, 2, 3, 4, 0, 1, 2, . . .

FIGURE 5

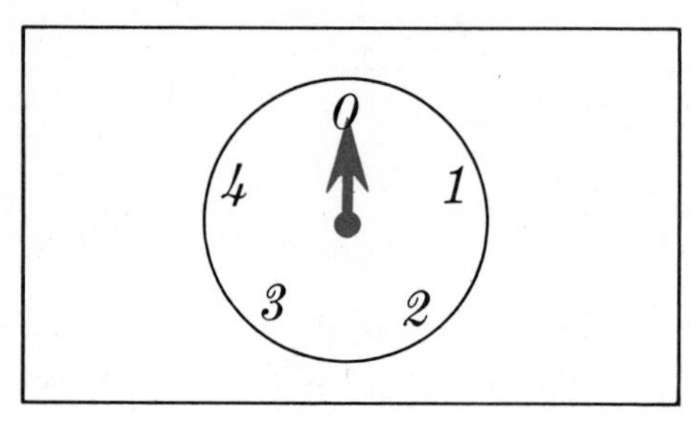

If it is now 1:00 on our 5-hour clock, then 3 hours from now it will be 4:00. Thus $1 + 3 = 4$. Suppose it is now 2:00 on the 5-hour clock. What time will it be 4 hours from now? To find out, we begin at 2 and count 4 units in a clockwise direction, ending up at 1. Therefore in this new system, $2 + 4 = 1$.

Similarly, we can verify that $4 + 3 = 2$ using the 5-hour clock. Continuing with this technique, we can construct a table of addition facts for a 5-hour clock using the set of elements $\{0, 1, 2, 3, 4\}$. See table 6.

TABLE 6

+	0	1	2	3	4
0	0	1	2	3	4
1	1	2	3	4	0
2	2	3	4	0	1
3	3	4	0	1	2
4	4	0	1	2	3

When we have a system such as the 5-hour or 12-hour clock that repeats itself in a cycle, we call it a **modular system.** The 5-hour clock is called the *modulo 5 system.* We abbreviate this to *mod 5 system.*

Recall that in order to convert a given number into a number in the 12-hour clock system, we divided it by 12 and recorded the remainder. Twelve did not affect the answer because it was the identity element. In order to convert a given number into a number in the 5-hour clock system (modulo 5), we divide it by 5 and record the remainder. For example, from table 6, we know that $4 + 4 = 3$ in the modulo 5 system, but we can also do this by finding the sum of 4 and 4 in ordinary arithmetic, 8, and converting it to a number in the modulo 5 system:

$$\begin{array}{r} 1 \\ 5\overline{)8} \\ \underline{5} \\ 3 \end{array}$$

When we divide 8 by 5, the remainder is 3, our desired result. Therefore, we can write

$$4 + 4 \equiv 3 \pmod 5$$

This is read "4 + 4 is equivalent to 3, mod 5."

Similarly, we can write $6 \equiv 1 \pmod 5$, $21 \equiv 1 \pmod 5$, and $36 \equiv 1 \pmod 5$. In each case, this indicates that 1 is the remainder when the numbers 6, 21, and 36 are divided by 5. In general terms,

$$a \equiv b \pmod m$$

means that a and b both have the same remainder when they are divided by m. We say that *a is equivalent to b mod m.*

EXAMPLE 1
Evaluate the following in the modulo 5 system.

a. $4 + 4$ b. $2 + 4$ c. $3 + 2$

Solution

a. Using table 6, we note that the sum of 4 and 4 is 3; that is, $4 + 4 \equiv 3 \pmod 5$.

b. According to the table, the sum of 2 and 4 is 1, or $2 + 4 \equiv 1 \pmod 5$.

c. Since $3 + 2$ is equivalent to 0, $3 + 2 \equiv 0 \pmod 5$.

Alternate solution

a. $4 + 4 = 8$; divide 5 into 8, and the remainder is 3. Hence, $4 + 4 \equiv 3 \pmod 5$.

b. $2 + 4 = 6$; divide 5 into 6, and the remainder is 1. Hence, $2 + 4 \equiv 1 \pmod 5$.

c. $3 + 2 = 5$; divide 5 into 5, and the remainder is 0. Hence $3 + 2 \equiv 0 \pmod 5$.

EXAMPLE 2
Evaluate the following in the modulo 7 system.

a. $4 + 5$ b. $4 + 4$ c. $6 + 5$ d. $4 + 3$

Solution

a. $4 + 5 = 9$; divide 7 into 9, and the remainder is 2. Therefore, $4 + 5 \equiv 2 \pmod 7$.

b. $4 + 4 = 8$; divide 7 into 8, and the remainder is 1. Therefore, $4 + 4 \equiv 1 \pmod{7}$.

c. $6 + 5 = 11$; divide 7 into 11, and the remainder is 4. Therefore, $6 + 5 \equiv 4 \pmod{7}$.

d. $4 + 3 = 7$; divide 7 into 7, and the remainder is 0. Therefore, $4 + 3 \equiv 0 \pmod{7}$.

EXAMPLE 3
Evaluate each of the following in the modulo 5 system.

a. $2 - 4$ b. $1 - 3$ c. $3 - 4$ d. $4 - 0$

Solution

a. In the mod 5 system, we can add 5 to a number and not change its identity. Therefore, $2 - 4 = 7 - 4 = 3$, so $2 - 4 \equiv 3 \pmod{5}$.

b. $1 - 3 = 6 - 3 = 3$, so $1 - 3 \equiv 3 \pmod{5}$

c. $3 - 4 = 8 - 4 = 4$, so $3 - 4 \equiv 4 \pmod{5}$

d. $4 - 0 \equiv 4 \pmod{5}$

Let's examine the properties of the modulo 5 mathematical system with the operation of addition. This system consists of the set of elements $\{0, 1, 2, 3, 4\}$. The addition operation on this set is shown in table 6 (page 288).

1. **Closure property:** All of the entries in the table are elements of the given set; there are no new elements appearing in the table. Therefore, the system satisfies the closure property.

2. **Identity property:** There is an element, 0, in the set of elements that does not change any element when it is added to that element:

   $$0 + 0 = 0, \quad 1 + 0 = 1, \quad 2 + 0 = 2, \quad 3 + 0 = 3, \quad 4 + 0 = 4$$

3. **Inverse property:** Each element in this system has an inverse, an element which when added to the given element results in the identity element, 0. Note that

   $$0 + 0 = 0, \quad 1 + 4 = 0, \quad 2 + 3 = 0, \quad 3 + 2 = 0, \quad 4 + 1 = 0$$

   Therefore, the inverse of 0 is 0, the inverse of 1 is 4, the inverse of 2 is 3, the inverse of 3 is 2, and the inverse of 4 is 1.

4. **Associative property:** Addition is associative if $(a + b) + c = a + (b + c)$. Trying an example, we have $(1 + 2) + 3 = 3 + 3 = 1$ and $1 + (2 + 3) = 1 + 0 = 1$, so that

$$(1 + 2) + 3 = 1 + (2 + 3)$$

Any other example for this system also works; hence, the system satisfies the associative property.

5. **Commutative property:** Addition is commutative if $a + b = b + a$. Trying an example, we have $2 + 4 = 1$ and $4 + 2 = 1$; therefore, $2 + 4 = 4 + 2$. Any other example for this system also works; hence, the system satisfies the commutative property.

The mathematical system (modulo 5) that consists of the set of elements $\{0, 1, 2, 3, 4\}$ and the operation of addition satisfies all of the properties listed. Since a group is composed of a set of elements together with a binary operation that satisfies the closure property, identity property, inverse property, and associative property, the modulo 5 system with the operation of addition forms a group. The fact that the system also satisfies the commutative property means that we have a commutative group under the operation of addition.

Now let's consider the modulo 5 system with the operation of multiplication. This system consists of the set of elements $\{0, 1, 2, 3, 4\}$ and the operation of multiplication. We first construct a multiplication table for mod 5. See table 7. We next determine what properties are satisfied under this system. The system is closed with respect to multiplication. It has an identity element, 1; that is, $a \times 1 = a$ for any a in the system. But, not every element has an inverse. In particular, there is no number in the system which when multiplied by zero will yield the identity element 1. Zero times a number is zero for every number in the system.

TABLE 7

×	0	1	2	3	4
0	0	0	0	0	0
1	0	1	2	3	4
2	0	2	4	1	3
3	0	3	1	4	2
4	0	4	3	2	1

EXAMPLE 4
Using the elements of the mod 5 system, find replacements for the question marks so that each of the following is true.

a. $4 + 3 \equiv ?\ (\text{mod } 5)$ b. $? + 4 \equiv 1\ (\text{mod } 5)$

c. $3 - ? \equiv 4\ (\text{mod } 5)$ d. $3 \times ? \equiv 1\ (\text{mod } 5)$

Solution

a. $4 + 3 = 7$; divide 5 into 7, and the remainder is 2. Hence, $4 + 3 \equiv 2\ (\text{mod } 5)$.

b. Since the mod 5 system contains the set of numbers $\{0, 1, 2, 3, 4\}$, we can try each one of these in place of the question mark. Starting with 0, we have: $0 + 4 = 4$; $1 + 4 = 0$; $2 + 4 = 6$, but $6 \equiv 1\ (\text{mod } 5)$. Therefore, our answer is 2.

c. Using the same technique, we have $3 - 0 = 3$, $3 - 1 = 2$, $3 - 3 = 0$, and $3 - 4 = 8 - 4 = 4$. Therefore the solution is 4.

d. We see that $3 \times 0 = 0$, $3 \times 1 = 3$, and $3 \times 2 = 6$. Since $6 \equiv 1\ (\text{mod } 5)$, the solution is 2.

EXAMPLE 5
Scott, an avid sports fan, collects football-player cards. Each day, he studies his collection of cards. On Monday, he divided his set of cards into piles of 5 with 2 left over; on Tuesday, he divided his set of cards into piles of 4 with 2 left over; and on Wednesday, he divided his set of cards into piles of 7 with 0 left over. If Scott has less than 50 cards in his collection, how many does he have?

Solution
There is no information given as to exactly how many cards Scott has, but we do know how many are in each pile on each day. Let x represent the number of cards. On Monday, each pile contained 5 cards with 2 left over. Therefore if 5 is divided into x, there is a remainder of 2; that is, $x \equiv 2\ (\text{mod } 5)$. On Tuesday, each pile contained 4 cards with 2 left over. Therefore if 4 is divided into x, there is a remainder of 2; so $x \equiv 2\ (\text{mod } 4)$. On Wednesday, each pile contained 7 cards with 0 left over. Hence, if 7 is divided into x, there is a remainder of 0; so $x \equiv 0\ (\text{mod } 7)$.

Now we must find a value that satisfies all three statements:

$$x \equiv 2\ (\text{mod } 5), \qquad x \equiv 2\ (\text{mod } 4), \qquad x \equiv 0\ (\text{mod } 7)$$

For the first statement, $x \equiv 2\ (\text{mod } 5)$, the set of possible replacements for x is $\{7, 12, 17, 22, 27, 32, 37, 42, 47\}$. We stop at 47 since

Scott has less than 50 cards. Now which of these numbers also satisfies the second statement, $x \equiv 2 \pmod{4}$? That is, which of these numbers have a remainder of 2 when divided by 4? These numbers are 22 and 42. Now which of these two numbers also satisfies the third statement, $x \equiv 0 \pmod{7}$? The remainder is zero when 42 is divided by 7, and therefore Scott had 42 cards in his collection.

EXERCISES FOR SECTION 6.4

1. Evaluate each sum on a 5-hour clock—that is, in the modulo 5 system.
   a. $1 + 3$ b. $2 + 3$
   c. $4 + 2$ d. $3 + 4$
   e. $(3 + 2) + 4$ f. $(4 + 3) + 4$

2. Evaluate each sum in the modulo 5 system.
   a. $3 + 3$ b. $4 + 4$
   c. $2 + 2$ d. $1 + 4$
   e. $3 + (3 + 4)$ f. $(2 + 2) + 2$

3. Find the equivalent of each number in the modulo 5 system.
   a. 33 b. 44 c. 55
   d. 342 e. 780 f. -8

4. Find the equivalent of each number in the modulo 5 system.
   a. 32 b. 41 c. 53
   d. 287 e. 2001 f. -17

5. Evaluate each difference in the modulo 5 system.
   a. $2 - 4$ b. $1 - 4$
   c. $1 - 3$ d. $3 - 4$
   e. $3 - (2 - 4)$ f. $2 - (3 - 4)$

6. Evaluate each difference in the modulo 5 system.
   a. $2 - 3$ b. $1 - 2$
   c. $4 - 4$ d. $(2 - 3) - 4$
   e. $(3 - 4) - 1$ f. $1 - (2 - 4)$

7. Evaluate each product in the modulo 5 system.
   a. 4×3 b. 2×4
   c. 3×2 d. $2 \times (4 \times 3)$
   e. $3 \times (4 + 2)$ f. $2 \times (3 - 4)$

8. Evaluate each product in the modulo 5 system.
   a. 4×4 b. 3×3
   c. 4×2 d. $3 \times (3 \times 3)$
   e. $2 \times (3 + 4)$ f. $3 \times (2 - 4)$

9. Given the set of elements $\{0, 1, 2, 3, 4, 5, 6\}$, construct a complete table of addition facts for the modulo 7 system.
   a. Does the closure property hold for this system?
   b. What is the identity element (if any) for the operation of addition in this system?
   c. Does the commutative property hold for this system?
   d. Verify that the associative property holds for one specific instance in this system.
   e. List the elements in this system that have an additive inverse, and list their inverses.
   f. Does this system form a group under the operation of addition?

10. Given the set of elements $\{0, 1, 2, 3, 4, 5, 6\}$, construct a complete table of multiplication facts for the modulo 7 system.
    a. Does the closure property hold for this system?
    b. What is the identity element (if any) for the operation of multiplication?
    c. Does the commutative property hold for this system?
    d. List the elements in this system that have a multiplicative inverse, and list their inverses.
    e. Does this system form a group under the operation of multiplication?

11. Determine whether each statement is true or false.
    a. $18 \equiv 3 \pmod 5$
    b. $22 \equiv 1 \pmod 5$
    c. $144 \equiv 4 \pmod 5$
    d. $33 \equiv 5 \pmod 7$
    e. $49 \equiv 0 \pmod 7$
    f. $99 \equiv 2 \pmod 7$

12. Determine whether each statement is true or false.
    a. $44 \equiv 9 \pmod 5$
    b. $27 \equiv 2 \pmod 5$
    c. $17 \equiv 3 \pmod 5$
    d. $140 \equiv 88 \pmod 7$
    e. $213 \equiv 12 \pmod 7$
    f. $1000 \equiv 55 \pmod 5$

13. Using the elements of the indicated modular system, find a replacement for each question mark so that each statement is true.
    a. $4 + ? \equiv 1 \pmod 5$
    b. $3 + ? \equiv 2 \pmod 5$
    c. $? + 4 \equiv 3 \pmod 5$
    d. $? + 3 \equiv 2 \pmod 7$
    e. $2 - ? \equiv 4 \pmod 6$
    f. $3 - ? \equiv 4 \pmod 7$

14. Using the elements of the indicated modular system, find a replacement for each question mark so that each statement is true.
    a. $2 \times ? \equiv 3 \pmod 7$
    b. $? \times 4 \equiv 1 \pmod 5$
    c. $? - 6 \equiv 2 \pmod 7$
    d. $1 - ? \equiv 4 \pmod 5$
    e. $2 \times ? \equiv 3 \pmod 9$
    f. $2 - ? \equiv 3 \pmod{12}$

15. Stan, a stock boy in a local supermarket, had to change the price on a set of soup cans on a particular shelf. In order to make the job easier, Stan decided to arrange the cans in stacks of 10, but when he did this he had 1 left over. Trying again, he arranged them in stacks of 7, but then he had 4 left over. Finally, he arranged them in stacks of 3, and there were none left over. If the shelf can only hold 100 cans, on how many cans did Stan have to change the price?

16. Irene is a cashier in a restaurant. After the rush hour, she began to tabulate the customers' checks. First, she arranged the checks in stacks of 5, and there was 1 left over. Next, Irene arranged the checks in stacks of 7, and there were 2 left over. Finally, she arranged them in stacks of 4, and there were 3 left over. If Irene had less than 100 checks, how many did she have?

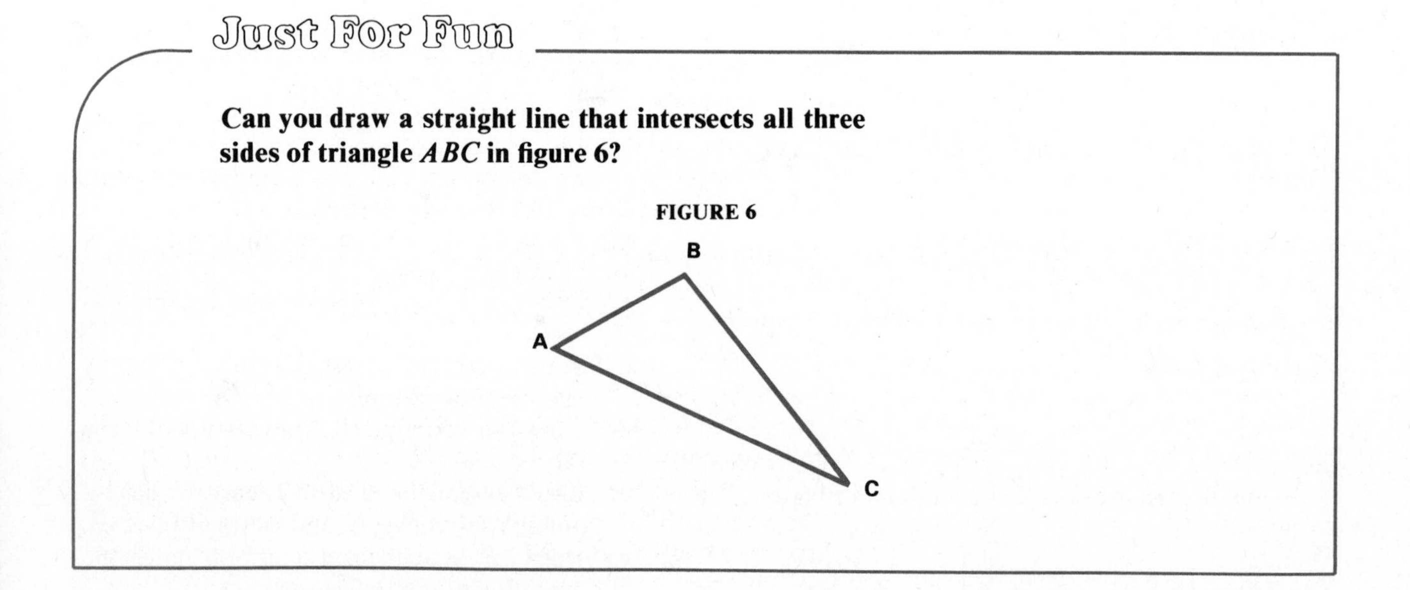

Just For Fun

Can you draw a straight line that intersects all three sides of triangle *ABC* in figure 6?

FIGURE 6
B
A
C

6.5 MATHEMATICAL SYSTEMS WITHOUT NUMBERS

Thus far in this chapter, we have examined mathematical systems such as clock arithmetic, the system of months, the week system, and various modular systems. Recall that a mathematical system consists of a set of elements together with one or more operations (rules) for combining elements of the set. The systems that we have considered so far have been based upon the use of numbers, but now we want to consider systems that are more abstract.

Consider the set of elements $\{A, B, C, D\}$, together with the operation $*$. We define this operation by means of table 8. The operation $*$ is a binary operation because we combine two elements of the given set to obtain each result. For example, to find the answer to $B * C$, we find the first element, B, in the vertical column under $*$, and the second element, C, in the top horizontal row following $*$. The answer is found where the row containing B and the column containing C intersect, at D. Hence, $B * C = D$. Similarly, $C * D = B$ and $D * C = B$.

TABLE 8

*	A	B	C	D
A	A	B	C	D
B	B	C	D	A
C	C	D	A	B
D	D	A	B	C

Let's examine the mathematical properties of this particular system, the set of elements $\{A, B, C, D\}$ and the operation $*$.

1. **Closure property:** All of the entries in the table are elements in the given set; there are no new elements appearing in the table. Therefore the system satisfies the closure property.

2. **Identity property:** There is an element, A, in the set that does not change any element when it is operated on together with that element. Note that $A * A = A$, $B * A = B$, $C * A = C$, $D * A = D$.

3. **Inverse property:** Each element in this system has an inverse, an element which when operated on with the given element results in the identity element A. Note that $A * A = A$, $B * D = A$, $C * C = A$, $D * B = A$. From these equations, we can see that the inverse of A is A, the inverse of B is D, the inverse of C is C, and the inverse of D is B.

4. **Associative property:** An operation is associative if the location of parentheses does not affect the answer. Trying an example, we have $(B * C) * D = D * D = C$, and $B * (C * D) = B * B = C$; hence $(B * C) * D = B * (C * D)$. Any other example for this system also works, and therefore the system satisfies the associative property.

5. **Commutative property:** An operation is commutative if it does not matter in what order you perform the operation. For an example in this system, we find that $B * C = D$ and $C * B = D$; therefore $B * C = C * B$. Any other example for this system also works, so the system satisfies the commutative property.

The abstract mathematical system consisting of the set of elements $\{A, B, C, D\}$ and the binary operation $*$ satisfies all of the properties listed. Therefore this system forms a group and, since it satisfies the commutative property, it forms a commutative group.

EXAMPLE 1

The following table defines the operation $\times$ for the set of elements {*odd, even*}.

$\times$	*odd*	*even*
odd	*odd*	*even*
even	*even*	*even*

Find each product using this table.

a. *odd* $\times$ *even* b. *odd* $\times$ *odd* c. *even* $\times$ *even*

Solution

a. *odd* $\times$ *even* = *even*

b. *odd* $\times$ *odd* = *odd*

c. *even* $\times$ *even* = *even*

EXAMPLE 2

Using the table in example 1, answer the following:

a. Is the set closed with respect to the operation $\times$? Why or why not?

b. What is the identity element (if any)?

c. Which elements of the set have an inverse? Name the inverse of each of these elements.

d. Does the commutative property hold for this system? Verify your answer.

e. Does this set form a group under the operation of $\times$?

Solution

a. Yes, the set is closed with respect to the operation $\times$. There are no new elements appearing in the table.

b. The identity element is *odd*. It does not change any of the elements in the given set when it is operated on with them: *odd* $\times$ *odd* = *odd*, *even* $\times$ *odd* = *even*.

c. The element *even* does not have an inverse. However, *odd* $\times$ *odd* = *odd*, so *odd* has an inverse, itself.

d. Yes, the commutative property does hold for this system. Checking the elements, we have *odd* $\times$ *even* = *even* and *even* $\times$ *odd* = *even*. Therefore, *odd* $\times$ *even* = *even* $\times$ *odd*. Note that *odd* $\times$ *odd* = *odd* $\times$ *odd* and *even* $\times$ *even* = *even* $\times$ *even*.

e. No, the set of elements {*odd, even*} with the operation $\times$, does not form a group. Not every element has an inverse, because *even* does not have an inverse.

EXAMPLE 3

Sergeant Gig, a drill instructor, drills his drill team daily. During the drills, he issues four commands: *right face, left face, about face,* and *as you were.* Let

r = *right face* a = *about face*
l = *left face* y = *as you were*

Sometimes Sergeant Gig gives two commands in succession, such as *right face* followed by *left face.* If a person followed these commands he would wind up in the original position, the position of *as you were.* If we let $\otimes$ stand for the operation *followed by,* then $r \otimes l = y$. From this information, evaluate the following:

a. $r \otimes r$ b. $a \otimes r$ c. $l \otimes a$

Solution

a. $r \otimes r$ means *right face* followed by *right face,* which would be the same as *about face.* Therefore $r \otimes r = a$.

b. $a \otimes r$ means *about face* followed by *right face,* which would be the same as *left face.* Therefore $a \otimes r = l$.

c. $l \otimes a$ means *left face* followed by *about face,* which would be the same as *right face*. Therefore $l \otimes a = r$.

Using the idea of drill commands in example 3, we can create a table to illustrate the results when any two commands are given. Using the notation r = *right face*, l = *left face*, a = *about face*, and y = *as you were*, and letting $\otimes$ stand for the operation *followed by*, we have a system consisting of a set of elements $\{r, l, a, y\}$ and an operation $\otimes$. Table 9 illustrates the operation $\otimes$ in this system.

TABLE 9

$\otimes$	r	l	a	y
r	a	y	l	r
l	y	a	r	l
a	l	r	y	a
y	r	l	a	y

From table 9, we can verify the exercises in example 3: $r \otimes r = a$, $a \otimes r = l$, and $l \otimes a = r$. Examining the entries in the table, we see that the set is closed with respect to the operation $\otimes$. The set also contains an identity element for the operation $\otimes$, the element y. Since y is the identity element, we can find the inverse of each element. The inverse of r is l, the inverse of l is r, a is its own inverse, and y is its own inverse. It can also be shown that the set satisfies the associative property for the operation $\otimes$. Hence, the set of elements $\{r, l, a, y\}$ with the operation $\otimes$ satisfies the closure property, identity property, inverse property, and associative property, and therefore forms a group.

It is interesting to note that many of the examples discussed in this chapter are different, but similar. We have discussed examples that contain different sets of elements which are combined using different operations, but they still satisfy the same properties of closure, identity, etc. This is one of the unique characteristics of mathematical systems.

EXERCISES FOR SECTION 6.5

For exercises 1–14, use table 10, which defines an operation : for the set of elements $\{P, Q, R, S\}$.

TABLE 10

:	P	Q	R	S
P	P	Q	R	S
Q	Q	R	S	P
R	R	S	P	Q
S	S	P	Q	R

Evaluate the following:

1. $P{:}Q$
2. $R{:}S$
3. $S{:}S$
4. $R{:}R$

5. $Q : R$
6. $R : P$
7. $R:(Q:S)$
8. $(S:Q):R$
9. $(P:Q):R$
10. $P:(Q:R)$
11. Is the set closed with respect to the operation : ? Why or why not?
12. What is the identity element (if any)?
13. Which elements of the set have an inverse? Name the inverse of each of these elements.
14. Does this set form a group under the operation : ?

Answer exercises 15–29 by using table 11, which defines an operation ⊙ for the elements of the set {$, ?, ¢}.

TABLE 11

⊙	$	?	¢
$	$	!	?
?	?	¢	!
¢	¢	?	$

Evaluate the following:

15. $ ⊙ ?
16. ? ⊙ $
17. ? ⊙ ¢
18. ¢ ⊙ ¢
19. ¢ ⊙ ?
20. $ ⊙ $
21. $ ⊙ (¢ ⊙ ?)
22. ($ ⊙ ¢) ⊙ ?
23. ¢ ⊙ (? ⊙ $)
24. (¢ ⊙ ?) ⊙ $
25. Is the set closed with respect to the operation ⊙? Why or why not?
26. What is the identity element (if any)?
27. Which elements of the set have an inverse? Name the inverse of each of these elements.
28. Is the commutative property satisfied for this system? Why or why not?
29. Does this set form a group under the operation ⊙?

Answer exercises 30–46 by using table 12, which defines an operation $*$ for the elements of the set $\{a, b, c, d, e\}$.

TABLE 12

$*$	a	b	c	d	e
a	a	b	c	d	e
b	b	c	d	e	a
c	c	d	e	a	b
d	d	e	a	b	c
e	e	a	b	c	d

Evaluate the following:

30. $b * a$
31. $c * d$
32. $a * d$
33. $d * a$
34. $e * e$
35. $d * d$
36. $c * a$
37. $c * c$
38. $c * (d * e)$
39. $(c * d) * e$
40. $(b * a) * d$
41. $b * (a * d)$
42. Is the set closed with respect to the operation $*$? Why or why not?
43. What is the identity element (if any)?
44. Which elements of the set have an inverse? Name the inverse of each of these elements.
45. Is the commutative property satisfied for this system? Why or why not?
46. Does this set form a group under the operation of $*$?
47. Given the set of counting numbers, $\{1, 2, 3, 4, \ldots\}$, and the operation of addition, answer the following:
    a. Is this set closed with respect to the operation of addition?
    b. What is the identity element (if any)?
    c. Which elements of the set have an inverse with respect to addition?
    d. Does this set form a group under the operation of addition?

FIGURE 7

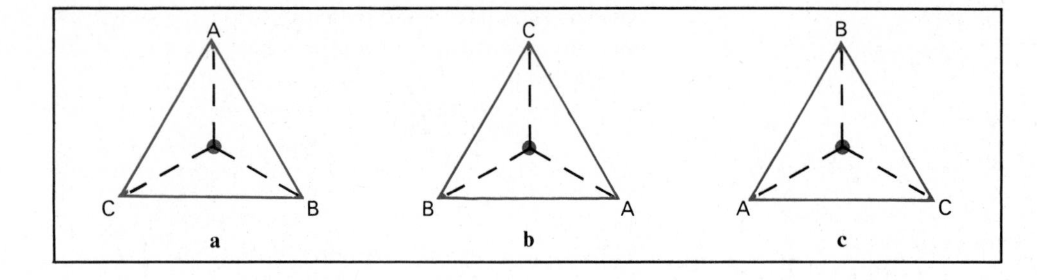

48. Given the set of counting numbers, $\{1, 2, 3, 4, \ldots\}$, and the operation of multiplication, answer the following:
    a. Is this set closed with respect to the operation of multiplication?
    b. What is the identity element (if any)?
    c. Which elements of the set have an inverse with respect to the operation of multiplication?
    d. Does this set form a group under the operation of multiplication?

49. Given the equilateral (all sides equal) triangle ABC, with a point in the center of the triangle so that the triangle may be rotated, as in figure 7a. If we rotate the triangle 120° clockwise (call it rotation a), the triangle changes position: A goes to B, B goes to C, and C goes to A. See figure 7b. If we rotate the triangle 240° clockwise (call it rotation b), the triangle again changes position: A goes to C, B goes to A, and C goes to B. See figure 7c. If we rotate the triangle 360° clockwise (call it rotation c), the triangle returns to its original position: A goes to A, B goes to B, C goes to C. See figure 7a.

Let's define an operation $*$ to mean "followed by." Therefore, $a * b$ would mean rotation a followed by rotation b, which is the same as a rotation of 360°, or rotation c. Hence, for this system $a * b = c$.

a. Complete the following table for this system:

*	a	b	c
a		c	
b			b
c	a		

b. Is this set closed with respect to the operation $*$?
c. What is the identity element (if any)?
d. Which elements of the set have an inverse? Name the inverse of each of these elements.
e. Does this set form a group under the operation of $*$?

FIGURE 8

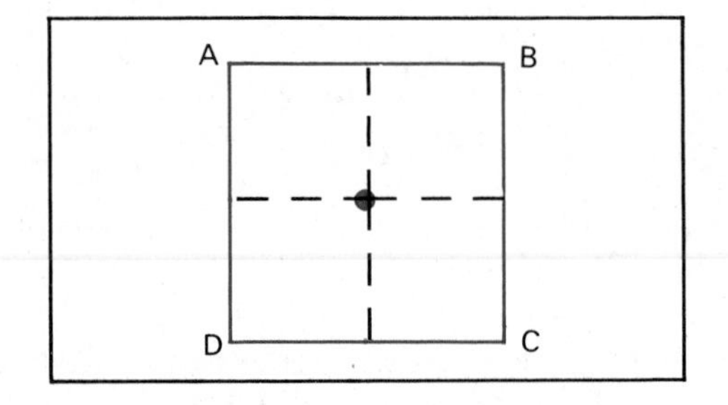

50. Given the square $ABCD$ (all sides equal), with a point in the center of the square so that the square can be rotated (see figure 8). Define the following rotations:

a = a clockwise rotation of 90°

b = a clockwise rotation of 180°

c = a clockwise rotation of 270°

d = a clockwise rotation of 360°

Let the operation $*$ mean "followed by."

a. Complete the following table for this system:

*	a	b	c	d
a		c		
b		d		
c			b	
d	a			

b. Is this set closed with respect to the operation $*$?
c. What is the identity element (if any)?
d. Which elements of the set have an inverse? Name the inverse of each of these elements.
e. Does the set form a group under the operation $*$?

Just For Fun

If a certain kind of egg sells for $1 per dozen, which costs more, $\frac{1}{2}$ dozen dozen eggs, or 6 dozen dozen eggs?

6.6 AXIOMATIC SYSTEMS

Throughout this chapter we have examined the basic structure of mathematical systems. We have considered a mathematical system to be a set of elements together with one or more operations (rules) for combining any two elements of the set. In this section we will examine more closely the structure of mathematics itself.

Regardless of the branch of mathematics we choose to examine, the various branches are similar in the way that they are constructed. We can compare the basic characteristics of mathematics to the basic characteristics of a game.

Most games have a vocabulary of special terms, some defined and some undefined. After a player acquires this vocabulary, he learns the rules of the games—that is, what moves he can make, and what moves he cannot make. Normally, he accepts these rules without question. For instance, in the game of baseball, one rule says that a runner must run the bases in a counterclockwise direction. When a youngster is first learning how to play baseball, he sometimes wants to run the bases in a clockwise direction. But when he is told that the rules state that a runner must run the bases the other way (counterclockwise), he accepts this. Similarly, we all accept the fact that a queen ranks higher than a jack in card games. Why? Because the rules say so!

In mathematics, the rules are called *axioms*. An **axiom** is a statement that is accepted as true without proof. Each mathematical system must have axioms that are consistent. They must not contradict one another, just as the rules of a game should not contradict one another.

After learning the undefined terms, defined terms, and rules of a game, we are ready to play. Once we have mastered the elementary moves of the game, we usually try more complicated moves using the rules (as in the game of chess). In mathematics, these new results that have evolved from the undefined terms, defined terms, and axioms (rules) are called *theorems.* **Theorems** are logical deductions that are made from undefined terms, defined terms, and axioms. Some theorems are even logical deductions from other theorems.

In essence, an axiomatic system consists of four main parts:

1. **undefined terms**
2. **defined terms (definitions)**
3. **axioms**
4. **theorems**

Undefined terms are necessary in an axiomatic system, as they are used to form a fundamental vocabulary with which other terms can be defined. Even though a term may be undefined, that does not mean that we do not know what it is. In a high school geometry course, the terms *point* and *line* are not defined terms, yet we know what they are. In chapter 1 of this text, *set* is an undefined term, but that does not prevent us from having an intuitive idea of what *set* means.

Definitions of defined terms may use undefined terms or terms that have been previously defined. Definitions should be concise, consistent, and not circular.

As we noted earlier, axioms are statements that are accepted as true without proof. In a given system, no axiom should contradict another; that is, the axioms must be consistent. Axioms are necessary in a system, since not everything can be proved. Axioms are needed to derive other statements. The derived statements are the theorems.

Let us now examine a proved theorem for a system that uses undefined terms, defined terms, and axioms. The following proof is found in a high school geometry course. The statement (theorem) to be proved is "If two straight lines intersect, then the vertical angles thus formed are equal."

We first construct a diagram representing this situation. See figure 9.

FIGURE 9

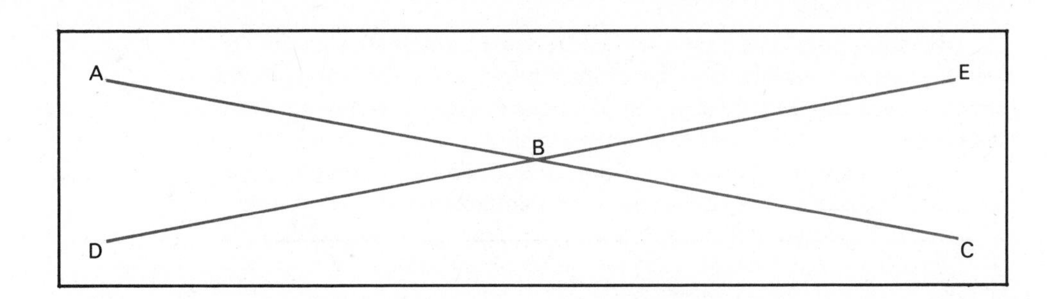

Figure 9 shows straight lines AC and DE intersecting at B and forming vertical angles ABD and EBC. We want to prove that the measure of angle ABD equals the measure of angle EBC, that is, $m\angle ABD = m\angle EBC$. The following is a formal proof.

Statements	Reasons
1. AC and DE are straight lines	1. Given
2. Angles ABC and DBE are straight angles	2. Definition of a straight angle
3. $m\angle ABC = m\angle DBE$	3. All straight angles are equal
4. $m\angle ABE = m\angle ABE$	4. Identity
5. $m\angle ABD = m\angle EBC$	5. If the same quantity is subtracted from two equal quantities, the remainders are equal

This proof uses undefined terms (straight lines), definitions (the definition of a straight angle), and axioms. The axioms appear in reason 3 (All straight angles are equal) and reason 5 (If the same quantity is subtracted from two equal quantities, the remainders are equal). The axioms used are consistent; that is, they do not contradict each other. The diagram in figure 9 could be considered a *model*—that is, a physical interpretation of the undefined terms which satisfies the axioms.

Let's examine another example, but one that is less familiar. We start with the following axioms, and the undefined terms *road, town,* and *stop sign.*

1. There is at least one road in the town.
2. Every stop sign is on exactly two roads.
3. Every road has exactly two stop signs on it.

We wish to prove that there is at least one stop sign in the town. It is helpful to construct a diagram (model) that satisfies all of the axioms. See figure 10.

FIGURE 10

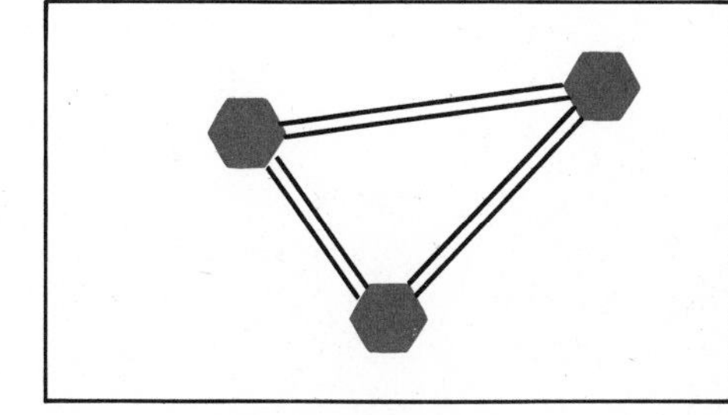

We want to prove that there is at least one stop sign in the town. Axiom 1 states that there must be at least one road in the town, and axiom 3 says that every road has exactly two stop signs on it. Hence, there must be at least one stop sign in the town.

Note that from this set of axioms we could have derived other conclusions, but we only derived the desired conclusion.

EXAMPLE 1

Given:

Axiom 1: There is exactly one road between any two traffic lights.

Axiom 2: For every road there exists a traffic light not on that road.

Axiom 3: There exist at least two traffic lights.

Prove: There exist at least three roads.

Solution

First we construct a model which satisfies all of the given axioms. Let ▣ represent a traffic light and ═══ represent a road. Our model appears in figure 11; this helps us understand the axioms.

FIGURE 11

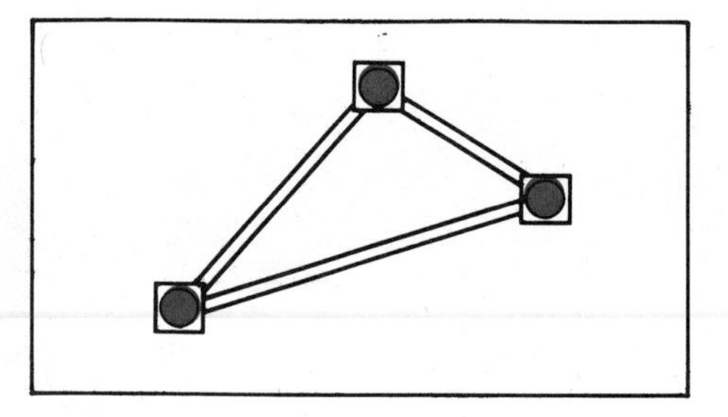

Axiom 3 tells us that there are at least two traffic lights, and axiom 1 states that there is exactly one road between any two traffic lights. So far we have one road. Now axiom 2 tells us that there has to be another traffic light not on the given road. But, if there is another traffic light, then there must be two more roads: axiom 1 tells us that there is exactly one road between any two traffic lights, and since there is a third traffic light, it must be connected to the other two lights by means of two roads. Therefore we have at least three roads.

EXERCISES FOR SECTION 6.6

1. Consider the following sets of axioms:

    I. There are at least two buildings on campus.
    II. There is exactly one sidewalk between any two buildings.
    III. Not all of the buildings are on the same sidewalk.

    If the capital letters A, B, C, etc., represent buildings and the lines represent sidewalks, which of the following models represent the given axiomatic system?

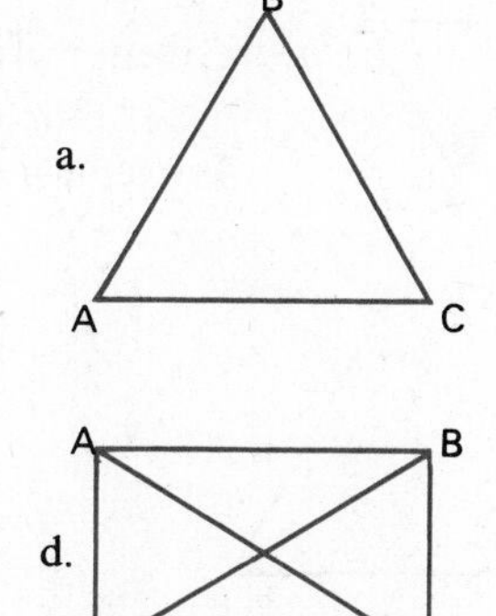

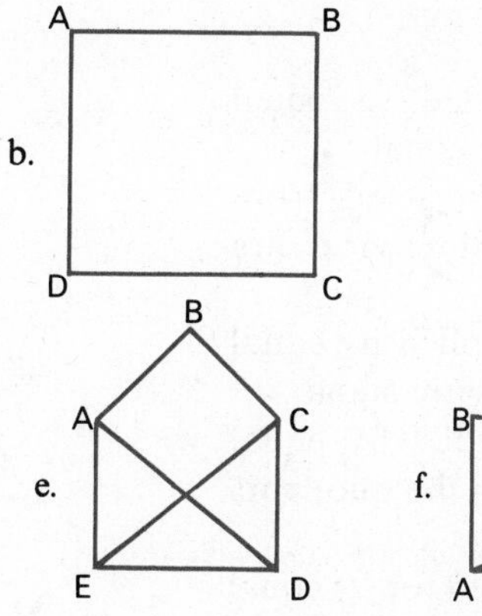

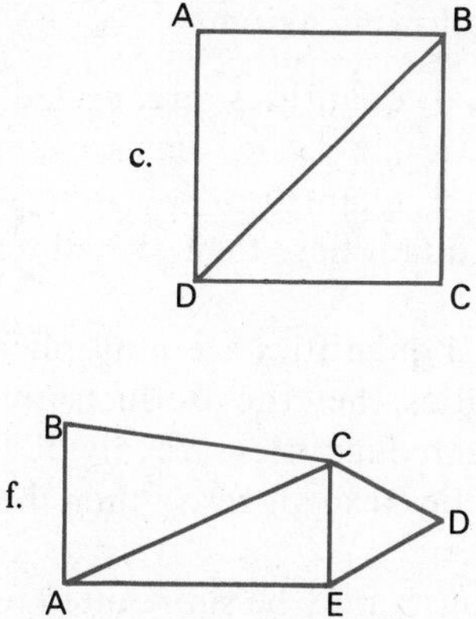

2. Consider the following set of axioms:

   I. There are exactly four bleeps.
   II. Each bleep is on a cleep.
   III. No bleep is on a cleep by itself.

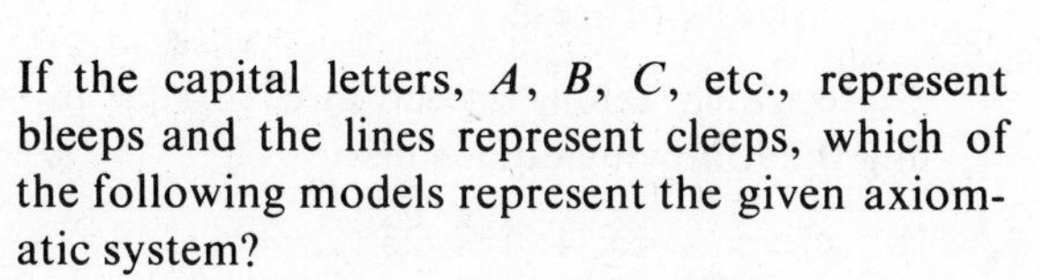

If the capital letters, *A*, *B*, *C*, etc., represent bleeps and the lines represent cleeps, which of the following models represent the given axiomatic system?

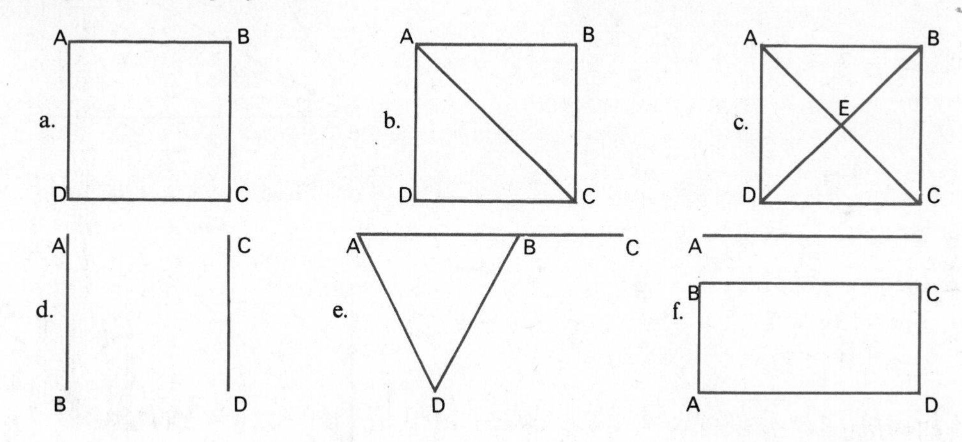

3. What is the basic structure of an axiomatic system?

4. Given the following axioms:

   I. There are at least two buildings on campus.
   II. There is exactly one sidewalk between any two buildings.
   III. Not all of the buildings are on the same sidewalk.

   Prove: There exist at least three buildings on campus.

5. Given the following axioms:

   I. There are exactly two points on each line.
   II. There is at least one line.
   III. For each pair of points, there is one and only one line containing them.
   IV. Corresponding to each line there is exactly one other line which has no point in common with it.

   Prove: There are at least four points.

6. Given the following axioms:

   I. If equal quantities are added to equal quantities, then the sums are equal.
   II. If equal quantities are subtracted from equal quantities, then the differences are equal.
   III. If equal quantities are multiplied by equal quantities, then the products are equal.
   IV. If equal quantities are divided by equal quantities (except zero), then the quotients are equal.
   V. A quantity may be substituted for its equal in any process.

   State which axiom is used in each step in solving the following equations.

   a. Given: $3x + 4 = 25$
      1. $3x = 21$
      2. $x = 7$

   b. Given: $2x - 7 = 15$
      1. $2x = 22$
      2. $x = 11$

   c. Given: $\frac{x}{2} - 4 = 3$
      1. $\frac{x}{2} = 7$
      2. $x = 14$

   d. Given: $\begin{cases} x + 3y = 7 \\ x + y = 5 \end{cases}$
      1. $2y = 2$
      2. $y = 1$
      3. $x + 1 = 5$
      4. $x = 4$

   e. Given: $\begin{cases} 2x + y = 8 \\ x - y = 4 \end{cases}$
      1. $3x = 12$
      2. $x = 4$
      3. $4 - y = 4$
      4. $4 = 4 + y$
      5. $0 = y$

7. Using the axioms given in exercise 6, prove the desired conclusion for each of the following:

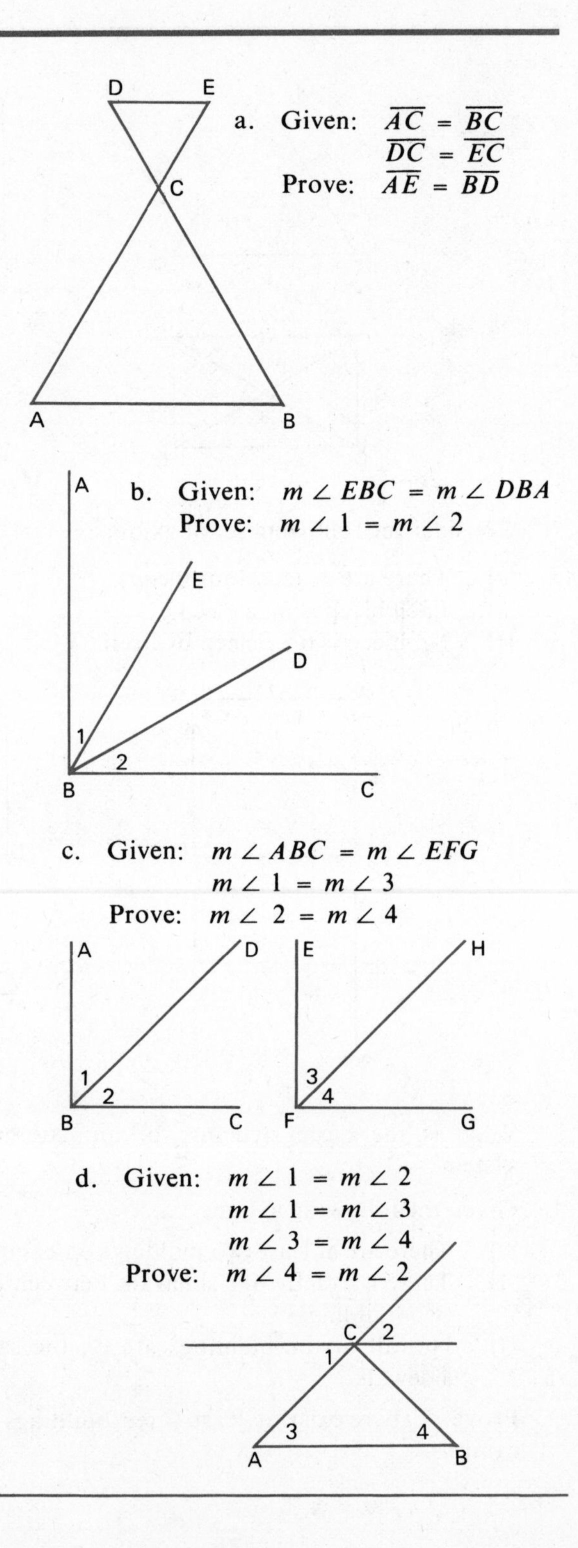

   a. Given: $\overline{AC} = \overline{BC}$, $\overline{DC} = \overline{EC}$
      Prove: $\overline{AE} = \overline{BD}$

   b. Given: $m\angle EBC = m\angle DBA$
      Prove: $m\angle 1 = m\angle 2$

   c. Given: $m\angle ABC = m\angle EFG$, $m\angle 1 = m\angle 3$
      Prove: $m\angle 2 = m\angle 4$

   d. Given: $m\angle 1 = m\angle 2$, $m\angle 1 = m\angle 3$, $m\angle 3 = m\angle 4$
      Prove: $m\angle 4 = m\angle 2$

Just For Fun

Benny wanted to plant some tomato plants in his garden. Altogether he had 12 tomato plants. He wanted to plant these in 6 rows, but with 4 tomato plants in each row. How did Benny do it?

6.7 SUMMARY

A mathematical system is composed of a nonempty set of elements together with one or more operations (rules) for combining any two elements of the set. A system is *closed* if when we operate on any two elements in the given set, the result is also a member of the given set. If there is an element e in the set that does not change any element when it is operated on together with that element, then e is the *identity element.* Each element in the system has an *inverse* if there exists an element which when operated on together with the given element results in the identity element. An operation is *associative* if the location of parentheses in a problem does not affect the answer. When a set of elements and an operation satisfy the closure property, identity property, inverse property, and associative property, we say that the set of elements forms a *group* under the given operation. A group is called a *commutative group* if it also satisfies the *commutative property*—that is, if $a * b = b * a$ for all elements a and b in the group.

Clock arithmetic is an example of a mathematical system, and can be used to illustrate the properties mentioned above. A system that repeats itself or is cyclical is called a *modular system.* If we have $a \equiv b \pmod{m}$, we say that *a is equivalent to b mod m.* This means that a and b both have the same remainder when they are divided by m.

Abstract mathematical systems can be thought of as mathematical systems without numbers. This is just an extension of our basic notion of a system. The elements in the set can be anything (they are usually letters), and the operation can be defined in any way (e.g., "followed by" in example 3 of section 6.5). An abstract system can also form a group, providing it satisfies the properties of a group.

An *axiomatic system* consists of *undefined terms, defined terms, axioms,* and *theorems.* A *model* for an axiomatic system is a physical interpretation or diagram of the undefined terms; it must be constructured so that all of the axioms of the system are satisfied. The axioms in a system must also be consistent; that is, they must not contradict each other. As we continue our study of mathematics, we will see that all branches of mathematics are similar in that they are based upon an axiomatic system.

Review Exercises for Chapter 6

1. Evaluate each of the following on a 12-hour clock.
   a. $7 + 7$ b. $9 + 4$
   c. $8 + 7$ d. $11 + 12$
   e. $10 + 10$ f. $9 + 8$

2. Evaluate each of the following on a 12-hour clock.
   a. $6 - 8$ b. $7 - 11$
   c. $3 - 7$ d. $2 - 12$
   e. $6 - 10$ f. $8 - 9$

3. Evaluate each of the following on a 12-hour clock.
   a. 4×9 b. 8×4
   c. 6×7 d. 9×3
   e. 6×5 f. 6×4

4. Evaluate each of the following on a 12-hour clock.
   a. $7 \times (8 + 6)$ b. $4 \times (3 - 6)$
   c. $4 - (6 - 8)$ d. $(4 - 6) - 8$
   e. $2 \times (6 - 8)$ f. $(2 \times 6) - 8$

5. Describe in your own words what a *mathematical system* is.

6. Given the set of whole numbers, $\{0, 1, 2, 3, \ldots\}$, tell if each sentence is true or false.
   a. The set is closed with respect to addition.
   b. The set is closed with respect to subtraction.
   c. The set is closed with respect to multiplication.
   d. The set is closed with respect to division.

7. True or false?
   a. The set of whole numbers, $\{0, 1, 2, 3, \ldots\}$, contains an identity element for the operation of addition.
   b. The set of whole numbers contains an additive inverse element for each element in the set.
   c. The 12-hour clock system contains an identity element for the operation of multiplication.
   d. The 12-hour clock system contains a multiplicative identity element for each element.

8. True or false: The set of whole numbers—
   a. is associative with respect to addition
   b. is commutative with respect to addition.

9. True or false: The set of whole numbers—
   a. is a group with respect to addition
   b. is a group with respect to multiplication.

For questions 10 and 11, evaluate the following given that 1 = *spring*, 2 = *summer*, 3 = *fall*, and 4 = *winter*. (Your answer should be in terms of a season.)

10. a. summer + summer b. winter + summer
    c. fall + winter d. summer − fall
    e. fall − winter f. spring − summer

11. a. fall × fall
    b. summer × fall
    c. fall × (fall + winter)
    d. fall × (spring − winter)
    e. winter × (fall − winter)
    f. summer × (winter + spring)

12. Find the equivalent to each of the following in the modulo 5 system.
    a. 42 b. 61 c. 89
    d. −32 e. −108 f. 2003

13. Using the elements of the indicated modular systems, find a replacement for each question mark so that the statement is true.
    a. $4 + ? \equiv 2 \pmod{5}$
    b. $2 - ? \equiv 4 \pmod{7}$
    c. $3 \times ? \equiv 2 \pmod{7}$
    d. $? \times 2 \equiv 3 \pmod{7}$
    e. $4 \times (3 + ?) \equiv 1 \pmod{5}$
    f. $2 \times (3 - ?) \equiv 4 \pmod{7}$

14. Carl the coin collector obtained 2 rolls of pennies from the bank. After examining the coins, he divided the pennies into 2 groups: those with mint marks, and those without mint marks. Working with the coins that had mint marks, Carl arranged these pennies in stacks of 5, with 2 left over. Next he arranged these coins in stacks of 7, with 1 left over. Finally, he arranged them in stacks of 3, with 0 left over. How many of the pennies had a mint mark?

15. The speedometer on a car indicates how fast a car travels; the odometer indicates how many miles a car travels. The odometer is an example of a modular system. Why? What modular system is used on the odometer of an ordinary car?

16. Answer the following by using table 13, which defines an operation * for the elements of the set {$, ¢, &, ?}.

TABLE 13

*	$	¢	&	?
$	¢	$	?	&
¢	$	¢	&	?
&	?	&	π	$
?	&	?	$	π

    a. $ * ¢ b. ¢ * ?
    c. & * & d. ? * ?
    e. ¢ * & f. & * ?
    g. ¢ * (& * $) h. & * (? * $)
    i. Is this set closed with respect to the operation * ? Why or why not?
    j. What is the identity element (if any)?
    k. Which elements of the set have an inverse?
    l. Does this set form a group under the operation * ?

17. Name the basic parts of an axiomatic system.

18. Given the following axioms:

    I. There are at least three squirrels.
    II. Each squirrel is in exactly one tree.
    III. No squirrel is in a tree by itself.
    IV. For every tree, there is a squirrel that is not in that tree.

    Prove: There exist at least four squirrels.

19. The sides of a triangle are represented by $3x + 4$, $2x + 8$, and $5x - 4$. If the perimeter of the triangle is 48, prove that the triangle is equilateral, that is, that all three sides are equal.

Just For Fun

Most proofs are done by means of deduction; that is, we proceed from the premises, step by step, to the conclusion. As we go from one step to the next, we must have a reason for each step to show that it follows logically. The following is an example of a proof that does not obey the rules: even though the derivation appears to be correct, it is not. Can you find the error?

Statements	Reasons
1. $a = b$	Given
2. $a^2 = ab$	Multiplying both sides by a
3. $a^2 - b^2 = ab - b^2$	Subtracting b^2 from both sides
4. $(a + b)(a - b) = b(a - b)$	Factoring both sides
5. $\dfrac{(a + b)(a - b)}{(a - b)} = \dfrac{b(a - b)}{(a - b)}$	Dividing both sides by $(a - b)$
6. $(a + b) = b$	Result of step 5
7. $b + b = b$	Recall that $a = b$ (step 1), so we may substitute b for a
8. $2b = b$	Combining $b + b$
9. $\dfrac{2b}{b} = \dfrac{b}{b}$	Dividing both sides by b
10. Therefore, $2 = 1$	Result of step 9 (this is the conclusion)

7 SYSTEMS OF NUMERATION

After studying this chapter, you will be able to do the following:

1. Express a counting number as an **Egyptian numeral,** and express an Egyptian numeral as a **base ten numeral**
2. Add and subtract Egyptian numerals
3. List the distinguishing characteristics of systems of numeration that use a **simple grouping, multiplicative grouping,** or **place-value system**
4. Identify systems of numeration that use simple grouping, multiplicative grouping, and place-value systems
5. Convert a numeral in any base other than ten to a base ten numeral by means of **expanded notation**
6. Convert a base ten numeral to any desired base by means of the **division algorithm**
7. Identify systems of numeration other than base ten to which we are exposed in everyday life
8. Add, subtract, and multiply in base five
9. Add, subtract, and multiply in base two.

7.1 INTRODUCTION

Regardless of the human culture we examine, the people in that culture need a system of counting. Man needs numbers: he must be able to count. No matter what type of job a person has, he or she must be able to count in order to cope with situations encountered in everyday living, even in such common activities as paying bills and making change. Man has needed some form of counting (numeration) system ever since he began to reason and to develop a civilization.

Anthropologists maintain that many of the primitive tribes of prehistoric times had some system for counting. The most primitive counting systems went only as high as three or four, with anything greater being described as "many," but many so-called primitive peoples had much more sophisticated systems.

Harold M. Lambert

Probably the first form of counting was done by matching the things to be counted (such as animals) with something else, such as fingers, toes, stones, or sticks. A pile of pebbles may have been matched, one-to-one, to the number of animals in a certain herd. A herdsman might have placed the pebbles in a pile as he was letting the animals out to graze, one pebble corresponding to each animal. When the animals returned, the herdsman would match a pebble from the pile with each animal. If, after all the animals had passed by, there were some pebbles still remaining that had not been matched, then the herdsman would know that there were some animals not accounted for, perhaps one, two, three, or even "many."

Man has invented many different ways to record numbers. The ancient Chinese recorded numbers by tying knots on a string, as did the Incas. Another way of recording numbers used in early history was to draw pictures or slashes in the dirt. Early man made slashes (marks) in the dirt; each mark represented an animal that he owned. Other civilizations recorded these marks on stones or pieces of clay, or made notches in a stick. Even today we still sometimes use tally marks to record numbers.

Tally marks such as /// are used to represent numbers. The mark /// represents the number three, while ////// represents the number six. Note that the tally marks are symbols used to represent numbers. All systems of numeration use symbols to represent numbers. In the Roman system of numeration, V represents the number five. You may recall that the symbols, I, II, III, IV, V, X, L, D, and so on, are called *Roman numerals.* Any symbol for a number is called a **numeral.** Symbols like V or 5 represent the number five; they are numerals, not numbers. Numbers exist in your mind, but when you put the symbol V or 5 on a piece of paper you are using a written symbol (numeral) to represent the number five.

The distinction between a number and a numeral is pointed out here because of the many different types of numerals used throughout history by different cultures. Do not become overly concerned with this distinction; we shall make use of it only when it is helpful to do so. Many times we shall use numerals just as we would numbers.

The distinction between a number and a numeral can be compared to the distinction between the letter O and the number zero (0). There is a distinct difference of meaning between the two, but we all know what a person means when he tells us that his phone number is "two-oh-eight-five," that is, 2085. Similarly, there is a distinction between a number and a numeral, but we shall only emphasize this distinction when it is helpful to do so.

7.2 SIMPLE GROUPING SYSTEMS

One of the oldest known systems of numeration is that of the Egyptians. The Egyptian culture was a fairly advanced one, and consequently they had an advanced system of numeration. The pyramids are proof of their technical ingenuity. These pyramids were tombs for Egyptian kings. Pictures called *hieroglyphics* were painted on the walls of these tombs to represent numbers.

The Egyptians used hieroglyphics as early as 3400 B.C. A single line or stroke represented items up to ten. After reaching ten, a new symbol is used to indicate a set of ten things. This is an example of *simple grouping.*

A simple grouping system is a system in which the position of a symbol does not affect the number represented, and in which a different symbol is used to indicate a certain number or group of things.

In the Egyptian system, special symbols were used to represent tens, hundreds, thousands, ten-thousands, hundred-thousands, and millions. That is, the Egyptians used a different symbol for each power of ten. A **power of ten** is ten raised to a number. For example, $10^1 = 10$ is the first power of ten, $10^2 = 10 \times 10 = 100$ is the second power of ten, and $10^3 = 10 \times 10 \times 10 = 1000$ is the third power of ten. Since the Egyptians used a different symbol for each power of ten, we say that their system had a **base** of ten. In the expression 10^3, 10 is the base and 3 is the exponent, and 1000, or 10^3, is the third power of ten.

Below are some of the Egyptian hieroglyphic numerals and their corresponding values.

Egyptian numerals	Name	Value	Power of ten
|	Stroke	1	10^0
∩	Heelbone	10	10^1
	Coiled Rope	100	10^2
	Lotus Flower	1000	10^3
	Pointed Finger	10,000	10^4
	Polywog	100,000	10^5
	Astonished Man	1,000,000	10^6

Since the Egyptians used a simple grouping system, the position of a symbol does not affect the number represented. That is, the numeral ∩|| represents 12, as do ||∩ and ∩ above ||. This is certainly not the case for other systems of numeration. For example, 23 does not equal 32 and 128 does not equal 821 in our system of numeration.

A simple grouping system can also be thought of as an "additive" system, since we must add the values of the symbols rather than concern ourselves with the position of the symbols. Sometimes the Egyptians wrote their numerals in left to right order, from largest to smallest; at other times the numerals are written in a right to left order. Regardless of the arrangement of the symbols, the value of the number represented is not affected. Both of the following pictures represent 234:

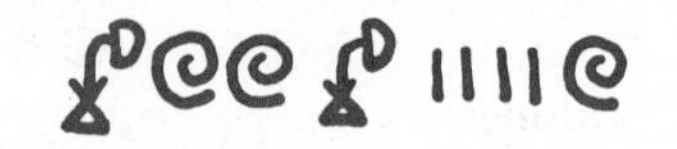

In order to evaluate the Egyptian numeral represented, we just add the values of the hieroglyphics, regardless of their position.

To add numbers that are expressed as Egyptian numerals, we group the same symbols together, and then simplify (rewrite) the expression. Consider the following addition problem:

In order to add these two numbers, we first group the same symbols together. Then to obtain our final answer, we rewrite ten of the strokes as a heelbone. This results in a string of ten heelbones, which we can rewrite as a coiled rope.

In order to subtract two numbers that are expressed as Egyptian numerals, we simply take away those symbols that are contained in both numbers. It may be necessary to rewrite some numerals in terms of other symbols, such as ∩ = ||||||||||. Consider the following subtraction problem:

Since we cannot subtract five strokes from three strokes, we rewrite one of the heelbones as ten strokes, and we have:

$$
\begin{array}{rll}
 & \text{𓍢 𓎆𓎆𓎆𓎆𓎆} & \text{𓏺𓏺𓏺𓏺𓏺𓏺𓏺𓏺𓏺𓏺 𓏺𓏺𓏺} \\
- & \text{𓍢 𓎆𓎆𓎆} & \text{𓏺𓏺𓏺𓏺𓏺} \\
\hline
 & \text{𓎆𓎆} & \text{𓏺𓏺𓏺𓏺𓏺𓏺𓏺𓏺}
\end{array}
$$

Remember that in the Egyptian system of computation, it is necessary to group the symbols together, and it is also sometimes necessary to rewrite some numerals in terms of other symbols.

EXAMPLE 1

Express the following as Egyptian numerals.

a. 13 b. 231 c. 13,423

Solution

a. 13 is 1 ten and 3 ones; that is, 𓎆 𓏺𓏺𓏺

b. 231 is 2 hundreds, 3 tens, and 1 one; that is, 𓍢𓍢 𓎆𓎆𓎆 𓏺

c. 13,423 is 1 ten thousand, 3 thousands, 4 hundreds, 2 tens, and 3 ones; that is, 𓂭 𓆼𓆼𓆼𓍢𓍢𓍢𓍢𓎆𓎆𓏺𓏺𓏺

EXAMPLE 2

Evaluate the following Egyptian numerals.

a. 𓍢𓍢𓎆𓎆𓎆𓎆 𓏺𓏺𓏺𓏺 b. 𓆼𓆼𓍢𓎆𓎆𓎆𓏺

c. 𓁨 𓂭𓍢𓎆𓏺𓏺

Solution

a. 𓍢𓍢𓎆𓎆𓎆𓎆𓏺𓏺𓏺𓏺 = 2 hundreds, 4 tens, and 4 ones; that is, $(2 \times 100) + (4 \times 10) + (4 \times 1) = 200 + 40 + 4 = 244$.

b. 𓆼𓆼𓍢𓎆𓎆𓎆𓏺 = 2 thousands, 1 hundred, 3 tens, and 1 one; that is, $(2 \times 1000) + (1 \times 100) + (3 \times 10) + (1 \times 1) = 2000 + 100 + 30 + 1 = 2131$.

c. 𓁨𓂭𓍢𓎆𓏺𓏺 = 1 million, 1 ten thousand, 1 hundred, 1 ten, and 2 ones; that is, $(1 \times 1{,}000{,}000) + (1 \times 10{,}000) + (1 \times 100) + (1 \times 10) + (2 \times 1) = 1{,}000{,}000 + 10{,}000 + 100 + 10 + 2 = 1{,}010{,}112$.

EXAMPLE 3
Add.

a. 𓎆𓎆𓎆𓏺𓏺 + 𓎆𓎆𓎆𓎆𓏺𓏺𓏺

b. 𓆼𓎆𓎆𓎆𓎆𓎆𓎆𓏺𓏺𓏺𓏺𓏺𓏺 + 𓆼𓎆𓎆𓎆𓏺𓏺𓏺𓏺

Solution

a. In order to add these two numbers, we simply combine the symbols.

𓎆𓎆𓎆𓏺𓏺 + 𓎆𓎆𓎆𓎆𓏺𓏺𓏺 = 𓎆𓎆𓎆𓎆𓎆𓎆𓎆𓏺𓏺𓏺𓏺𓏺

b. In order to add these two numbers, we first group the same symbols together and then simplify.

𓆼𓎆𓎆𓎆𓎆𓎆𓎆 𓏺𓏺𓏺𓏺𓏺𓏺 + 𓆼𓎆𓎆𓎆 𓏺𓏺𓏺𓏺

= 𓆼𓆼 𓎆𓎆𓎆𓎆𓎆𓎆𓎆𓎆𓎆𓏺𓏺𓏺𓏺𓏺𓏺𓏺𓏺𓏺𓏺

= 𓆼𓆼 𓎆𓎆𓎆𓎆𓎆𓎆𓎆𓎆𓎆𓎆𓏺

= 𓆼𓆼𓍢𓏺

EXAMPLE 4
Subtract.

a. 𓎆𓎆𓎆𓎆𓏺𓏺𓏺 − 𓎆𓎆𓎆𓏺𓏺

b. 𓍢𓎆𓎆𓎆𓏺𓏺 − 𓎆𓎆𓎆𓎆 𓏺𓏺𓏺𓏺𓏺𓏺𓏺

Solution

a. In order to subtract two numbers that are expressed as Egyptian numerals, we subtract (or take away) those symbols that are contained in both numerals.

𓎆𓎆𓎆𓎆𓏺𓏺𓏺 − 𓎆𓎆𓎆𓏺𓏺 = 𓎆𓏺

b. This subtraction cannot be performed immediately; we must first rewrite the first number, since it does not contain 7 ones or 4 tens, as such.

𓍢𓎆𓎆𓎆𓏺𓏺 − 𓎆𓎆𓎆𓎆𓏺𓏺𓏺𓏺𓏺𓏺𓏺

= 𓎆𓎆𓎆𓎆𓎆𓎆𓎆𓎆𓎆𓎆𓎆𓎆𓏺𓏺𓏺𓏺𓏺𓏺𓏺𓏺𓏺𓏺𓏺𓏺 − 𓎆𓎆𓎆𓎆𓏺𓏺𓏺𓏺𓏺𓏺𓏺

= 𓎆𓎆𓎆𓎆𓎆𓎆𓎆𓎆𓏺𓏺𓏺𓏺𓏺

EXERCISES FOR SECTION 7.2

1. Express each number with an Egyptian numeral.
   a. 18 b. 23 c. 34
   d. 102 e. 201 f. 1132
2. Express each number with an Egyptian numeral.
   a. 13 b. 22 c. 124
   d. 1492 e. 13,213 f. 22,123
3. Evaluate each Egyptian numeral.
   a. 𓎆𓎆𓏺𓏺 b. 𓆼𓍢𓍢𓎆𓏺𓏺𓏺
   c. 𓆼𓍢𓏺 d. 𓂭𓆼𓆼𓍢𓍢𓎆𓏺𓏺
   e. 𓎆𓎆𓆼𓍢𓍢𓏺𓏺 f. 𓏺𓎆𓏺𓆼𓍢𓆼𓍢
4. Evaluate each Egyptian numeral.
   a. 𓎆𓎆𓏺𓏺𓎆 b. 𓍢𓍢𓎆𓏺𓎆
   c. 𓆼𓏺𓏺𓎆 d. 𓆼𓆼𓍢𓍢𓏺𓏺𓏺𓏺
   e. 𓂭𓆼𓏺𓍢𓆼 f. 𓏺𓎆𓍢𓆼𓂭
5. Add the following:
   a. 𓎆𓎆𓎆𓏺 + 𓎆𓎆𓏺𓏺𓏺
   b. 𓍢𓍢𓎆𓎆𓎆𓏺𓏺𓏺𓏺𓏺 + 𓍢𓎆𓏺𓏺𓏺𓏺𓏺
   c. 𓍢𓍢𓎆𓎆𓎆𓏺𓏺 + 𓍢𓎆𓎆𓎆𓎆𓏺𓏺
   d. 𓆼𓆼𓍢𓍢𓍢
   \+ 𓍢𓍢𓍢𓍢𓍢𓍢𓍢
6. Add the following:
   a. 𓏺𓏺𓏺𓏺𓏺𓏺 + 𓏺𓏺𓏺𓏺𓏺
   b. 𓍢𓎆𓎆𓎆𓎆𓎆𓏺𓏺𓏺𓏺 + 𓏺𓏺𓏺𓏺𓏺𓎆
   c. 𓎆𓎆𓎆𓎆 + 𓎆𓎆𓎆𓎆𓎆𓎆𓎆
   d. 𓆼𓍢𓍢𓍢𓍢𓍢𓍢 + 𓆼𓍢𓍢𓍢𓍢
7. Perform each indicated subtraction.
   a. 𓎆𓎆𓎆𓏺 − 𓎆𓎆𓏺𓏺𓏺𓏺𓏺
   b. 𓍢𓎆𓎆𓏺𓏺 − 𓎆𓎆𓎆𓏺𓏺𓏺
   c. 𓍢𓍢𓎆𓎆𓎆𓎆𓏺𓏺 − 𓍢𓎆𓎆𓏺
   d. 𓆼𓆼𓍢𓍢 − 𓍢𓎆𓎆𓏺𓏺𓏺
8. Perform each indicated subtraction.
   a. 𓎆𓏺 − 𓏺𓏺𓏺𓏺𓏺
   b. 𓎆𓎆 − 𓎆𓏺𓏺𓏺𓏺𓏺𓏺𓏺
   c. 𓍢𓎆𓏺 − 𓎆𓎆𓎆𓎆𓎆𓎆𓏺𓏺
   d. 𓆼𓍢 − 𓍢𓍢𓎆𓎆𓎆𓎆𓏺𓏺𓏺𓏺𓏺𓏺

Just For Fun

How big is a million? Do you really have any idea? Suppose you were to start counting to a million, counting 1, 2, 3, 4, 5, ... , one number per second. Assuming that you could keep counting one number per second without stopping, how long would it take for you to reach 1,000,000?

7.3 MULTIPLICATIVE GROUPING SYSTEMS

One of the numeration systems developed by the Greeks used letters to represent numbers; that is, letters were used as numerals. Listed below are the Greek numerals together with their corresponding values.

Greek numerals	Values
I (iota)	1
Γ (gamma)	5
Δ (delta)	10
H (eta)	100
X (chi)	1000
M (mu)	10,000

One thing that the Greeks did differently from the other systems of numeration is to create symbols for multiples of 5. Since the Greeks had no symbol for 50, they thought of 50 as 5 tens and wrote it as Γ^{Δ}. Similarly, 500 was thought of as 5 hundreds and written as Γ^{H}. It follows that $\Gamma^{X} = 5000$ and $\Gamma^{M} = 50{,}000$.

In order to express 1984 in Greek numerals, we first think of 1984 as 1000 + 900 + 80 + 4. But 900 = 500 + 400 and 80 = 50 + 30. Therefore, 1984 = 1000 + 500 + 400 + 50 + 30 + 4, or

$$1984 = \mathrm{X}\Gamma^{\mathrm{H}}\mathrm{HHHH}\Gamma^{\Delta}\Delta\Delta\Delta\mathrm{IIII}$$

Note that the number is represented by using multiples of 5; this enabled the Greeks to use fewer symbols to express a number. The use of multiples of 5 is an example of *multiplicative grouping*.

A multiplicative grouping system is a system that uses certain symbols for numbers in a basic group, together with a second symbol or notation to represent numbers that are multiples of the basic group.

In the Greek system, 5 is the basic group. The symbol for 5, Γ, is used together with other symbols to represent numbers that are multiples of 5.

EXAMPLE 1
Express each number with a Greek numeral.

a. 12   b. 56   c. 88   d. 167   e. 1776

Solution

a. 12 is 1 ten and 2 ones: ΔII

b. 56 is 5 tens, 1 five, and 1 one: 𐅄ΓI

c. 88 is 1 fifty, 3 tens, 1 five, and 3 ones: 𐅄ΔΔΔΓIII

d. 167 is 1 hundred, 1 fifty, 1 ten, 1 five, and 2 ones: H𐅄ΔΓII

e. 1776 is 1 thousand, 1 five hundred, 2 hundreds, 1 fifty, 2 tens, 1 five, and 1 one: X𐅅HH𐅄ΔΔΓI

EXAMPLE 2
Evaluate each Greek numeral.

a. 𐅄ΓII

b. 𐅅HH𐅄I

c. 𐅆XX𐅅H𐅄ΔII

Solution

a. 𐅄 is 5 tens, or 50, Γ is 5, and II is 2; hence, 𐅄ΓII = 50 + 5 + 2 = 57.

b. 𐅅 = 500, HH = 200, 𐅄 = 50, and I = 1; hence, 𐅅HH𐅄I = 500 + 200 + 50 + 1 = 751.

c. 𐅆 = 5000, XX = 2000, 𐅅 = 500, H = 100, 𐅄 = 50, Δ = 10, and II = 2; hence, 𐅆XX𐅅H𐅄ΔII = 5000 + 2000 + 500 + 100 + 50 + 10 + 2 = 7662

Joseph Needham, Cambridge University Press

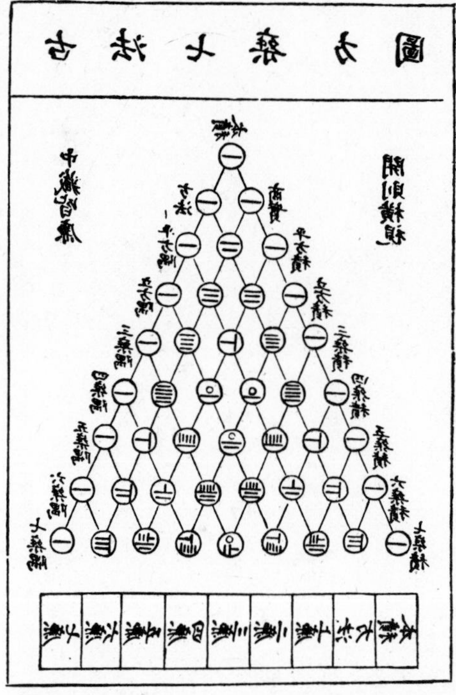

This excerpt from a fourteenth-century Chinese mathematical treatise shows an array of numbers known in the West as *Pascal's triangle*.

The Greek system of numeration uses six symbols. The system is repetitive and it uses multiples of five. The use of multiples of five in the Greek numeration system is an example of multiplicative grouping.

The Chinese-Japanese system of numeration also involves multiplicative grouping. But this system differs from the Greek system in that it uses multiples of 10, 100, and 1000 in its grouping system. Before we examine the multiplicative grouping of this system, we will first acquaint ourselves with the characteristics of the Chinese-Japanese system. One of the most important things to remember is that the system uses vertical instead of horizontal writing. Listed below are the Chinese-Japanese numerals together with their corresponding values.

Chinese-Japanese numerals	Values	Chinese-Japanese numerals	Values
一	1	七	7
二	2	八	8
三	3	九	9
四	4	十	10
五	5	百	100
六	6	千	1000

Since the Chinese-Japanese system of numeration is a system that uses multiplicative grouping, a number such as 2347 is thought of as 2 thousands, 3 hundreds, 4 tens, and 7. Two thousands is written as 二千, 3 hundreds is written as 三百, and 4 tens is written as 四十. Hence 2347 is written as shown in the margin.

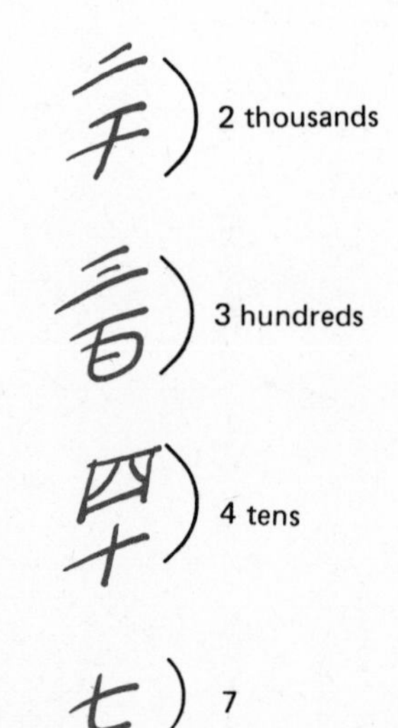

EXAMPLE 3

Express each number with a Chinese-Japanese numeral.

a. 12 b. 56 c. 88 d. 167 e. 2776

Solution

a. 十二 b. 五十六 c. 八十八 d. 百六十七 e. 二千七百七十六

EXAMPLE 4

Evaluate each Chinese-Japanese numeral.

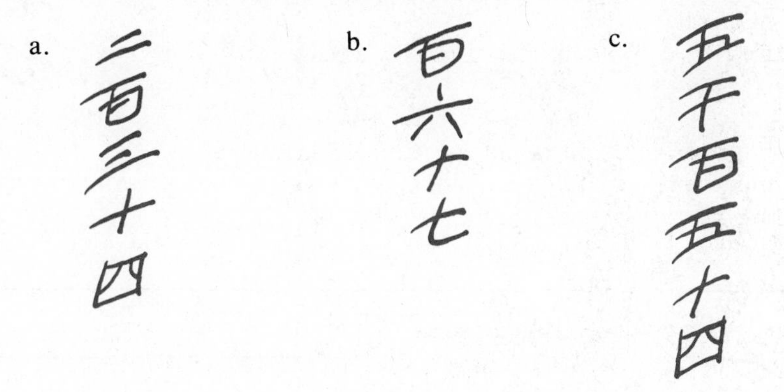

Solution

a. Since the Chinese-Japanese system of numeration uses multiplicative grouping, we have 2 hundreds, 3 tens, and 4: $(2 \times 100) + (3 \times 10) + 4 = 234$.

b. We have 1 hundred, 6 tens, and 7: $100 + (6 \times 10) + 7 = 167$.

c. This numeral contains 5 thousands, 1 hundred, 5 tens, and 4: $(5 \times 1000) + 100 + (5 \times 10) + 4 = 5154$.

EXERCISES FOR SECTION 7.3

1. Express each number with a Greek numeral.
   a. 18 b. 23 c. 34
   d. 44 e. 187 f. 598

2. Express the following as Greek numerals.
   a. 16 b. 54 c. 147
   d. 1492 e. 2194 f. 6875

3. Evaluate each Greek numeral.
   a. ΓII b. ΔI c. ΔΓII
   d. ΗΔΔΓII e. ΗΓ^{Δ}ΓΙ f. $\Gamma^{H}\Gamma^{\Delta}$ΔΓΙ

4. Evaluate the following Greek numerals.
   a. ΔΓII b. ΗΔΓII c. ΜΧΔ
   d. Γ^{H}ΗΓ^{Δ}ΓΙ e. Γ^{X}ΧΓ^{H}ΗΙ f. $\Gamma^{X}\Gamma^{H}\Gamma^{\Delta}$Γ

5. Express each number with a Chinese-Japanese numeral.
   a. 18 b. 23 c. 34
   d. 46 e. 234 f. 477

6. Express the following as Chinese-Japanese numerals.
   a. 16 b. 54 c. 147
   d. 897 e. 3473 f. 4176

7. Evaluate each Chinese-Japanese numeral.
   a. 二十 b. 七百五 c. 十三 d. 四十九

8. Evaluate the following Chinese-Japanese numerals.
   a. 五百二十七 b. 三百十五 c. 千百十 d. 八千二百三十六

Just For Fun

An inventor was trying to invent a substance that would dissolve anything. Finally, one day he declared that he had a box of the dissolving substance. Was he telling the truth?

7.4 PLACE-VALUE SYSTEMS

Thus far in our discussion of systems of numeration, we have examined a simple grouping system of numeration—the Egyptian system—and two multiplicative grouping systems—the Greek and the Chinese-Japanese. We shall now examine two systems of numeration that use a *place-value system*—the Babylonian system and the Hindu-Arabic system.

A place-value system is a system in which the position of a symbol matters; that is, the value that any symbol represents depends upon the position it occupies within the numeral.

The Babylonian system of numeration uses only two symbols to represent numbers, ▼ to represent one and ◀ to represent ten. The Babylonians pressed the end of a stick into a clay tablet in order to write their numbers. They used these two symbols, ▼ = 1 and ◀ = 10, to write any number up to 60. That is, ▼ = 1, ▼▼ = 2, ▼▼▼ = 3, and so on. They used the principle of addition to write numbers. For example, to record the number 7, the Babylonians would write 7 ones:

7 = ▼▼▼▼ ▼▼▼

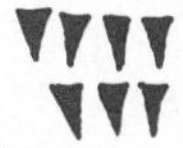

Since ◀ = 10, we can express 11 as 1 ten and 1 one:

11 = ◀▼ , 12 = ◀▼▼ , 13 = ◀▼▼▼ , and so on

Forty-three can be expressed as 4 tens and 3 ones:

The Babylonians used a place-value system, in which the position of a symbol is important. The symbols for ten were always placed to the left of the symbols for one. Therefore to represent the number 57 in the Babylonian system, we use 5 tens and 7 ones arranged as follows:

In order to represent a number greater than 60, such as 85, the Babylonians used a *sexagesimal* system, a system based on 60. The number 85 was thought of as one 60 and 25, that is, as $(1 \times 60) + 25$. To indicate this, the Babylonians placed a symbol for one, ▼, to the left of the numeral for 25:

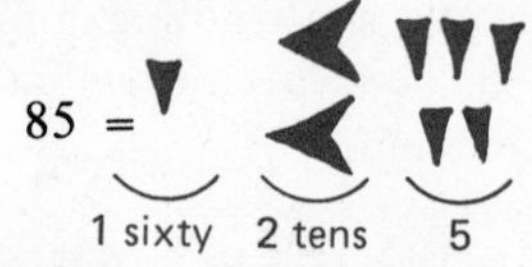

In our system, a number such as 1984 is read "one thousand, nine hundred, eighty-four." In other words, 1984 is composed of 1 thousand, 9 hundreds, 8 tens, and 4 ones. Symbolically, we can write this as $(1 \times 10^3) + (9 \times 10^2) + (8 \times 10^1) + (4 \times 1)$. Our system is based on powers of ten; hence it is called a **decimal** system of numeration. (The word *decimal* is derived from the Latin word *decem,* which means ten.) The Babylonian system is based on powers of sixty; hence it is called a **sexagesimal** system. Therefore, a Babylonian numeral such as

is interpreted as $(2 \times 60^2) + (21 \times 60^1) + 32$, or $(2 \times 3600) + (21 \times 60) + 32 = 7200 + 1260 + 32 = 8492$.

EXAMPLE 1

Express the following as Babylonian numerals.

a. 12 b. 42 c. 56 d. 88 e. 147

Solution

a. 12 is 1 ten and 2 ones; hence 12 =

b. 42 is 4 tens and 2 ones; hence 42 =

c. 56 is 5 tens and 6 ones; hence 56 =

d. 88 is 1 sixty and 28, or 1 sixty, 2 tens, and 8 ones; hence 88 =

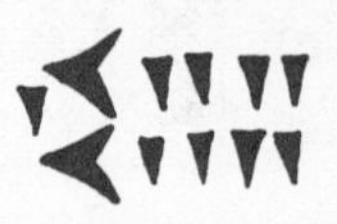

e. 147 is 2 sixties, 2 tens, and 7 ones; hence 147 =

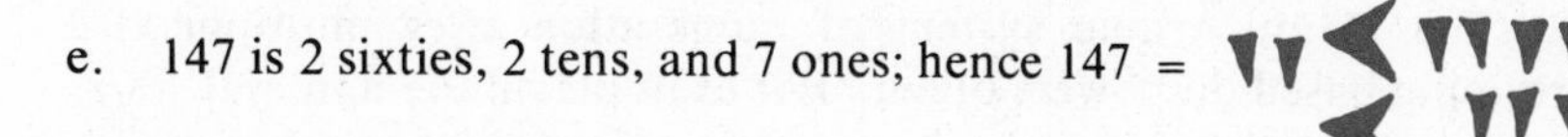

EXAMPLE 2

Evaluate each Babylonian numeral.

Solution

a. In this expression, we have 1 ten and 3 ones; $10 + 3 = 13$

b. Here we have 4 tens and 4 ones; $(4 \times 10) + 4 = 44$

c. In this expression, we have 1 sixty, 3 tens, and 2 ones; $(1 \times 60) + (3 \times 10) + 2 = 60 + 30 + 2 = 92$.

The Babylonian system of numeration uses only two symbols, and the symbols can be repeated for numbers up to sixty. After that, the numbers are expressed in powers of sixty. The numeral represents 1 sixty, 2 tens, and 2 ones; that is, $(1 \times 60) + (2 \times 10) + 2 = 82$.

The system of numerals we use today is called the **Hindu-Arabic system.** That is, we use Hindu-Arabic numerals to express numbers. The symbols that we use are 0, 1, 2, 3, 4, 5, 6, 7, 8, 9. These symbols had their beginning in India; it was the Arabs who were responsible for making their existence known in Europe. The Arabs did their calculations with these new numerals and then exposed the Europeans to their new techniques.

One of the most significant contributions of the Hindu numerals was zero. Zero evolved from a need for a placeholder, because it was important to distinguish between numerals such as 501 and 51. The transition to Hindu-Arabic numerals was a slow process, and it was not until the end of the sixteenth century that the changeover was fairly complete.

Recall that the Egyptian system of numeration uses a different symbol for each power of ten, but it has no place value. That is, the positions of the symbols do not affect the value of the number. Our system of numeration also uses powers of ten, but it does have **place value:** the position that a symbol has within a numeral is important.

The Hindu-Arabic system of numeration uses multiplicative grouping based on powers of ten. For example, in the numeral 1978, the 8 represents 8 ones, the 7 represents 7 tens, the 9 represents 9 hundreds, and the 1 represents 1 thousand. One, ten, one hundred, and one thousand are all powers of ten:

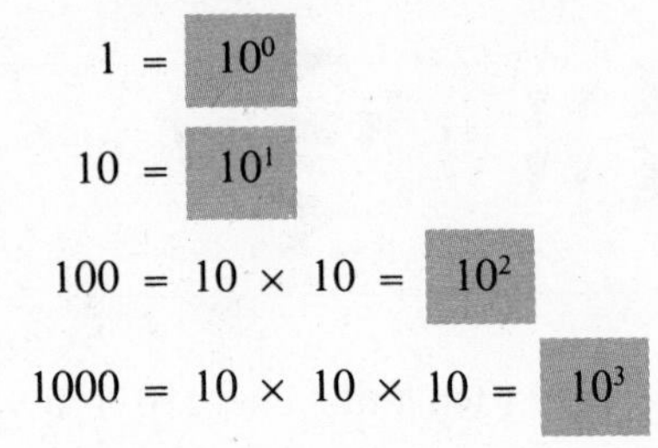

The number 1978 is the result of combining multiples of these powers of ten:

$$\begin{array}{r} 8 = 8 \times 10^0 \\ 70 = 7 \times 10^1 \\ 900 = 9 \times 10^2 \\ \underline{1000 = 1 \times 10^3} \\ 1978 \end{array}$$

Notice that the positions of the numerals are important. The only way that we know that the 8 in 1978 represents 8 ones is by its position; in 1987, the 8 represents 8 tens. Thus we say that the Hindu-Arabic system of numeration is a place-value system.

When we write a number in terms of powers of ten, we are writing it in **expanded notation.** In expanded notation,

$$1978 = (1 \times 10^3) + (9 \times 10^2) + (7 \times 10^1) + (8 \times 10^0)$$

When expressing a number in expanded notation, it is convenient to start with the ones place and proceed from right to left.

EXAMPLE 3

Write each number in expanded notation.

a. 123　　b. 2347　　c. 2003

Solution

a. The Hindu-Arabic numeral 123 is composed of 1 hundred, 2 tens, and 3 ones. Therefore, we have $(1 \times 100) + (2 \times 10) + (3 \times 1)$. Rewriting this using powers of ten, we have $(1 \times 10^2) + (2 \times 10^1) + (3 \times 10^0)$. (Recall that $10^0 = 1$. Any number raised to the zero power, except zero, equals one.)

b. $2347 = (2 \times 1000) + (3 \times 100) + (4 \times 10) + (7 \times 1) = (2 \times 10^3) + (3 \times 10^2) + (4 \times 10^1) + (7 \times 10^0)$

c. $2003 = (2 \times 10^3) + (0 \times 10^2) + (0 \times 10^1) + (3 \times 10^0)$. Note that there are no tens or hundreds, but we still must indicate this in expanded notation, as the zeros are placeholders, and 2003 is not the same as 23.

EXAMPLE 4

Write each of the following as Hindu-Arabic numerals in base ten (decimal) notation.

a. one hundred eighty-seven

b. two thousand three hundred forty-one

c. one thousand two

d. $(3 \times 10^3) + (2 \times 10^2) + (1 \times 10^1) + (0 \times 10^0)$

e. $(4 \times 10^3) + (2 \times 10^0)$

Solution

a. 187

b. 2341

c. 1002 (Note that we did not write "one thousand *and* two." In mathematics, the word *and* is used to indicate the position of the decimal point, as in one hundred three and two-tenths, which is 103.2.)

d. $(3 \times 10^3) + (2 \times 10^2) + (1 \times 10^1) + (0 \times 10^0)$
$= (3 \times 1000) + (2 \times 100) + (1 \times 10) + (0 \times 1)$
$= 3000 + 200 + 10 + 0 = 3210$

e. $(4 \times 10^3) + (2 \times 10^0) = (4 \times 1000) + (2 \times 1) = 4000 + 2 = 4002$ (Note that the hundreds and tens places were omitted in the expanded notation, but we were able to obtain the correct result by proceeding in an orderly manner.)

EXERCISES FOR SECTION 7.4

1. Express each number with a Babylonian numeral.
   a. 18 b. 23 c. 82
   d. 102 e. 349 f. 864

2. Express the following as Babylonian numerals.
   a. 34 b. 44 c. 93
   d. 201 e. 423 *f. 3674

3. Evaluate each Babylonian numeral.
   a. ◀ ▼▼▼ b. ◀◀ ▼▼
   c. ◀◀◀ ▼ d. ▼ ◀◀ ▼▼▼
4. Evaluate the following Babylonian numerals.
   a. ▼ ◀◀◀◀ ▼ b. ▼▼ ◀◀ ▼▼
   c. ▼ ◀ ▼ d. ▼ ◀◀ ▼
5. Write each number in expanded notation.
   a. 243 b. 378 c. 1234
   d. two thousand fifty-one
   e. ten thousand four hundred one
6. Write each number in expanded notation.
   a. 345 b. 1776 c. 19,876
   d. three thousand four hundred fifty-six
   e. twelve thousand nine hundred three
7. Write each of the following in base ten (decimal) notation.
   a. two hundred forty
   b. two thousand three hundred eleven
   c. one thousand seven hundred seventy-six
   d. $(4 \times 10^3) + (2 \times 10^2) + (1 \times 10^1) + (3 \times 10^0)$
   e. $(2 \times 10^2) + (0 \times 10^1) + (4 \times 10^0)$
   f. $(4 \times 10^4) + (3 \times 10^2) + (1 \times 10^0)$
8. Write each of the following in base ten (decimal) notation.
   a. three hundred forty-five
   b. forty thousand two
   c. one million one
   d. $(5 \times 10^3) + (3 \times 10^2) + (2 \times 10^1) + (1 \times 10^0)$
   e. $(2 \times 10^4) + (3 \times 10^2) + (4 \times 10^1) + (5 \times 10^0)$
   f. $(7 \times 10^5) + (8 \times 10^3) + (9 \times 10^1)$

Just For Fun

Look at the following equation:

XI + I = X

Can you make it a correct statement without adding, crossing out, or changing anything?

7.5 NUMERATION IN BASES OTHER THAN TEN

Most of the time we group items by tens, but it is not uncommon to group items in some other manner. We group things such as socks and mittens by twos. Another common grouping is by twelves. How do you buy doughnuts and eggs? We purchase items such as these by the dozen, that is, in groups of twelve. Three dozen doughnuts is 3 twelves, or 36, doughnuts.

Consider the instructor who is ordering supplies and orders a *gross* of chalk. One *gross* is a dozen dozen, or 12 twelves. Therefore, the instructor ordered 144 pieces of chalk. Also, there are twelve inches in a foot, twelve hours in one complete cycle of the clock, and twelve months in a year.

Suppose we have $2\frac{1}{2}$ dozen doughnuts. How many doughnuts do we have? We have 2 dozen and $\frac{1}{2}$ of a dozen, or $(2 \times 12) + 6 = 24 + 6 = 30$. Since we are grouping by dozens, we could have written this as 2 dozen + 6. But, we also could have said 2 dozen + 6 ones, and this is the same as $(2 \times 12) + (6 \times 1)$. Since any number (except zero) raised to the zero power is one, we can write this as $(2 \times 12^1) + (6 \times 12^0)$. This expanded notation indicates that we are grouping by twelves. In decimal notation, we grouped by tens and hence were working in base ten. Now we are grouping by twelves and therefore we can say that we are working in base twelve. Hence we have

Vandermark/Stock, Boston

$$(2 \times 12^1) + (6 \times 12^0) = 26_{\text{twelve}}$$

When we write numerals in some base other than base ten, we must indicate what base we are working with. We do this by using a subscript. The numeral

$$47_{\text{twelve}}$$

indicates that we are grouping by twelves, and for this particular example we have 4 twelves and 7 ones—that is, $(4 \times 12^1) + (7 \times 12^0)$, or $(4 \times 12) + (7 \times 1) = 55$.

EXAMPLE 1
Change to base ten notation.

a. 42_{twelve} b. 30_{twelve} c. 234_{twelve}

Solution

a. In order to change a numeral to base ten notation, we first write the numeral in expanded notation:

$$\begin{aligned} 42_{\text{twelve}} &= (4 \times 12^1) + (2 \times 12^0) = (4 \times 12) + (2 \times 1) \\ &= 48 + 2 = 50 \end{aligned}$$

b. $30_{\text{twelve}} = (3 \times 12^1) + (0 \times 12^0) = (3 \times 12) + (0 \times 1) = 36 + 0 = 36$

c. In order to write numerals in expanded notation, we start from the right (note that this is the units place), and proceed to the left in successive higher powers of the indicated base. Therefore,

$234_{\text{twelve}} = (2 \times 12^2) + (3 \times 12^1) + (4 \times 12^0) = (2 \times 144) + (3 \times 12) + (4 \times 1) = 288 + 36 + 4 = 328.$

Many people believe that the reason we normally group items by tens is that man has ten fingers. But man could just as easily group items by fives, since he has five fingers on one hand. The *base five system* is a system of numeration that groups items by fives. Consider the numeral 13: in base ten, this is 1 ten and 3 ones, but in base five, it is 2 fives and 3 ones. Therefore $13 = 23_{\text{five}}$. If we are given a numeral in base five notation, we can convert it to base ten by writing the base five numeral in expanded notation. Recall that in order to write numerals in expanded notation, we start from the right, which is the units place and represents the zero power of the indicated based. Then we proceed to the left in successive higher powers of the indicated base. Therefore we have

$$23_{\text{five}} = (2 \times 5^1) + (3 \times 5^0) = (2 \times 5) + (3 \times 1) = 10 + 3 = 13$$

Next, let us convert 342_{five} to base ten notation:

$$\begin{aligned} 342_{\text{five}} &= (3 \times 5^2) + (4 \times 5^1) + (2 \times 5^0) \\ &= (3 \times 25) + (4 \times 5) + (2 \times 1) \\ &= 75 + 20 + 2 \\ &= 97 \end{aligned}$$

Remember that when we want to convert from a given base to base ten, we use expanded notation.

EXAMPLE 2

Write each number in base ten notation.

a. 40_{five} b. 121_{five} c. 203_{five}

Solution

a. We first write the numeral in expanded notation, then simplify:

$$\begin{aligned} 40_{\text{five}} &= (4 \times 5^1) + (0 \times 5^0) = (4 \times 5) + (0 \times 1) \\ &= 20 + 0 = 20 \end{aligned}$$

b. $$\begin{aligned} 121_{\text{five}} &= (1 \times 5^2) + (2 \times 5^1) + (1 \times 5^0) \\ &= (1 \times 25) + (2 \times 5) + (1 \times 1) \\ &= 25 + 10 + 1 = 36 \end{aligned}$$

c. $$\begin{aligned} 203_{\text{five}} &= (2 \times 5^2) + (0 \times 5^1) + (3 \times 5^0) \\ &= (2 \times 25) + (0 \times 5) + (3 \times 1) \\ &= 50 + 0 + 3 = 53 \end{aligned}$$

Thus far we have only considered changing numerals in a given base to base ten notation. We should also be able to express base ten numerals in terms of other bases. Suppose we want to express 13 as a numeral in base five. Since base five groups items by fives, we determine how many fives are contained in 13. There are 2 fives and 3 ones. Therefore,

$$13 = 23_{\text{five}}$$

There is a convenient rule that enables us to convert from base ten to any given base. We can convert any number from base ten to another base by recording the remainders of successive divisions. We stop dividing when we obtain a quotient of zero. Using the last example, we can illustrate the procedure involved. We wish to convert 13 to base five, so we divide 13 by 5 and record the remainders:

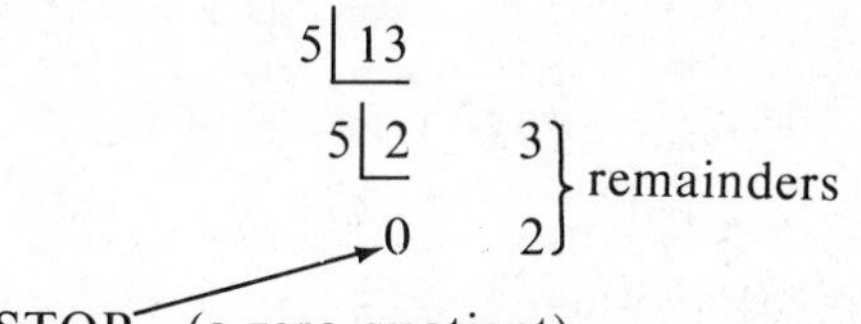

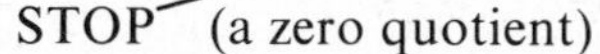

The answer is determined by reading the remainders from *bottom to top*. Therefore, $13 = 23_{\text{five}}$.

Consider the number 97: let's express it with a base five numeral. Performing the successive divisions, we have:

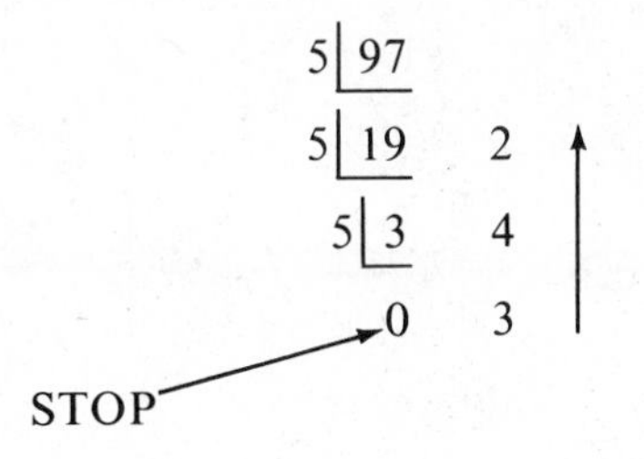

Reading the remainders from bottom to top, we have

$$97 = 342_{\text{five}}$$

We can check the answer by converting the base five numeral to base ten: $342_{\text{five}} = (3 \times 5^2) + (4 \times 5^1) + (2 \times 5^0) = (3 \times 25) + (4 \times 5) + (2 \times 1) = 75 + 20 + 2 = 97$. The answer checks.

EXAMPLE 3

Express each number with base five notation.

a. 43 b. 147 c. 520

Solution

In each case, we perform successive divisions by five, the new base, and record the remainders for each division. We determine the answer by reading the remainders from bottom to top.

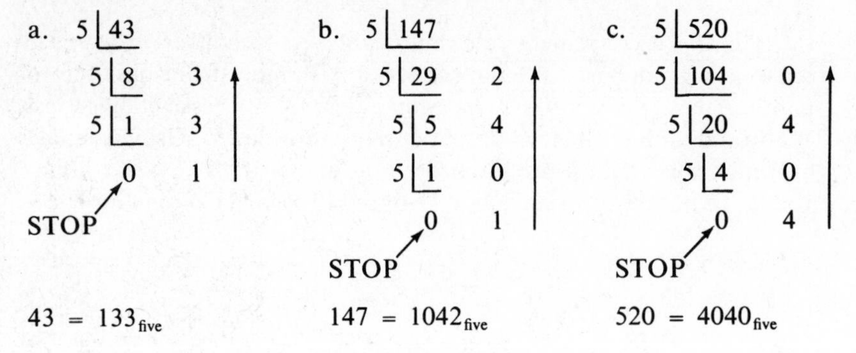

$43 = 133_{five}$ $147 = 1042_{five}$ $520 = 4040_{five}$

EXAMPLE 4

Change the following to base twelve notation.

a. 43 b. 100 c. 520

Solution

In each case, we shall apply our handy rule and perform successive divisions. Remember that the answer is determined by reading the remainders from bottom to top. Since we are converting to base twelve, we shall divide by twelve.

a. 12 | 43
12 | 3 7
0 3
STOP

b. 12 | 100
12 | 8 4
0 8
STOP

c. 12 | 520
12 | 43 4
12 | 3 7
0 3
STOP

$43 = 37_{twelve}$ $100 = 84_{twelve}$ $520 = 374_{twelve}$

Suppose you buy 22 of something (eggs, doughnuts, or bagels, for example). You have purchased 1 dozen plus 10 more. How can we express this in base twelve notation? We cannot say $22 = 110_{twelve}$. Why? Because if we evaluate 110_{twelve}, we have $(1 \times 12^2) + (1 \times 12^1) + (0 \times 12^0)$, or $(1 \times 144) + (1 \times 12) + (0 \times 1) = 144 + 12 + 0 = 156$, which is not 22!

In our decimal system of numeration we use ten symbols, 0, 1, 2, 3, 4, 5, 6, 7, 8, 9. Consequently we should use twelve symbols in the base twelve numeration system. Since we already have ten symbols we can borrow from the base ten system, let us agree to use the additional symbols T and E in the base twelve system of numeration. Let T stand for ten and E stand for eleven. Then the twelve symbols that we shall use in base twelve are 0, 1, 2, 3, 4, 5, 6, 7, 8, 9, T, E. Keep in mind that, in the base ten system of numeration, 11 represents 1 ten and 1 one, or eleven. But in the base twelve system of numeration, T represents eleven, and 11_{twelve} represents 1 twelve and 1 one, or thirteen.

EXAMPLE 5
Change to base ten notation.

a. 40_{twelve} b. $4T_{\text{twelve}}$ c. $ET2_{\text{twelve}}$

Solution

a. $40_{\text{twelve}} = (4 \times 12^1) + (0 \times 12^0) = (4 \times 12) + (0 \times 1) = 48 + 0 = 48$

b. $4T_{\text{twelve}} = (4 \times 12^1) + (T \times 12^0) = (4 \times 12) + (10 \times 1) = 48 + 10 = 58$

c. $$\begin{aligned} ET2_{\text{twelve}} &= (E \times 12^2) + (T \times 12^1) + (2 \times 12^0) \\ &= (E \times 144) + (T \times 12) + (2 \times 1) \\ &= (11 \times 144) + (10 \times 12) + (2 \times 1) \\ &= 1584 + 120 + 2 = 1706 \end{aligned}$$

The base twelve system of numeration is a fairly common system. It is also called the **duodecimal** system of numeration, which indicates that it is a system of numeration with twelve as its base, as opposed to the decimal system of numeration, which has ten as its base.

EXERCISES FOR SECTION 7.5

1. In each example, items are grouped by twelve. Perform the indicated operations.

a.
```
  4 years 7 months
+ 2 years 9 months
```

c.
```
  13 feet 11 inches
+ 11 feet 10 inches
```

e.
```
  2 gross  3 dozen 8 units
+ 4 gross 11 dozen 6 units
```

b.
```
  7 years 3 months
- 4 years 10 months
```

d.
```
  14 feet 6 inches
- 10 feet 7 inches
```

f.
```
  5 gross  9 dozen 3 units
- 2 gross 11 dozen 7 units
```

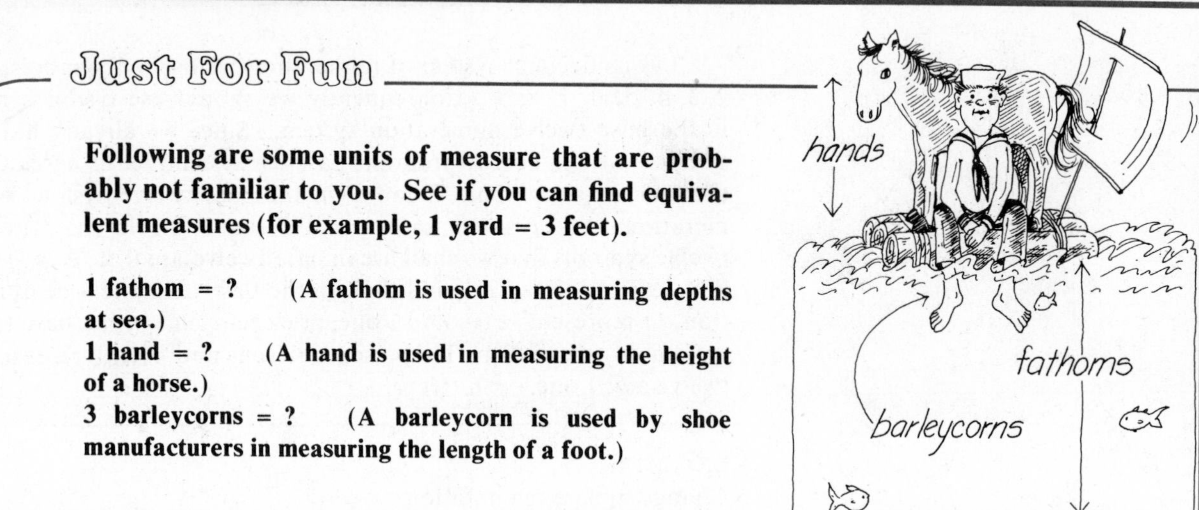

Just For Fun

Following are some units of measure that are probably not familiar to you. See if you can find equivalent measures (for example, 1 yard = 3 feet).

1 fathom = ? (A fathom is used in measuring depths at sea.)
1 hand = ? (A hand is used in measuring the height of a horse.)
3 barleycorns = ? (A barleycorn is used by shoe manufacturers in measuring the length of a foot.)

2. In each example, items are grouped by twelve. Perform the indicated operations. (*Note:* 1 great gross = 12 gross)

a.
```
  3 years 9 months
+ 2 years 7 months
```

b.
```
  8 years 2 months
- 3 years 9 months
```

c.
```
  3 gross 3 dozen 9 units
+ 5 gross 8 dozen 4 units
```

d.
```
  7 gross 2 dozen 5 units
- 3 gross 9 dozen 7 units
```

e.
```
  3 great gross  2 gross 4 dozen
+ 4 great gross 10 gross 8 dozen
```

f.
```
  3 great gross 2 gross 4 dozen
- 1 great gross 3 gross 7 dozen
```

3. Change to base ten notation.
   a. 32_{five} b. 12_{five} c. 14_{five}
   d. 123_{five} e. 203_{five} f. 2031_{five}

4. Change to base ten notation.
   a. 13_{five} b. 44_{five} c. 231_{five}
   d. 304_{five} e. 100_{five} f. 4021_{five}

5. Change to base five notation.
   a. 6 b. 19 c. 38
   d. 3 e. 121 f. 497

6. Change to base five notation.
   a. 9 b. 27 c. 4
   d. 243 *e. 2003 *f. 3421

7. Change to base ten notation.
   a. 47_{twelve} b. 28_{twelve} c. 59_{twelve}
   d. 124_{twelve} e. 347_{twelve} f. 1001_{twelve}

8. Change to base ten notation.
   a. 42_{twelve} b. 54_{twelve} c. 99_{twelve}
   d. 137_{twelve} e. 243_{twelve} f. 2001_{twelve}

9. Change to base twelve notation.
   a. 42 b. 53 c. 60
   d. 137 e. 234 f. 876

10. Change to base twelve notation.
   a. 49 b. 66 c. 118
   d. 341 *e. 3421 *f. 5736

11. Change to base ten notation.
 a. $2E_{\text{twelve}}$ b. TE_{twelve}
 c. $T2E_{\text{twelve}}$ d. $E2T_{\text{twelve}}$
12. Change to base ten notation.
 a. $T3_{\text{twelve}}$ b. ET_{twelve}
 c. $3E4_{\text{twelve}}$ d. $TE5_{\text{twelve}}$
13. An instructor ordered the following classroom supplies: a gross of pencils, a great gross of chalk (1 great gross is 12 gross), 3 dozen note-books, $\frac{1}{2}$ dozen pens, and 1 grade book. What is the total number of items ordered?
14. Jake the bagel maker prepared the following order: He made 6 dozen salt bagels, 3 dozen poppy bagels, 6 dozen dozen rye bagels, a gross of plain bagels, and a gross of onion bagels. How many bagels did Jake make?
15. a. What numeral is 47_{twelve} in base five?
 b. What numeral is 34_{five} in base twelve?

7.6 BASE FIVE ARITHMETIC

Thus far we have converted base ten numerals to base twelve numerals and also to base five numerals. We also have converted base twelve and base five numerals to base ten numerals. In a manner of speaking, we have learned to count in some base other than base ten. The next step is to perform some arithmetic operations in a base system other than base ten. A convenient base to work with is base five. Using conventional Hindu-Arabic numerals, the base five system involves the numerals 0, 1, 2, 3, 4. Recall that the numeral 14_{five} is composed of 1 five and 4 ones; hence $14_{\text{five}} = 9$ in base ten. For the sake of convenience, we shall now write all numerals in base five with the numeral 5 as a subscript. Therefore, 14_{five} is the same as 14_5. Recall that if no subscript is written—that is, if no base is indicated—then it is understood that the numeral is written in base ten notation.

Now suppose we want to perform the following addition: $14_5 + 24_5$. At first glance, you might want to say that the answer is 38. But 38 is an impossible answer! Why? For one reason, the base five system of numeration uses only the digits 0, 1, 2, 3, 4, and therefore we cannot have an 8 in our answer.

Let us try again. One way to find an answer to the problem $14_5 + 24_5$ is to convert each of the base five numerals to base ten, add the base ten numerals, and then convert the answer back to base five.

$$14_5 = 9 \quad \text{and} \quad 24_5 = 14$$
$$9 + 14 = 23 \quad \text{and} \quad 23 = 43_5$$

Therefore,

$$14_5 + 24_5 = 43_5$$

This may seem like the best method, but it isn't—especially if we want to add three- and four-digit numerals such as 1223_5 and 4223_5. One thing that would help us to add numerals in base five is a table of addition facts. Table 1 is a table of addition facts for base five.

TABLE 1. Base five addition table

+	0_5	1_5	2_5	3_5	4_5
0_5	0_5	1_5	2_5	3_5	4_5
1_5	1_5	2_5	3_5	4_5	10_5
2_5	2_5	3_5	4_5	10_5	11_5
3_5	3_5	4_5	10_5	11_5	12_5
4_5	4_5	10_5	11_5	12_5	13_5

The entries in this table may seem odd at first. Let us check a few of them so we can see that they do make sense. How does $3_5 + 4_5 = 12_5$? If we add $3 + 4$ in base ten, we get 7. Since 7 is composed of 1 five and 2 ones, $7 = 12_5$, and so $3_5 + 4_5 = 12_5$. Similarly, $4_5 + 4_5 = 13_5$, because $4 + 4 = 8$ in base ten, and 8 is composed of 1 five and 3 ones; hence $8 = 13_5$. Let us solve an addition problem:

$$\begin{array}{r} 23_5 \\ +\ \underline{34_5} \end{array}$$

We start with the units place: $3_5 + 4_5 = 12_5$. Writing the 2 and carrying the 1, we have

$$\begin{array}{r} {\scriptstyle 1} \\ 23_5 \\ +\ \underline{34_5} \\ 2_5 \end{array}$$

Next we add the fives: $1_5 + 2_5 + 3_5 = 3_5 + 3_5 = 11_5$. Therefore the completed problem is

$$\begin{array}{r} 23_5 \\ +\ \underline{34_5} \\ 112_5 \end{array}$$

We can check our work in base ten:

$$\begin{aligned} 23_5 &= (2 \times 5^1) + (3 \times 5^0) = (2 \times 5) + (3 \times 1) = 10 + 3 = 13 \\ +\ \underline{34_5} &= (3 \times 5^1) + (4 \times 5^0) = (3 \times 5) + (4 \times 1) = 15 + 4 = \underline{19} \\ 112_5 &= (1 \times 5^2) + (1 \times 5^1) + (2 \times 5^0) \\ &= (1 \times 25) + (1 \times 5) + (2 \times 1) = 25 + 5 + 2 \qquad = 32 \end{aligned}$$

Let us try the preceding problem without using the addition table. In order to add 23_5 and 34_5, we can combine 3 + 4 as we normally would; that is, 3 + 4 = 7. Remember that the numerals 0, 1, 2, 3, 4 have the same meaning in base five as they do in base ten. Our only problem is the 7, but 7 is composed of 1 five and 2 ones, and therefore $3_5 + 4_5 = 12_5$. We write the 2 and carry the 1 as before:

$$\begin{array}{r} {\scriptstyle 1} \\ 23_5 \\ +\ 34_5 \\ \hline 2_5 \end{array}$$

We now add 1 + 2 + 3 = 6, but in base five notation $6 = 11_5$ (1 five and 1 one). Hence $23_5 + 34_5 = 112_5$.

EXAMPLE 1

Find the sum of 234_5 and 341_5.

Solution

Starting with the units place, 4 + 1 = 5, but in base five, $5 = 10_5$ (1 five and no ones). Placing the 0 and carrying the 1, we have 1 + 3 + 4 = 8; but in base five, $8 = 13_5$ (1 five and 3 ones). Placing the 3 and carrying the 1, we now have 1 + 2 + 3 = 6, and in base five, $6 = 11_5$ (1 five and 1 one). Our addition is complete.

$$\begin{array}{r} 234_5 \\ +\ 341_5 \\ \hline 1130_5 \end{array}$$

EXAMPLE 2

Find the sum of 133_5 and 341_5.

Solution

Starting with the units place, 3 + 1 = 4. This is the same for base five as for base ten, because the numerals 0, 1, 2, 3, 4 have the same meaning in base five as they do in base ten. Next we proceed to the fives place: 3 + 4 = 7, but in base five, $7 = 12_5$ (1 five and 2 ones). Placing the 2 and carrying the 1, we now have 1 + 1 + 3 = 5, and in base five, $5 = 10_5$ (1 five and no ones). Our addition is complete.

$$\begin{array}{r} 133_5 \\ +\ 341_5 \\ \hline 1024_5 \end{array}$$

EXAMPLE 3

Find the sum of 342_5 and 324_5.

Solution

$$\begin{array}{r} 342_5 \\ +\ 324_5 \\ \hline 1221_5 \end{array}$$

Check:

$$\begin{aligned}
342_5 &= (3 \times 5^2) + (4 \times 5^1) + (2 \times 5^0) \\
&= (3 \times 25) + (4 \times 5) + (2 \times 1) = 75 + 20 + 2 = 97 \\
+\ 324_5 &= (3 \times 5^2) + (2 \times 5^1) + (4 \times 5^0) \\
&= (3 \times 25) + (2 \times 5) + (4 \times 1) = 75 + 10 + 4 = \underline{89} \\
1221_5 &= (1 \times 5^3) + (2 \times 5^2) + (2 \times 5^1) + (1 \times 5^0) \\
&= (1 \times 125) + (2 \times 25) + (2 \times 5) + (1 \times 1) \\
&= 125 + 50 + 10 + 1 = 186
\end{aligned}$$

Another arithmetic operation that goes hand in hand with addition is the operation of subtraction. Before we try a subtraction problem in base five, let's review subtraction in base ten. Suppose we wish to subtract 248 from 735, that is, 735 − 248. We set the problem up as shown below. Note that the parts of the problem have been labeled.

$$\begin{array}{rl} 735 & \textit{minuend} \\ -\ 248 & \textit{subtrahend} \\ \hline \end{array}$$

The answer is usually called the *difference*, but it is also sometimes called the *remainder*. Performing the subtraction, we have

$$\begin{array}{rrr} {}^{6} & {}^{12} & {}^{15} \\ \not{7} & \not{3} & 5 \\ -\ 2 & 4 & 8 \\ \hline 4 & 8 & 7 \end{array}$$

Notice that we must rename 3 tens, 5 ones as 2 tens, 15 ones. Seven hundreds, 2 tens are then renamed as 6 hundreds, 12 tens. The numbers in small type over the minuend are shown only to indicate the new arrangement of the number in order to facilitate the subtraction.

Now let us try a subtraction problem in base five. Consider the following:

$$\begin{array}{r} 42_5 \\ -\ 13_5 \\ \hline \end{array}$$

We note that we cannot subtract 3 from 2, so we must borrow from the 4. But what do we borrow? Since we are in base five, we borrow 1 five. So we now have 1 five and 2 ones in the units place, that is, 12_5. But this is the same as 7, and 3 from 7 is 4. Now we have only 3 in the fives place, and 1 from 3 is 2. This process is illustrated below.

$$\begin{array}{rcc} & \overset{3}{\not{4}} & {}^{1}2_5 \\ - & 1 & 3_5 \\ \hline & 2 & 4_5 \end{array}$$

Remember to indicate the base with which you are working in your answer. If no subscript is written, then it is understood that the numeral is written in base ten notation.

Let us check our answer to the preceding problem by translating it into base ten.

$$\begin{array}{rlllr} 42_5 = (4 \times 5^1) + (2 \times 5^0) & = (4 \times 5) + (2 \times 1) & = 20 + 2 = & 22 \\ -\ 13_5 = (1 \times 5^1) + (3 \times 5^0) & = (1 \times 5) + (3 \times 1) & = 5 + 3 = & -\ 8 \\ \hline 24_5 = (2 \times 5^1) + (4 \times 5^0) & = (2 \times 5) + (4 \times 1) & = 10 + 4 = & 14 \end{array}$$

Since $22 - 8 = 14$, the answer checks.

Now let us try another problem. Consider

$$\begin{array}{r} 431_5 \\ -\ 132_5 \\ \hline \end{array}$$

Since we cannot subtract 2 from 1, we borrow 1 from the 3 in the fives place, leaving a 2 in the fives place and giving us 11_5 in the units place. We know 11_5 is the same as 6, and 2 from 6 is 4. So far we have:

$$\begin{array}{rccc} & 4 & \overset{2}{\not{3}} & {}^{1}1_5 \\ - & 1 & 3 & 2_5 \\ \hline & & & 4_5 \end{array} \qquad 11_5 = 6 \quad \text{so} \quad 11_5 - 2_5 = 4_5$$

We cannot subtract 3 from 2 in the fives place, so again we borrow. This time we borrow 1 from the 4 in the next place (the twenty-fives). This means we have borrowed 1 twenty-five, or 5 fives, which gives us a total of 7 fives, and 3 fives from 7 fives is 4 fives. We can also think of this as subtracting 3_5 from 12_5, or 3 from 7, which gives us a 4 in the fives place. Now we have only 3 in the twenty-fives place, and 1 from 3 is 2. The completed subtraction process is illustrated as follows:

$$\begin{array}{rrrr} & {}^{3}\cancel{4} & {}^{12}\cancel{3} & {}^{1}1_5 \\ - & 1 & 3 & 2_5 \\ \hline & 2 & 4 & 4_5 \end{array} \qquad 12_5 = 7 \quad \text{so} \quad 12_5 - 3_5 = 4_5$$

Table 1, the base five addition table, may be helpful to you in subtraction problems. Remember that when you borrow, you are borrowing a number in the indicated base, not a ten. In these examples we are borrowing fives and twenty-fives. For problems in other bases, you may be borrowing sevens, fours, etc.

EXAMPLE 4

Subtract 24_5 from 33_5.

Solution

Since we cannot subtract 4 from 3, we borrow 1 from the fives place, which gives us 1 five and 3 ones in the units place, or $13_5 = 8$; hence 4_5 from 13_5 is 4_5. We are left with a 2 in the fives place, and 2 from 2 is zero.

$$\begin{array}{rrr} & {}^{2}\cancel{3} & {}^{1}3_5 \\ - & 2 & 4_5 \\ \hline & & 4_5 \end{array} \qquad \text{or} \qquad \begin{array}{rr} & 33_5 \\ - & 24_5 \\ \hline & 4_5 \end{array}$$

EXAMPLE 5

Subtract 234_5 from 433_5.

Solution

We cannot subtract 4 from 3, so we borrow 1 from the fives place. This gives 1 five and 3 ones in the units place, or $13_5 = 8$, and 4 from 8 is 4. In the fives place, we cannot subtract 3 from 2, so we borrow 1 from the 4 in the twenty-fives place. One twenty-five and 2 fives gives us 7 fives, and 3 fives from 7 fives is 4 fives. We are left with a 3 in the twenty-fives place, and 2 from 3 is 1.

$$\begin{array}{rrrr} & {}^{3}\cancel{4} & {}^{12}\cancel{3} & {}^{1}3_5 \\ - & 2 & 3 & 4_5 \\ \hline & 1 & 4 & 4_5 \end{array} \qquad \text{or} \qquad \begin{array}{rr} & 433_5 \\ - & 234_5 \\ \hline & 144_5 \end{array}$$

We shall also examine the operation of multiplication in base five. The procedure for multiplication in base five (or in any other base) is the same as that in base ten. Table 2 is a table of multiplication facts for multiplying numerals in base five.

TABLE 2. Base five multiplication table

$\times$	0_5	1_5	2_5	3_5	4_5
0_5	0_5	0_5	0_5	0_5	0_5
1_5	0_5	1_5	2_5	3_5	4_5
2_5	0_5	2_5	4_5	11_5	13_5
3_5	0_5	3_5	11_5	14_5	22_5
4_5	0_5	4_5	13_5	22_5	31_5

We see that $3_5 \times 3_5 = 14_5$, since $3 \times 3 = 9$ and 9 is equal to 1 five and 4 ones. Similarly, $4_5 \times 3_5 = 22_5$; $4 \times 3 = 12$, but 12 is 2 fives and 2 ones.

We must master single-digit multiplication before we can proceed to other examples. To help in an example such as 4_5 times 4_5, just remember that $4 \times 4 = 16$, and 16 equals 3 fives and 1 one; therefore $4_5 \times 4_5 = 31_5$.

Consider the following multiplication problem:

$$\begin{array}{r} 21_5 \\ \times \quad 31_5 \\ \hline 21_5 \\ 113_5 \\ \hline 1201_5 \end{array}$$

Since the procedure for multiplication is the same for any base, we first multiply 21_5 by 1_5: $21_5 \times 1_5 = 21_5$. Now we multiply by the next digit, that is, $21_5 \times 3_5$. To do this, we multiply each digit of 21_5 by 3_5: $3 \times 1 = 3$ and $3 \times 2 = 6$. However, in base five, $6 = 11_5$; hence $21_5 \times 3_5 = 113_5$. Note that the partial product is indented just as in base ten. Next we find the sum of the partial products:

$$\begin{array}{r} 21_5 \\ +\ 113_5 \\ \hline 1201_5 \end{array}$$

Note that in the fives place $2 + 3 = 5$, but in base five, $5 = 10_5$, so we write the 0 and carry the 1 to the next place, adding it to the 1 already there.

Let us try another example. Consider

$$\begin{array}{r} 433_5 \\ \times \quad 2_5 \\ \hline \end{array}$$

One way to do this problem is shown below. Note that partial products are used here, and each one is indented to indicate the powers of five involved.

$$\begin{array}{r} 433_5 \\ \times \quad 2_5 \\ \hline 11_5 \\ 11_{5} \\ 13_{5} \\ \hline 1421_5 \end{array}$$

We can do this problem in another way, but we will have to do some work mentally. The solution to the problem could have appeared as

$$\begin{array}{r} 433_5 \\ \times \quad 2_5 \\ \hline 1421_5 \end{array}$$

To do the multiplication in the shortened form, we first multiply the 3 in the units place by 2: $2 \times 3 = 6 = 11_5$. Therefore we write 1 and carry 1. In the fives place, we again have 2×3 which equals 11_5, but we carried 1 from the units place, so we have $11_5 + 1_5 = 12_5$. Hence we write the 2 and carry the 1. In the twenty-fives place, we have 2×4 which equals 13_5, and we carried a 1 from the fives place, so $13_5 + 1_5 = 14_5$ and our multiplication is complete.

EXAMPLE 6

Find the product

$$\begin{array}{r} 342_5 \\ \times \quad 23_5 \\ \hline \end{array}$$

Solution

First, we multiply 342_5 by 3_5: $3 \times 2 = 6$, but $6 = 11_5$, so we write 1 and carry 1. We have $3 \times 4 = 12 = 22_5$, and since we carried a 1, we have $22_5 + 1_5 = 23_5$. We write the 3 and carry the 2. $3 \times 3 = 9 = 14_5$, and since we carried a 2, we have $14_5 + 2_5 = 21_5$. Therefore our partial product ($342_5 \times 3_5$) appears in the problem as

$$\begin{array}{r} 342_5 \\ \times \quad 23_5 \\ \hline 2131_5 \end{array}$$

Now we are ready to multiply by the next digit, 2: $2 \times 2 = 4 = 4_5$; we write the 4 and proceed. We have $2 \times 4 = 8$, but $8 = 13_5$, so we write the 3 and carry the 1. $2 \times 3 = 6 = 11_5$ and since we carried a 1, we have $11_5 + 1_5 = 12_5$. Now the problem looks like

$$\begin{array}{r} 342_5 \\ \times \quad 23_5 \\ \hline 2131_5 \\ 1234_5 \\ \hline \end{array}$$

Note that the partial product is indented to indicate the powers of five involved. Next, we find the sum of the partial products, and the completed problem is

$$\begin{array}{r} 342_5 \\ \times \quad 23_5 \\ \hline 2131_5 \\ 1234_5 \\ \hline 20021_5 \end{array}$$

EXAMPLE 7
Multiply 33_5 by 4_5.

Solution
In doing this problem, we can use table 2 to find $4_5 \times 3_5 = 22_5$. We write 2 and carry 2. We again have $4_5 \times 3_5 = 22_5$, but we carried a 2, so we have $22_5 + 2_5 = 24_5$. Our multiplication is complete.

Another way to do this problem is the following: $4 \times 3 = 12$, but $12 = 22_5$, so we write 2 and carry 2; again, $4 \times 3 = 12 = 22_5$, and since we carried a 2, we have $22_5 + 2_5 = 24_5$.

$$\begin{array}{r} 33_5 \\ \times \quad 4_5 \\ \hline 242_5 \end{array}$$

EXAMPLE 8
Multiply 44_5 by 23_5.

Solution
According to table 2, $3_5 \times 4_5 = 22_5$; we write 2 and carry 2. Again, we have $3_5 \times 4_5 = 22_5$, and since we carried a 2, we have $22_5 + 2_5 = 24_5$. Therefore the partial product is 242_5. We find the next partial product as follows: $2_5 \times 4_5 = 13_5$, so we write the 3 and carry the 1; again, we have $2_5 \times 4_5 = 13_5$, and since we carried a 1, we have

$13_5 + 1_5 = 14_5$. This partial product is 143_5. We next find the sum of the partial products and obtain the final result, 2222_5.

$$\begin{array}{r} 44_5 \\ \times \quad 23_5 \\ \hline 242_5 \\ 143_5 \\ \hline 2222_5 \end{array}$$

A true test of understanding of the process of computation in base five is the operation of division. We divide in base five in the same way that we divide in base ten. In fact, we can think in base ten, but we must write our computation and answer in base five. Consider the problem 123_5 divided by 2_5. We first set up this problem just as we would in base ten, but remember that we are working in base five. Therefore we have

$$2_5 \overline{)\,123_5}$$

At first glance, we might want to say that 2 divides 12 six times; but this is not the case. We are dividing 2_5 into 12_5. Besides, we cannot have a 6 for an answer since base five only uses the numerals 0, 1, 2, 3, 4. Since we are dividing 2_5 into 12_5, we can think of this as dividing 2 into $7 = 12_5$. Two divides 7 three times, and $2_5 \times 3_5 = 11_5$. Thus we have

$$\begin{array}{r} 3 \\ 2_5 \overline{)\,123_5} \\ 11_5 \\ \hline \end{array}$$

Next, we subtract and bring down the next digit. The problem now appears as

$$\begin{array}{r} 3 \\ 2_5 \overline{)\,123_5} \\ 11_5 \\ \hline 13_5 \end{array}$$

Now we must divide 2_5 into 13_5. We can think of this as dividing 2 into 8, which equals 4, and $2_5 \times 4_5 = 13_5$. We have completed the division, and there is no remainder. The completed problem and check appear as follows:

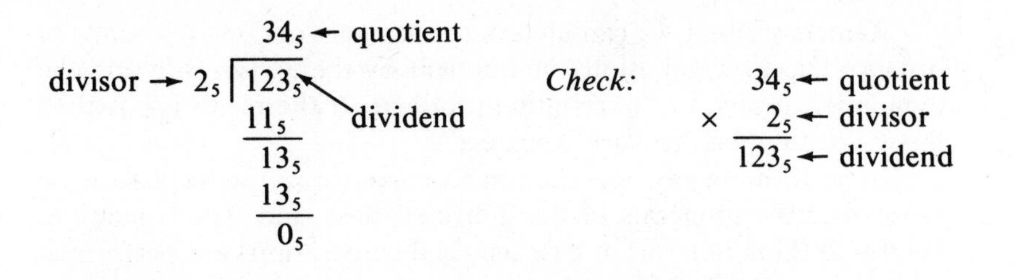

EXAMPLE 9

Divide 141_5 by 2_5.

Solution

First we divide 2_5 into 14_5 (think of this as 2 into 9), which equals 4. As we do in base ten, we now multiply $4_5 \times 2_5 = 13_5$; we then subtract this from 14_5 in the dividend. We bring down the next digit, 1. Now we must divide 2_5 into 11_5 (think of this as 2 into 6), which equals 3, and $3_5 \times 2_5 = 11_5$. We have completed the division, and there is no remainder.

$$\begin{array}{r|l} & \ \ 43_5 \\ \hline 2_5 & 141_5 \\ & \underline{13_5} \\ & \ \ 11_5 \\ & \ \ \underline{11_5} \\ & \ \ \ \ 0_5 \end{array} \qquad \text{Recall that } \begin{cases} 1_5 \times 2_5 = 2_5 \\ 2_5 \times 2_5 = 4_5 \\ 3_5 \times 2_5 = 11_5 \\ 4_5 \times 2_5 = 13_5 \end{cases}$$

EXAMPLE 10

Divide 234_5 by 4_5.

Solution

Dividing 4_5 into 23_5 (think of this as 4 into 13), we get 3. Multiplying $3_5 \times 4_5$, we obtain 22_5, and subtract this from 23_5 in the dividend. After bringing down the next digit, we divide 4_5 into 14_5 (think of this as 4 into 9), which equals 2, and $2_5 \times 4_5 = 13_5$. Subtracting this from 14_5, we obtain a remainder of 1_5, and the division is complete.

$$\begin{array}{r|l} & \ \ 32_5 \\ \hline 4_5 & 234_5 \\ & \underline{22_5} \\ & \ \ 14_5 \\ & \ \ \underline{13_5} \\ & \ \ \ \ 1_5 \end{array} \qquad \text{remainder: } 1_5$$

Remember that we can always check our work in division. In order to check, we multiply the quotient by the divisor and add the remainder, if any, to the resulting product. If the result is equal to the dividend, then the work is correct.

In performing any operation in base five (or any other base), we must write the numerals in the indicated base, but it is sometimes helpful to think in terms of base ten, as the operations are performed in the same manner in every base.

EXERCISES FOR SECTION 7.6

1. Perform the following additions in base five. Check your work by converting to base ten.

   a. $13_5 + 23_5$ b. $14_5 + 22_5$ c. $23_5 + 32_5$

   d. $123_5 + 124_5$ e. $231_5 + 222_5$ f. $343_5 + 112_5$

2. Perform the following additions in base five. Check your work by converting to base ten.

   a. $34_5 + 34_5$ b. $42_5 + 24_5$ c. $34_5 + 11_5$

   d. $1213_5 + 2312_5$ e. $3142_5 + 2233_5$ f. $4241_5 + 1204_5$

3. Perform the following subtractions in base five. Check your work by converting to base ten.

   a. $23_5 - 14_5$ b. $22_5 - 13_5$ c. $12_5 - 4_5$

   d. $321_5 - 231_5$ e. $231_5 - 132_5$ f. $411_5 - 122_5$

4. Perform the following subtractions in base five. Check your work by converting to base ten.

   a. $32_5 - 23_5$ b. $22_5 - 14_5$ c. $11_5 - 3_5$

   d. $434_5 - 332_5$ e. $4211_5 - 1232_5$ f. $3212_5 - 2233_5$

5. Perform the following multiplications in base five. Check your work by converting to base ten.

   a. $231_5 \times 3_5$ b. $432_5 \times 2_5$ c. $432_5 \times 4_5$

   d. $231_5 \times 21_5$ e. $324_5 \times 23_5$ f. $432_5 \times 34_5$

6. Perform the following multiplications in base five. Check your work by converting to base ten.

   a. $343_5 \times 2_5$ b. $434_5 \times 3_5$ c. $231_5 \times 21_5$

   d. $234_5 \times 42_5$ e. $434_5 \times 34_5$ f. $434_5 \times 234_5$

7. Perform the following divisions in base five. Check your work by converting to base ten.

   a. $2_5 \overline{)11_5}$ b. $3_5 \overline{)22_5}$ c. $4_5 \overline{)31_5}$

   d. $2_5 \overline{)32_5}$ e. $3_5 \overline{)212_5}$ f. $4_5 \overline{)103_5}$

8. Perform the following divisions in base five. Check your work by converting to base ten.

   a. $2_5 \overline{)124_5}$ b. $3_5 \overline{)343_5}$ c. $4_5 \overline{)342_5}$

   d. $3_5 \overline{)1234_5}$ e. $11_5 \overline{)243_5}$ f. $13_5 \overline{)341_5}$

Just For Fun

What three words in the English language are pronounced the same as 4?

7.7 BINARY NOTATION AND OTHER BASES

Arithmetic operations in other bases are performed in the same manner as in base ten or base five. The base two system of numeration (also called *binary notation*) is of particular interest because it has some useful applications, particularly in the area of computers. **Binary notation** is a system of numeration that groups items by twos. Therefore, the only numerals used in this system are 0 and 1. We can illustrate this if we examine a numeral in base two notation. Consider 10_2: writing this in expanded notation, we have $10_2 = (1 \times 2^1) + (0 \times 2^0) = (1 \times 2) + (0 \times 1) = 2 + 0 = 2$. How would we express 3 in base two? Three is composed of 1 two and 1 one; hence $3 = 11_2$.

The binary system is unique in that it uses only two symbols, 0 and 1, to represent any number. This is important for computers, because these machines consist of a large number of electrical switches. Each of these switches, like a light switch, has only two possible positions: on or off. The position of a switch can be changed in nanoseconds. (A *nanosecond* is one-billionth of a second.) This is because electricity travels at a rate of approximately 186,000 miles per second, or about 1 foot per nanosecond. If a switch is on, then it represents the numeral 1; if it is off, it represents the numeral 0. Using only these two symbols, a computer can perform thousands of calculations per second and thereby save thousands of hours of human worktime. Information is fed into most computers in base ten notation and then converted to base two. The machine performs the necessary base two calculations by turning many switches on and off.

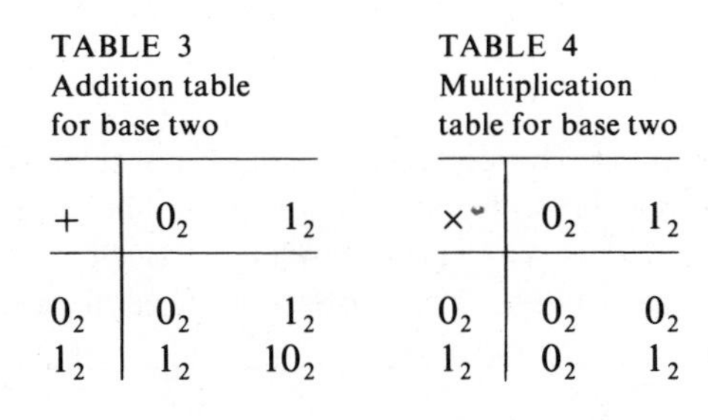

TABLE 3
Addition table for base two

+	0_2	1_2
0_2	0_2	1_2
1_2	1_2	10_2

TABLE 4
Multiplication table for base two

×	0_2	1_2
0_2	0_2	0_2
1_2	0_2	1_2

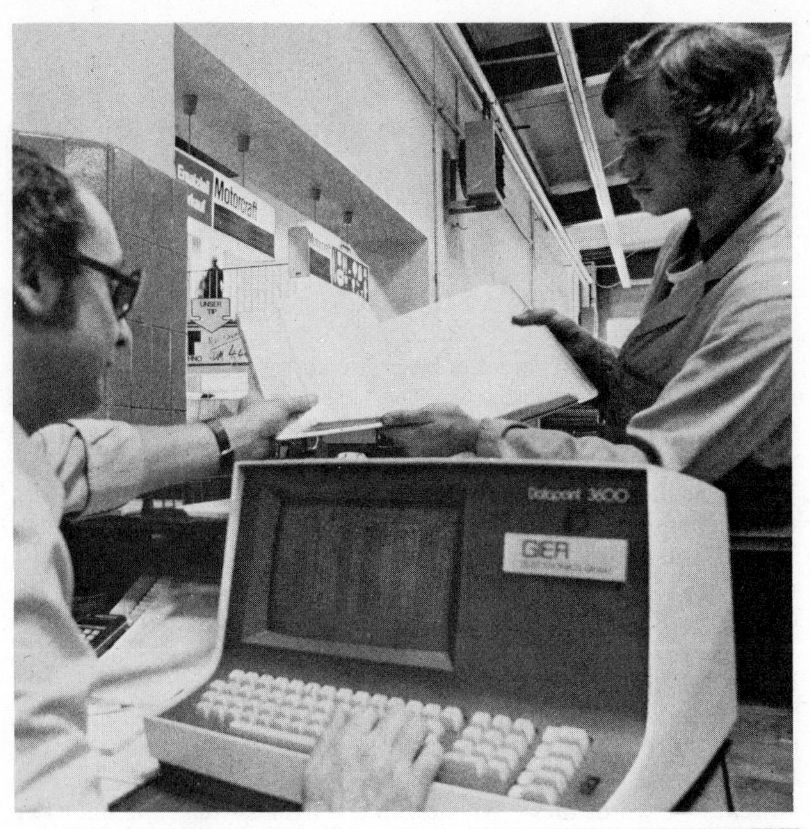

This computer performs all of its internal calculations in base two. The results are converted to base ten for display on the monitor screen.

TRW

To convert a base ten numeral to base two, we perform successive divisions by 2, the new base, and record the remainders for each separate division. The answer is determined by reading the remainders from the bottom to the top.

To express 7 as a base two numeral, we perform the division in the following manner:

$$\begin{array}{r|l l} 2 & \underline{7} & \\ 2 & \underline{3} & 1 \uparrow \\ 2 & \underline{1} & 1 \\ & 0 & 1 \end{array} \qquad 7 = 111_2$$

We can check our answer by converting 111_2 back to base ten. We can do this by writing 111_2 in expanded notation: $111_2 = (1 \times 2^2) + (1 \times 2^1) + (1 \times 2^0) = (1 \times 4) + (1 \times 2) + (1 \times 1) = 4 + 2 + 1 = 7$. Note that since we are now working in base two, each place value is a power of 2: 2^0, 2^1, 2^2, 2^3, and so on.

EXAMPLE 1
Change to base ten notation.

a. 101_2 b. 1101_2

Solution

a. In order to change a numeral to base ten notation, we must write the numeral in expanded notation.

$$\begin{aligned} 101_2 &= (1 \times 2^2) + (0 \times 2^1) + (1 \times 2^0) \\ &= (1 \times 4) + (0 \times 2) + (1 \times 1) \\ &= 4 + 0 + 1 \\ &= 5 \end{aligned}$$

b.
$$\begin{aligned} 1101_2 &= (1 \times 2^3) + (1 \times 2^2) + (0 \times 2^1) + (1 \times 2^0) \\ &= (1 \times 8) + (1 \times 4) + (0 \times 2) + (1 \times 1) \\ &= 8 + 4 + 0 + 1 \\ &= 13 \end{aligned}$$

EXAMPLE 2
Change to base two notation.

a. 9 b. 15

Solution
We perform successive divisions by 2 and record the remainders for each division. The answer is determined by reading the remainders from bottom to top.

a.
$$\left.\begin{array}{rlc} 2 & \underline{\big|\,9} & \\ 2 & \underline{\big|\,4} & 1 \\ 2 & \underline{\big|\,2} & 0 \\ 2 & \underline{\big|\,1} & 0 \\ & \;\;0 & 1 \end{array}\right\uparrow$$
$$9 = 1001_2$$

b.
$$\left.\begin{array}{rlc} 2 & \underline{\big|\,15} & \\ 2 & \underline{\big|\,7} & 1 \\ 2 & \underline{\big|\,3} & 1 \\ 2 & \underline{\big|\,1} & 1 \\ & \;\;0 & 1 \end{array}\right\uparrow$$
$$15 = 1111_2$$

In order to perform addition in base two, we must remember the four addition facts listed in table 3. We must also remember the "carrying" process. Consider the following addition problem:

$$\begin{array}{r} 111_2 \\ +\ \underline{110_2} \end{array}$$

We start with the units place: $0 + 1 = 1$, and this is the answer for both base two and base ten, because the numerals 0 and 1 have the same meaning in base two and base ten. The sum of 0 and 1 is 1, and there is nothing to carry. We proceed to the twos place: 1 + 1 = 2, but in base two, 2 = 10_2. We write the 0 and carry the 1. Next we have 1 + 1 + 1, which equals 3; in base two, 3 = 11_2 (1 two and 1 one). The addition is complete, as shown below.

$$\begin{array}{r} 111_2 \\ +\ 110_2 \\ \hline 1101_2 \end{array}$$

EXAMPLE 3
Find the sum of 1010_2 and 1011_2.

Solution

$$\begin{array}{r} 1010_2 \\ +\ 1011_2 \\ \hline 10101_2 \end{array}$$

Check:

$$\begin{aligned} 1010_2 &= (1 \times 2^3) + (0 \times 2^2) + (1 \times 2^1) + (0 \times 2^0) \\ &= (1 \times 8) + (0 \times 4) + (1 \times 2) + (0 \times 1) \\ &= 8 + 0 + 2 + 0 && = 10 \\ +\ 1011_2 &= (1 \times 2^3) + (0 \times 2^2) + (1 \times 2^1) + (1 \times 2^0) \\ &= (1 \times 8) + (0 \times 4) + (1 \times 2) + (1 \times 1) \\ &= 8 + 0 + 2 + 1 && = \underline{11} \\ 10101_2 &= (1 \times 2^4) + (0 \times 2^3) + (1 \times 2^2) + (0 \times 2^1) + (1 \times 2^0) \\ &= (1 \times 16) + (0 \times 8) + (1 \times 4) + (0 \times 2) + (1 \times 1) \\ &= 16 + 0 + 4 + 0 + 1 && = 21 \end{aligned}$$

EXAMPLE 4
Subtract 111_2 from 1011_2.

Solution
In the units and twos place, we subtract 1 from 1 and obtain 0. But in the fours place, we cannot subtract 1 from 0. Therefore, we borrow 1 from the eights place. One eight gives us 2 fours, and 1 four from 2 fours is 1 four. The subtraction is complete.

$$\begin{array}{r} 1011_2 \\ -\ 111_2 \\ \hline 100_2 \end{array}$$

Multiplication in base two does not present much of a problem, providing we can add the partial products, because we only have to remember the four multiplication facts listed in table 4: $0 \times 0 = 0$, $0 \times 1 = 0$, $1 \times 0 = 0$, and $1 \times 1 = 1$. Consider the following multiplication problem:

$$\begin{array}{r} 101_2 \\ \times \quad 11_2 \\ \hline 101_2 \\ 101_2 \\ \hline 1111_2 \end{array}$$

Since the procedure for multiplication is the same for any base, we first multiply 101_2 by 1_2, which equals 101_2. Now we multiply by the next digit, that is, $101_2 \times 1_2$, which equals 101_2. Note that the partial product is indented, as in base ten. Next we find the sum of the partial products. For this example, we do not have to carry anything.

EXAMPLE 5
Multiply 110_2 by 11_2.

Solution

$$\begin{array}{r} 110_2 \\ \times \quad 11_2 \\ \hline 110_2 \\ 110_2 \\ \hline 10010_2 \end{array}$$

The only problem that occurs here is in the adding of the partial products:

$$\begin{array}{r} 110_2 \\ + \; 110_2 \\ \hline 10010_2 \end{array}$$

Note that in the fours place we have $1 + 1$, which equals 10_2, so we place the 0 and carry the 1. This again gives $1 + 1$, which is 10_2. Since our addition is completed, we write down this sum to give us the final answer, 10010_2.

EXAMPLE 6
Multiply 101_2 by 101_2.

Solution

The only difference between the two solutions below is that the first solution indicates the multiplication by 0 in the second partial product, while the second solution actually shows the multiplication by 0.

$$\begin{array}{r} 101_2 \\ \times \quad 101_2 \\ \hline 101_2 \\ 1010_2 \\ \hline 11001_2 \end{array} \qquad \text{or} \qquad \begin{array}{r} 101_2 \\ \times \quad 101_2 \\ \hline 101_2 \\ 000_2 \\ 101_2 \\ \hline 11001_2 \end{array}$$

In this section we have examined arithmetic operations in base two, while in the preceding section we thoroughly examined arithmetic operations in base five. We could have used other bases instead of base two and base five, but these two base systems are good examples to work with as they provide good illustrative examples of the principles involved when working with a given base system. Regardless of the base system being considered, the arithmetic principles involved are basically the same.

If we wish to convert a base ten numeral to a different base, we perform successive divisions by the numeral of the new base, and record the remainders for each separate division. The answer is determined by reading the remainders from the bottom to the top. For example, if we wanted to express 34 as a numeral in base six, we would divide 34 by 6:

$$\begin{array}{rrl} 6\,\underline{|\,34} & & \uparrow \\ 6\,\underline{|\,5} & 4 & | \\ 0 & 5 & | \end{array} \qquad 34 = 54_6 \qquad \text{(5 sixes and 4 ones)}$$

If we want to convert a numeral in some other base to a base ten numeral, then we use the concept of expanded notation. To convert 342_8 to a base ten numeral, for example, we would write 342_8 as $(3 \times 8^2) + (4 \times 8^1) + (2 \times 8^0)$ and proceed to evaluate it as $(3 \times 64) + (4 \times 8) + (2 \times 1) = 192 + 32 + 2 = 226$.

In order to convert numerals from one base to another base, where neither base is base ten, it is best to convert first to base ten and then convert this result to the desired base. For example, suppose we want to convert 54_6 to base eight. First we convert 54_6 to base ten:

$$54_6 = (5 \times 6^1) + (4 \times 6^0) = (5 \times 6) + (4 \times 1) = 30 + 4 = 34$$

Now we convert 34 to base eight by performing successive divisions:

$$\begin{array}{r|ll} 8 & 34 & \\ 8 & 4 & 2 \uparrow \\ & 0 & 4 \end{array} \qquad 34 = 42_8$$

Therefore, $54_6 = 42_8$.

EXAMPLE 7

Convert 354_6 to base five.

Solution

First we convert 354_6 to base ten:

$$\begin{aligned} 354_6 &= (3 \times 6^2) + (5 \times 6^1) + (4 \times 6^0) \\ &= (3 \times 36) + (5 \times 6) + (4 \times 1) \\ &= 108 + 30 + 4 = 142 \end{aligned}$$

Now we have $354_6 = 142$, and we can convert 142 to a base five numeral.

$$\begin{array}{r|ll} 5 & 142 & \\ 5 & 28 & 2 \uparrow \\ 5 & 5 & 3 \\ 5 & 1 & 0 \\ & 0 & 1 \end{array}$$

Therefore, $354_6 = 1032_5$.

Check:

$$\begin{aligned} 354_6 &= (3 \times 6^2) + (5 \times 6^1) + (4 \times 6^0) \\ &= (3 \times 36) + (5 \times 6) + (4 \times 1) \\ &= 108 + 30 + 4 = 142 \\ 1032_5 &= (1 \times 5^3) + (0 \times 5^2) + (3 \times 5^1) + (2 \times 5^0) \\ &= (1 \times 125) + (0 \times 25) + (3 \times 5) + (2 \times 1) \\ &= 125 + 0 + 15 + 2 = 142 \end{aligned}$$

EXAMPLE 8

Convert 322_4 to base seven.

Solution

$322_4 = (3 \times 4^2) + (2 \times 4^1) + (2 \times 4^0) = (3 \times 16) + (2 \times 4) + (2 \times 1) = 48 + 8 + 2 = 58$

$$
\begin{array}{r|l l}
7 & 58 & \\
7 & 8 & 2 \uparrow \\
7 & 1 & 1 \\
 & 0 & 1
\end{array}
\qquad 58 = 112_7
$$

Therefore, $322_4 = 112_7$.

EXAMPLE 9

Which is greater, 211_3 or 10111_2?

Solution

We will convert each of the numerals to base ten and compare the results.

$$
\begin{aligned}
211_3 &= (2 \times 3^2) + (1 \times 3^1) + (1 \times 3^0) \\
&= (2 \times 9) + (1 \times 3) + (1 \times 1) = 18 + 3 + 1 = 22 \\
10111_2 &= (1 \times 2^4) + (0 \times 2^3) + (1 \times 2^2) + (1 \times 2^1) + (1 \times 2^0) \\
&= (1 \times 16) + (0 \times 8) + (1 \times 4) + (1 \times 2) + (1 \times 1) \\
&= 16 + 0 + 4 + 2 + 1 = 23
\end{aligned}
$$

Since $211_3 = 22$ and $10111_2 = 23$, $10111_2 > 211_3$.

EXERCISES FOR SECTION 7.7

1. Change to base ten notation.
   a. 10_2 b. 11_2 c. 101_2
   d. 111_2 e. 1011_2 f. 11011_2
2. Change to base ten notation.
   a. 100_2 b. 110_2 c. 1101_2
   d. 1010_2 e. 10110_2 f. 11111_2
3. Change to binary notation.
   a. 5 b. 7 c. 8
   d. 11 e. 17 f. 23
4. Change to binary notation.
   a. 6 b. 9 c. 13
   d. 21 e. 25 f. 33
5. Perform the following additions in base two. Check your work by converting to base ten.
   a. $10_2 + 11_2$ b. $11_2 + 11_2$ c. $10_2 + 10_2$
   d. $110_2 + 10_2$ e. $100_2 + 101_2$ f. $110_2 + 110_2$
6. Perform the following additions in base two. Check your work by converting to base ten.
   a. $110_2 + 11_2$ b. $101_2 + 10_2$ c. $111_2 + 110_2$
   d. $111_2 + 111_2$ e. $1110_2 + 1011_2$ f. $1011_2 + 1111_2$
7. Perform the following subtractions in base two. Check your work by converting to base ten.
   a. $10_2 - 1_2$ b. $11_2 - 10_2$ c. $101_2 - 10_2$
   d. $101_2 - 11_2$ e. $111_2 - 101_2$ f. $1011_2 - 101_2$

8. Perform the following subtractions in base two. Check your work by converting to base ten.

a. $11_2 - 1_2$ b. $111_2 - 10_2$ c. $111_2 - 101_2$

d. $1010_2 - 101_2$ *e. $1001_2 - 11_2$ *f. $1001_2 - 110_2$

9. Perform the following multiplications in base two. Check your work by converting to base ten.

a. $11_2 \times 11_2$ b. $11_2 \times 10_2$ c. $101_2 \times 11_2$

d. $110_2 \times 11_2$ e. $101_2 \times 111_2$ f. $101_2 \times 101_2$

10. Perform the following multiplications in base two. Check your work by converting to base ten.

a. $111_2 \times 10_2$ b. $101_2 \times 10_2$ c. $111_2 \times 100_2$

d. $1001_2 \times 11_2$ e. $1011_2 \times 101_2$ *f. $111_2 \times 111_2$

11. Convert 321_6 to base five.

12. Convert 434_5 to base four.

13. Convert 101101_2 to base seven.

14. Convert 243_5 to base eight.

15. Convert 111_5 to base two.

16. Indicate whether each statement is true or false.

a. $111_2 > 12_5$ b. $321_4 < 10111_2$

c. $1011_2 = 21_5$ d. $110_2 = 11_5$

17. Indicate whether each statement is true or false.

a. $41_8 < 46_7$ b. $32_8 > 1111_2$

c. $421_5 = 111$ d. $213_4 < 211_5$

*18. Perform each operation in the indicated base.

a. $23_6 + 34_6$ b. $352_7 + 405_7$

c. $41_7 - 23_7$ d. $241_6 - 42_6$

CONCHY
James Childress
THE FUTURE IS DOOMED TO A LANGUAGE OF NUMBERS...
AND WHAT WILL BECOME OF US WHO REFUSE TO SPEAK THE DIGITAL ALPHABET? WHAT PROBLEMS WILL WE FACE WITH THE COMING OF SUCH A COMMUNICATIONS GAP?
4!
4 WHAT?
BLAP!
© Field Enterprises, Inc., 1975
6-12

CONCHY by James Childress. © Field Enterprises, Inc., 1975. Courtesy of Field Newspaper Syndicate.

Just For Fun

Three Yankee fans and three Dodger fans have to ride an elevator up to the top floor (there are no steps between the ground floor and the top), but the elevator will hold no more than two people. The Dodger fans always start an argument if they are left in a situation where they outnumber the Yankee fans, but they are fine if they are left alone or if they are with the same or a greater number of Yankee fans. How do they all get up to the top floor, using the elevator, without any arguments?

7.8 SUMMARY

The introduction to systems of numeration presented here should enable you to understand some of the basic concepts of numeration. You should be aware that this chapter has merely been an introduction to some of the systems of numeration that have been developed by man. Man has needed some form of numeration ever since he was able to reason and began to develop a civilization.

One of the oldest known systems of numeration is that of the Egyptians (approximately 3400 B.C.). The Egyptians used a simple grouping system, grouping items by tens. They used a different symbol for each power of ten.

When we say that the Egyptians used a simple grouping system, we mean that the position of a symbol in an Egyptian numeral does not affect the number represented.

The Greek system of numeration differs from other systems of numeration in that it uses special symbols for multiples of 5. For example, since they had no symbol for 50, they thought of 50 as 5 tens, and similarly 500 was thought of as 5 one hundreds. This concept enabled the Greeks to use fewer symbols to express a number. The use of multiples of 5 is an example of *multiplicative grouping*. Another system of numeration that involves multiplicative grouping is the Chinese-Japanese system of numeration. This system arranges numbers vertically instead of horizontally. It also uses positional notation and groups items in powers of ten.

The Babylonian system of numeration uses only two symbols to represent numerals. This system also uses a *place-value* system: the position of a symbol matters. The Babylonian system of numeration is based on powers of sixty, and is called a sexagesimal system.

Our decimal system of numeration uses *Hindu-Arabic* numerals to express numbers. The decimal system of numeration groups by powers of ten. It also uses a place-value system: the position that a symbol has in a decimal numeral affects its meaning.

Even though we use the decimal system of numeration, it is not uncommon for us to group items in some other manner, such as pairs (twos) or dozens. When we group items by some number other than ten, we say that we are working with a *base* different from ten. When we write numerals in a base other than base ten, we must indicate what base we are working with. We do this by using a subscript to indicate the base. An example is 23_5, which indicates that we are grouping by fives.

In order to change a numeral in some other base to base ten notation, we write the numeral in expanded notation. For example, $23_5 = (2 \times 5^1) + (3 \times 5^0) = (2 \times 5) + (3 \times 1) = 10 + 3 = 13$. We can also convert any number from base ten to another base by performing successive divisions by the new base and recording the remainders for each division; we stop dividing when we get a quotient of zero. The answer in the new base is found by reading the remainders from bottom to top.

In this chapter, we performed arithmetic operations in base five and base two. Our main purpose in doing arithmetic in these bases is to enlarge our understanding of what actually occurs when we work in base ten. When we perform any operation in any base other than base ten, we must write the numerals in the indicated base; but it is sometimes helpful to think in terms of base ten, since the arithmetic operations are performed in the same manner in every base.

Review Exercises for Chapter 7

1. Express the following as Egyptian numerals.
   a. 32 b. 211 c. 1111

2. Express the following Egyptian numerals as base ten numerals.

   a. 𓆼𓍢𓍢𓎆𓎆𓎆𓏺𓏺𓏺

   b. 𓂭𓆼𓆼𓎆𓎆𓏺𓏺

3. Perform the indicated operations. Express your answers in Egyptian numerals.

   a. 𓎆𓎆𓏺𓏺𓏺 + 𓎆𓎆𓎆𓏺𓏺

   b. 𓍢𓎆𓎆𓎆𓏺𓏺𓏺 + 𓍢𓎆𓏺𓏺𓏺𓏺𓏺𓏺𓏺

   c. 𓍢𓎆𓎆𓏺𓏺 − 𓎆𓎆𓎆𓏺𓏺𓏺

   d. 𓆼 − 𓍢𓎆𓎆

4. List the distinguishing characteristics of—
   a. a simple grouping system
   b. a multiplicative grouping system
   c. a place-value system.

5. Name a system of numeration that uses—
   a. simple grouping
   b. multiplicative grouping
   c. place value.

6. Write in expanded notation.
   a. 345 b. 342_5 c. 10111_2

7. Convert to base ten.
   a. 47_{12} b. 34_5 c. 241_5
   d. 36_8 e. 10101_2 f. $2TE_{12}$

8. Convert to the indicated base.
   a. $45 = \_\_\_\__5$ b. $51 = \_\_\_\__5$
   c. $32 = \_\_\_\__{12}$ d. $33 = \_\_\_\__2$
   e. $63 = \_\_\_\__2$ f. $34 = \_\_\_\__{12}$

9. Name at least two systems of numeration, other than base ten, to which we are exposed in everyday life and give examples of how each is used.

10. Perform the indicated operations in base five.
   a. $314_5 + 221_5$ b. $231_5 - 132_5$
   c. $231_5 \times 34_5$ d. $2_5 \overline{)124_5}$

11. Perform the indicated operations in base five.
   a. $23_5 + 22_5$ b. $231_5 \times 3_5$
   c. $22_5 - 13_5$ d. $123_5 \times 4_5$

12. Perform the indicated operations in base two.
   a. $111_2 - 11_2$ b. $1011_2 - 101_2$
   c. $111_2 \times 10_2$ d. $1011_2 \times 101_2$

13. Perform the indicated operations in base two.
   a. $101_2 + 110_2$ b. $101_2 + 101_2$
   c. $1101_2 + 1101_2$ d. $101_2 - 10_2$

14. Convert each of the following to the indicated base.
   a. $321_5 = \_\_\_\__6$ b. $42_5 = \_\_\_\__3$
   c. $10111_2 = \_\_\_\__5$ d. $77_8 = \_\_\_\__2$

Just For Fun

Can you plant 10 tomato plants in 5 rows with 4 plants in each row?

20
0
20
40

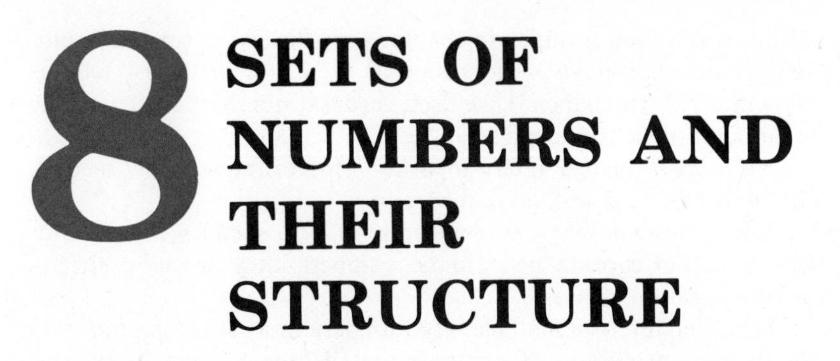

8 SETS OF NUMBERS AND THEIR STRUCTURE

After studying this chapter, you will be able to do the following:

1. Determine whether a natural number is **prime** or **composite**
2. Find the **prime factors** of a given composite number
3. Find the **greatest common divisor** and the **least common multiple** for a given pair of numbers
4. Add, subtract, and multiply integers
5. Add, subtract, multiply, and divide rational numbers
6. Express a rational number as a decimal
7. Express a **terminating decimal** or a **repeating nonterminating decimal** as a quotient of integers
8. Identify an **irrational number** as a nonterminating and nonrepeating decimal
9. Identify the set of **real numbers** as the union of the sets of rational and irrational numbers.

8.1 INTRODUCTION

Numbers, like many other things, can be described or classified in a variety of ways. It is not unusual to classify an item in a certain category and then later reclassify it differently to describe it better.

If a person is holding a playing card from a standard deck of cards, we don't know much about the card: it is one of fifty-two possibilities. If the person tells us that it is a heart, then we know that it is one of thirteen possibilities. If we are told that it is also a picture card, then we know it is one of three possibilities—king, queen, or jack of hearts. Finally, the person may identify it completely by telling us that the card is the king of hearts.

Once an item has been placed in a general category, we can better describe it by placing it in more specific categories. This can also be done with numbers. That is, numbers can be classified in

H. Armstrong Roberts

various ways. For example, in chapter 1 we discussed cardinal numbers. A *cardinal number*—for example, 1, 2, 10, or 2001—tells us "how many." In chapter 1 we used cardinal numbers to determine the number of elements in a set.

An *ordinal number* refers to order—for example, first, second, 14th, and 123rd are ordinal numbers.

Your phone number, student number, and social security number are neither cardinal nor ordinal numbers: they are used strictly for identification purposes.

In this chapter we shall classify numbers in another manner and examine the properties of numbers that belong to these particular categories.

8.2 NATURAL NUMBERS—PRIMES AND COMPOSITES

When man first started to count he did not begin with zero. He began with 1 and then one more, 2, and proceeded in a like manner. He counted in this manner:

$$1, 2, 3, 4, 5, 6, 7, 8, 9, \ldots$$

That is, he began with 1 and continued to count. The set of numbers $\{1, 2, 3, 4, \ldots\}$ is often referred to as the set of **counting numbers.** More formally, they are called the set of **natural numbers.** The first natural number is 1, sometimes called the unit number.

Any natural number can be expressed as the product of two or more natural numbers. For example, $6 = 3 \times 2$, $8 = 4 \times 2$, $3 = 3 \times 1$, and $1 = 1 \times 1$. The numbers that are multiplied together to form a number are the **factors** of the number. Since $6 = 3 \times 2$, 3 and 2 are factors of 6. A factor of a number divides the given number with zero remainder. Every natural number is divisible by itself and 1. Some natural numbers are divisible only by themselves and 1. These natural numbers are called *prime numbers.*

A prime number is any natural number greater than one that is divisible only by itself and one.

Two, 3, 5, 7, and 11 are the first five prime numbers. Note that 1 is not a prime number. The first prime number, 2, is unique in that it is the only even prime number. Any other even natural number such as 10, 200, or 484 is divisible by 2 and therefore cannot be prime, because a prime number is divisible only by itself and 1.

If a natural number other than 2 is not prime, then it is called a **composite number** since it can be composed of other factors. Any composite number can be expressed as a product of prime factors. For example, $6 = 2 \times 3$, $12 = 2 \times 2 \times 3$, and $4 = 2 \times 2$. In fact, if we disregard order, every composite natural number can be expressed as a product of prime factors in one and only one way. ($6 = 2 \times 3$ or $6 = 3 \times 2$, but these are the same if we disregard order.) If we phrase this statement more formally, we have:

Every natural number greater than 1 is either a prime or can be expressed as a product of prime factors. Except for the order of the factors, this can be done in one and only one way.

This statement is the *fundamental theorem of arithmetic.* Since every natural number except 1 is either a prime number or can be expressed as a product of primes, the concept of a prime number is an important one.

How can we tell if a number is prime? There is no quick and easy solution; no formula exists for finding primes. Suppose we want to determine if 29 is prime. How do we go about it? In order to determine whether a number is prime, we check divisors to see if they divide it. The first divisor to check is 2: 29 is not divisible by 2. Try 3: 29 is not divisible by 3. How about 4? We need not test 4 because it is a composite number ($4 = 2 \times 2$) and contains smaller divisors (2) which have already been tried. Next we try 5: 29 is not divisible by 5. We do not need to test 6. Why? Because $6 = 3 \times 2$ and neither 3 nor 2 divide 29. Try 7: 29 is not divisible by 7. How far do we keep testing? All the way to 29? No; in fact, we should have stopped at 6, since $6 \times 6 = 36$ and $36 > 29$. If 6 or a number greater than 6 divided 29, then the quotient would be less than 6 and would also be a factor of 29. But we have already tested all possible factors less than 6, and they all failed. Therefore, our conclusion is that 29 is prime.

To determine if a number is prime, we need only test the prime divisors $\{2, 3, 5, 7, 11, \ldots\}$ up to the largest natural number whose square is less than or equal to the number we are testing.

Remember, we do not have to check composite divisors, since a composite number can be expressed as a product of prime factors.

EXAMPLE 1
Is 43 prime?

Solution
Yes, 43 is not divisible by 2, 3, 5, or 7. We need not check any other divisors, since $7^2 = 49$ and $49 > 43$.

EXAMPLE 2
Is 91 prime?

Solution
No. We need not check past 10, since $10^2 = 100$ and $100 > 91$. In fact, we only have to check through 7, since 8, 9, and 10 are all composite numbers. Two does not divide 91, 3 does not divide 91, and 5 does not divide 91. But 7 does divide 91: $91 = 7 \times 13$. Hence, 91 is not prime.

EXAMPLE 3
Is 1001 prime?

Solution
No. The primes less than 100 are 2, 3, 5, 7, 11, 13, 17, 19, 23, 29, 31, 37, 41, 43, 47, 53, 59, 61, 67, 71, 73, 79, 83, 89, and 97. We do not have to check past 37, since $37^2 = 1369$ and $1369 > 1001$. In fact, we only have to check through 31, since the other natural numbers up to 37 are composite. We see that 2 does not divide 1001, 3 does not divide 1001, and 5 does not divide 1001. But 7 does divide 1001: $1001 = 7 \times 143$. Therefore, 1001 is not prime.

EXAMPLE 4
Is 2003 prime?

Solution
Yes. We need not check past 45, since $45^2 = 2025$ and $2025 > 2003$. None of the prime numbers 2, 3, 5, 7, 11, 13, 17, 19, 23, 29, 31, 37, 41, or 43 divides 2003. This is far enough to check, since 44 and 45 are composite numbers. Therefore 2003 is prime.

The following is a list of rules for divisibility by certain numbers. These rules may aid you in determining whether a given number is prime, or in finding the prime factors of a number.

1. A natural number is divisible by 2 if the natural number is an even number. (For example, 2 divides 2754, since 2754 ends in 4 and 4 is even.)
2. A natural number is divisible by 3 if the sum of the digits is divisible by 3. (For example, 3 divides 2754, because $2 + 7 + 5 + 4 = 18$ and 18 is divisible by 3.)
3. A natural number is divisible by 5 if the last digit on the right is 0 or 5. (For example, 5 divides 1350 and 234,795.)
4. A natural number is divisible by 9 if the sum of the digits is divisible by 9. (For example, 9 divides 2754, because $2 + 7 + 5 + 4 = 18$ and 18 is divisible by 9.)

5. A natural number is divisible by 11 if the difference between the sum of the digits in the odd places and the sum of the digits in the even places is 0 or divisible by 11. (For example, 368,610 is divisible by 11 since the difference of the sum of the digits in the odd places, $6 + 6 + 0 = 12$, and the sum of the digits in the even places, $3 + 8 + 1 = 12$, is 0.)

EXAMPLE 5

Test each of the following numbers for divisibility by 2, 3, 5, 9, and 11.

a. 330 b. 410 c. 369 d. 1331

Solution

a. 330 is divisible by 2, 3, 5, and 11.

b. 410 is divisible by 2 and 5.

c. 369 is divisible by 3 and 9.

d. 1331 is divisible by 11.

There exists an interesting and ancient technique for finding primes. It is called the **sieve of Eratosthenes** and was invented by a Greek scholar, Eratosthenes (276–194 B.C.), who was also head of the famous library in Alexandria. We shall use the numbers from 1 to 100 to illustrate Eratosthenes' method.

~~1~~	(2)	(3)	~~4~~	(5)	~~6~~	(7)	~~8~~	~~9~~	~~10~~
(11)	~~12~~	(13)	~~14~~	~~15~~	~~16~~	(17)	~~18~~	(19)	~~20~~
~~21~~	~~22~~	(23)	~~24~~	~~25~~	~~26~~	~~27~~	~~28~~	(29)	~~30~~
(31)	~~32~~	~~33~~	~~34~~	~~35~~	~~36~~	(37)	~~38~~	~~39~~	~~40~~
(41)	~~42~~	(43)	~~44~~	~~45~~	~~46~~	(47)	~~48~~	~~49~~	~~50~~
~~51~~	~~52~~	(53)	~~54~~	~~55~~	~~56~~	~~57~~	~~58~~	(59)	~~60~~
(61)	~~62~~	~~63~~	~~64~~	~~65~~	~~66~~	(67)	~~68~~	~~69~~	~~70~~
(71)	~~72~~	(73)	~~74~~	~~75~~	~~76~~	~~77~~	~~78~~	(79)	~~80~~
~~81~~	~~82~~	(83)	~~84~~	~~85~~	~~86~~	~~87~~	~~88~~	(89)	~~90~~
~~91~~	~~92~~	~~93~~	~~94~~	~~95~~	~~96~~	(97)	~~98~~	~~99~~	~~100~~

We exclude 1, since 1 is not prime. Eratosthenes determined that 2 is prime, and also that every second number (beginning with 2) is not prime, since it has 2 as a factor. Therefore we cross out 4, 6, 8, 10, ... Similarly, 3 is a prime number, and every third number (beginning with 3) is not prime since it has 3 as a factor. Therefore, we cross out 6, 9, 12, ... Note that some of these numbers, such as 6, 12, and 18, have already been crossed out. The next number to consider is 5: it is prime, so we cross out every fifth number.

We can continue this process, but for how long? We do not have to go past 11, since $11^2 = 121$ and $121 > 100$. In fact, after determining that 7 is prime and crossing out every seventh number, we have a list of all the prime numbers less than 100. This process is called the sieve of Eratosthenes because instead of crossing out the numbers he punched them out with a sharp stick to form a "sieve."

Since every natural number (except 1) is either a prime number or a composite number, the concept of a prime number is an important one. If a number is not prime, then it is composite. A composite number can be expressed as a product of its prime factors in one and only one way, disregarding order. Is 60 prime? No, it is a composite number. What are its prime factors? To determine the prime factors of 60 in a systematic manner, we first try 2; 2 is a factor of 60, since $60 = 2 \times 30$. Next we try to factor 30, and 2 is a factor of 30: $30 = 2 \times 15$. We now have $60 = 2 \times 2 \times 15$. What are the factors of 15 (if any)? $15 = 3 \times 5$ and 3 and 5 are both prime. Hence, $60 = 2 \times 2 \times 3 \times 5$. We have expressed 60 as a product of its prime factors.

Another way to determine the prime factors of 60 is by successive divisions by prime divisors. This division technique is the same as we used in the last chapter, but now we must always have remainders of zero. In order to use the technique of successive divisions, we first determine the smallest prime that will divide into 60, which is 2. We divide 2 into 60 and continue in this manner until we reach a quotient that is prime. This indicates that no more divisions are necessary. This technique is illustrated below:

$$\begin{array}{r|l} 2 & \underline{60} \\ 2 & \underline{30} \\ 3 & \underline{15} \\ & \;\;5 \end{array}$$

Since 5 is prime, no more divisions are necessary. Therefore,

$$60 = 2 \times 2 \times 3 \times 5$$

Note that we could also express this as $60 = 2 \times 3 \times 5 \times 2$. This is not a different factorization, but merely a rearrangement of the original one. A composite number can be expressed as a product of its prime factors in one and only one way, *disregarding order:* we still have two factors of 2, one factor of 3, and one factor of 5.

EXAMPLE 6
Determine the prime factors of 345.

Solution
345 is not divisible by 2 since 345 is not even. But 345 is divisible by 3, since the sum of the digits $3 + 4 + 5 = 12$ is divisible by 3. So we have

$$\begin{array}{r|l} 3 & \underline{345} \\ & 115 \end{array}$$

Since 115 ends in 5, it is divisible by 5.

$$\begin{array}{r|l} 5 & \underline{115} \\ & 23 \quad \text{STOP: 23 is prime} \end{array}$$

Putting the two steps together, we have

$$\begin{array}{r|l} 3 & \underline{345} \\ 5 & \underline{115} \\ & 23 \end{array}$$

and $345 = 3 \times 5 \times 23$.

EXAMPLE 7
Determine the prime factors of 4830.

Solution

$$\begin{array}{r|l} 2 & \underline{4830} \\ 3 & \underline{2415} \\ 5 & \underline{805} \\ 7 & \underline{161} \\ & 23 \quad \text{STOP: 23 is prime} \end{array}$$

$4830 = 2 \times 3 \times 5 \times 7 \times 23$

EXAMPLE 8
Determine the prime factors of 900.

Solution

$$\begin{array}{r|l} 2 & 900 \\ 2 & 450 \\ 3 & 225 \\ 3 & 75 \\ 5 & 25 \\ & 5 \end{array}$$

STOP: 5 is prime

$900 = 2 \times 2 \times 3 \times 3 \times 5 \times 5$

(*Note*: We also could have said $900 = 2 \times 3 \times 5 \times 3 \times 2 \times 5$ and still be correct, as this is still the same set of prime factors. We can also write this as $900 = 2^2 \times 3^2 \times 5^2$.)

EXERCISES FOR SECTION 8.2

1. Determine whether the number used in each of the following is used as a cardinal number, an ordinal number, or an identification number. (*Hint:* See section 8.1.)
   a. He shot 82 for a round of golf.
   b. Julie was third in line.
   c. My social security number is 089-20-4944.
   d. Ben's phone number is 727-3100.
   e. Janie plays first singles on the tennis team.
   f. Pam bowled 157 for her only score.

2. Determine whether each number is prime or composite.
   a. 97 b. 89 c. 1
   d. 243 e. 741 f. 1955

3. Determine whether each number is prime or composite.
   a. 101 b. 103 c. 323
   d. 2007 e. 1003 f. 4,159,731

4. Determine the prime factors for each number.
   a. 36 b. 72 c. 216
   d. 475 e. 625 f. 147

5. Find the prime factors of each number.
   a. 234 b. 213 c. 891
   d. 1331 e. 902 f. 7429

6. Two is the only even prime number; all of the other primes are odd, and hence any two consecutive odd primes must differ by at least 2, as do 3 and 5, and 5 and 7. Prime numbers that differ by exactly 2 are called *twin primes*. Find three pairs of twin primes other than those already mentioned.

Just For Fun

A perfect number is one that is equal to the sum of its divisors, excluding itself. An example of a perfect number is 6: its divisors are 1, 2, 3, and 6; we exclude 6, and 1 + 2 + 3 = 6. Therefore, 6 is perfect. Can you discover some more perfect numbers? (*Hint:* The next perfect number is less than 50.)

8.3 GREATEST COMMON DIVISOR AND LEAST COMMON MULTIPLE

Now that we are able to determine the prime factors of a given natural number, we shall use this process in determining some other properties of natural numbers. One important concept we will need is that of the *greatest common divisor.*

> **The greatest common divisor (G.C.D.) of two natural numbers is the greatest (largest) natural number that divides a given pair of natural numbers with remainders of zero.**

Consider the two natural numbers 32 and 40. We list the set of divisors of each number:

32: $\{1, 2, 4, 8, 16, 32\}$
40: $\{1, 2, 4, 5, 8, 10, 20, 40\}$

If we find the intersection of these two sets of divisors, $\{1, 2, 4, 8, 16, 32\} \cap \{1, 2, 4, 5, 8, 10, 20, 40\}$, we have $\{1, 2, 4, 8\}$. Since 8 is the greatest number in the intersection, it is the greatest common divisor of 32 and 40.

When we are given two prime numbers, then their greatest common divisor is 1. But there are pairs of composite numbers whose greatest common divisor is also 1. Consider the two natural numbers 24 and 25. The sets of divisors of these numbers are

24: $\{1, 2, 3, 4, 6, 8, 12, 24\}$
25: $\{1, 5, 25\}$

The intersection of these two sets of divisors is $\{1, 2, 3, 4, 6, 8, 12, 24\} \cap \{1, 5, 25\} = \{1\}$. Therefore 1 is the greatest common divisor of 24 and 25. Two numbers whose G.C.D. is 1 are said to be **relatively prime.**

Do we always have to list the sets of divisors to determine the greatest common divisor? The answer is no. We can do it in a more efficient manner. Let's consider our original example, 32 and 40, and find the prime factors (divisors) of each number:

$$\begin{array}{r|l} 2 & 32 \\ \hline 2 & 16 \\ \hline 2 & 8 \\ \hline 2 & 4 \\ \hline & 2 \end{array} \qquad \begin{array}{r|l} 2 & 40 \\ \hline 2 & 20 \\ \hline 2 & 10 \\ \hline & 5 \end{array}$$

$$32 = 2 \times 2 \times 2 \times 2 \times 2 \qquad 40 = 2 \times 2 \times 2 \times 5$$

Now examine the two sets of prime factors and determine the factors common to both sets. Note that both sets of prime divisors contain $2 \times 2 \times 2$. Therefore the greatest common divisor of 32 and 40 is $2 \times 2 \times 2$, or 8.

Let's try another example. What is the greatest common divisor of 30 and 45? First we find the prime factors of each:

$$\begin{array}{r|l} 2 & 30 \\ \hline 3 & 15 \\ \hline & 5 \end{array} \qquad \begin{array}{r|l} 3 & 45 \\ \hline 3 & 15 \\ \hline & 5 \end{array}$$

$$30 = 2 \times 3 \times 5 \qquad 45 = 3 \times 3 \times 5$$

Examining the two sets of prime factors (divisors), we see that the intersection is 3×5. Therefore, the greatest common divisor of 30 and 45 is 3×5, or 15.

EXAMPLE 1

Find the greatest common divisor of 8 and 12.

Solution

First we find the prime factors of 8 and 12:

$$\begin{array}{r|l} 2 & 8 \\ \hline 2 & 4 \\ \hline & 2 \end{array} \quad 8 = 2 \times 2 \times 2 \qquad \begin{array}{r|l} 2 & 12 \\ \hline 2 & 6 \\ \hline & 3 \end{array} \quad 12 = 2 \times 2 \times 3$$

Next, we examine the intersection of the two sets of prime factors and we note that 2×2 is common to both sets; hence $2 \times 2 = 4$ is the G.C.D.

EXAMPLE 2

Find the greatest common divisor of 8 and 15.

Solution

We first find the prime factors of 8 and 15.

$$\begin{array}{r|l} 2 & 8 \\ \hline 2 & 4 \\ \hline & 2 \end{array} \qquad 8 = 2 \times 2 \times 2 \qquad\qquad \begin{array}{r|l} 3 & 15 \\ \hline & 5 \end{array} \qquad 15 = 3 \times 5$$

Note that the two sets of prime factors have no elements in common; their intersection is empty. When this occurs, the G.C.D. for the two numbers is 1, so the numbers are relatively prime. Note that we did not list 1 as a prime factor for either number because 1 is not a prime number.

EXAMPLE 3

Find the greatest common divisor of 342 and 380.

Solution

Finding the prime factors of each number, we have

$$\begin{array}{r|l} 2 & 342 \\ \hline 3 & 171 \\ \hline 3 & 57 \\ \hline & 19 \end{array} \qquad 342 = 2 \times 3 \times 3 \times 19$$

$$\begin{array}{r|l} 2 & 380 \\ \hline 2 & 190 \\ \hline 5 & 95 \\ \hline & 19 \end{array} \qquad 380 = 2 \times 2 \times 5 \times 19$$

Examining the intersection of the two sets of prime factors, we see that 2×19 is common to both; hence the G.C.D. of 342 and 380 is $2 \times 19 = 38$.

One application of the greatest common divisor is in the reduction of fractions. We can simplify, or reduce, fractions if we determine the greatest common divisor of both the numerator and denominator. Suppose we are asked to reduce the fraction

$$\frac{65}{91}$$

We can do this in the following manner. We find the greatest common divisor of both the numerator and denominator; that is, we find the G.C.D. of 65 and 91:

$$\begin{array}{r|l} 5 & 65 \\ \hline & 13 \end{array} \qquad 65 = 5 \times 13 \qquad\qquad \begin{array}{r|l} 7 & 91 \\ \hline & 13 \end{array} \qquad 91 = 7 \times 13$$

The G.C.D. of 65 and 91 is 13. We now rewrite the original fraction.

$$\frac{65}{91} = \frac{\cancel{13} \times 5}{\cancel{13} \times 7} = \frac{5}{7}$$

The numerator and denominator of the reduced fraction are relatively prime; that is, their G.C.D. is 1.

EXAMPLE 4

Reduce $\frac{130}{455}$ to lowest terms

Solution

We first find the G.C.D. of 130 and 455.

$$\begin{array}{r|l} 2 & 130 \\ \hline 5 & 65 \\ \hline & 13 \end{array} \qquad 130 = 2 \times 5 \times 13 \qquad \begin{array}{r|l} 5 & 455 \\ \hline 7 & 91 \\ \hline & 13 \end{array} \qquad 455 = 5 \times 7 \times 13$$

The G.C.D. of 130 and 455 is 5 × 13. Now we rewrite the original fraction.

$$\frac{130}{455} = \frac{\cancel{(5 \times 13)} \times 2}{\cancel{(5 \times 13)} \times 7} = \frac{2}{7}$$

EXAMPLE 5

Reduce $\frac{310}{460}$ to lowest terms.

Solution

We first find the G.C.D. of 310 and 460.

$$\begin{array}{r|l} 2 & 310 \\ \hline 5 & 155 \\ \hline & 31 \end{array} \qquad 310 = 2 \times 5 \times 31 \qquad \begin{array}{r|l} 2 & 460 \\ \hline 2 & 230 \\ \hline 5 & 115 \\ \hline & 23 \end{array} \qquad 460 = 2 \times 2 \times 5 \times 23$$

The G.C.D. of 310 and 460 is 2 × 5, and we can rewrite the original fraction.

$$\frac{310}{460} = \frac{\cancel{(2 \times 5)} \times 31}{\cancel{(2 \times 5)} \times 2 \times 23} = \frac{31}{2 \times 23} = \frac{31}{46}$$

If we want to find the greatest common divisor for three or more numbers, we extend the process for finding the G.C.D. for two numbers: we find the prime factors that are common to all the sets of prime factors. For example, what is the G.C.D. of 24, 36, and 48? We first find the prime factors of each number.

$$24 = 2 \times 2 \times 2 \times 3$$
$$36 = 2 \times 2 \times 3 \times 3$$
$$48 = 2 \times 2 \times 2 \times 2 \times 3$$

Now, what is the intersection of the three sets of prime factors? Upon inspection, we see that $2 \times 2 \times 3$ is common to all three sets. Therefore the G.C.D. of 24, 36, and 48 is $2 \times 2 \times 3$, or 12.

When we are given two natural numbers, we can now find their greatest common divisor. Another concept that goes hand in hand with the G.C.D. is the *least common multiple.*

The least common multiple (L.C.M.) of two natural numbers is the smallest (least) natural number that is a multiple of each of the two given numbers.

The least common multiple can also be thought of as the smallest (least) natural number that is divisible by both of the given numbers. Four is a multiple of 2, as is 6, 8, 10, and so on, because each of these numbers has 2 as a factor. The multiples of 3 are: {3, 6, 9, 12, 15, . . .}. Listing the sets of multiples for 2 and 3, we have

2: {2, 4, 6, 8, 10, . . .}
3: {3, 6, 9, 12, 15, . . .}

Upon inspection, we note that 6 is the least common multiple (L.C.M.) of 2 and 3. Also, it is the smallest number that is divisible by both of the given numbers, 2 and 3.

Let's consider another example, the L.C.M. of 10 and 12. Listing the sets of multiples for 10 and 12, we have

10: {10, 20, 30, 40, 50, 60, . . .}
12: {12, 24, 36, 48, 60, 72, . . .}

Inspection of these two sets of multiples indicates that 60 is the least common multiple of 10 and 12. It is the smallest number that is divisible by both of the given numbers, 10 and 12. Can we do this problem another way? The answer is yes. First find the greatest common divisor of 10 and 12.

$$2\,\underline{|\,10}\qquad 10 = 2 \times 5$$
$$\quad 5$$

$$2\,\underline{|\,12}$$
$$2\,\underline{|\,6}\qquad 12 = 2 \times 2 \times 3$$
$$\quad 3$$

The only prime factor that 10 and 12 have in common is 2. Therefore, the G.C.D. of 10 and 12 is 2. Note that $10 \times 12 = 120$. If we divide 120 by the G.C.D. of 10 and 12, that is, by 2, we have $120 \div 2 = 60$, and 60 is the least common multiple (L.C.M.) of 10 and 12.

In general, we can determine the least common multiple (L.C.M.) of two natural numbers by dividing the product of the two numbers by their greatest common divisor (G.C.D.)

EXAMPLE 6

Find the least common multiple of 8 and 12.

Solution

Using the general rule for finding the L.C.M., we first find the G.C.D. of 8 and 12, which is 4 (see example 1). Next we determine the product of 8 and 12: $8 \times 12 = 96$. Finally, the L.C.M. of 8 and 12 is

$$\frac{8 \times 12}{4} = \frac{96}{4} = 24$$

EXAMPLE 7

Find the L.C.M. of 48 and 72.

Solution

First we find the G.C.D. of 48 and 72.

$$2\,\underline{|\,48}$$
$$2\,\underline{|\,24}$$
$$2\,\underline{|\,12}\qquad 48 = 2 \times 2 \times 2 \times 2 \times 3$$
$$2\,\underline{|\,6}$$
$$\quad 3$$

$$2\,\underline{|\,72}$$
$$2\,\underline{|\,36}$$
$$2\,\underline{|\,18}\qquad 72 = 2 \times 2 \times 2 \times 3 \times 3$$
$$3\,\underline{|\,9}$$
$$\quad 3$$

The G.C.D. of 48 and 72 is $2 \times 2 \times 2 \times 3 = 24$. Next we find the product of 48 and 72, $48 \times 72 = 3456$. The L.C.M. of 48 and 72 is

$$\frac{48 \times 72}{24} = \frac{3456}{24} = 144$$

or

$$\text{L.C.M. of 48 and 72 is } \frac{48}{24} \times 72 = 2 \times 72 = 144$$

When do we use the least common multiple? It is usually used in combining fractions. Suppose we want to add $\frac{2}{5}$ and $\frac{1}{6}$.

$$\frac{2}{5} + \frac{1}{6}$$

We cannot add these two fractions as they are represented here. Before we can add or subtract two fractions, they must have a common denominator; when they have a common denominator, we can add or subtract the numerators. Since $\frac{2}{5}$ and $\frac{1}{6}$ do not have a common denominator, we must rewrite them so that they do. In doing this, we usually use the **least common denominator,** which is the least common multiple of the given denominators. The least common multiple of 5 and 6 is 30. Therefore we now have

$$\frac{2}{5} + \frac{1}{6} = \frac{12}{30} + \frac{5}{30} = \frac{17}{30}$$

EXAMPLE 8

Add $\frac{1}{4}$ and $\frac{2}{9}$.

Solution

Before we can add these two fractions, they must have a common denominator. A common denominator is the L.C.M. of 4 and 9. Since 4 and 9 are relatively prime, their G.C.D. is 1. Therefore the L.C.M. is

$$\frac{4 \times 9}{1} = 36$$

$$\frac{1}{4} + \frac{2}{9} = \frac{9}{36} + \frac{8}{36} = \frac{17}{36}$$

Subtract $\frac{1}{6}$ from $\frac{2}{9}$.

Solution

We must first rewrite the fractions so that they have a common denominator, the L.C.M. of 9 and 6. First we find the G.C.D. of 9 and 6, which is 3. Therefore, the L.C.M. is

$$\frac{9 \times 6}{3} = \frac{54}{3} = 18.$$

$$\frac{2}{9} - \frac{1}{6} = \frac{4}{18} - \frac{3}{18} = \frac{1}{18}$$

EXERCISES FOR SECTION 8.3

1. Find the greatest common divisor for each of the following:
   a. 8 and 14 b. 14 and 28
   c. 15 and 24 d. 52 and 78
   e. 111 and 267 f. 24, 48, and 60

2. Find the greatest common divisor for each of the following:
   a. 8 and 15 b. 21 and 25
   c. 48 and 72 d. 234 and 470
   e. 801 and 999 f. 60, 90, and 210

3. Reduce each fraction to lowest terms.
   a. $\frac{30}{36}$ b. $\frac{42}{54}$ c. $\frac{39}{65}$
   d. $\frac{120}{180}$ e. $\frac{294}{304}$ f. $\frac{195}{390}$

4. Reduce each fraction to lowest terms.
   a. $\frac{16}{24}$ b. $\frac{15}{28}$ c. $\frac{28}{72}$
   d. $\frac{50}{273}$ e. $\frac{213}{450}$ f. $\frac{115}{450}$

5. Find the least common multiple for each of the following:
   a. 8 and 14 b. 14 and 28
   c. 15 and 24 d. 52 and 78
   e. 66 and 90 f. 111 and 267

6. Find the least common multiple for each of the following:
   a. 8 and 15 b. 21 and 25
   c. 48 and 72 d. 13 and 17
   e. 234 and 470 f. 801 and 999

7. Perform the indicated operation and reduce your answers to lowest terms.
   a. $\frac{3}{4} + \frac{2}{9}$ b. $\frac{5}{9} + \frac{1}{12}$ c. $\frac{1}{6} + \frac{3}{4}$
   d. $\frac{7}{8} - \frac{1}{12}$ e. $\frac{10}{11} - \frac{4}{5}$ f. $\frac{13}{15} - \frac{3}{20}$

8. Perform the indicated operations and reduce your answers to lowest terms.
   a. $\frac{7}{15} + \frac{1}{40}$ b. $\frac{4}{9} + \frac{5}{36}$ c. $\frac{2}{7} + \frac{3}{11}$
   d. $\frac{7}{9} - \frac{1}{5}$ e. $\frac{5}{18} - \frac{1}{24}$ f. $\frac{7}{11} - \frac{2}{5}$

Just For Fun

It has been stated that "Every even number greater than 2 can be expressed as the sum of two prime numbers." For example, 6 = 3 + 3, 8 = 3 + 5, 12 = 7 + 5, and 14 = 11 + 3. This statement is called Goldbach's conjecture. It is a conjecture because it has never been proved. Show that Goldbach's conjecture is true for all even numbers except 2, up to and including 30.

8.4 INTEGERS

In section 8.2, we introduced the set of natural numbers, {1, 2, 3, 4, . . .}. The first natural number is 1. This is the first number that man used in counting. The use of zero did not come about until approximately 700 A.D. Zero was first used to indicate position, distinguishing between numbers like 32 and 302. Later, zero began to be used as a starting point in counting. The set of natural numbers was expanded to include zero, forming the set {0, 1, 2, 3, 4, . . .}. This set is called the set of **whole numbers** to indicate that it is different from the set of natural numbers, which does not include zero.

$$\{1, 2, 3, 4, \ldots\} = \text{natural numbers}$$
$$\{0, 1, 2, 3, 4, \ldots\} = \text{whole numbers}$$

Recall that 0 is the identity element for the operation of addition: it does not change the identity of a number when it is added to that number. That is, $1 + 0 = 1$, $2 + 0 = 2$, $3 + 0 = 3$, and so on. Recall also that -3 is the additive inverse of 3, since $-3 + 3 = 0$, and -1 is the additive inverse of 1, since $-1 + 1 = 0$. The set of numbers consisting of the set of whole numbers and their additive inverses is called the set of **integers**; that is,

$$\{\ldots, -4, -3, -2, -1, 0, 1, 2, 3, 4, \ldots\} = \text{integers}$$

It should be noted that a number such as 2 may be classified as a natural number, a whole number, or an integer, since it is an element of all three of these sets. But a number such as -2 can only be classified as an integer, and not as a natural number or a whole number. We can also classify -2 as a negative integer, while 2 is a positive integer. Note that 0 is neither positive nor negative, but it is an integer.

It is possible to picture the set of integers, $\{\ldots, -4, -3, -2, -1, 0, 1, 2, 3, 4, \ldots\}$, on a **number line,** as shown in figure 1. First,

FIGURE 1

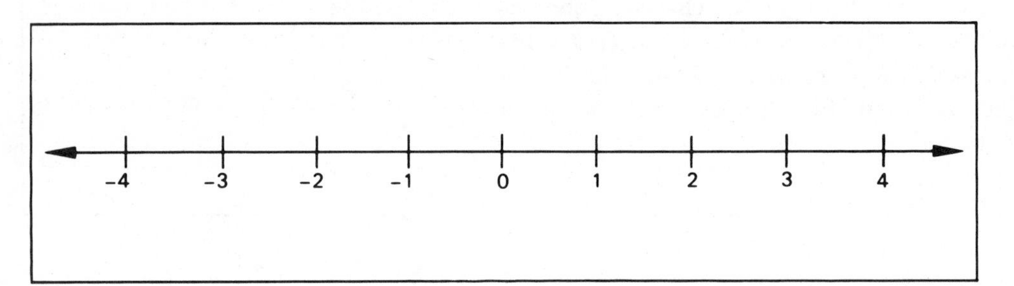

we draw a line, pick a point on the line, and label it 0. Next mark off equal units to the right and left of 0. Label the end points of the intervals to the right of 0 with the positive integers, and those to the left of 0 with the negative integers.

Ronald Oriti

This desert area has a negative altitude because it is below sea level.

The number line is a picture of the integers and their relationship to each other. The number line may be extended indefinitely in either direction, that is to the left or right of zero. Those numbers associated with points to the left of zero are called the negative integers, and those numbers associated with points to the right of zero are called the positive integers. Remember that zero is an integer, but it is neither positive nor negative.

Using the number line, we can determine the order of the integers. Three is greater than 2 ($3 > 2$) since 3 is to the right of 2 on the number line. Similarly, 4 is greater than 1 ($4 > 1$) since 4 is to the right of 1 on the number line. For the same reason, 0 is greater than -1 ($0 > -1$) and -1 is greater than -4 ($-1 > -4$). Other observations that we might make are

$$4 > 0, \qquad 1 > 0, \qquad 1 > -1, \qquad 2 > -2$$

Instead of stating that 3 is greater than 2 ($3 > 2$), we could have said that 2 is less than 3 ($2 < 3$), since 2 is to the left of three on the number line. Similarly, $1 < 4$ (1 is less than 4) since 1 is to the left of 4 on the number line. We can also say that

$$0 < 4, \qquad 0 < 1, \qquad -1 < 1, \qquad -2 < 2$$

Prior to our discussion of the set of integers, we only used the minus sign ($-$) to indicate subtraction. Now we are also using it to label negative integers. The number 3 is 3 units to the right of 0 on the number line and its "opposite," negative 3 (-3), is 3 units to the left of 0 on the number line. The opposite of 4 is negative 4 (-4), and the opposite of 1 is negative 1 (-1). Note that the sum of any number and its opposite is 0: $3 + (-3) = 0$, $4 + (-4) = 0$, and $1 + (-1) = 0$. It should also be noted that the opposite of 0 is 0: $0 + 0 = 0$. Recall that we can also describe -3 as the additive inverse of 3, since $3 + (-3) = 0$. Similarly, -1 is the additive inverse of 1, since $1 + (-1) = 0$.

How do we combine integers? Consider the sum of the positive integers 2 and 3. We know that $2 + 3 = 5$. However, let us work this problem on the number line. To find the solution to $2 + 3$ on the number line, we start at 0 and proceed to the right two units to 2. Since we want to add 3, we proceed three more units to the right; this brings us to 5. Hence, $2 + 3 = 5$. This process is illustrated in figure 2.

FIGURE 2 $2 + 3 = 5$

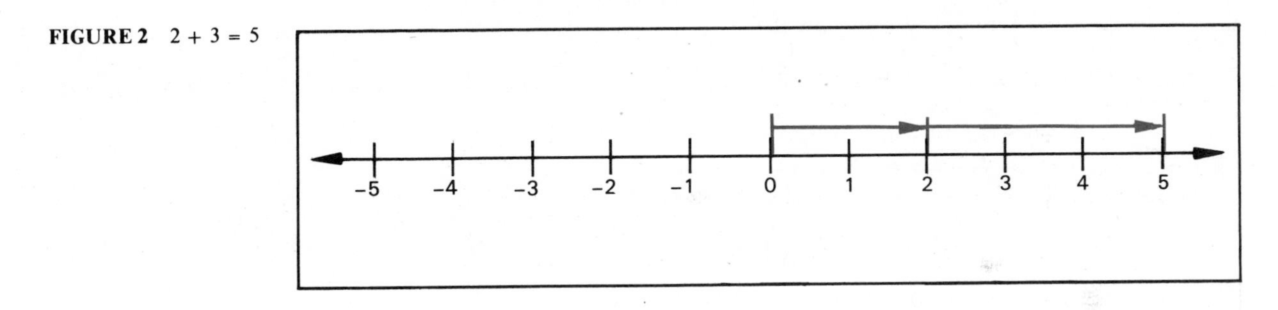

Now let's combine a positive and a negative integer, such as $2 + (-3)$, on a number line. We start at 0 and proceed two units to the right to 2. Since we want to add -3 to 2, we move three units to the *left* from 2. We end up at -1. Therefore $2 + (-3) = -1$. This is illustrated in figure 3

We could do this problem in another way: -3 can be expressed as $(-2) + (-1)$; then the original problem becomes $2 + (-3) = 2 + (-2) + (-1)$. But (-2) is the opposite, or additive inverse, of 2 and

FIGURE 3 $2 + (-3) = -1$

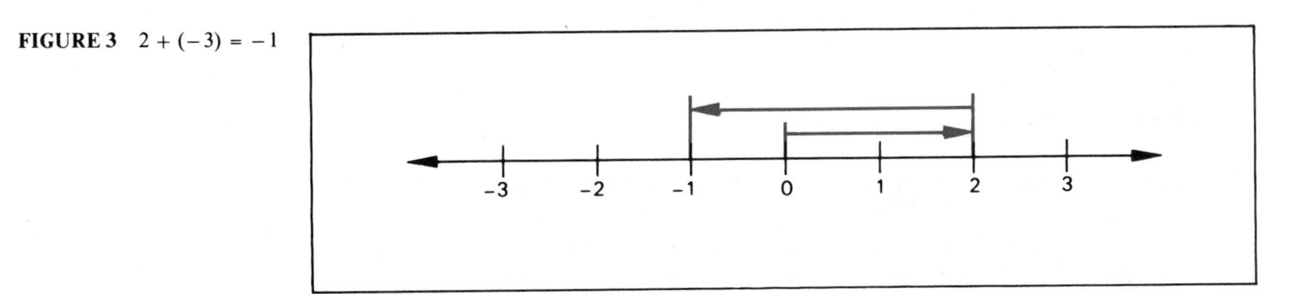

therefore $2 + (-2) + (-1) = 0 + (-1) = -1$. Summarizing these steps we have

$$2 + (-3) = 2 + (-2) + (-1) = 0 + (-1) = -1$$

We are not advocating one technique over the other. The method that you best understand is the one to use.

EXAMPLE 1
Evaluate $1 + (-4)$.

Solution

$$1 + (-4) = -3$$

On the number line in figure 4, we start at 0 and proceed one unit to the right to 1. Next, we move four units in a negative direction—that is, to the left—from 1. We end up at -3.

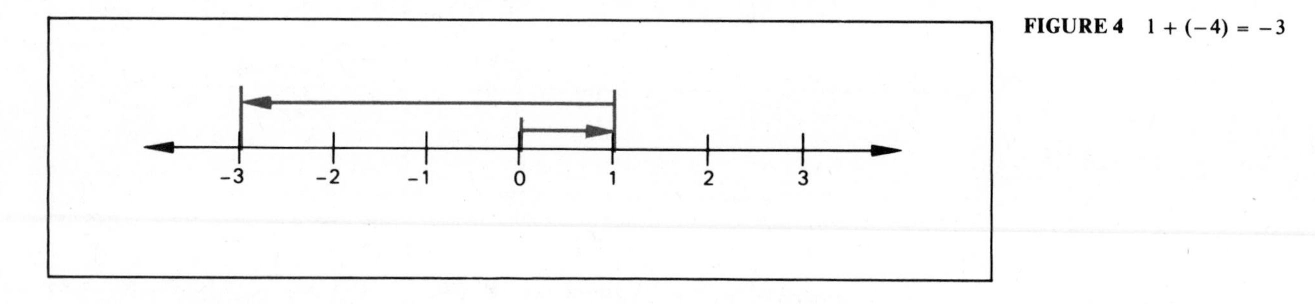

FIGURE 4 $1 + (-4) = -3$

Alternate solution

$$1 + (-4) = 1 + (-1) + (-3) = 0 + (-3) = -3$$

EXAMPLE 2
Evaluate $-2 + 5$.

Solution

$$-2 + 5 = 3$$

We start at 0 on the number line in figure 5 and move two units to the left to -2. Next we move five units to the right (positive direction) from -2; we end up at 3.

FIGURE 5 $-2 + 5 = 3$

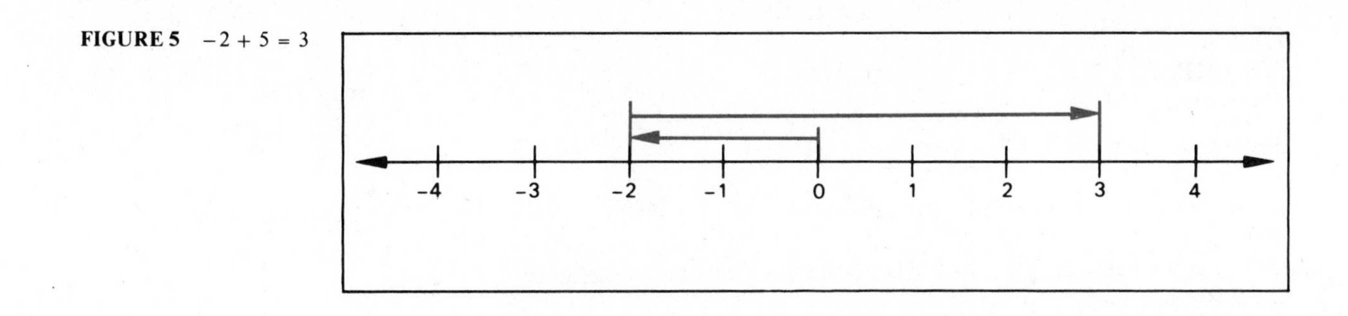

Alternate solution

$$-2 + 5 = -2 + 2 + 3 = 0 + 3 = 3$$

If we add two positive integers, then their sum will be a positive integer. If we add two negative integers, then their sum will be a negative integer. If we add a positive integer and a negative integer, then their sum may be a positive integer, a negative integer, or zero.

The problem $8 - 5$ has the same answer as $8 + (-5)$. Similarly, $7 - 4 = 7 + (-4)$ and $3 - 1 = 3 + (-1)$. This seems to indicate that for integers a subtraction problem can be thought of as an addition problem. That is, subtracting integers is the same as adding the opposite of the second integer to the first integer.

You probably do not need this rule for problems like $8 - 5$ and $7 - 4$. But what about $5 - 8$? According to the above rule, we can think of $5 - 8$ as $5 + (-8)$. Now we have a problem similar to the ones that we did in examples 1 and 2. Using the number line, as in figure 6, $5 + (-8) = -3$.

FIGURE 6 $5 + (-8) = -3$

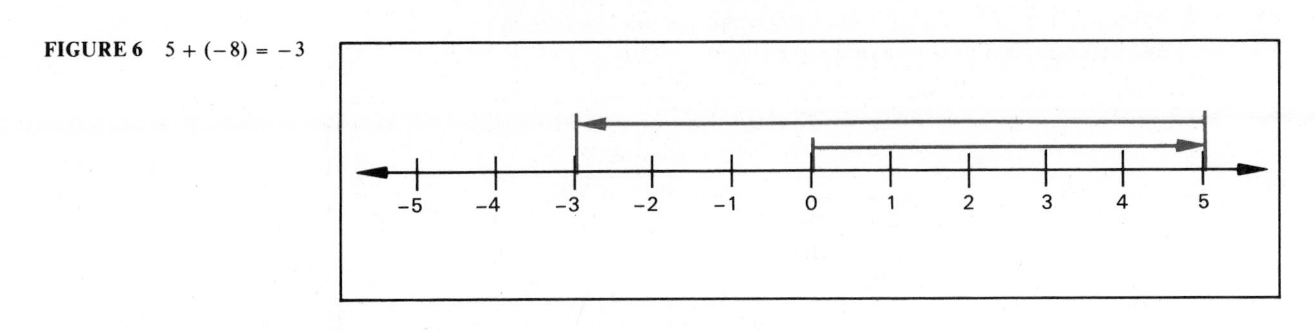

Using an alternate method,

$$5 - 8 = 5 + (-8) = 5 + (-5) + (-3) = 0 + (-3) = -3$$

EXAMPLE 3
Evaluate $3 - 4$.

Solution

$$3 - 4 = -1$$

$3 - 4$ is the same as $3 + (-4)$. Using the number line as in figure 7, we have $3 + (-4) = -1$.

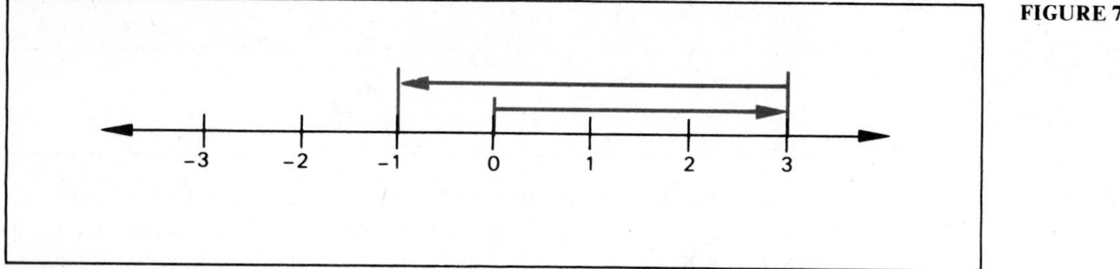

FIGURE 7 $3 + (-4) = -1$

Alternate solution

$$3 - 4 = 3 + (-4) = 3 + (-3) + (-1) = 0 + (-1) = -1$$

EXAMPLE 4
Evaluate $-3 - 2$.

Solution

$$-3 - 2 = -5$$

$-3 - 2$ is the same as $-3 + (-2)$. Using the number line, as in figure 8, we have $-3 + (-2) = -5$. Note that when we add two negative integers their sum is a negative integer.

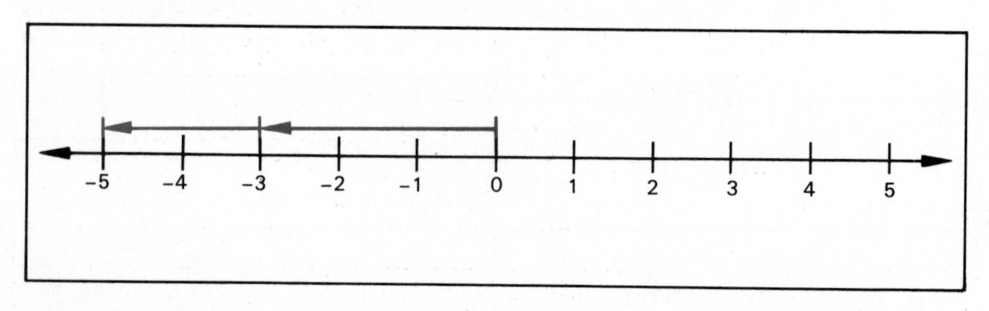

FIGURE 8 $-3 + (-2) = -5$

Alternate solution

$$-3 - 2 = -3 + (-2) = -(3 + 2) = -5$$

$p \rightarrow q \equiv$

$\sim q \rightarrow \sim p \equiv$

De Morgan's law

$\sim(p \wedge q) \equiv$

$\sim(p \vee q) \equiv$

$p \rightarrow q$

converse

inverse

contrapositive

negation

EXAMPLE 5

Evaluate $-2 - (-3)$.

Solution

$$-2 - (-3) = 1$$

Subtracting integers is the same as adding the opposite of the second integer to the first integer. Therefore, $-2 - (-3)$ is the same as $-2 + 3$. (Note that 3 is the opposite of -3.) Using the number line, we see that $-2 + 3 = 1$, as shown in figure 9.

FIGURE 9 $-2 + 3 = 1$

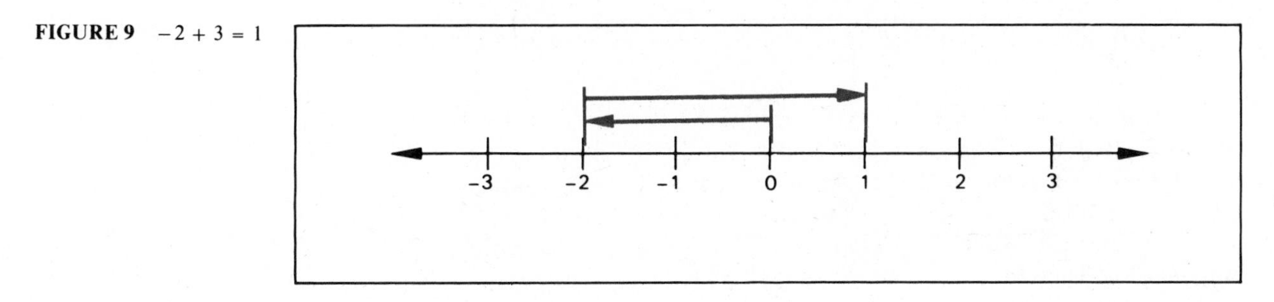

Alternate solution

$$-2 - (-3) = -2 + 3 = -2 + 2 + 1 = 0 + 1 = 1$$

EXAMPLE 6
Evaluate $-5 - (-3)$.

Solution

$$-5 - (-3) = -2$$

This problem is similar to example 5: $-5 - (-3)$ is the same as $-5 + 3$. By means of the number line in figure 10, we see that $-5 + 3 = -2$.

FIGURE 10 $-5 + 3 = -2$

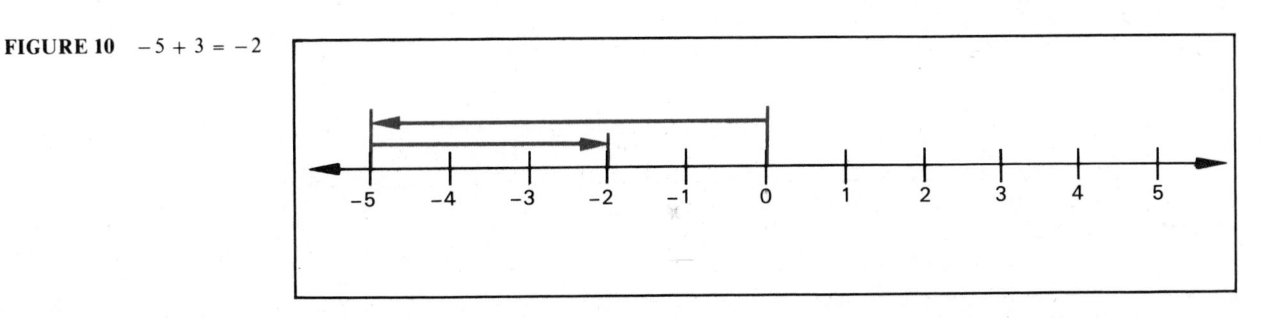

Alternate solution

$-5 - (-3) = -5 + 3 = -2 + (-3) + 3 = -2 + 0 = -2$

Once we have mastered addition and subtraction of integers, we can proceed to the operation of multiplication. We already know how to multiply the natural numbers: for example, $4 \times 2 = 8$, $5 \times 3 = 15$, and $7 \times 5 = 35$. These problems are also examples of multiplying positive integers. One observation that we can make from these examples is the following:

A positive integer multiplied by a positive integer yields a positive integer.

But, what happens when we multiply a positive integer by a negative integer? Consider the example $2 \times (-4)$. What is the answer? We can think of $2 \times (-4)$ as $(-4) + (-4)$, which equals -8. Therefore, $2 \times (-4) = -8$. Let's try another example, $3 \times (-5)$; this can also be expressed as $(-5) + (-5) + (-5)$, which equals -15. Therefore, $3 \times (-5) = -15$.

Whenever we multiply a positive integer by a negative integer, the answer is always a negative integer.

Before we consider the product of two negative integers, let us consider the *distributive law:* It states that

$$a \times (b + c) = a \times b + a \times c$$

for all integers a, b, and c. In terms of a specific example,

$$3 \times (2 + 5) = 3 \times 2 + 3 \times 5.$$

More formally, we call this the **distributive law (or property) for multiplication over addition.** When we are given an expression such as $3 \times (2 + 5)$, we can evaluate it in two different ways:

$$3 \times (2 + 5) = 3 \times 7 = 21$$

or

$$3 \times (2 + 5) = 3 \times 2 + 3 \times 5 = 6 + 15 = 21$$

Similarly, $(-2) \times (5 + 6) = (-2) \times 11 = -22$, or

$$(-2) \times (5 + 6) = (-2) \times 5 + (-2) \times 6 = -10 + (-12) = -22$$

Consider the example $2 \times (-4 + 4)$. We can also evaluate this in two ways:

$$2 \times (-4 + 4) = 2 \times (0) = 0$$

or

$$2 \times (-4 + 4) = 2 \times (-4) + 2 \times 4 = -8 + 8 = 0$$

Now consider the example $(-2) \times (-4 + 4)$. Evaluating this both ways, we have

$$(-2) \times (0) = 0 \quad \text{and} \quad (-2) \times (-4) + (-2) \times 4 = ?$$

Here we run into the problem of multiplying two negative integers. But now we can determine the answer. Since $a \times (b + c) = a \times b + a \times c$ for all integers a, b, and c, $(-2) \times (-4 + 4)$ must have the same answer regardless of which way we evaluate it. We already know that

$$(-2) \times (-4 + 4) = (-2) \times (0) = 0$$

Consequently,

$$(-2) \times (-4 + 4) = (-2) \times (-4) + (-2) \times 4$$

must also equal 0. We already know that $(-2) \times 4 = -8$; we must determine the answer to $(-2) \times (-4)$. We also know that when the answer to $(-2) \times (-4)$ is added to -8, the final answer must be 0. What number added to -8 will give an answer of 0? The opposite, or additive inverse, of -8: namely, 8. Therefore, $(-2) \times (-4)$ must equal 8. We list the steps again:

$$\begin{aligned}(-2) \times (-4 + 4) &= (-2) \times (-4) + \underbrace{(-2) \times 4} = 0 \\ &= \underbrace{(-2) \times (-4)} + \quad (-8) \quad = 0 \\ &= \quad\quad 8 \quad\quad + \quad (-8) \quad = 0\end{aligned}$$

Whenever we multiply two negative integers, the product is always positive.

The following is a summary of the rules we have developed for multiplication of integers:

1. **The product of two positive integers is positive. (For example, $2 \times 3 = 6$.)**

2. **The product of a positive integer and a negative integer is negative. (For example, $2 \times (-3) = -6$.)**
3. **The product of two negative integers is positive. (For example, $(-2) \times (-3) = 6$.)**
4. **The product of any integer and zero is zero. (For example, $(-2) \times 0 = 0$, $100 \times 0 = 0$.)**

EXAMPLE 7

Evaluate each of the following:

a. 4×3 b. $4 \times (-3)$ c. $(-4) \times (-3)$ d. 4×0

Solution

a. $4 \times 3 = 12$. The product of two positive integers is positive.

b. $4 \times (-3) = -12$. The product of a positive integer and a negative integer is negative.

c. $(-4) \times (-3) = 12$. The product of two negative integers is positive.

d. $4 \times 0 = 0$. The product of any integer and zero is zero.

EXAMPLE 8

Evaluate each of the following:

a. $7 \times (-8)$ b. $(-3) \times 9$

c. $(-9) \times (-4)$ d. $(-6) \times (-6)$

Solution

a. $7 \times (-8) = -56$ b. $(-3) \times 9 = -27$

c. $(-9) \times (-4) = 36$ d. $(-6) \times (-6) = 36$

Every integer can be classified as either an even integer or an odd integer. An integer is even if it can be expressed as 2 times another integer. For example, 6 is an even integer because $6 = 2 \times 3$. Similarly, -2 is even because $-2 = 2 \times (-1)$, and 0 is even since $0 = 2 \times 0$. In general, any even integer may be expressed in the form $2 \times n$, $2 \cdot n$, or $2(n)$, where n represents some integer.

If an integer is not even, then it must be odd. If $2 \cdot n$ represents an even integer, then adding 1 more to it will make it odd. Therefore $2 \cdot n + 1$ represents an odd integer. Since 6 is even, adding 1 more will yield an odd integer, 7: $7 = 2 \cdot 3 + 1$. Another odd integer is -5; $-5 = 2 \cdot (-3) + 1$.

O'Brien/Stockmarket

Some interesting observations can be made about odd and even integers. For example, what happens when you add two even integers? Considering a few examples, we see that $2 + 2 = 4$, $2 + 4 = 6$, $8 + 10 = 18$, and $12 + 6 = 18$. It appears that the sum of two even integers is also even. Are you sure? Considering some more examples, we see that $6 + 6 = 12$, $8 + 12 = 20$, and so on.

You are probably convinced that the sum of two even integers is even, but have we proved it? No, we have just observed what happens for a certain set of examples. We have not examined all of the possibilities, and therefore we cannot be sure that it is always true. If we can show that it is true for *any* two even integers, then we can be certain that the sum of any two specific even integers is even. Any even integer can be expressed as 2 times another integer, so let's consider the even integers $2 \cdot n$ and $2 \cdot k$, where n and k are also integers. Then we have

$$(2 \cdot n) + (2 \cdot k)$$

But

$$(2 \cdot n) + (2 \cdot k) = 2 \cdot (n + k)$$

by means of the distributive property. Note that $(n + k)$ is an integer, since the sum of any two integers is an integer. Therefore, $2 \cdot (n + k)$ is an even integer, since 2 times any integer is an even integer, and we have shown that the sum of any two even integers is also an even integer.

What happens when you add two odd integers? Considering $3 + 5 = 8$, $1 + 3 = 4$, and $5 + 7 = 12$, it appears that the sum of two odd integers is an even integer. Let's examine what happens when we add *any* two odd integers. Consider the odd integers $2 \cdot n + 1$ and $2 \cdot k + 1$: adding them, we have

$$(2 \cdot n + 1) + (2 \cdot k + 1) = 2 \cdot n + 2 \cdot k + 2 = 2 \cdot (n + k + 1)$$

by means of the distributive property. Note that $(n + k + 1)$ is an integer, since $n + k$ is an integer and the sum of 1 and an integer is still an integer. Therefore, $2 \cdot (n + k + 1)$ is an even integer, since 2 times any integer is an even integer.

EXAMPLE 9

Prove that the product of any two even integers is an even integer.

Solution

Any two even integers may be expressed as $2 \cdot n$ and $2 \cdot k$. Consequently, we have

$$(2 \cdot n) \cdot (2 \cdot k) = 2 \cdot (2 \cdot n \cdot k)$$

We see that $(2 \cdot n \cdot k)$ represents an integer, since the product of three integers is an integer. Therefore, $2 \cdot (2 \cdot n \cdot k)$ is an even integer. Hence, the product of any two even integers is an even integer.

EXERCISES FOR SECTION 8.4

1. Evaluate each of the following:
   a. $2 + (-3$ b. $4 + (-7)$
   c. $5 + (-8)$ d. $-3 + 5$
   e. $-7 + 8$ f. $-9 + 12$

2. Evaluate each of the following:
   a. $3 - 5$ b. $6 - 9$
   c. $-8 - 6$ d. $6 - (-5)$
   e. $-2 - (-1)$ f. $-7 - (-3)$

3. Evaluate each of the following:
   a. $-7 - (-2)$ b. $-10 - (-10)$
   c. $12 - (-13)$ d. $6 \times (-5 + 5)$
   e. $7 \times (-5)$ f. $(-8 + 5) \times (-3)$

4. Evaluate each of the following:
   a. $-8 + 8$ b. $-13 - (-12)$
   c. $-6 - (8 + 3)$ d. $(-8) \times 9$
   e. $(5 - 3) \times (-2)$ f. $(-2 - 1) \times (-3)$

5. Replace each question mark with $=$, $>$, or $<$ to make the sentence true.
   a. 11 ? 3 b. 0 ? 1
   c. 0 ? -1 d. $-2 + 3$? $3 - 2$
   e. $2 + (-3)$? $1 \times (-1)$
   f. $(-2)(3)$? $(-2)(-3)$

6. Replace each question mark with $=$, $>$, or $<$ to make the sentence true.
   a. -1 ? -2 b. -10 ? -3
   c. $-1 - (-1)$? 0 d. $-3 + 2$? $2 - (-3)$
   e. $-1 - (-2)$? $-3 + 4$
   f. $(-1)(-3)$? $-3 + (-1)$

7. Classify each integer as odd or even and express it in the form $2 \cdot n + 1$ or $2 \cdot n$. (*Example:* 7 is an odd integer; $7 = 2 \cdot 3 + 1$.)
   a. 10 b. 15 c. 21 d. -5

8. Classify each integer as odd or even and express it in the form $2 \cdot n + 1$ or $2 \cdot n$. (*Example:* 6 is an even integer; $6 = 2 \cdot 3$.)
   a. 12 b. 17 c. -6 d. 101

9. One of Goldbach's conjectures is that any odd number greater than 7 can be expressed as the sum of three odd primes (for example, $9 = 3 + 3 + 3$). Show that this conjecture is true for all positive integers from 11 up to an including 29.

10. An elevator in the Empire State Building started at the 35th floor, rose 7 floors, descended 12 floors, rose 4 floors, descended 11 floors, descended 3 floors, and then rose 15 floors and stopped. At what floor did the elevator stop? Assuming that there are 15 feet between floors, how far did the elevator travel?

11. If Aristotle was born in 384 B.C. and Euclid was born in 365 B.C., who was born first?

12. Prove that the product of any two odd integers is an odd integer.
13. Prove that the sum of an odd integer and even integer is an odd integer.
14. When you multiply an even integer by an odd integer, is the answer even or odd? Prove your answer.

B.C. by permission of Johnny Hart and Field Enterprises, Inc.

Just For Fun

A numismatist (coin collector) was examining a collection of coins. In this collection he discovered a coin dated 384 B.C. What is your conclusion regarding this coin?

8.5 RATIONAL NUMBERS

Thus far we have examined the set of natural numbers, the set of whole numbers, and the set of integers. All of these can be shown on a number line like that in figure 11.

But what about the intervals between the numbers on the number line? Do any other numbers belong in these intervals? The answer is yes. Consider the number $\frac{3}{4}$: it is greater than 0 and less than 1, so it belongs in the interval between 0 and 1 as shown in figure 11. What kind of number is it? It is not a natural number, it is not a whole number, and it is not an integer. It is a *rational number*.

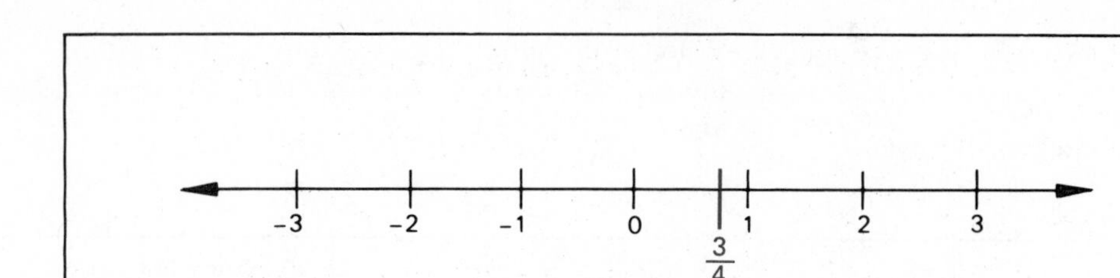

FIGURE 11

A **rational number** is a number that can be expressed in the form $\frac{a}{b}$, where a and b are integers and $b \neq 0$ (we cannot divide by zero). In other words, a rational number is any number that can be expressed as the quotient of two integers. A rational number like $\frac{3}{4}$ is commonly referred to as a fraction. Remember that both the numerator and denominator of the fraction must be integers, and that the denominator of the fraction cannot be 0.

Is the number 4 a rational number? The answer is yes, since a rational number is any number that can be expressed as the quotient of two integers, and

$$4 = \frac{4}{1}$$

In fact, using this idea we can see that any integer can be expressed as a quotient of two integers:

$$7 = \frac{7}{1}, \quad 10 = \frac{10}{1}, \quad -4 = \frac{-4}{1}, \quad -8 = \frac{-8}{1}, \quad \text{and} \quad 0 = \frac{0}{1}$$

Every integer is also a rational number, but remember that not all rational numbers are integers. (The fraction $\frac{3}{4}$ is an example of a rational number that is not an integer.)

A rational number may be expressed in many different ways. Recall that 4 is a rational number since $4 = \frac{4}{1}$, but we could also express 4 in other ways:

$$4 = \frac{4}{1} = \frac{8}{2} = \frac{16}{4} = \frac{32}{8} = \cdots$$

Note that $\frac{4}{1} \times \frac{2}{2} = \frac{8}{2}$, $\frac{16}{4} = \frac{8}{2} \times \frac{2}{2}$, and so on.

For any fraction $\frac{a}{b}$, if k is any number other than zero,

$$\frac{a \times k}{b \times k} = \frac{a}{b}$$

This rule is helpful in reducing, or simplifying, fractions. Given the fraction $\frac{15}{25}$, we can reduce it by factoring both the numerator and denominator into prime factors:

$$\frac{15}{25} = \frac{3 \times 5}{5 \times 5}$$

Applying the rule, we have

$$\frac{15}{25} = \frac{3 \times \cancel{5}}{5 \times \cancel{5}} = \frac{3}{5}$$

Note that we eliminated the factors that were common to both the numerator and denominator.

How do we know that our answer is correct? Does $\frac{15}{25} = \frac{3}{5}$? An expression such as this can be verified by cross multiplying to see if the products are equal.

$$15 \times 5 \stackrel{?}{=} 25 \times 3$$
$$75 = 75$$

You may recall that the expression $\frac{15}{25} = \frac{3}{5}$ can be thought of as a proportion. For any proportion to be true, the product of the *means* must equal the product of the *extremes*. More formally, we say that

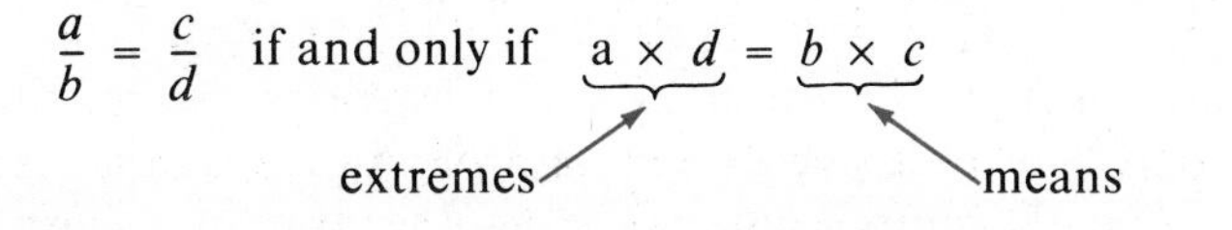

$\frac{a}{b} = \frac{c}{d}$ if and only if $a \times d = b \times c$
extremes
means

EXAMPLE 1
Does $\frac{3}{11} = \frac{9}{33}$?

Solution
The answer is yes, since $3 \times 33 = 9 \times 11$, that is, $99 = 99$.

Alternate solution
We reduce $\frac{9}{33}$ and see if the result is $\frac{3}{11}$:

$$\frac{9}{33} = \frac{\cancel{3} \times 3}{\cancel{3} \times 11} = \frac{3}{11}$$

EXAMPLE 2
Reduce $\frac{42}{54}$.

Solution

$$\frac{42}{54} = \frac{\not{2} \times \not{3} \times 7}{\not{2} \times \not{3} \times 3 \times 3} = \frac{7}{3 \times 3} = \frac{7}{9}$$

Check: $\frac{42}{54} = \frac{7}{9}$ if and only if $42 \times 9 = 7 \times 54$
$378 = 378$

How do we combine rational numbers? Since 2 and 3 are rational numbers and $2 + 3 = 5$, we already know how to combine some rationals. But consider the problem of adding the rational numbers $\frac{1}{5} + \frac{2}{3}$. In section 8.3 we discussed the process of adding and subtracting fractions. However those problems were considered only with regard to the use of the least common multiple. Let's state a general rule for adding any two rational numbers:

If $\frac{a}{b}$ and $\frac{c}{d}$ are rational numbers, then

$$\frac{a}{b} + \frac{c}{d} = \frac{ad + bc}{bd}$$

Therefore, for the example $\frac{1}{5} + \frac{2}{3}$, we have the following:

$$\frac{1}{5} + \frac{2}{3} = \frac{1 \times 3 + 5 \times 2}{5 \times 3} = \frac{3 + 10}{15} = \frac{13}{15}$$

Recall that in order to add two fractions, we rewrite them so that they have the same denominator, and then we add the numerators. The above procedure is just another way of doing this. Since the two given fractions do not have the same denominator, we could have found a common denominator by finding the least common multiple of 5 and 3, which is 15. This would result in the same answer as above:

$$\frac{1}{5} + \frac{2}{3} = \frac{3}{15} + \frac{10}{15} = \frac{13}{15}$$

EXAMPLE 3
Add $\frac{1}{3} + \frac{2}{5}$.

Solution
Using the rule for addition, we have

$$\frac{1}{3} + \frac{2}{5} = \frac{1 \times 5 + 3 \times 2}{3 \times 5} = \frac{5 + 6}{15} = \frac{11}{15}$$

EXAMPLE 4
Add $\frac{2}{6} + \frac{3}{9}$.

Solution

$$\frac{2}{6} + \frac{3}{9} = \frac{2 \times 9 + 6 \times 3}{6 \times 9} = \frac{18 + 18}{54} = \frac{36}{54} = \frac{\not{2} \times 2 \times \not{3} \times \not{3}}{\not{2} \times 3 \times \not{3} \times \not{3}} = \frac{2}{3}$$

How do we subtract two rational numbers? Consider the problem $\frac{2}{3} - \frac{1}{5}$. The expression $-\frac{1}{5}$ is equivalent to $\frac{1}{-5}$ and also to $\frac{-1}{5}$. Hence the problem $\frac{2}{3} - \frac{1}{5}$ is the equivalent to $\frac{2}{3} + \frac{-1}{5}$, which turns out to be an addition problem similar to those that we have been considering. Therefore

$$\frac{2}{3} - \frac{1}{5} = \frac{2}{3} + \frac{-1}{5} = \frac{2 \times 5 + 3 \times (-1)}{15} = \frac{10 + (-3)}{15} = \frac{7}{15}$$

EXAMPLE 5
Subtract $\frac{4}{5} - \frac{1}{3}$.

Solution

$$\frac{4}{5} - \frac{1}{3} = \frac{4}{5} + \frac{-1}{3} = \frac{4 \times 3 + 5 \times (-1)}{5 \times 3} = \frac{12 + (-5)}{15} = \frac{7}{15}$$

Now that we have examined the operations of addition and subtraction for rational numbers, we next examine multiplication. Most students feel that multiplication is the easiest operation to perform with fractions. In order to multiply two fractions, we simply multiply numerator times numerator and denominator times denominator.

If $\frac{a}{b}$ and $\frac{c}{d}$ are rational numbers, then

$$\frac{a}{b} \times \frac{c}{d} = \frac{a \times c}{b \times d}$$

If we want to find the product of $\frac{3}{5}$ and $\frac{2}{7}$, we simply multiply the numerators together to find the numerator of the product and multiply the denominators together to find the denominator of the product. Therefore

$$\frac{3}{5} \times \frac{2}{7} = \frac{3 \times 2}{5 \times 7} = \frac{6}{35}$$

Consider the problem $\frac{5}{18} \times \frac{6}{25}$. We can do this problem in the same manner as the previous example; that is:

$$\frac{5}{18} \times \frac{6}{25} = \frac{5 \times 6}{18 \times 25} = \frac{30}{450} = \frac{\not{2} \times \not{3} \times \not{5}}{\not{2} \times 3 \times \not{3} \times \not{5} \times 5} = \frac{1}{15}$$

Or we can make the problem a little easier by simplifying it before performing the actual multiplication:

$$\frac{5}{18} \times \frac{6}{25} = \frac{5}{2 \times 3 \times 3} \times \frac{2 \times 3}{5 \times 5} = \frac{\not{5} \times \not{2} \times \not{3}}{\not{2} \times 3 \times \not{3} \times \not{5} \times 5} = \frac{1}{15}$$

Note that we eliminated the factors that are common to both the numerator and denominator.

EXAMPLE 6
Multiply $\frac{4}{9} \times \frac{2}{5}$.

Solution

$$\frac{4}{9} \times \frac{2}{5} = \frac{4 \times 2}{9 \times 5} = \frac{8}{45}$$

EXAMPLE 7
Multiply $\frac{7}{16} \times \frac{40}{42}$.

Solution

$$\frac{7}{16} \times \frac{40}{42} = \frac{7}{2 \times 2 \times 2 \times 2} \times \frac{2 \times 2 \times 2 \times 5}{2 \times 3 \times 7}$$

$$= \frac{\not{7} \times \not{2} \times \not{2} \times \not{2} \times 5}{\not{2} \times \not{2} \times \not{2} \times 2 \times 2 \times 3 \times \not{7}} = \frac{5}{2 \times 2 \times 3} = \frac{5}{12}$$

Division of fractions can be defined in terms of multiplication.

$$\frac{a}{b} \div \frac{c}{d} = \frac{a}{b} \times \frac{d}{c}$$

You may recall a rule that you learned previously: "In order to divide two fractions, invert the divisor and multiply." Why does this work? Consider the problem $\frac{2}{3} \div \frac{1}{2}$. According to the rule,

$$\frac{2}{3} \div \frac{1}{2} = \frac{2}{3} \times \frac{2}{1} = \frac{2 \times 2}{3 \times 1} = \frac{4}{3}$$

Another way to look at this problem is

$$\frac{\frac{2}{3}}{\frac{1}{2}}$$

This is a **complex fraction,** since the numerator or (inclusive *or*) denominator of the fraction is also a fraction. The complex fraction would no longer be complex if the denominator were 1. In order to convert this denominator to 1, we must multiply $\frac{1}{2}$ by its *reciprocal,* $\frac{2}{1}$. But, if we multiply the denominator by $\frac{2}{1}$, we must multiply the numerator by $\frac{2}{1}$. Therefore,

$$\frac{\frac{2}{3}}{\frac{1}{2}} = \frac{\frac{2}{3} \times \frac{2}{1}}{\frac{1}{2} \times \frac{2}{1}} = \frac{\frac{2 \times 2}{3 \times 1}}{\frac{1 \times 2}{2 \times 1}} = \frac{\frac{4}{3}}{\frac{2}{2}} = \frac{\frac{4}{3}}{1} = \frac{4}{3}$$

In general terms, we have

$$\frac{\frac{a}{b}}{\frac{c}{d}} = \frac{\frac{a}{b} \times \frac{d}{c}}{\frac{c}{d} \times \frac{d}{c}} = \frac{\frac{a}{b} \times \frac{d}{c}}{1} = \frac{a}{b} \times \frac{d}{c}$$

From the illustrative example, we can see that in order to divide two rational numbers, we multiply the first rational number (the *dividend*) by the multiplicative inverse of the second rational number (the *divisor*).

For any rational numbers $\frac{a}{b}$ and $\frac{c}{d}$,

$$\frac{a}{b} \div \frac{c}{d} = \frac{a}{b} \times \frac{d}{c} \qquad b \neq 0, c \neq 0$$

EXAMPLE 8
Divide: $\frac{9}{11} \div \frac{5}{4}$

Solution

$$\frac{9}{11} \div \frac{5}{4} = \frac{9}{11} \times \frac{4}{5} = \frac{9 \times 4}{11 \times 5} = \frac{36}{55}$$

EXAMPLE 9
Divide: $\frac{6}{7} \div \frac{9}{14}$

Solution

$$\frac{6}{7} \div \frac{9}{14} = \frac{6}{7} \times \frac{14}{9} = \frac{2 \times 3}{7} \times \frac{2 \times 7}{3 \times 3} = \frac{2 \times \cancel{3} \times 2 \times \cancel{7}}{\cancel{7} \times 3 \times \cancel{3}} = \frac{4}{3}$$

Alternate solution

$$\frac{6}{7} \div \frac{9}{14} = \frac{6}{7} \times \frac{14}{9} = \frac{84}{63} = \frac{2 \times 2 \times \cancel{3} \times \cancel{7}}{3 \times \cancel{3} \times \cancel{7}} = \frac{4}{3}$$

Note that the only difference between the two solutions in example 9 is that in the first solution, the prime factors are determined first and those common to both the numerator and denominator are eliminated before the answer is determined.

EXERCISES FOR SECTION 8.5

1. Reduce each fraction to lowest terms.
   a. $\frac{6}{16}$ b. $\frac{8}{72}$ c. $\frac{81}{129}$ d. $\frac{54}{448}$
2. Reduce each fraction to lowest terms.
   a. $\frac{24}{72}$ b. $\frac{4}{9}$ c. $\frac{484}{576}$ d. $\frac{775}{1325}$
3. Perform the indicated operations.
   a. $\frac{4}{5} + \frac{1}{7}$ b. $\frac{2}{3} + \frac{1}{4}$ c. $\frac{1}{9} + \frac{1}{8}$
   d. $\frac{9}{11} - \frac{2}{3}$ e. $\frac{13}{16} - \frac{4}{5}$ f. $\frac{8}{9} - \frac{1}{3}$
4. Perform the indicated operations.
   a. $\frac{4}{11} + \frac{5}{13}$ b. $\frac{6}{11} + \frac{2}{15}$ c. $\frac{8}{33} + \frac{3}{16}$
   d. $\frac{9}{11} - \frac{1}{3}$ e. $\frac{13}{16} - \frac{3}{5}$ f. $\frac{12}{33} - \frac{4}{11}$
5. Perform the indicated operations.
   a. $\frac{4}{5} \times \frac{2}{7}$ b. $\frac{3}{11} \times \frac{4}{5}$ c. $\frac{8}{13} \times \frac{4}{7}$
   d. $\frac{4}{5} \div \frac{2}{7}$ e. $\frac{3}{11} \div \frac{4}{9}$ f. $\frac{8}{13} \div \frac{4}{7}$

6. Perform the indicated operations.
   a. $\frac{6}{11} \times \frac{2}{9}$ b. $\frac{8}{9} \times \frac{3}{4}$ c. $\frac{7}{14} \times \frac{22}{14}$
   d. $\frac{8}{33} \div \frac{4}{11}$ e. $\frac{8}{9} \div \frac{2}{3}$ f. $\frac{12}{33} \div \frac{4}{11}$

7. Determine whether each statement is true or false. (*Hint:* Convert the fractions under consideration to fractions with the same denominator.)
   a. $\frac{4}{7} > \frac{2}{3}$ b. $\frac{3}{4} < \frac{7}{8}$ c. $\frac{4}{11} > \frac{3}{7}$
   d. $\frac{4}{9} < \frac{16}{36}$ e. $\frac{5}{11} > \frac{11}{5}$ f. $\frac{6}{7} < \frac{8}{8}$

8. Determine whether each statement is true or false. (See exercise 7.)
   a. $\frac{4}{3} > \frac{5}{4}$ b. $\frac{6}{4} < \frac{8}{9}$ c. $\frac{8}{9} > \frac{6}{5}$
   d. $\frac{8}{33} < \frac{4}{11}$ e. $\frac{8}{9} < \frac{2}{3}$ f. $\frac{12}{33} = \frac{4}{11}$

9. Determine whether each statement is true or false.
   a. Every rational number is an integer.
   b. Every integer is a rational number.
   c. Every rational number is a natural number.
   d. Every natural number is a rational number.
   e. Every rational number is a whole number.

10. Determine whether each statement is true or false.
   a. Every whole number is a rational number.
   b. Every whole number is an integer.
   c. Every integer is a whole number.
   d. The rationals are a subset of the integers.
   e. The integers are a subset of the rationals.

Just For Fun

If you double $\frac{1}{4}$ of a certain fraction and multiply it by that fraction, the answer is $\frac{1}{8}$. What is the fraction?

8.6 RATIONAL NUMBERS AND DECIMALS

Before we continue our discussion regarding what other numbers belong on the number line, let's review the topic of *decimals*. **Decimals** are fractions that have a power of ten, such as 10, 100, 1000, or 10,000, for their denominator. The word decimal comes from the Latin word *decem,* which means *ten.* The fraction $\frac{3}{10}$ is represented by the decimal 0.3.

The *decimal point* is a period (.) which appears just to the left of the tenths place in the decimal. Some other examples of fractions expressed as decimals are:

$$\frac{4}{10} = 0.4, \quad \frac{31}{100} = 0.31, \quad \frac{471}{1000} = 0.471$$

The decimal 0.4 is read as "4 tenths," 0.31 is read as "31 hundredths," and 0.471 is read as "471 thousandths." The example below indicates the names of the places for decimals.

thousands	hundreds	tens	units		tenths	hundredths	thousandths	ten-thousandths
4	3	4	1	.	2	1	3	4
Integers					Decimals			

In order to change any fraction in the form $\frac{a}{b}$ to a decimal, we divide the denominator into the numerator.

$$\begin{array}{r} 0.2 \\ 5\overline{)1.0} \\ \underline{1\,0} \\ 0 \end{array} \qquad \frac{1}{5} = 0.2 \qquad\qquad \begin{array}{r} 0.375 \\ 8\overline{)3.000} \\ \underline{24} \\ 60 \\ \underline{56} \\ 40 \\ \underline{40} \\ 0 \end{array} \qquad \frac{3}{8} = 0.375$$

Texas Instruments, Inc.

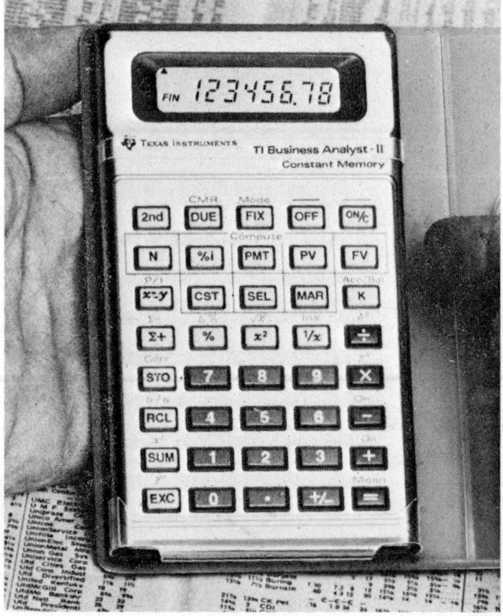

Decimals such as 0.2 and 0.375 are called **terminating decimals** since at some point in the division a remainder of zero is obtained; that is, when we change the fraction to a decimal, the division terminates.

Not all fractions can be expressed as terminating decimals. For example, consider the rational number $\frac{1}{3}$. Converting $\frac{1}{3}$ to a decimal, we have:

$$\begin{array}{r} 0.3333\ldots \\ 3\overline{)1.0000} \\ \underline{9} \\ 10 \\ \underline{9} \\ 10 \\ \underline{9} \\ 10 \\ \underline{9} \\ 1 \end{array}$$

The division does not terminate: we will never obtain a remainder of zero, but the remainder of 1 will keep reappearing at regular intervals.

Instead of writing the decimal expression for $\frac{1}{3}$ as 0.3333 . . . , we can express it in a more convenient and efficient way by placing a bar over the 3, which indicates that the 3 repeats endlessly, that is, $0.333\ldots = 0.\overline{3}$. In the same manner, $0.121212\ldots = 0.\overline{12}$. In this case the digits 12 repeat endlessly, so we place a bar over both the 1 and the 2. Decimals such as $0.\overline{3}$ and $0.\overline{12}$ are called **repeating nonterminating decimals.**

EXAMPLE 1
Express $\frac{5}{8}$ as a decimal.

Solution

$$\begin{array}{r|l} & 0.625 \\ \hline 8 & 5.000 \\ & \underline{4\,8} \\ & \;\;20 \\ & \;\;\underline{16} \\ & \;\;\;40 \\ & \;\;\;\underline{40} \\ & \;\;\;\;0 \end{array} \qquad \frac{5}{8} = 0.625$$

EXAMPLE 2
Express $\frac{4}{9}$ as a decimal.

Solution

$$\begin{array}{r|l} & 0.44\ldots \\ \hline 9 & 4.00 \\ & \underline{3\,6} \\ & \;\;40 \\ & \;\;\underline{36} \\ & \;\;\;4 \end{array} \qquad \frac{4}{9} = 0.\overline{4}$$

EXAMPLE 3
Express $\frac{3}{7}$ as a decimal.

Solution

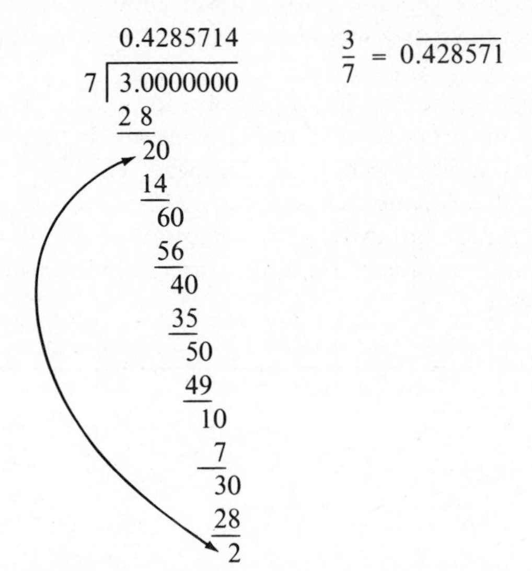

Note that we place a bar over the six digits $0.\overline{428571}$. The last remainder of 2 is a repeat of a remainder that we had previously. Therefore the pattern of division will repeat, and the digits 428571 will repeat endlessly.

In examining the set of rational numbers, we learned how to express a fraction as a decimal. For example, $\frac{5}{8} = 0.625$, $\frac{4}{9} = 0.\overline{4}$, and $\frac{3}{7} = 0.\overline{428571}$. The next thing to consider is whether, given the decimal expression, we can find its equivalent fraction. We already know that $0.\overline{3} = \frac{1}{3}$ and $0.25 = \frac{1}{4}$, but what about other examples we might encounter?

To convert a terminating decimal to a fraction, we simply omit the decimal point and supply the proper denominator. For example,

$$0.3 = \frac{3}{10}, \qquad 0.25 = \frac{25}{100} = \frac{1}{4}, \qquad 0.125 = \frac{125}{1000} = \frac{1}{8}$$

This technique does not work for repeating decimals, so we will need to develop another method for repeating decimals. Consider the decimal $0.3\overline{3}$. Let $x = 0.3\overline{3}$. Then $10x = 3.3\overline{3}$. (Multiplying a number by 10 moves the decimal point one place to the right, multiplying by 100 moves the decimal point two places to the right, and so on.) Thus far we have

$$\begin{aligned} 10x &= 3.3\overline{3} \\ x &= 0.3\overline{3} \\ \hline \end{aligned}$$

The decimal points are lined up in the same position.

Now subtract x from $10x$ and $0.3\overline{3}$ from $3.3\overline{3}$:

$$\begin{aligned} 10x &= 3.3\overline{3} \\ x &= 0.3\overline{3} \\ \hline 9x &= 3 \end{aligned}$$

All of the repeating 3s are subtracted from repeating 3s.

Next divide both sides of the resulting equation by 9:

$$\frac{9x}{9} = \frac{3}{9}$$

$$x = \frac{3}{9} = \frac{1}{3};$$

$$\text{therefore} \quad 0.3\overline{3} = \frac{1}{3}$$

Let's try another example. Suppose we wish to convert the repeating decimal $0.\overline{13}$ to a fraction. Let $x = 0.\overline{13}$; since the digits repeat in cycles of two, we multiply both sides of the equation by 100. If $x = 0.\overline{13}$, then $100x = 13.\overline{13}$. We subtract, which gives

$$\begin{aligned} 100x &= 13.\overline{13} \\ x &= 0.\overline{13} \\ \hline 99x &= 13 \end{aligned}$$

Dividing both sides of the equation by 99, we have

$$\frac{99x}{99} = \frac{13}{99}$$

$$x = \frac{13}{99};$$

$$\text{therefore} \quad 0.\overline{13} = \frac{13}{99}$$

EXAMPLE 4

Express $0.\overline{25}$ as a quotient of integers (that is, convert $0.\overline{25}$ to a fraction).

Solution

Let $x = 0.\overline{25}$ and multiply both sides of the equation by 100 (two digits repeating). Subtracting, we have

$$\begin{aligned} 100x &= 25.\overline{25} \\ x &= 0.\overline{25} \\ \hline 99x &= 25 \end{aligned}$$

Dividing both sides by 99,

$$\frac{99x}{99} = \frac{25}{99}$$

$$x = \frac{25}{99}$$

$$\text{therefore} \quad 0.\overline{25} = \frac{25}{99}$$

EXAMPLE 5

Express $3.\overline{162}$ as a quotient of integers.

Solution

Let $x = 3.\overline{162}$. Multiplying both sides of the equation by 1000 (three digits repeating) and subtracting, we have

$$\begin{aligned} 1000x &= 3162.\overline{162} \\ x &= 3.\overline{162} \\ \hline 999x &= 3159 \end{aligned}$$

Dividing both sides by 999,

$$\frac{999x}{999} = \frac{3159}{999}$$

$$x = \frac{3159}{999} = \frac{117}{37}$$

$$\text{therefore} \quad 3.\overline{162} = \frac{117}{37}$$

EXAMPLE 6

Express $2.14\overline{27}$ as a quotient of integers.

Solution

Let $x = 2.14\overline{27}$, and multiply both sides of the equation by 100 (two digits repeating). We place the decimal points in the same position and subtract:

$$\begin{aligned} 100x &= 214.27\overline{27} \\ x &= 2.14\overline{27} \\ \hline 99x &= 212.13 \end{aligned}$$

Dividing both sides by 99,

$$\frac{99x}{99} = \frac{212.13}{99}$$

But we are not finished. We were supposed to express $2.14\overline{27}$ as a quotient of two integers, and 212.13 is not an integer—it is a decimal.

In our earlier discussion of rational numbers (section 8.5), we noted that

$$\frac{a}{b} = \frac{a \times k}{b \times k}$$

Using this idea, we multiply $\frac{212.13}{99}$ by $\frac{100}{100}$ (we use 100, since we have two decimal places). Therefore

$$x = \frac{212.13}{99} \times \frac{100}{100} = \frac{21213}{9900}$$

$$= \frac{2357 \times \not{3} \times \not{3}}{1100 \times \not{3} \times \not{3}} = \frac{2357}{1100}$$

$$\text{therefore} \quad 2.14\overline{27} = \frac{2357}{1100}$$

Thus far, we have examined the set of natural numbers, the set of whole numbers, the set of integers, and the set of rational numbers. All of these can be shown on a number line. For example, figure 12 is a number line showing the integers zero and 1 and the rational numbers $\frac{1}{2}$, $\frac{14}{16}$, and $\frac{15}{16}$.

FIGURE 12

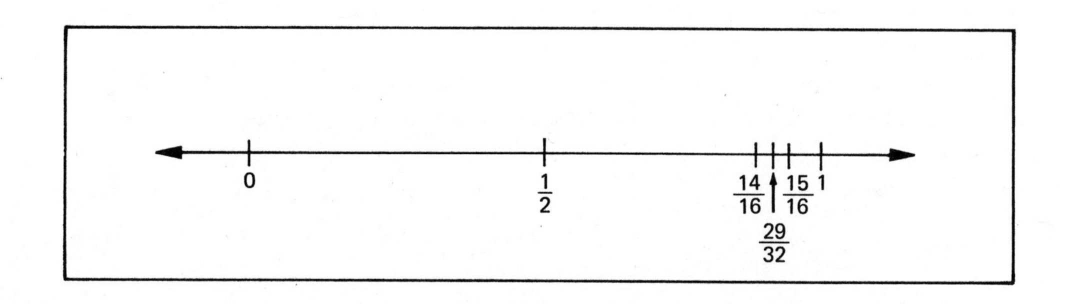

If two rational numbers are indicated on a number line, can we find other numbers that fit between them? The answer is yes. As an example, we will find a number that lies between $\frac{14}{16}$ and $\frac{15}{16}$ on the line in figure 12. To do this, we convert $\frac{14}{16}$ to $\frac{28}{32}$ and $\frac{15}{16}$ to $\frac{30}{32}$. Now we can see that the number $\frac{29}{32}$ is between the two given fractions. This process can be continued indefinitely: it is possible to find a rational number between any two given rational numbers. This process may seem to "fill up" the number line, but this is not the case, since there is always room for one more number. In fact, between any two rational numbers, there is always another rational number. This particular property is called the **density property of rational numbers.** We can also say that the rational numbers are **dense**. Note that the density property does not hold for all kinds of numbers. The natural numbers are not dense, because there is no natural number between the natural numbers 3 and 4.

We have seen one example of how to find a rational number between two given rational numbers, but how do we do it for *any* two rational numbers? One way to do this is to find the arithmetic mean of the two given rational numbers. In general terms, if we let x equal the number we are seeking, and we let $\frac{a}{b}$ and $\frac{c}{d}$ equal the given rational numbers, then

$$x = \frac{1}{2} \times \left(\frac{a}{b} + \frac{c}{d}\right)$$

Using $\frac{14}{16}$ and $\frac{15}{16}$ from our previous discussion, we have

$$x = \frac{1}{2} \times \left(\frac{14}{16} + \frac{15}{16}\right)$$

$$x = \frac{1}{2} \times \left(\frac{29}{16}\right)$$

$$x = \frac{29}{32}$$

Let's consider another example, finding a rational number between $\frac{2}{7}$ and $\frac{1}{3}$. Using the given formula, we have

$$x = \frac{1}{2} \times \left(\frac{2}{7} + \frac{1}{3}\right)$$

$$x = \frac{1}{2} \times \left(\frac{6+7}{21}\right) = \frac{1}{2} \times \left(\frac{13}{21}\right)$$

$$x = \frac{13}{42}$$

We can check that $\frac{13}{42}$ is between $\frac{2}{7}$ and $\frac{1}{3}$ by rewriting $\frac{2}{7}$ as $\frac{12}{42}$ and $\frac{1}{3}$ as $\frac{14}{42}$; it is clear that $\frac{13}{42}$ is between $\frac{12}{42}$ and $\frac{14}{42}$.

EXAMPLE 7
Find a rational number between $\frac{5}{7}$ and $\frac{6}{7}$.

Solution
Finding the arithmetic mean of the given numbers, we have

$$x = \frac{1}{2} \times \left(\frac{5}{7} + \frac{6}{7}\right)$$

$$x = \frac{1}{2} \times \left(\frac{11}{7}\right)$$

$$x = \frac{11}{14}$$

Check:

$$\frac{5}{7} = \frac{10}{14} \quad \text{and} \quad \frac{6}{7} = \frac{12}{14}$$

$$\frac{10}{14} < \frac{11}{14} \quad \text{and} \quad \frac{11}{14} < \frac{12}{14}$$

EXAMPLE 8
Find a rational number between $\frac{1}{5}$ and $\frac{1}{9}$.

Solution

$$x = \frac{1}{2} \times \left(\frac{1}{5} + \frac{1}{9}\right)$$

$$x = \frac{1}{2} \times \left(\frac{9+5}{45}\right) = \frac{1}{\not{2}} \times \left(\frac{\overset{7}{\not{14}}}{45}\right)$$

$$x = \frac{7}{45}$$

Check:

$$\frac{1}{9} = \frac{5}{45} \quad \text{and} \quad \frac{1}{5} = \frac{9}{45}$$

$$\frac{5}{45} < \frac{7}{45} \quad \text{and} \quad \frac{7}{45} < \frac{9}{45}$$

Note that there are other rational numbers that are also between $\frac{1}{5}$ and $\frac{1}{9}$. However, using the formula $x = \frac{1}{2} \times \left(\frac{a}{b} + \frac{c}{d}\right)$, we found the arithmetic mean, which is exactly halfway between the two given numbers.

In summary, whenever we are given two rational numbers $\frac{a}{b}$ and $\frac{c}{d}$, we can find another rational number (call it x) such that $\frac{a}{b} < x$ and $x < \frac{c}{d}$, or $\frac{a}{b} > x$ and $x > \frac{c}{d}$. This particular property is called the *density property of rational numbers.* The rational numbers are dense, but the integers are not. For example, there is no integer between the integers 1 and 2.

EXERCISES FOR SECTION 8.6

1. Express each fraction as a decimal.
   a. $\frac{3}{8}$ b. $\frac{5}{16}$ c. $\frac{2}{3}$
   d. $\frac{7}{33}$ e. $\frac{1}{11}$ f. $\frac{15}{37}$

2. Express each fraction as a decimal.
   a. $\frac{5}{8}$ b. $\frac{15}{16}$ c. $\frac{2}{33}$
   d. $\frac{7}{11}$ e. $\frac{19}{37}$ f. $\frac{1}{7}$

3. Express each decimal as a quotient of integers, in simplest form.
   a. 0.45 b. 0.035 c. $0.\overline{6}$
   d. $0.\overline{12}$ e. $0.\overline{134}$ f. $2.1\overline{78}$

4. Express each decimal as a quotient of integers, in simplest form.
   a. 0.125 b. 0.0025 c. $0.\overline{7}$
   d. $0.\overline{34}$ e. $6.2\overline{81}$ f. $0.\overline{9}$

5. Find a rational number between each pair of rational numbers.
   a. $\frac{1}{2}, \frac{1}{3}$ b. $\frac{1}{3}, \frac{1}{4}$ c. $\frac{1}{4}, \frac{1}{5}$
   d. $\frac{2}{3}, \frac{7}{8}$ e. $\frac{3}{4}, \frac{9}{11}$ f. $\frac{7}{11}, \frac{15}{16}$

6. Find a rational number between each pair of rational numbers.
   a. $\frac{2}{9}, \frac{3}{9}$ b. $\frac{4}{7}, \frac{5}{7}$ c. $\frac{3}{7}, \frac{4}{11}$
   d. $\frac{3}{5}, \frac{7}{9}$ e. $\frac{4}{9}, \frac{11}{12}$ f. $\frac{7}{13}, \frac{9}{17}$

Just For Fun

Name a four-letter word that ends with *e n y*.

8.7 IRRATIONAL NUMBERS AND THE SET OF REAL NUMBERS

We have seen that any rational number can be expressed as a decimal, and that this decimal will be a terminating decimal or a repeating nonterminating decimal. For example, $\frac{5}{8}$ and $\frac{1}{11}$ are rational numbers and, expressing each as a decimal, we have

$$\frac{5}{8} = 0.625 \qquad \text{(a terminating decimal)}$$

$$\frac{1}{11} = 0.\overline{09} \qquad \text{(a repeating nonterminating decimal)}$$

Are all decimals either terminating or repeating nonterminating decimals? The answer is no. We can construct a decimal that does not terminate, yet does not repeat, as follows: choose any digit and write it after the decimal point, then write a zero; repeat the digit, then write 2 zeros; repeat the digit, then write 3 zeros; repeat the digit, then write 4 zeros; repeat the digit, and so on. Using this idea we construct decimals like

0.101001000100001000001 . . .
0.202002000200002000002 . . .
0.909009000900009000009 . . .

For each of these decimals, we have no repeating cycle as we did for

$\frac{1}{3}$ and $\frac{3}{7}$. Regardless of how far we extend these decimals, there will be no repeating set of digits. These decimals have a pattern, but no repeating cycle. Decimals that are *nonterminating* and *nonrepeating* are called **irrational numbers.**

Probably the most famous irrational number is *pi* (π). The value of π for the first fifty decimal places is

$$\pi = 3.14159265358979323846264338327950288419716939937510$$

You will note that there is no repeating sequence of digits for this decimal.

The formula for finding the circumference of a circle says that to find the circumference of a circle, we multiply its diameter times *pi;* that is, $C = \pi d$. Therefore π is the ratio of the circumference of a circle to its diameter. In other words, to find the value of π, you must divide the circumference of a circle by its diameter. You will obtain a nonrepeating nonterminating decimal. Computers have been used to find the value of π well beyond the fifty decimal places shown above, but no one has ever reached the end, or ever will.

There are many other examples of irrational numbers—numbers which, when they are expressed as decimals, are nonterminating and nonrepeating. Some other examples are:

2.718281824 . . .
0.12122122212222122222 . . .
$\sqrt{2}$, $\sqrt{3}$, $\sqrt{5}$, $\sqrt{6}$, $\sqrt{7}$

A **perfect square** is a number that is the product of an integer times itself. For example, 4 is a perfect square because $4 = 2 \cdot 2$. Some other examples of perfect squares are 1, 9, 16, 25, and 36. The square root of any positive integer that is *not* a perfect square is an irrational number. Numbers such as $\sqrt{2}$, $\sqrt{3}$, $\sqrt{5}$, $\sqrt{6}$, and $\sqrt{7}$ are irrational because they are nonterminating and nonrepeating when expressed as decimals. Each is a square root of a nonnegative number that is not a perfect square.

EXAMPLE 1

Classify each number as rational or irrational.

a. $0.\overline{13}$ b. 0.131131113 . . .

c. $\sqrt{11}$ d. $\sqrt{49}$

Solution

a. $0.\overline{13}$ is a rational number. It is a nonterminating decimal, but it is repeating.

b. 0.131131113 . . . is an irrational number. It is a decimal that is nonterminating and nonrepeating.

c. $\sqrt{11}$ is an irrational number. The square root of a nonnegative number that is not a perfect square is an irrational number.

d. $\sqrt{49}$ is a rational number because 49 is a perfect square. In fact, $\sqrt{49} = 7$.

It is interesting to note that the set of rational numbers and the set of irrational numbers are disjoint sets; that is, their intersection is empty. If we take the union of these two sets, then we have all of the numbers on the number line. The union of the set of rational numbers and the set of irrational numbers yields a set of numbers that is called the set of **real numbers.** Therefore, any rational or irrational number is a real number. Note that a rational number may be expressed as either a terminating decimal or a repeating nonterminating decimal, and an irrational number may be expressed as a nonrepeating nonterminating decimal. Hence, every decimal is a real number and every real number may be expressed as a decimal.

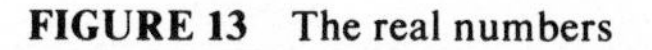

FIGURE 13 The real numbers

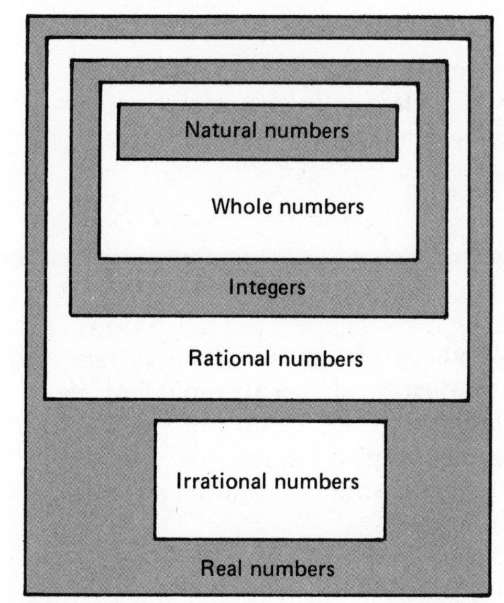

Recall that we began our discussion of the classification of numbers with the set of natural numbers. Including zero with the set of natural numbers produced the set of whole numbers. The set of integers was composed of the set of whole numbers and their additive inverses. The set of integers and the set of fractions yielded the set of rational numbers. Finally, came the set of irrational numbers, and the union of the set of rationals and the set of irrationals yielded the set of real numbers. This process is illustrated by the following diagram.

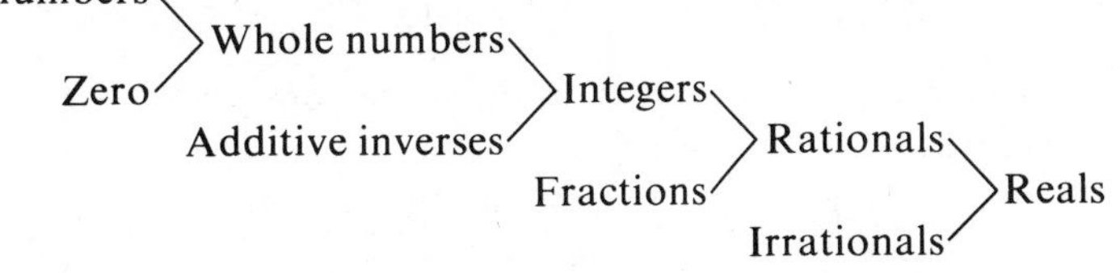

Figure 13 illustrates the relationship of these sets of numbers to the set of real numbers.

EXAMPLE 2

Let

$$R = \{\text{real numbers}\}, \qquad I = \{\text{integers}\},$$
$$Y = \{\text{rational numbers}\}, \qquad Z = \{\text{irrational numbers}\}$$

Find—

a. $R \cap I$ b. $I \cap Y$

c. $Y \cap Z$ d. $Y \cup Z$

Solution

We can find these sets by examining figure 13.

a. $R \cap I = I$ (the set of integers is common to both sets)

b. $I \cap Y = I$ (the set of integers is common to both sets)

c. $Y \cap Z = \emptyset$ (the set of rational numbers and the set of irrational numbers have no elements in common)

d. $Y \cup Z = R$ (the set of rational numbers together with the set of irrational numbers forms the set of real numbers)

EXERCISES FOR SECTION 8.7

1. Classify each number as rational or irrational.
   a. $\frac{1}{3}$ b. -2 c. $0.\overline{3}$
   d. $\sqrt{2}$ e. $\sqrt{3}$ f. $\sqrt{4}$
2. Classify each number as rational or irrational.
   a. 0 b. π
   c. $\sqrt{9}$ d. 2.343343334 . . .
   e. 1.010010001 . . . f. 3.14
3. Determine whether each of the following can be represented by a terminating decimal, a repeating decimal, or a nonterminating nonrepeating decimal.
   a. $\frac{3}{4}$ b. $\frac{2}{3}$ c. $\sqrt{2}$
   d. $\sqrt{25}$ e. $\frac{1}{7}$ f. π
4. Determine whether each of the following can be represented by a terminating decimal, a repeating decimal, or a nonterminating nonrepeating decimal.
   a. $\frac{1}{8}$ b. $\frac{3}{7}$ c. $\sqrt{3}$
   d. $\sqrt{\frac{9}{16}}$ e. $\sqrt{99}$ f. $\frac{1}{11}$
5. Determine whether each sentence is true or false.
   a. A real number is either a rational or irrational number.
   b. A real number is positive, negative, or zero.
   c. A repeating nonterminating decimal is a rational number.
   d. A nonrepeating nonterminating decimal is a rational number.

e. A terminating decimal is an irrational number.

6. Determine whether each sentence is true or false.
   a. The intersection of the set of rational numbers and the irrational numbers is not empty.
   b. The union of the set of rational numbers and the set of irrational numbers is the set of real numbers.
   c. All rational numbers are real numbers.
   d. All real numbers are rational numbers.
   e. Every real number can be expressed as a terminating decimal or a repeating decimal.

7. Let R = {real numbers}, I = {integers}, Y = {rational numbers}, and Z = {irrational numbers}. Determine whether each of the following is true or false.
   a. $R \cap Y = I$
   b. $I \cup Y = Z$
   c. $Y \subset R$
   d. $R \subset Z$
   e. $Y \cap Z = R$
   f. $Y \cup Z = R$

Just For Fun

If *pi* is an irrational number (a number that is a nonrepeating nonterminating decimal), then why is it that we are usually *told* that pi (π) is $\frac{22}{7}$, when $\frac{22}{7} = 3.\overline{142857}$?

8.8 SUMMARY

Numbers, like many other things, can be classified in a variety of ways. The first set of numbers that we discussed in this chapter was the set of natural numbers, that is, the set of numbers $\{1, 2, 3, 4, \ldots\}$. Every natural number is divisible by itself and 1. Some natural numbers are divisible only by themselves and 1. These natural numbers are called *prime numbers.* A *prime number* is any natural number greater than 1 that is divisible only by itself and 1. The numbers 2, 3, 5, and 7 are examples of prime numbers. If a natural number is not prime, then it is called a *composite number.* Disregarding order, every composite natural number can be expressed as a product of prime factors in one and only one way.

The *greatest common divisor* (G.C.D.) of a pair of natural numbers is the greatest natural number that divides both of the given numbers. Two numbers whose G.C.D. is 1 are said to be relatively prime. The *least common multiple* (L.C.M.) of two natural numbers is the least natural number that is a multiple of each of the given numbers. In other words, the L.C.M. is the least number that is divisible by both of the given numbers.

Expanding the set of natural numbers to include zero gives us the set of numbers $\{0, 1, 2, 3, 4, \ldots\}$. This is called the set of *whole numbers.* The set of numbers consisting of the set of whole numbers and their additive inverses is called the set of *integers*; that is, the set of integers is the set $\{\ldots, -3, -2, -1, 0, 1, 2, 3, \ldots\}$. It should be noted that the set of natural numbers and the set of whole numbers are subsets of the integers. We can use the number line to determine the order of the integers. For example, $-1 > -3$ since -1 is to the right of -3 on the number line.

In combining integers, we should remember the following:

1. If we add two positive integers, the sum is a positive integer.
2. If we add two negative integers, the sum is a negative integer.
3. If we add a positive integer and a negative integer, the sum may be positive or negative.

The product of two positive integers is positive, as is the product of two negative integers. The product of a positive integer and a negative integer is negative. An integer is *even* if it can be expressed in the form $2 \cdot n$, where n represents an integer. An integer is *odd* if it can be expressed in the form $2 \cdot n + 1$.

A *rational number* is a number that can be expressed as a quotient of integers, that is, in the form $\frac{a}{b}$, where a and b are integers and $b \neq 0$. Rational numbers can also be classified as either terminating decimals or repeating nonterminating decimals. Decimals that are nonterminating and nonrepeating are called *irrational numbers.* Some examples of irrational numbers are π, $\sqrt{2}$, $\sqrt{3}$, and 0.1010010001... If we take the union of the set of rational numbers and the set of irrational numbers, then we have all of the numbers that belong on the number line. This set of numbers is called the set of *real numbers.*

Review Exercises for Chapter 8

1. Determine whether each number is prime or composite.
   a. 99 b. 97 c. 83
   d. 431 e. 657 f. 10,101

2. Determine the prime factors of each natural number.
   a. 78 b. 111 c. 475
   d. 147 e. 903 f. 1111

3. Find the greatest common divisor for each pair of numbers.
   a. 30 and 48 b. 42 and 55
   c. 48 and 72 d. 66 and 90
   e. 111 and 231 f. 342 and 612

4. Find the least common multiple for each pair of numbers.
   a. 30 and 48 b. 42 and 55
   c. 48 and 72 d. 66 and 90
   e. 111 and 231 f. 342 and 612

5. Evaluate each of the following:
   a. $3 + (-7)$ b. $4 + (-5)$
   c. $-8 + 7$ d. $-7 - (-3)$
   e. $13 - (-3)$ f. $-8 - 5$

6. Evaluate each of the following:
   a. $(-3) \times (-2)$
   b. $(-4) \times 4$
   c. $(-2) \times (-3) \times (-4)$
   d. $(-2 + 5) \times (-3)$
   e. $(1 - 5) \times 2$
   f. $(-2 - 6) \times (-2 - 3)$

7. Perform the indicated operation.
   a. $\frac{1}{3} + \frac{2}{5}$ b. $\frac{3}{7} + \frac{4}{9}$ c. $\frac{2}{11} + \frac{1}{13}$
   d. $\frac{4}{7} - \frac{1}{3}$ e. $\frac{3}{5} - \frac{1}{2}$ f. $\frac{7}{8} - \frac{2}{5}$

8. Perform the indicated operation.
   a. $\frac{1}{3} \times \frac{2}{5}$ b. $\frac{3}{7} \times \frac{4}{9}$ c. $\frac{2}{11} \times \frac{3}{5}$
   d. $\frac{2}{3} \div \frac{4}{7}$ e. $\frac{3}{5} \div \frac{1}{2}$ f. $\frac{6}{11} \div \frac{3}{5}$

9. Express each of the following as a decimal.
   a. $\frac{7}{8}$ b. $\frac{7}{16}$ c. $\frac{2}{11}$
   d. $\frac{13}{37}$ e. $\frac{5}{13}$ f. $\frac{3}{7}$

10. Express each of the following as a quotient of integers.
    a. 0.75 b. 0.213 c. 3.14
    d. $0.\overline{46}$ e. $2.\overline{49}$ f. $4.\overline{123}$

11. Which of the following numbers are rational and which are irrational
    a. 2.1 b. $2.\overline{1}$
    c. $\sqrt{5}$ d. 3.141141114 . . .
    e. 2.121121112 . . . f. 3.1415926

12. Determine whether each statement is true or false.
    a. Every real number is an irrational number.
    b. Every real number is a rational number.
    c. Every real number is either a rational number or an irrational number.
    d. Every rational number is a real number.
    e. Every irrational number is a real number.
    f. The union of the sets of rational and irrational numbers is the set of real numbers.

13. Answer yes or no to tell whether each of the following numbers is—

    I. a natural number

    II. a whole number

    III. an integer

    IV. a rational number

    V. an irrational number

    VI. a real number.

    a. 6 b. -1 c. 0
    d. 2.89 e. $1.\overline{34}$ f. π
    g. $\sqrt{16}$ h. $\sqrt{5}$ i. $\frac{2}{3}$

Just For Fun

The value of π has been computed to hundreds of decimal places by modern computers. The value of π to seven decimal places is

$$\pi = 3.1415926\ldots$$

The following diagram shows a trick for remembering these digits:

3	1	4	1	5	9	2	6
May	I	have	a	large	container	of	coffee?

How does this trick work?

9 AN INTRODUCTION TO ALGEBRA

After studying this chapter, you will be able to do the following:

1. Find the solution sets of some simple **open sentences**
2. **Graph** the solution sets of inequalities in one variable on the number line
3. Solve equations in one variable
4. Translate word problems into mathematical equations, and solve verbal problems involving one variable
5. Find at least three solutions for a **linear equation** in two unknowns, and graph a linear equation in two unknowns on the **Cartesian plane**
6. Graph a **parabola** of the form $y = ax^2 + bx + c$ on the Cartesian plane
7. Graph an inequality in two variables on the Cartesian plane
8. Solve a **linear-programming problem** by translating the word problem into mathematical inequalities and equations, and use graphing and corner-point evaluation to determine the values of the variables that will provide the desired maximum or minimum value for a given expression.

9.1 INTRODUCTION

What is algebra? Basically, algebra is the area of mathematics that generalizes the facts of arithmetic. In arithmetic we encounter expressions such as $3 + 2 = 5$, while in algebra we encounter an expression such as $x + 2 = y$. In algebra, letters are used to denote numbers or a certain set of numbers. An arithmetic expression is one such as $4 + 2$, while an algebraic expression is one such as $4x + 2$, or $3x + 2y$. The operations in algebra are similar to those in arithmetic, that is, they are addition, subtraction, multiplication, and division.

It appears that algebra had its basic beginnings in Egypt in approximately 1700 B.C. The Greeks did some work with algebra, as they used it to solve equations; but it was René Descartes (1596–

Le Jeune/Stockmarket

1650) who introduced the concept of using a letter of the alphabet to represent the unknown quantity in an equation.

Letters used in equations are called **variables.** They are symbols that represent an unknown member of a set. The symbols x and y are variables in the expression $x + y = y + x$. In this chapter, we shall examine the basic properties of algebra and its relationship to some of the topics covered in previous chapters.

9.2 OPEN SENTENCES AND THEIR GRAPHS

You may recall that in chapter 2 we encountered statements such as the following:

February has 30 days.
$3 + 2 = 1$
Gerald Ford was president of the United States.
$5 + 4 + 3 + 2 + 1 = 14$

Each of these statements is either true or false (but not both). We can tell whether one of these statements is true or false upon reading the statement.

Consider the sentence

$$x + 1 = 3$$

Is this sentence true or false? We cannot answer that question yet. The sentence is neither true nor false until we replace x with a number. Sentences like $x + 1 = 3$, which cannot be classified as true or false until we replace x by a number, are called **open sentences.** Note that any real number can be substituted for x in the open sentence $x + 1 = 3$. The sentence will be true or false depending on the value we substitute for x. Since x represents an unknown number, we can call x a variable.

If we replace x by 4 in the open sentence $x + 1 = 3$, we have $4 + 1 = 3$, which is a false statement. If we replace x by 2, we have $2 + 1 = 3$, which is a true statement. Therefore, a solution to the open sentence $x + 1 = 3$ is 2. Since it can be shown that 2 is the only solution to $x + 1 = 3$, we call 2 the *solution set* for the open sentence. The **solution set** of an open sentence is the set of numbers that make the sentence true when they are substituted for x.

If we replace the equal sign in the open sentence $x + 1 = 3$ with the symbol $>$, we have

$$x + 1 > 3$$

an open sentence of *inequality*. The sentence $x + 1 > 3$ is read as "$x + 1$ is greater than 3." Note that if we replace x by 2 in this statement, we have $2 + 1 > 3$, which is not a true statement. Suppose we restrict our replacements for x to the set of integers. What will the solution set be? The integers 3, 4, 5, and so on will make the sentence true. Therefore the solution set is $\{3, 4, 5, \ldots\}$.

Some other examples of open sentences of inequality are:

$x > 2$	x is greater than 2
$x < 2$	x is less than 2
$x \geq 2$	x is greater than or equal to 2
$x \leq 2$	x is less than or equal to 2

It is important to realize that an open sentence may have different solution sets, depending on the restrictions placed on the set of permissible replacements, the **replacement set.** For instance, the equation $x + 1 = 0$ has no solution set (the solution is the empty set, $\emptyset$) if we restrict the set of permissible replacements to the set of natural numbers. But if the replacement set is the set of integers, then the solution set for $x + 1 = 0$ is $\{-1\}$.

Consider the following sentence:

$$x + 1 < 5$$

If the replacement set is the set of natural numbers, then the solution set is $\{1, 2, 3\}$. If the replacement set is the set of whole numbers, then the solution set is $\{0, 1, 2, 3\}$. If the replacement set is the set of integers, then the replacement set is $\{\ldots, -2, -1, 0, 1, 2, 3\}$. If the replacement set is the set of real numbers, then the solution set is {all real numbers less than 4}.

EXAMPLE 1

Find the solution set for $x + 5 = 9$, where x is any real number.

Solution

There is only one real number that will make this sentence true. Since $4 + 5 = 9$, the solution set is $\{4\}$.

EXAMPLE 2

Find the solution set for $x + 2 \geq 7$, where x is an integer.

Solution

The sentence is read as "$x + 2$ is greater than or equal to 7," and the integers that satisfy this sentence are 5, 6, 7, and so on. Hence, the solution set is $\{5, 6, 7, \ldots\}$. Note that 5 is a member of this set because the sentence reads "greater than *or* equal to."

EXAMPLE 3
Find the solution set for $x + 2 < 3$, where x is a natural number.

Solution
We want the sum $(x + 2)$ of two natural numbers to be less than 3. There are no natural numbers that we may substitute for x to make this statement true; hence the solution set is the empty set, ϕ.

EXAMPLE 4
Find the solution set of $x + 2 < 3$, where x is an integer.

Solution
Now we are considering a different replacement set than in example 3. There are integers that can be added to 2 to yield a sum less than 3, namely, 0, -1, -2, and so on. Therefore, the solution set is $\{\ldots, -2, -1, 0\}$.

Once we find the solution set for an open sentence, we can make a "picture" of the solution set. That is, we can **graph** the solution set. In chapter 8 we introduced the concept of a *number line*. First, we drew a horizontal line, picked a point on the line, and labeled it zero. Next, we marked off equal units to the right and left of zero. We labeled the end points of the intervals to the right of zero with the positive integers, and those to the left of zero with the negative integers. The number line can be extended indefinitely in either direction. See figure 1.

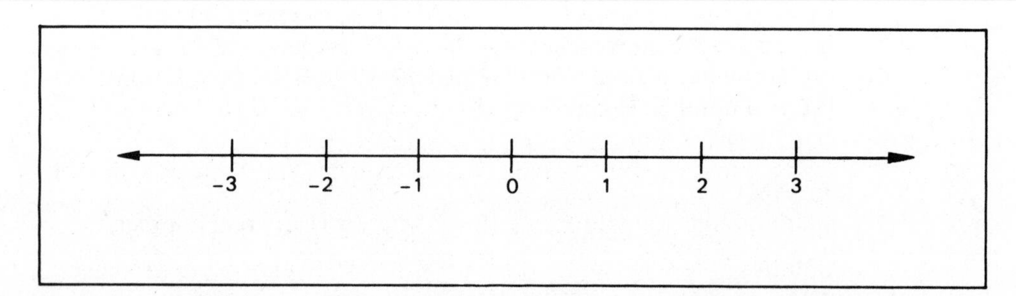

FIGURE 1

Recall that each integer corresponds to a particular point on the number line. Also, each point on the number line corresponds to a real number. The set of real numbers is composed of the set of rational numbers and the set of irrational numbers. It is not physically possible to label every point on the number line, but we do know that there exists a point $\frac{3}{4}$ unit from 0, a point -1.5 units from 0, and points $\sqrt{3}$ units and π units from 0. There is a point on the number line that corresponds to every rational or irrational number, and there is a rational or irrational number that corresponds to every point on the number line. Hence we can use the number line to represent the set of real numbers.

How can we picture these numbers on the real number line? We can represent individual numbers on the number line by marking a heavy solid dot on the line at the point that corresponds to that number. For example, figure 2 is a graph of the number 1.

FIGURE 2

FIGURE 3

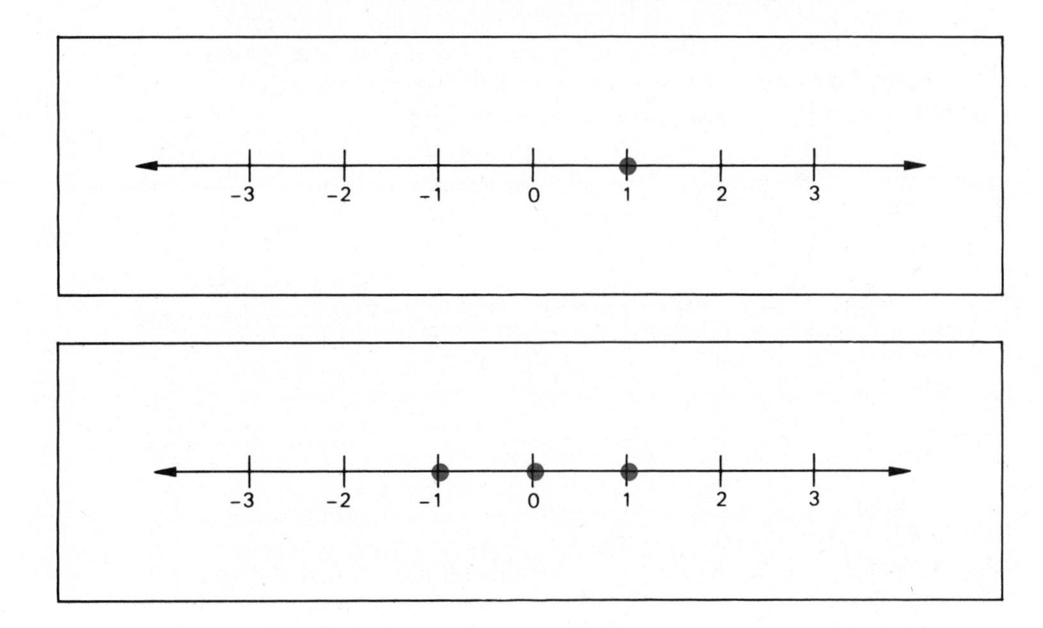

Figure 3 shows the graph of the set of integers $\{-1, 0, 1\}$.

Note that we can also call figure 3 the graph of the integers between -2 and 2. Figure 3 is the correct graph because "between -2 and 2" means that we do not include the integers -2 and 2.

Suppose we want a graph that represents all of the real numbers between -2 and 2. This graph would have to include all of the integers, rational numbers, and irrational numbers between -2 and 2. We indicate this by using hollow dots at -2 and 2 and drawing a solid bar between the hollow dots as shown in figure 4.

FIGURE 4

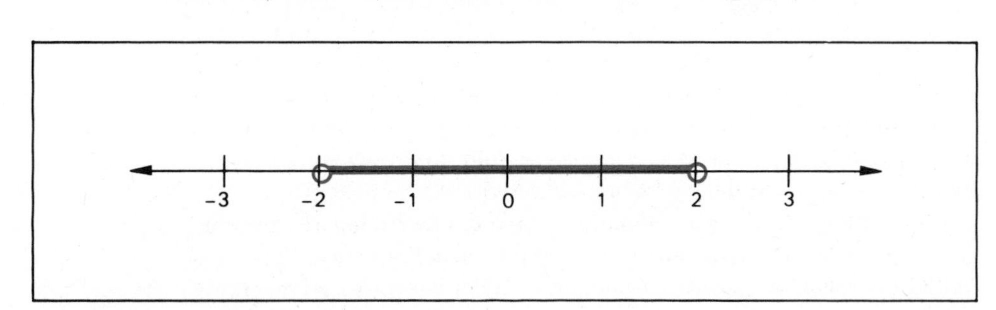

The fact that the dots at points -2 and 2 are not solid indicates that these points are not included in our solution set. Speaking in terms of algebra and the set of real numbers, figure 4 is a graph of all real numbers x such that $-2 < x$ and $x < 2$. We can shorten this expression to $-2 < x < 2$.

Now consider the algebraic sentence $-2 \leq x \leq 2$, read "-2 is less than or equal to x, and x is less than or equal to 2." If we are to represent the solution set of this sentence on the number line, where x is any real number, then we must include the endpoints. Consequently, they would be colored in, as shown in figure 5.

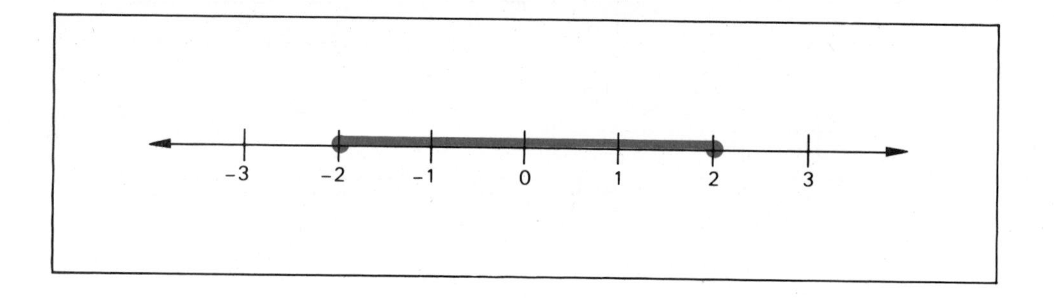

FIGURE 5

Next, let's consider the open sentence $x + 2 < 3$, where x is a real number. This statement is true for any real number that is less than 1, that is, $x < 1$. We can represent this on the number line as shown in figure 6.

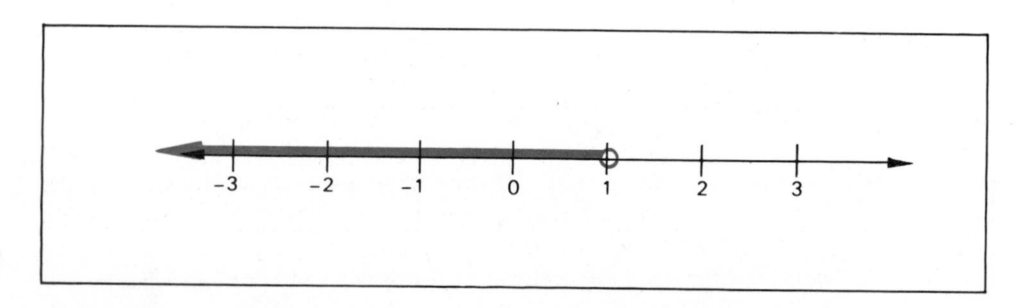

FIGURE 6

The solid arrow indicates that the solution set extends to the left indefinitely. Note that we have $x < 1$ (x is less than 1) and therefore we do not include 1 in the solution. If we had the statement $x \leq 1$ (x is less than or equal to 1), then we would have included 1 in the solution, and 1 would have been marked with a solid dot.

The graph in figure 6 is sometimes called a **half line.** If the point at 1 were included, then the graph would be called a **ray.** Figure 4 depicts an **open line segment,** while figure 5 depicts a **line segment.**

EXAMPLE 5
Graph the solution of $x - 2 > 1$, where x is a real number.

Solution
In order for the open sentence $x - 2 > 1$ to be true, x must be a number greater than 3, that is, $x > 3$. The solution is the half line shown in figure 7.

FIGURE 7

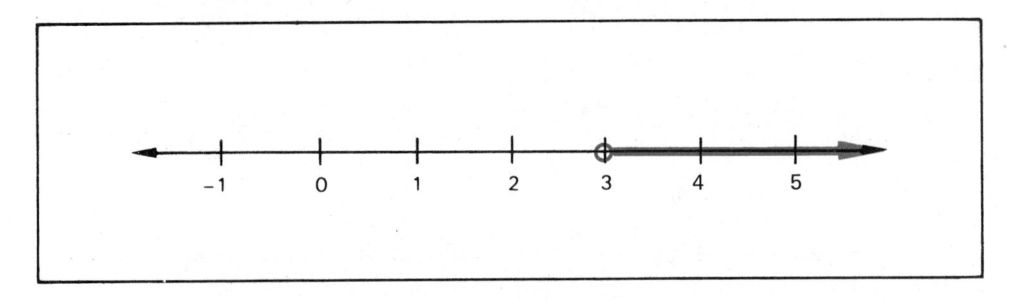

EXAMPLE 6
Graph the solution of $x + 2 \leq 1$, where x is a real number.

Solution
In order for the open sentence $x + 2 \leq 1$ to be true, x must be a number less than or equal to negative 1, that is, $x \leq -1$. The solution is the ray shown in figure 8.

FIGURE 8

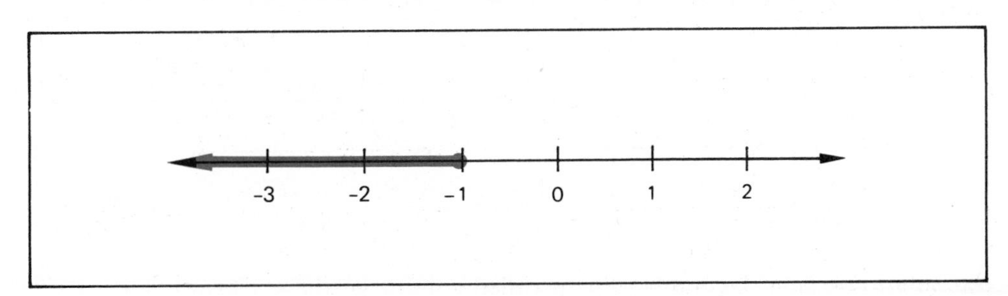

EXAMPLE 7
Graph the solution of $x + 2 > x$, where x is a real number.

Solution
Note that there are no restrictions on x in order for the open sentence $x + 2 > x$ to be true. Regardless of the number with which we replace x, we have a true statement. For example, $1 + 2 > 1$, $0 + 2 > 0$, and $-2 + 2 > -2$. Therefore, the solution set is the set of all real numbers. The graph of the set of all real numbers is a **line,** as shown in figure 9.

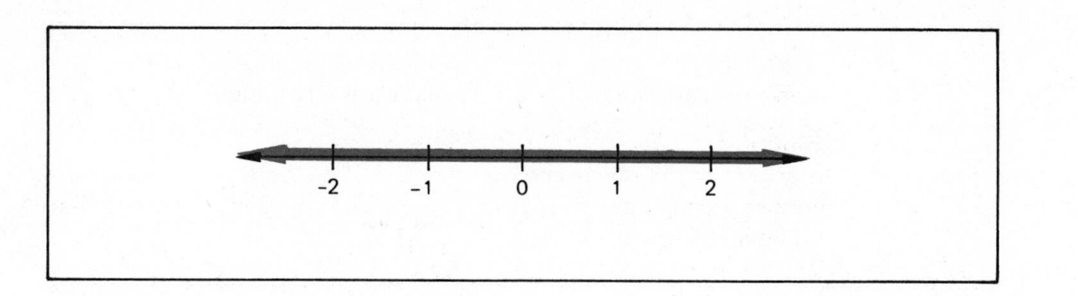

FIGURE 9

EXAMPLE 8

Graph the solution of the open sentence $-1 < x < 3$, where x is a real number.

Solution

The sentence $-1 < x < 3$ is read as "-1 is less than x and x is less than 3." The solution set for this sentence is the set of real numbers between -1 and 3. The graph of the solution set is the open line segment shown in figure 10.

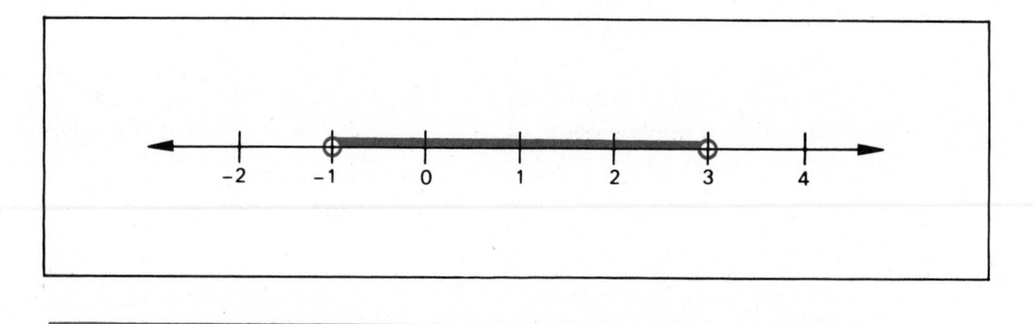

FIGURE 10

In this section we discussed open sentences, that is, sentences that are neither true nor false. A solution of an open sentence is a member of the solution set for the open sentence, and a "picture" of the solution set on the number line is called a graph of the solution set.

EXERCISES FOR SECTION 9.2

For exercises 1–16, find the solution set for each open sentence. Unless otherwise noted, the replacement set is the set of real numbers.

1. $x + 2 = 5$
2. $x + 1 = 5$
3. $x + 1 = 0$
4. $x + 4 = 5$
5. $x + 4 = 5 - 2$ (x is a whole number)
6. $x - 3 = 4 - 1$ (x is a whole number)
7. $x + 2 < 7$ (x is a natural number)
8. $x + 3 \leq 5$ (x is a natural number)

9. $x + 4 > 5$ 10. $x - 3 > -5$

11. $0 < x < 3$ (x is an integer)

12. $-1 \leq x \leq 3$ (x is an integer)

13. $0 \leq x < 5$ (x is an integer)

14. $0 < x \leq 5$ (x is an integer)

15. $-2 < x \leq 3$ (x is an integer)

16. $-2 \leq x \leq 2$ (x is a whole number)

For exercises 17–28, graph the solution set for each open sentence on the number line. In each case, the replacement set is the set of real numbers.

17. $x + 3 = 4$ 18. $x - 1 = 0$

19. $x < 4$ 20. $x \geq 2$

21. $x + 2 \geq 0$ 22. $x - 2 > 1$

23. $2 < x < 4$ 24. $-1 \leq x \leq 3$

25. $-2 < x \leq 1$ 26. $-1 \leq x < 1$

27. $x + 1 > x$ 28. $x + 1 < x$

Just For Fun

Can you name 100 different words that do not contain the letters *A*, *B*, *C*, or *Q*? (*Hint:* What is one of the first things you do with numbers?) Time limit: 3 minutes.

9.3 ALGEBRAIC NOTATION

In the preceding section, we examined open sentences such as $x + 3 = 4$ and $x - 2 > 1$. The open sentences that we have encountered up to now have contained expressions involving only x, not $2x$, $3x$, and so on. You should be aware that if we wish to express the product of a number such as 2 and the variable y, we can do this by writing any of the following:

$$(2)(y), \quad (2)y, \quad 2(y), \quad 2 \cdot y, \quad \text{or} \quad 2y$$

Note that we avoid the use of $2 \times y$, since the symbol "×" is easily confused with x, which is reserved for use as a variable. In algebra, the most common method of expressing the product 2 and y is to write $2y$. Similarly, $5x$ is a method of expressing "5 times x," and $7z$ means "7 times z." But the product of 2 and 3 must *not* be written as 23. We use the above method for expressing a product only when we are working with variables.

If we have an expression such as $2y + 3y$, we can simplify this expression by writing $2y + 3y = 5y$. This can be illustrated in the following manner:

$$2y + 3y = (2 + 3)y = 5y$$

In the second step, the y has been factored out of $2y + 3y$ using the distributive property; we then add $(2 + 3)$ and obtain 5. We can also think of this as combining two like quantities: if we have 2 yaks plus 3 yaks, we will obtain 5 yaks; that is, 2 yaks + 3 yaks = 5 yaks.

An expression such as $5z - 2z$ can also be simplified:

$$5z - 2z = (5 - 2)z = 3z$$

We can also think of this as 5 zebras minus 2 zebras, resulting in 3 zebras; that is, 5 zebras − 2 zebras = 3 zebras.

It should be noted that when we wish to write $1x$, we usually do this by writing x by itself. Similarly, $1y$ is written as y. Therefore, an expression such as

$$3x + x$$

is the indicated sum of $3x$ and $1x$, so that $3x + x = 4x$. Also, $4y - y = 3y$ and $x + x = 2x$.

If we have $3x$ and we wish to double the amount, we can multiply $3x$ by 2, that is,

$$2(3x) = 3x + 3x = 6x$$

The following are some other examples of indicated multiplications and the resulting products.

$$3(4x) = 12x, \qquad 2(6y) = 12y, \qquad 4(2z) = 8z$$

$$9(z) = 9(1z) = 9z, \qquad \tfrac{1}{2}(4y) = 2y$$

The last multiplication example also illustrates another operation. If we multiply a quantity by $\frac{1}{2}$, then this is the same as dividing the quantity by 2. In each of the examples above, we are only working with the numbers, not the variables. For example, $2x + 3x = 5x$, $3y - y = 2y$, and $3(2z) = 6z$. The same holds true for the operation of division. If we have $9x$ and divide that quantity by a number such as 3, the quotient is $3x$:

$$9x \div 3 = \frac{9x}{3} = 3x$$

The following are some other examples of indicated divisions and the resulting quotients:

$$\frac{5y}{5} = y, \quad \frac{10z}{5} = 2z, \quad \frac{6x}{2} = 3x$$

$$\frac{3y}{1} = 3y, \quad \frac{8x}{2} = 4x$$

EXERCISES FOR SECTION 9.3

Simplify each of the following algebraic expressions.

1. $2x + 3x$
2. $4y + 5y$
3. $2x + x$
4. $4x - 2x$
5. $3y - y$
6. $5z - 3z$
7. $2(4x)$
8. $3(2z)$
9. $2(7y)$
10. $8z \div 4$
11. $10x \div 2$
12. $2y \div 1$
13. $\frac{4x}{2}$
14. $\frac{16z}{4}$
15. $\frac{15x}{3}$
16. $x + 3x + 4$
17. $2y + 2 + 3y$
18. $4z - 2z + 3$
19. $8y + 2y + 4$
20. $8x - x + 3$
21. $2z - z + 2$
22. $2x + 3 - x - 2$
23. $4y - 2 - y + 3$
24. $6z + 2 - z - 3$
25. $4(3x) + 2$
26. $6(2x + 3)$
27. $3(x - 2)$
28. $2(4y - 3) + 2y$
29. $5(2z) + 3 - z$
30. $y + 3(4 - y)$
31. $0(3x - 4) + 1$
32. $3(4y) - 4(2y)$
33. $2(6z) - 3(4z)$

Just For Fun.

How many triangles are there in figure 11?

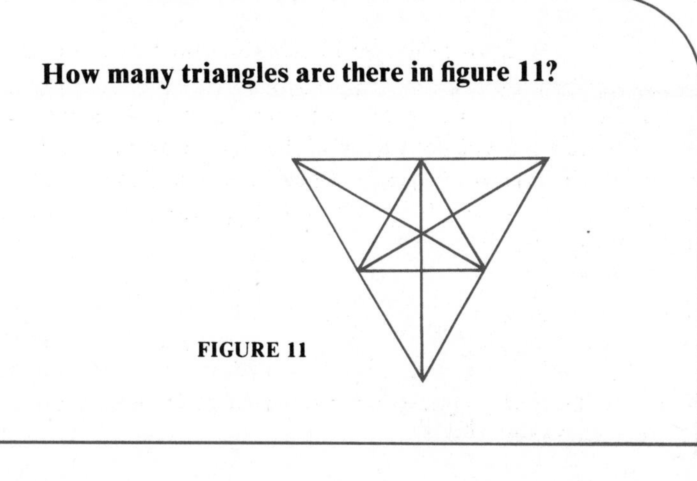

FIGURE 11

9.4 MORE OPEN SENTENCES

Thus far in our discussion of open sentences, we have only considered sentences such as $x - 2 > 1$ and $x + 2 = 3$. It is not too difficult to find replacements for x such that the given sentences will be true, as we can usually do this by observation or by trial and error. That is, we can replace x by a number and see if the resulting statement is true. But how about a sentence such as

$$4x - 2 = x + 7?$$

This open sentence is an equation since it is a statement of equality. We could probably find the solution to the equation $4x - 2 = x + 7$ by trial and error, but there does exist a more efficient method. By using certain techniques, we can systematically solve many equations in one unknown.

Before we generate specific techniques for solving equations, let's make sure that we understand the following axioms, or rules, that we will use in solving equations:

Axiom 1

If the same quantity or two equal quantities are added to two equal quantities, then the sums are equal.

As an example of axiom 1, we have:

$x - 1 = 2$	given
$+ 1 = + 1$	adding 1 to both sides
$x - 1 + 1 = 2 + 1$	sums are equal
$x = 3$	

Axiom 2

If the same quantity or two equal quantities are subtracted from two equal quantities, then the remainders (differences) are equal.

The following example illustrates axiom 2.

$x + 4 = 5$	given
$- 4 = - 4$	subtracting 4 from both sides
$x + 4 - 4 = 5 - 4$	remainders are equal
$x = 1$	

Axiom 3

If equal quantities are multiplied by equal quantities, then the products are equal.

As an example of axiom 3, we have:

$\frac{x}{2} = 4$	given
$\times \quad 2 = 2$	multiplying both sides by 2
$\frac{x}{2} \cdot 2 = 4 \cdot 2$	products are equal
$x = 8$	

Axiom 4

If equal quantities are divided by equal quantities (other than zero), then the quotients are equal.

To illustrate axiom 4, we have:

$3x = 9$	given
$\div \quad 3 = 3$	dividing both sides by 3
$\frac{3x}{3} = \frac{9}{3}$	quotients are equal
$x = 3$	

One thing to keep in mind in solving any equation is that an equation is like a balance scale, and in order to keep the equation in balance, whatever we do to one side of the equation, we must do to the other side of the equation. Our ultimate goal is to get x by itself on one side of the equal sign. It does not matter whether we wind up with x on the left side or the right side of the equal sign, as long as it is by itself.

Now let's solve the equation $4x - 2 = x + 7$ using the axioms whenever necessary. We must get x by itself on one side of the equation. We will first eliminate the -2 from the left side. We can do this by adding 2 to the left side; but if we add 2 to the left side, then, by axiom 1, we must also add 2 to the right side. Hence we have

$$4x - 2 + 2 = x + 7 + 2$$

which simplifies to

$$4x = x + 9$$

Remember that we must get all of the x's on one side of the equal sign. We can eliminate x from the right side, $x + 9$, by subtracting x from it. But if we subtract x from the right side, we must subtract x from the left side in order to maintain the balance (axiom 2). Therefore we have

$$4x - x = x - x + 9$$

or

$$3x = 9$$

Since we want to solve for x, we must divide the expression $3x$ (3 times x) by 3. Therefore we divide both sides by 3 (axiom 4).

$$\frac{3x}{3} = \frac{9}{3}$$
$$x = 3$$

We can check our answer by replacing x by 3 in the original equation to see if the left side of the equation equals the right side. Replacing x by 3, we have

$$4(3) - 2 \stackrel{?}{=} 3 + 7$$
$$12 - 2 \stackrel{?}{=} 10$$
$$10 = 10$$

The solution checks.

Below is the solution of the equation $4x - 2 = x + 7$ without the discussion. We list the reasons for each step of the solution.

$4x - 2 = x + 7$	given
$4x - 2 + 2 = x + 7 + 2$	axiom 1: adding 2 to both sides
$4x = x + 9$	combining like terms
$4x - x = x - x + 9$	axiom 2: subtracting x from both sides
$3x = 9$	combining like terms
$\frac{3x}{3} = \frac{9}{3}$	axiom 4: dividing both sides by 3
$x = 3$	simplifying

EXAMPLE 1

Solve $4x - 7 = x + 5$ for x.

Solution

$4x - 7 = x + 5$	given
$4x - 7 + 7 = x + 5 + 7$	axiom 1: adding 7 to both sides
$4x = x + 12$	combining like terms
$4x - x = x - x + 12$	axiom 2: subtracting x from both sides
$3x = 12$	combining like terms
$\frac{3x}{3} = \frac{12}{3}$	axiom 4: dividing both sides by 3
$x = 4$	simplifying

EXAMPLE 2

Solve $2y + 6 = 6y - 10$ for y.

Solution

$2y + 6 = 6y - 10$	given
$2y + 6 + 10 = 6y - 10 + 10$	axiom 1: adding 10 to both sides
$2y + 16 = 6y$	combining like terms
$2y - 2y + 16 = 6y - 2y$	axiom 2: subtracting $2y$ from both sides
$16 = 4y$	combining like terms
$\frac{16}{4} = \frac{4y}{4}$	axiom 4: dividing both sides by 4
$4 = y$	simplifying

EXAMPLE 3

Solve $\frac{3z}{4} + 2 = 8$ for z.

Solution

$\frac{3z}{4} + 2 = 8$	given
$\frac{3z}{4} + 2 - 2 = 8 - 2$	axiom 2: subtracting 2 from both sides
$\frac{3z}{4} = 6$	combining like terms
$\frac{3z}{4} \cdot 4 = 6 \cdot 4$	axiom 3: multiplying both sides by 4
$3z = 24$	simplifying
$\frac{3z}{3} = \frac{24}{3}$	axiom 4: dividing both sides by 3
$z = 8$	simplifying

EXERCISES FOR SECTION 9.4

Solve each of the following equations in the system of real numbers.

1. $x + 3 = 5$
2. $x - 4 = 2$
3. $6 = x - 2$
4. $4 = x + 1$
5. $2y = 12$
6. $3z = 12$
7. $2y - 2 = 10$
8. $3x + 4 = 16$
9. $2x = x + 5$
10. $z - 4 = 2z$
11. $3y = 2y + 4$
12. $3y - 4 = 4y$
13. $2x + 2 = x + 3$
14. $3y - 3 = 2y + 3$
15. $4x + 3 = 2x + 9$
16. $5x - 3 = 2x + 3$
17. $3y - 3 = y + 3$
18. $2y + 6 = 5y - 3$
19. $\frac{x}{2} = 5$
20. $\frac{y}{3} = 12$
21. $\frac{x}{4} + 1 = 15$
22. $\frac{z}{2} - 1 = 5$
23. $\frac{x}{2} = x - 1$
24. $\frac{y}{3} + y = 12$
25. $\frac{z}{4} + z = 5$
26. $\frac{x}{3} = 2 - x$
27. $\frac{x}{2} - 1 = x - 3$
28. $\frac{z}{5} + 2 = 1 - z$
29. $\frac{x}{2} + 3 = x + 2$
30. $\frac{y}{3} - 1 = 2y + 1$

Just For Fun

You are given a pizza pie and you have to share it with your friends. The problem is that you have to cut the pizza into eight equal pieces, but you can only make three cuts. How do you do it?

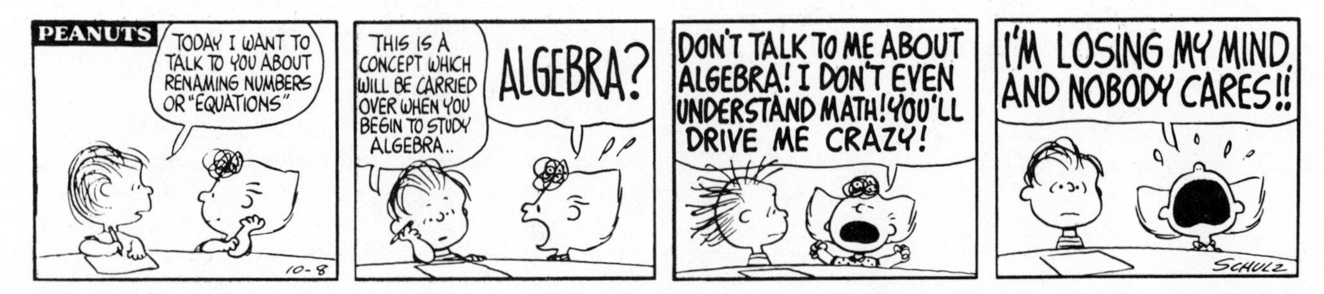

© 1965 United Feature Syndicate, Inc.

9.5 PROBLEM SOLVING

One of the oldest applications of algebra is solving word problems. As early as 2000 B.C., the Egyptians worked on word problems. You can probably solve some word problems by means of simple arithmetic, but most word problems require the use of algebra in order to find the solution in a systematic manner, as opposed to trial and error.

Consider the following sentence:

What number when decreased by 5 equals 15?

This sentence is an example of a word problem. What is the number? We can find the number if we can translate the sentence into an equation. In translating word problems into algebra, we must pay strict attention to what the words say. For instance, in the given sentence, the phrase "what number" is the key phrase; it tells us that we are looking for a certain number. Therefore we let x equal the number. (We could have just as easily let some other variable such as n or y equal the number.) Next we have the phrase "decreased by 5." What does this mean? Exactly what it says; that is, "decreased by 5" is the same as "minus 5," or "-5." The last part of the sentence, "equals 15," can be expressed as "$=15$." Therefore, the equation for the sentence

What number when decreased by 5 equals 15?

is

$$x - 5 = 15$$

Now we solve the equation:

$$\begin{aligned} x - 5 &= 15 \\ x - 5 + 5 &= 15 + 5 \\ x &= 20 \end{aligned}$$

As a check, observe that $20 - 5 = 15$.

Whenever we attempt to solve a word problem, it is necessary to translate the given sentence into an equation. The first thing we must do is let the unknown be represented by a variable such as x, y, z, m, or n. Next we must translate the given relationship with the unknown into an equation. The next step is to solve the equation in order to find the value of the unknown. The following are some examples of translating word phrases into equations.

EXAMPLE 1

Write an equation to illustrate "What number increased by 2 is 5?"

Solution

Let x equal the unknown. The phrase "increased by 2" means "plus 2," or "+2." Another word for "is" is "equals." Therefore, we have

$$x + 2 = 5$$

EXAMPLE 2

Write an equation to illustrate "The sum of what number and twice that number equals 9?"

Solution

Let n equal the unknown. Twice a number can be expressed as $2n$. Therefore we have

$$n + 2n = 9$$

EXAMPLE 3

Write an equation to illustrate "The product of what number and 3 is 12?"

Solution

Let y equal the unknown. The product of y and 3 can be expressed as $3y$. Therefore we have

$$3y = 12$$

EXAMPLE 4

Write an equation to illustrate "Two less than four times what number is 34?"

Solution

Let x equal the unknown. The phrase "two less" indicates that we subtract 2, but from what? We subtract 2 from "four times what number," that is, $4x$. Therefore we have

$$4x - 2 = 34$$

EXAMPLE 5

Write an equation to illustrate "One-half of what number and 5 more is 12?"

Solution

Let n equal the unknown. One-half of a number can be expressed as $\frac{1}{2}n$ or $\frac{n}{2}$. The phrase "and 5 more" indicates that we add 5 to $\frac{n}{2}$. Therefore we have

$$\frac{n}{2} + 5 = 12$$

EXAMPLE 6

A rectangular swimming pool is 6 feet longer than twice its width w. If the perimeter of the pool is 120 feet, write an equation to find its dimensions.

Solution

This is stated in a little more formal language, but we treat it the same as the other examples. We are already given that the width is w, so the length l can be expressed as $2w + 6$. Using the fact that the perimeter of a rectangle is $2w + 2l$, we express the perimeter as

$$2w + 2(2w + 6) = 120$$

EXAMPLE 7

The sum of two consecutive odd integers is 40. Write an equation for this problem.

Solution

Let x equal the first odd integer. The next (consecutive) odd integer is $x + 2$. (It cannot be $x + 1$, because if x is odd, then $x + 1$ would be even.) Therefore the sum of the two consecutive odd integers is $x + (x + 2)$. Hence we have

$$x + (x + 2) = 40$$

EXAMPLE 8

Scott is 5 years older than Steve, and the sum of their ages is 27. Write an equation for this problem

Solution

Let x equal Steve's age. Then Scott's age is $x + 5$ because he is 5 years older. Since the sum of their ages is 27, we have

$$x + x + 5 = 27$$

Remember that in order to solve a verbal problem it is necessary to do the following:

1. Represent the unknown in terms of a variable
2. Translate the given relationship with the unknown into an equation
3. Solve the equation to find the value of the unknowns.

Now that we have had some practice in translating verbal problems into algebraic expressions, let us complete the process and solve the problem. Consider the following problem:

The sum of two consecutive integers is 99. Find the value of the smaller integer.

The first thing we must do is represent the unknown in terms of a variable. Let x equal the first integer; then $x + 1$ will equal the next consecutive integer. Translating the given relationship with the unknown into an equation, we have

$$x + (x + 1) = 99$$

Next we solve the equation:

$2x + 1 = 99$	combining like terms
$2x + 1 - 1 = 99 - 1$	subtracting 1 from both sides
$2x = 98$	
$\frac{2x}{2} = \frac{98}{2}$	dividing both sides by 2
$x = 49$	(the solution)

We can check our solution to see if it is correct. If the first integer is 49, then the next consecutive integer is $x + 1$, that is, $49 + 1$, or 50. Is the sum of 49 and 50 equal to 99? Yes, $49 + 50$ does equal 99. Our answer checks, so the solution is correct.

EXAMPLE 9

Scott is 2 years older than Joe and the sum of their ages is 18. What are the ages of Scott and Joe?

Solution

Let x = Joe's age; then $x + 2$ = Scott's age. The sum of their ages is 18. Therefore we have the equation

$$x + x + 2 = 18$$

Solving the equation, we have

$2x + 2 = 18$	combining like terms
$2x + 2 - 2 = 18 - 2$	subtracting 2 from both sides
$2x = 16$	
$\frac{2x}{2} = \frac{16}{2}$	dividing both sides by 2
$x = 8$	Joe's age
$x + 2 = 8 + 2 = 10$	Scott's age.

EXAMPLE 10

Two less than 4 times what number is 34?

Solution

Let x = the number. Two less than 4 times x is 34 gives us the equation

$$4x - 2 = 34$$

Next, we solve the equation:

$4x - 2 + 2 = 34 + 2$	adding 2 to both sides
$4x = 36$	
$\frac{4x}{4} = \frac{36}{4}$	dividing both sides by 4
$x = 9$	(the solution)

EXAMPLE 11

The width of a rectangle is 2 feet less than its length. If the perimeter is 32 feet, find the dimensions of the rectangle.

Solution

Let x = the length of the rectangle; then $x - 2$ = the width of the rectangle. Since the perimeter is equal to the sum of the lengths of all the sides, we have the equation

$$x + x - 2 + x + x - 2 = 32$$

Solving the equation,

$4x - 4 = 32$	combining like terms
$4x - 4 + 4 = 32 + 4$	adding 4 to both sides
$4x = 36$	
$\frac{4x}{4} = \frac{36}{4}$	dividing both sides by 4
$x = 9$	length of the rectangle
$x - 2 = 9 - 2 = 7$	width of the rectangle

EXAMPLE 12

The sum of two angles of a triangle is 90 degrees. If one angle is 10 degrees more than three times the smaller angle, find the angles.

Solution

Let y = the smaller angle; then $3y + 10$ = the larger angle. Since the sum of the two angles is 90 degrees, we have the equation

$$y + 3y + 10 = 90$$

Solving the equation,

$4y + 10 = 90$	combining like terms
$4y + 10 - 10 = 90 - 10$	subtracting 10 from both sides
$4y = 80$	
$\frac{4y}{4} = \frac{80}{4}$	dividing both sides by 4
$y = 20$	the smaller angle
$3y + 10 = 3 \cdot 20 + 10 = 60 + 10 = 70$	the larger angle

EXAMPLE 13

Pam has 85 cents in her change purse. If there are only nickels and dimes in her change purse, and she has 13 coins altogether, how many nickels and dimes does she have?

Solution

Let x = number of dimes; then $13 - x$ = the number of nickels. Now if Pam has x dimes in her purse, then the value of these dimes in cents is $10x$. For instance, if she had 3 dimes, then the value is $10 \cdot 3$, or 30 cents. Similarly, the value of the nickels in Pam's purse is $5(13 - x)$. The total value of the coins in her purse is 85 cents, and hence we have the equation

$$10x + 5(13 - x) = 85$$

Solving the equation,

$10x + 65 - 5x = 85$	distributive property
$5x + 65 = 85$	combining like terms
$5x + 65 - 65 = 85 - 65$	subtracting 65 from both sides
$5x = 20$	
$\frac{5x}{5} = \frac{20}{5}$	dividing both sides by 5
$x = 4$	number of dimes
$13 - x = 13 - 4 = 9$	number of nickels

Check: $x = 4$, and $13 - x = 9$, so it checks that Pam has 13 coins. But we must also check the value of the coins; that is, $4 \cdot 10 = 40$ (the value of the dimes), $9 \cdot 5 = 45$ (the value of the nickels), and $40 + 45 = 85$ cents (the total value of the coins). Our solution is verified.

EXAMPLE 14

The sum of three consecutive even integers is 30. Find the integers.

Solution

Let z = the first even integer; then the next consecutive even integer is $z + 2$. The third consecutive even integer is 2 more than the second, that is, $z + 2 + 2$, or $z + 4$. Hence, we have

$$\begin{aligned} z &= \text{first even integer} \\ z + 2 &= \text{second consecutive even integer} \\ z + 4 &= \text{third consecutive even integer} \end{aligned}$$

Since the sum of these integers is 30, we have the equation

$$z + (z + 2) + (z + 4) = 30$$

Solving the equation,

$$\begin{aligned} 3z + 6 &= 30 && \text{combining like terms} \\ 3z + 6 - 6 &= 30 - 6 && \text{subtracting 6 from both sides} \\ 3z &= 24 \\ \frac{3z}{3} &= \frac{24}{3} && \text{dividing both sides by 3} \\ z &= 8 && \text{the first even integer} \\ z + 2 &= 10 && \text{the second consecutive even integer} \\ z + 4 &= 12 && \text{the third consecutive even integer} \end{aligned}$$

Check: The three consecutive even integers are 8, 10, and 12. Their sum is $8 + 10 + 12$, or 30. This satisfies the original problem and shows that the solution is correct.

In order to solve a verbal problem, perform the following steps: First, let the unknown be represented by a variable such as x or y. Next, translate the given relationships with the unknown into an equation. Finally, solve the equation in order to find the value of the unknown. You should check your answer to make sure that your solution is correct. This is done by checking to see if the answer satisfies the original problem (not the equation).

EXERCISES FOR SECTION 9.5

Solve each of the following problems. Only an algebraic solution will be accepted.

1. Three more than a certain number is 10. Find the number.
2. Four less than a certain number is 10. Find the number.
3. The sum of a certain number and 5 is 12. Find the number.
4. Two more than twice a certain number is 6. Find the number.
5. The sum of two consecutive integers is 15. Find the numbers.
6. Mary is 5 years older than Tom and the sum of their ages is 51. How old is each?
7. Julia is 7 years younger than Lewis and the sum of their ages is 53. How old is each?
8. Seven less than 4 times a number is 41. Find the number.
9. The sum of 5 times a number and 8 is 63. Find the number.
10. The sum of one-half of a number and 5 is 12. Find the number.
11. The width of a rectangle is 3 feet less than its length. If the perimeter of the rectangle is 50 feet, find the dimensions of the rectangle.
12. The length of a rectangle is 4 metres longer than its width. If the perimeter of the rectangle is 60 metres, find the dimensions of the rectangle.
13. The length of a rectangle is 2 metres longer than twice its width. If the perimeter of the rectangle is 100 metres, find the dimensions of the rectangle.
14. The sum of two consecutive integers is 101. Find the numbers.
15. The sum of two consecutive odd integers is 40. Find the numbers.
16. The sum of two consecutive even integers is 38. Find the numbers.
17. The sum of three consecutive integers is 93. Find the numbers.
18. The sum of three consecutive odd integers is 123. Find the numbers.
19. David has $2.25 in dimes and quarters in his pocket. If he has twice as many dimes as quarters, how many of each type of coin does he have?
20. Daniel, a newspaper carrier, has $2.90 in nickels, dimes, and quarters. If he has 3 more nickels than dimes and twice as many dimes as quarters, how many of each type of coin does he have?
21. In a collection of 60 coins the number of quarters is one-third the number of dimes and the number of nickels is 10 less than twice the number of dimes. How many of each type of coin is in the collection?
22. A rectangular garden is enclosed by 460 feet of fencing. If the length of the garden is 10 feet less than 3 times the width, find the dimensions of the garden.
23. Joe emptied his bank which contained only nickels, dimes, and quarters. Joe discovered that he had the same number of quarters and dimes and 5 more nickels than quarters. If the total value of the coins was $4.25, how many of each type of coin was in the bank?
24. The width of a rectangle is 3 inches more than one-half of its length. If the perimeter is 60

inches, find the length and width of the rectangle.

25. Two angles of a triangle are equal and the third angle is 20 degrees less than twice one of the equal angles. Find the number of degrees in each angle of the triangle. (*Hint:* The sum of the angles of a triangle is 180 degrees.)

Just For Fun

Addie has two coins whose total is 35 cents, and yet one of them is not a dime. What are the two coins?

9.6 LINEAR EQUATIONS IN TWO VARIABLES

Thus far in our discussion of open sentences and equations, we have dealt with sentences that contained only one unknown, or variable. But these are not the only type of open sentences that exist. Consider the equation

$$x + 2y = 10$$

This is an equation in two unknowns, or two variables. If we replace x by 2, that is,

$$2 + 2y = 10$$

we still have an open sentence, but now we can solve the equation $2 + 2y = 10$ for y. When $x = 2$ and $y = 4$, the open sentence $x + 2y = 10$ is true. This means that $x = 2$ and $y = 4$ is a *solution* to the equation $x + 2y = 10$.

Rather than write $x = 2$ and $y = 4$, we can denote this in another manner, namely, by means of the ordered pair (2, 4). (Recall our previous discussion of ordered pairs in section 1.7.) We say *ordered pair* because the order in which the numbers appear is important. By convention, the first number in an ordered pair is always a value for x, and the second number is always a value for y. We know that the ordered pair (2, 4) is a solution to the equation $x + 2y = 10$: when we replace x by 2 and y by 4, the resulting statement is true; that is, $2 + 2(4) = 10$, or $2 + 8 = 10$. Note that the ordered pair (4, 2) is not a solution to the equation $x + 2y = 10$; that is, $4 + 2(2) \neq 10$.

An ordered pair is always of the form (x, y); that is, the x value is listed first and the y value is listed second. The x value is formally called the **abscissa** and the y value is called the **ordinate,** but in our discussion we shall refer to them as the x and y values.

It should be noted that the ordered pair (2, 4) is not the only ordered pair that will make the open sentence $x + 2y = 10$ a true statement; it is not the only solution for the given equation. In fact, there are infinitely many ordered pairs in the solution set of the equation $x + 2y = 10$.

Any equation of the form

$$Ax + By = C$$

where A, B, and C are real numbers is called a **linear equation.** Thus $x + 2y = 10$ is a linear equation. It is called a linear equation because when we graph such an equation in the Cartesian plane, we get a straight line.

We mentioned before that there are infinitely many ordered pairs in the solution set of the equation $x + 2y = 10$. We already have the ordered pair (2, 4); now let's obtain some more ordered pairs that are in the solution set. One way to do this is to let x be any value we choose and then solve for y. Suppose $x = 0$. Then we have

$$\begin{aligned} & x + 2y = 10 \\ \text{let } x = 0: \quad & 0 + 2y = 10 \\ & \frac{2y}{2} = \frac{10}{2} \qquad \text{dividing both sides by 2} \\ & y = 5 \end{aligned}$$

Hence when $x = 0$ and $y = 5$, we have another solution to the equation $x + 2y = 10$. We can therefore say that the ordered pair (0, 5) is a solution.

To find another solution for the equation $x + 2y = 10$, let $x = 4$. Then we have

$$\begin{aligned} & x + 2y = 10 \\ \text{let } x = 4: \quad & 4 + 2y = 10 \\ & 4 + 2y - 4 = 10 - 4 \qquad \text{subtracting 4 from both sides} \\ & 2y = 6 \\ & \frac{2y}{2} = \frac{6}{2} \qquad \text{dividing both sides by 2} \\ & y = 3 \end{aligned}$$

Hence (4, 3) is a solution.

Thus far, we have three ordered pairs that satisfy the linear equation $x + 2y = 10$, namely, (2, 4), (0, 5), and (4, 3).

In order to find solutions for any linear equation, select a value for x and substitute it for x in the equation; then solve the resulting equation for y. We can also find a solution for a linear equation by selecting a value for y, substituting it for y in the equation, and then solving the resulting equation for x.

EXAMPLE 1

Find three solutions for $3x + 4y = 15$.

Solution

a. We let $x = 1$. Then we have

$$\begin{aligned} 3x + 4y &= 15 \\ \text{let } x = 1: \quad 3(1) + 4y &= 15 \\ 3 + 4y &= 15 \\ 3 + 4y - 3 &= 15 - 3 && \text{subtracting 3 from both sides} \\ 4y &= 12 \\ \frac{4y}{4} &= \frac{12}{4} && \text{dividing both sides by 4} \\ y &= 3 \end{aligned}$$

Therefore (1, 3) is a solution.

b. Let $x = 5$; then

$$\begin{aligned} 3(5) + 4y &= 15 \\ 15 + 4y &= 15 \\ 15 + 4y - 15 &= 15 - 15 && \text{subtracting 15 from both sides} \\ 4y &= 0 \\ \frac{4y}{4} &= \frac{0}{4} && \text{dividing both sides by 4} \\ y &= 0 \end{aligned}$$

Therefore (5, 0) is a solution.

c. Let $x = 0$; then

$$\begin{aligned} 3(0) + 4y &= 15 \\ 0 + 4y &= 15 \\ 4y &= 15 \\ \frac{4y}{4} &= \frac{15}{4} && \text{dividing both sides by 4} \\ y &= \frac{15}{4} \end{aligned}$$

Therefore $(0, \frac{15}{4})$ is a solution.

The third solution obtained in example 1, $(0, \frac{15}{4})$, contains a fraction. However, as we shall see in the next section, it is advantageous to obtain integral solutions, that is, solutions that contain only integers. We can avoid obtaining a fractional solution if we are careful about the values that we choose for x. For instance, in example 1, if $x = -3$, then $y = 6$, and $(-3, 6)$ is a solution for $3x + 4y = 15$.

EXAMPLE 2
Find three solutions for $2x - 3y = 12$.

Solution

a. Selecting a value for x, we let $x = 0$. Hence we have

$$\begin{aligned} 2x - 3y &= 12 \\ \text{let } x = 0: \quad 2(0) - 3y &= 12 \\ 0 - 3y &= 12 \\ -3y &= 12 \\ \frac{-3y}{-3} &= \frac{12}{-3} \qquad \text{dividing both sides by } -3 \\ y &= -4 \end{aligned}$$

Therefore $(0, -4)$ is a solution.

b. Let $x = 3$; then

$$\begin{aligned} 2(3) - 3y &= 12 \\ 6 - 3y &= 12 \\ 6 - 3y - 6 &= 12 - 6 \qquad \text{subtracting 6 from both sides} \\ -3y &= 6 \\ \frac{-3y}{-3} &= \frac{6}{-3} \qquad \text{dividing both sides by } -3 \\ y &= -2 \end{aligned}$$

Therefore $(3, -2)$ is a solution.

c. Let us try a different approach, selecting a value for y. Let $y = 0$; then

$$\begin{aligned} 2x - 3(0) &= 12 \\ 2x - 0 &= 12 \\ 2x &= 12 \\ \frac{2x}{2} &= \frac{12}{2} \qquad \text{dividing both sides by 2} \\ x &= 6 \end{aligned}$$

Therefore $(6, 0)$ is a solution. Note that even though we have selected a value for y first, we list the x value first and the y value second when we write the solution as an ordered pair.

Instead of listing the solutions to the equation $2x - 3y = 12$ in example 2 as the ordered pairs $(0, -4)$, $(3, -2)$, and $(6, 0)$, we can list the solutions as a **table of values,** as shown below.

$2x - 3y = 12$

x	y
0	−4
3	−2
6	0

EXERCISES FOR SECTION 9.6

Find three solutions for each of the following equations.

1. $x + y = 5$
2. $x + y = 8$
3. $x - y = 3$
4. $x - y = 1$
5. $2x + y = -6$
6. $2x - y = 3$
7. $3x - y = -10$
8. $2x + 3y = -6$
9. $3x - 2y = 8$
10. $5x + 3y = 15$
11. $3x + 5y = -15$
12. $x + y = 0$
13. $x - y = 0$
14. $x - 2y = 0$
15. $x + 2y = 0$
16. $x + 3y = 13$
17. $2x - 3y = -11$
18. $3x - 5y = -9$
19. $-2x - 3y = 7$
20. $-3x - 2y = -3$
21. $-5x - 4y = -12$

Just For Fun

In the "old, old days" students had to know the following kinds of measure: the tierce, hogshead, pipe, butt, and tun. These are similar kinds of measure. What do they measure?

9.7 GRAPHING EQUATIONS

Now that we are able to find solution sets for linear equations, our next goal is to illustrate these solution sets by means of a graph. We can do this on a grid called the **Cartesian plane,** named after the French mathematician-philosopher René Descartes.

In order to develop an understanding of the Cartesian plane and how to graph ordered pairs on it, consider the map of Anytown in figure 12. Assume that we are at the intersection of Main Street (east-west) and Euclid Avenue (north-south). In order to get to the town hall, we must go 3 blocks east and then 2 blocks north; to get to the school, we must go 2 blocks east and then 3 blocks south. Similarly, to get to the hospital, we must go 3 blocks west and then 3 blocks north, and to get to the library we must go 1 block west and then 3 blocks south.

FIGURE 12

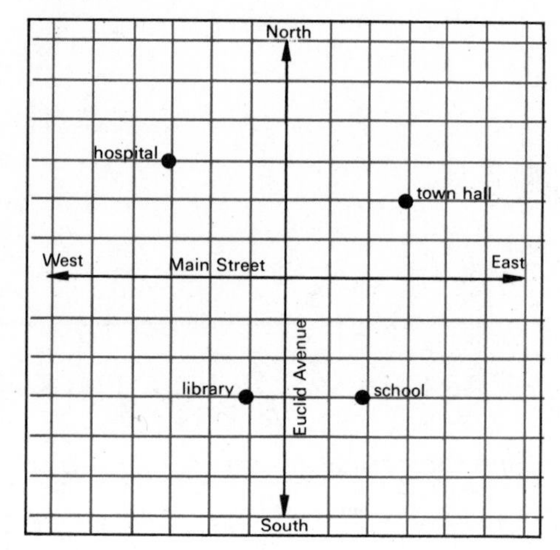

We can use this same idea to graph, or **plot,** ordered pairs. In chapter 8, we used a horizontal number line like the line representing the east-west Main Street in figure 12. Recall that zero[1] was in the middle of the line, the positive numbers were to the right (east) of zero, and the negative numbers were to the left (west) of zero. We shall use such a horizontal line and call it the **x–number line,** or the **x-axis.** Now we construct another number line perpendicular to the x-axis (like the north-south street Euclid Avenue in figure 12) and passing through zero. We mark off the numbers on this new line as we did for the x-number line. The numbers above the x-axis (north) are positive, and those below the x-axis (south) are negative. This vertical number line is called the **y–number line,** or **y-axis.** The Cartesian plane is based upon the x- and y-axes. They intersect at the common point where $x = 0$ and $y = 0$. This particular point is represented by the ordered pair $(0, 0)$ and is called the **origin.** See figure 13.

Instead of giving directions to the town hall in figure 12 by saying "go 3 blocks east and then 2 blocks north," we can now say "go 3 units in the positive x-direction, and then 2 units in the positive y-direction." We can shorten this even more by writing the ordered pair $(3, 2)$. If we are asked to plot the ordered pair $(3, 2)$ on the Cartesian plane, we start at the origin and move 3 units in the positive x-direction, and then 2 units in the positive y-direction. See figure 14.

Using this idea, we can graph the ordered pairs $(2, -3)$, $(-3, 3)$, and $(-1, -3)$ on the Cartesian plane shown in figure 14. In order to graph an ordered pair, we always start at the origin $(0, 0)$ and proceed from there. The point represented by $(2, -3)$ is obtained by moving 2 units in a positive x-direction and then 3 units in a negative y-direction. The point represented by $(-3, 3)$ is obtained by starting at the origin and moving 3 units in a negative x-direction and then 3 units in a positive y-direction. To plot the point represented by $(-1, -3)$, we start at the origin and move 1 unit in a negative x-direction and then 3 units in a negative y-direction.

In order to obtain any point on the plane, we start at the origin and then move in a positive or negative x-direction (east-west, right-left); next we move in a positive or negative y-direction (north-south, up-down). The first number in an ordered pair is the x-value and the second number is the y-value. These two numbers

FIGURE 13

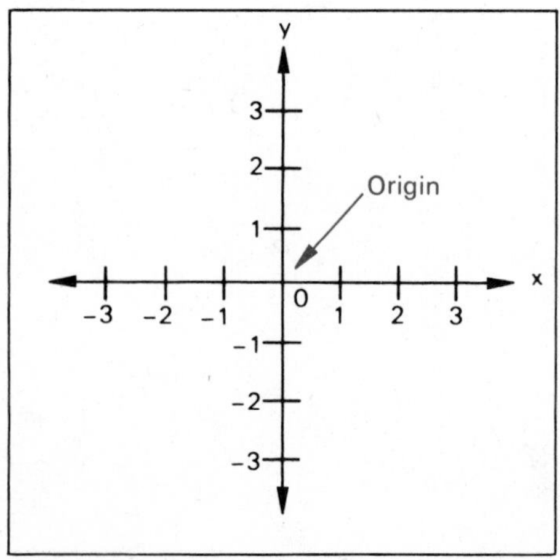

FIGURE 14

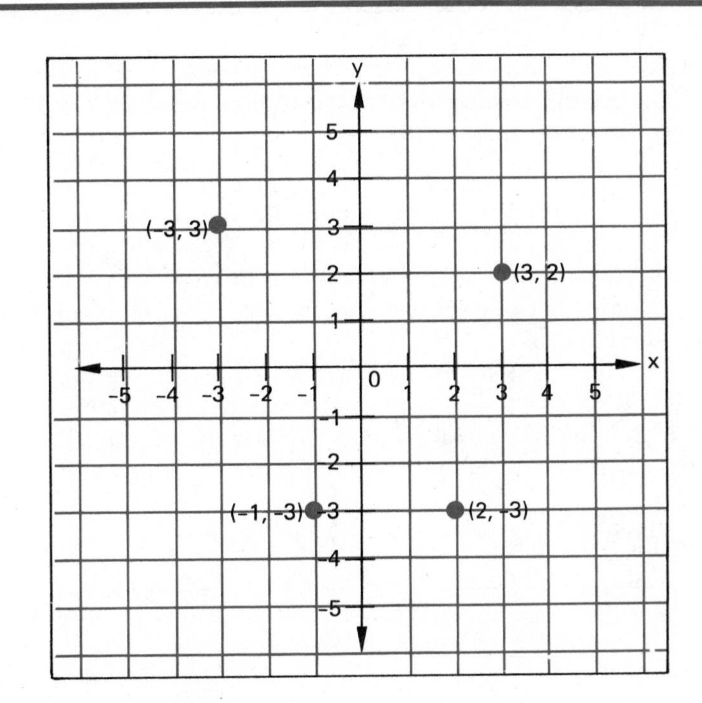

FIGURE 15

are called the **coordinates** of the point; hence the Cartesian plane is also called the **Cartesian coordinate system.**

EXAMPLE 1

Locate the points corresponding to the ordered pairs $A(1,2)$ and $B(2,-1)$.

Solution

The given ordered pairs are labeled A and B to aid us in our discussion. Point A, corresponding to the ordered pair $(1,2)$, is found by moving 1 unit in a positive x-direction from the origin and then 2 units in a positive y-direction. Point B, corresponding to $(2,-1)$, is found by moving 2 units in a positive x-direction from the origin and then 1 unit in a negative y-direction. Points A and B are shown in figure 15.

FIGURE 16

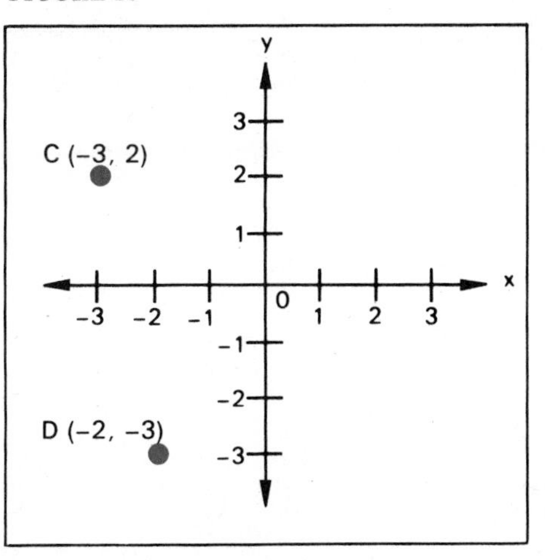

EXAMPLE 2

Locate the points corresponding to the ordered pairs $C(-3,2)$ and $D(-2,-3)$.

Solution

Point C, $(-3,2)$, is found by moving 3 units in a negative x-direction from the origin and then 2 units in a positive y-direction. Point D, $(-2,-3)$, is found by moving 2 units in a negative x-direction from the origin and then 3 units in a negative y-direction. See figure 16.

EXAMPLE 3

Locate the points corresponding to the ordered pairs $A(0, 2)$, $B(3, 0)$, $C(0, -1)$, $D(-2, 0)$, and $E(0, 0)$.

Solution

Each of the given ordered pairs contains 0 as one of its values. Point $A(0, 2)$ is found by moving 0 units in the x-direction from the origin and then 2 units in a positive y-direction. Point $B(3, 0)$ is found by moving 3 units in a positive x-direction from the origin and 0 units in the y-direction.

Point $C(0, -1)$ is found by moving 0 units in the x-direction from the origin and then 1 unit in the negative y-direction. Point $D(-2, 0)$ is found by moving 2 units in a negative x-direction from the origin and then 0 units in the y-direction. Point $E(0, 0)$ is the origin.

Points A, B, C, D, and E are shown in figure 17.

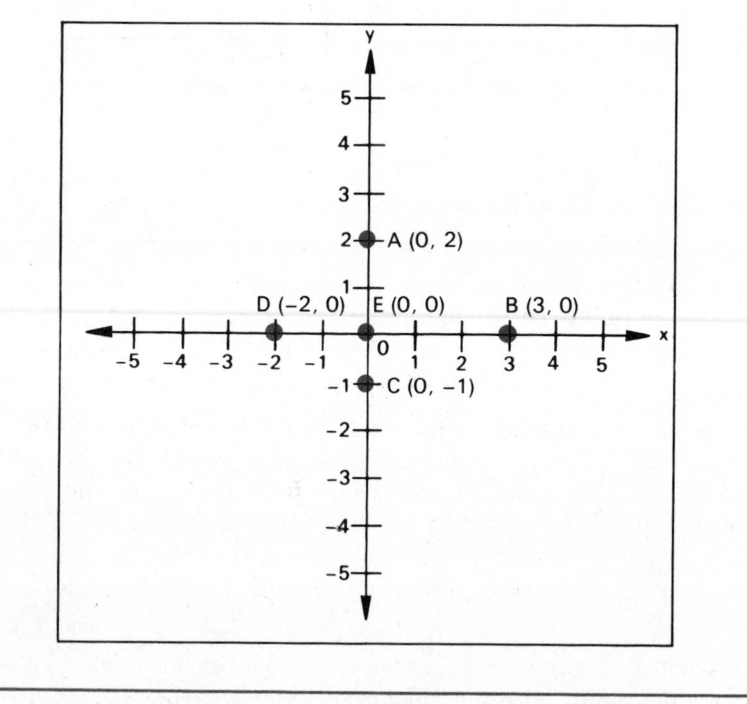

FIGURE 17

Since each ordered pair represents a point on the plane, let us now locate some ordered pairs that are solutions to the linear equations that we discussed in the previous section.

Consider the equation

$$x + 2y = 10$$

FIGURE 18

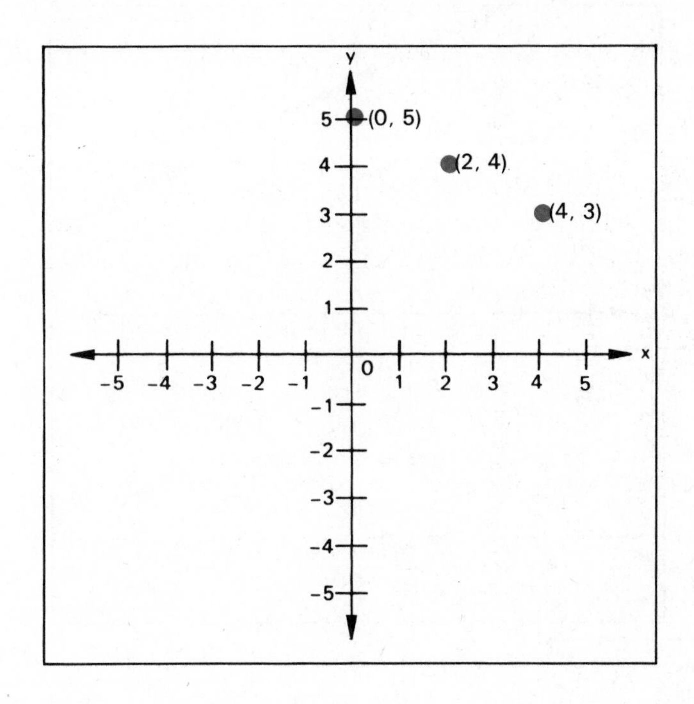

Three ordered pairs that satisfy this equation are $(4, 3)$, $(2, 4)$, and $(0, 5)$. We can plot all of these ordered pairs in the Cartesian plane, as shown in figure 18. Note that these points tend to form a pattern; that is, they appear to lie on the same path. If we connect these points, we see that all of these points do in fact lie on the same straight line. This line, shown in figure 19, is a graph of the equation $x + 2y = 10$.

The fact that the graph of this type of equation—that is, an equation of the form $Ax + By = C$, where A, B, and C are real numbers—is a straight line is the reason that such equations are called **linear equations.**

If the graph of a linear equation is always a straight line, then why did we use three points to determine the graph of $x + 2y = 10$? It is true that two points determine a line, that is, through two given points one and only one straight line can be drawn. We use three points because the third point is a check. We locate the third point to check that all three points lie on the same straight line. If one of the points is not on the line, then we must go back and check for an error in determining the solutions to the linear equation.

EXAMPLE 4
Graph $x + y = 5$.

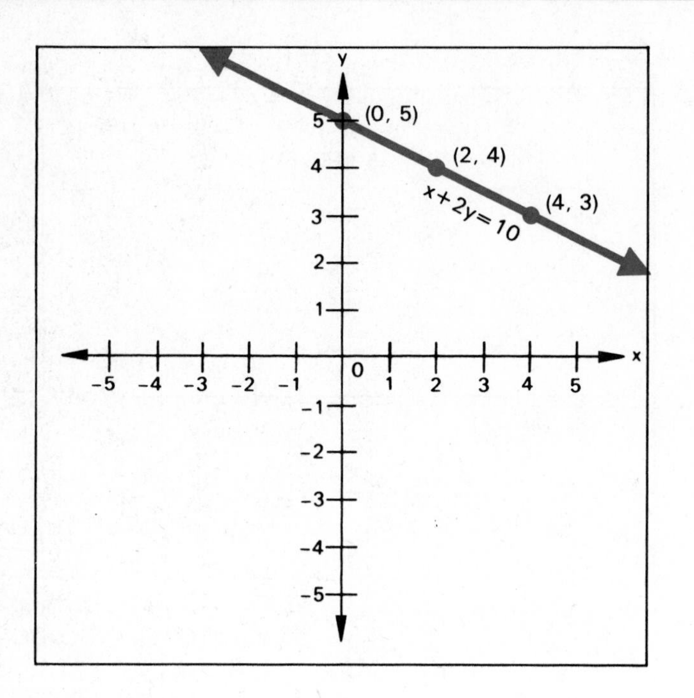

FIGURE 19

Solution

In order to graph $x + y = 5$, we must first find three ordered pairs that are solutions to the equation.

a. Selecting a value for x, we let $x = 1$. Hence, we have

$$\begin{aligned} & x + y = 5 \\ \text{let } x = 1: \quad & 1 + y = 5 \\ & y = 4 \end{aligned}$$

and (1, 4) is a solution.

b. Let $x = 4$; then

$$\begin{aligned} 4 + y &= 5 \\ y &= 1 \end{aligned}$$

and (4, 1) is a solution.

c. Let $x = 2$; then

$$\begin{aligned} 2 + y &= 5 \\ y &= 3 \end{aligned}$$

and (2, 3) is a solution.

Now that we have three different ordered pairs, (1, 4), (4, 1), and (2, 3), that are solutions to the equation $x + y = 5$, we plot these points and draw the line that contains them, as shown in figure 20.

FIGURE 20

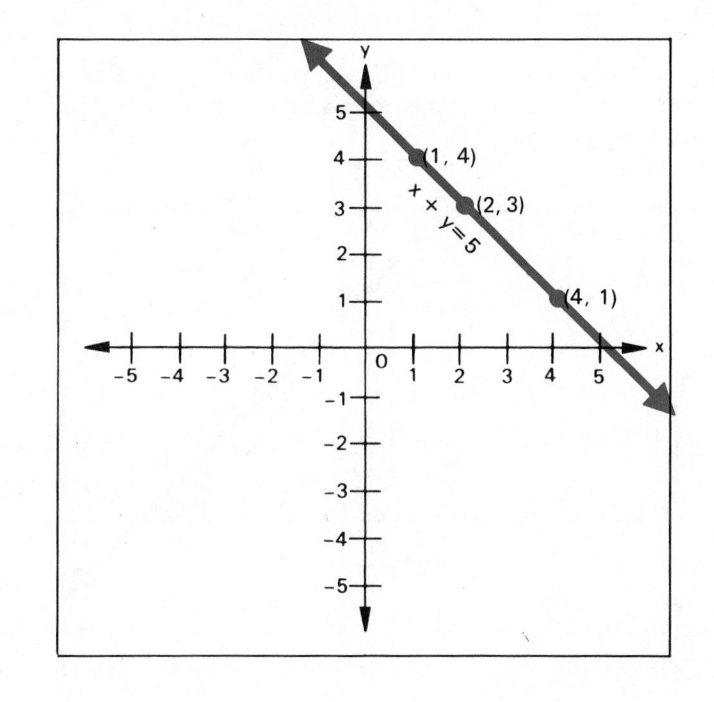

When we graph a linear equation, it is often convenient to find the points where the graph crosses each axis. Note that $y = 0$ at the point where a line crosses the x-axis, and $x = 0$ at the point where a line crosses the y-axis. The point at which a line crosses the x-axis is called the **x-intercept,** and, similarly, the point at which it crosses the y-axis is called the **y-intercept.** In order to find the intercepts of the equation in example 4, $x + y = 5$, we proceed as follows:

$$x + y = 5$$

let $x = 0$: $\quad 0 + y = 5$

$$y = 5 \qquad (0, 5) \text{ is a solution}$$

The y-intercept is (0, 5). Now let $y = 0$; then

$$x + 0 = 5$$

$$x = 5 \qquad (5, 0) \text{ is a solution}$$

The x-intercept is (5, 0).

If we want to graph $x + y = 5$ by using the intercepts, we should find a third point as a check:

let $x = 2$: $\quad 2 + y = 5$

$y = 3 \qquad (2, 3)$ is a solution

Using this point and the intercepts, we graph $x + y = 5$ as in figure 21. Note that we get the same graph as in figure 20, although we used different ordered pairs.

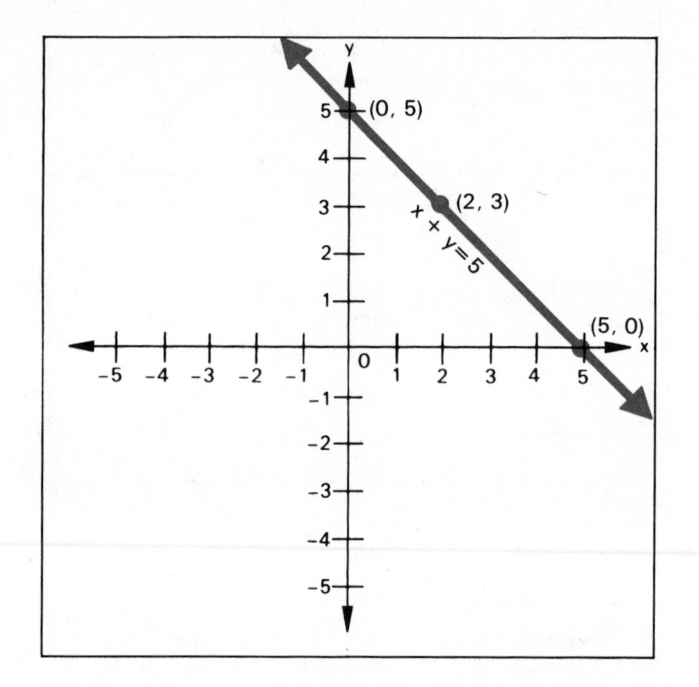

FIGURE 21

Remember that there are infinitely many solutions to a linear equation and that each solution (ordered pair) is a point on the line. All the solutions of the equation are points on the line, and all the points on the line are solutions of the equation. In graphing an equation, it is sometimes convenient to find the intercepts because they contain zeros, and this simplifies computation.

EXAMPLE 5

Graph $2x - y = 4$.

Solution

First we shall find the intercepts. Therefore the value we select for x is 0.

$$2x - y = 4$$

let $x = 0$: $2(0) - y = 4$

$$0 - y = 4$$

$$-y = 4$$

$$-1(-y) = -1(4)$$ multiplying both sides by -1 to make the y-term positive

$$y = -4$$

The ordered pair $(0, -4)$ is a solution; $(0, -4)$ is also the y-intercept. Next, let $y = 0$; then

$$2x - 0 = 4$$

$$2x = 4$$

$$x = 2$$

The ordered pair $(2, 0)$ is a solution; $(2, 0)$ is also the x-intercept. Selecting a third point as a check, we let $x = 3$; then

$$2(3) - y = 4$$

$$6 - y = 4$$

$$-y = -2$$

$$-1(-y) = -1(-2)$$ multiplying both sides by -1 to make the y-term positive

$$y = 2$$

The pair $(3, 2)$ is a solution.

Using these points, we graph $2x - y = 4$ in figure 22.

FIGURE 22

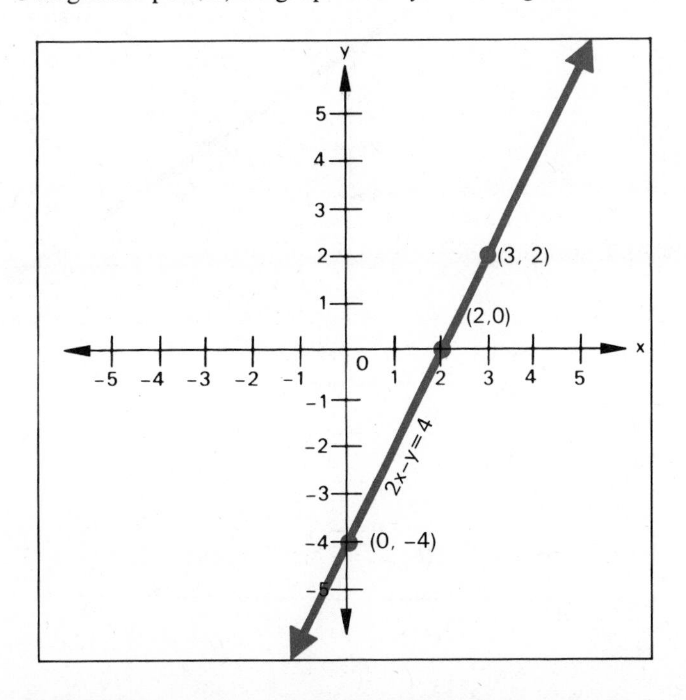

EXAMPLE 6

Graph $2x + 3y = 12$.

Solution

We find the intercepts first.

$$\begin{aligned} & & 2x + 3y &= 12 \\ \text{let } x = 0: & & 2(0) + 3y &= 12 \\ & & 0 + 3y &= 12 \\ & & 3y &= 12 \\ & & y &= 4 \end{aligned}$$

The y-intercept is $(0, 4)$.

$$\begin{aligned} \text{let } y = 0: & & 2x + 3(0) &= 12 \\ & & 2x + 0 &= 12 \\ & & 2x &= 12 \\ & & x &= 6 \end{aligned}$$

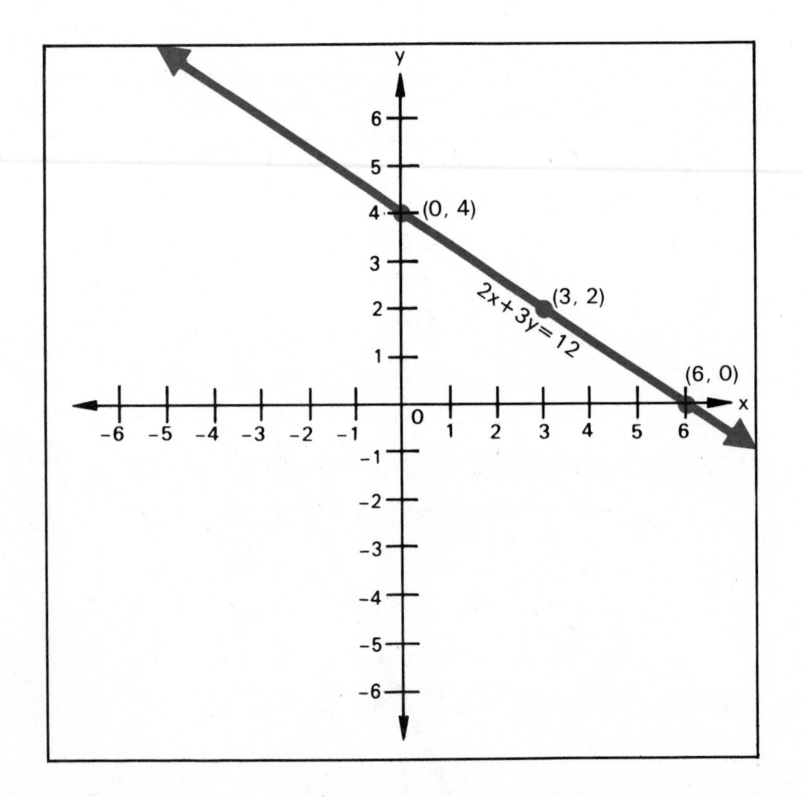

FIGURE 23

x	y
0	4
6	0
3	2

The x-intercept is $(6, 0)$.

We then select a third point.

$$\begin{aligned} \text{let } x = 3: \quad 2(3) + 3y &= 12 \\ 6 + 3y &= 12 \\ 3y &= 6 \\ y &= 2 \end{aligned}$$

The point $(3, 2)$ is a solution.

Using these points, we graph $2x + 3y = 12$ in figure 23. Note that we can also list the solutions to the equation $2x + 3y = 12$ in a table of values such as shown next to figure 23.

When we graph a linear equation, it is usually convenient to find the points where the graph crosses each axis, that is, the x-intercept and the y-intercept. But if letting x or y be zero produces an intercept with a fractional value, then it is better to find solutions to the equation that contain only integers. For example, consider the equation $3x + 4y = 15$. If $x = 0$, then $y = \frac{15}{4}$, and $(0, \frac{15}{4})$ is the y-intercept. It will be more difficult to locate this point accurately on the Cartesian plane than a pair of integers such as $(-3, 6)$. Therefore, it is better to use $(-3, 6)$ than $(0, \frac{15}{4})$ as one of the ordered pairs used to graph $3x + 4y = 15$.

EXERCISES FOR SECTION 9.7

1. Locate the points corresponding to the given ordered pairs on a Cartesian plane.

   a. $(4, 3)$ b. $(2, 5)$ c. $(3, -1)$
   d. $(-2, 4)$ e. $(0, 0)$ f. $(-2, -3)$
   g. $(3, -2)$ h. $(-1, 5)$ i. $(-4, 0)$

For exercises 2–22, graph the given equation on a Cartesian plane.

2. $x + y = 5$
3. $x + y = 8$
4. $x - y = 3$
5. $x - y = 1$
6. $2x + y = 6$
7. $2x - y = 4$
8. $3x - y = -12$
9. $2x + 3y = -6$
10. $3x - 2y = 12$
11. $5x + 3y = 15$
12. $3x - 5y = -15$
13. $x + y = 0$
14. $x - y = 0$
15. $x - 2y = 0$
16. $x + 2y = 0$
17. $2x + 3y = 5$
18. $3x - 2y = 7$
19. $x - 2y = 3$
20. $-2x - 3y = 6$
21. $-x - 2y = -4$
22. $-2x + y = -8$

Just For Fun

A bicycle dealer was asked how many bikes he had in stock. He answered, "If one-half, one-third, and one-quarter of the number of bikes were added together, they would make 13." How many bikes did he have in stock?

9.8 GRAPHING $y = ax^2 + bx + c$

In the preceding section, we graphed linear equations in two variables. The graphs of these equations were straight lines. In this section, we shall examine graphs of equations of the form $y = ax^2 + bx + c$. An equation of this form is called a **quadratic equation** because it contains a term, ax^2, whose exponent is 2—that is, a **second-degree** term.

A **relation** is defined to be a set of ordered pairs. For example, $\{(2, 3), (4, 5)\}$ is a relation, as is $\{(0, 1), (2, 3), (5, 7)\}$. A **function** is a relation in which no two ordered pairs have the same first coordinate and different second coordinates. Consider, for example, the equation $y = x^2$. Some of the ordered pairs that satisfy this equation are $(0, 0)$, $(1, 1)$, $(-1, 1)$, $(2, 4)$, and $(-2, 4)$. Note that no two of these ordered pairs have the same first coordinate and different second coordinates. Thus the set of ordered pairs (x, y) that satisfy $y = x^2$ is a function.

Now consider $x = y^2$. Some of the ordered pairs that satisfy this equation are $(0, 0)$, $(1, 1)$, $(4, 2)$, and $(4, -2)$. Some of these ordered pairs do have the same first coordinate and different second coordinates. Hence, the set of ordered pairs (x, y) that satisfy $x = y^2$ is not a function. In this section, we shall consider only *quadratic functions*. **Quadratic functions** are sets of ordered pairs (x, y) that satisfy equations of the general form $y = ax^2 + bx + c$.

Consider $y = x^2$. This is a quadratic equation of the form $y = ax^2 + bx + c$: in this case, $a = 1$, $b = 0$, and $c = 0$. To graph $y = x^2$, we first find some ordered pairs that satisfy the given equation. Recall that the ordered pairs that satisfy $y = x^2$ are a function. Therefore, for every value of x, there is a unique value of y for which $y = x^2$. For example, if $x = 0$, then $y = (0)^2 = 0$. Similarly, if $x = 1$, then $y = (1)^2 = 1$, and if $x = -1$, then $y = (-1)^2 = 1$. If

$x = 2$, $y = 4$ and if $x = -2$, $y = 4$. These points can be listed in a table of values as follows:

x	-2	-1	0	1	2
y	4	1	0	1	4

If we plot these points and connect them by means of a smooth curve, we obtain the curve shown in figure 24. Note that the graph cannot be a straight line, because $y = x^2$ is not a linear equation.

FIGURE 24

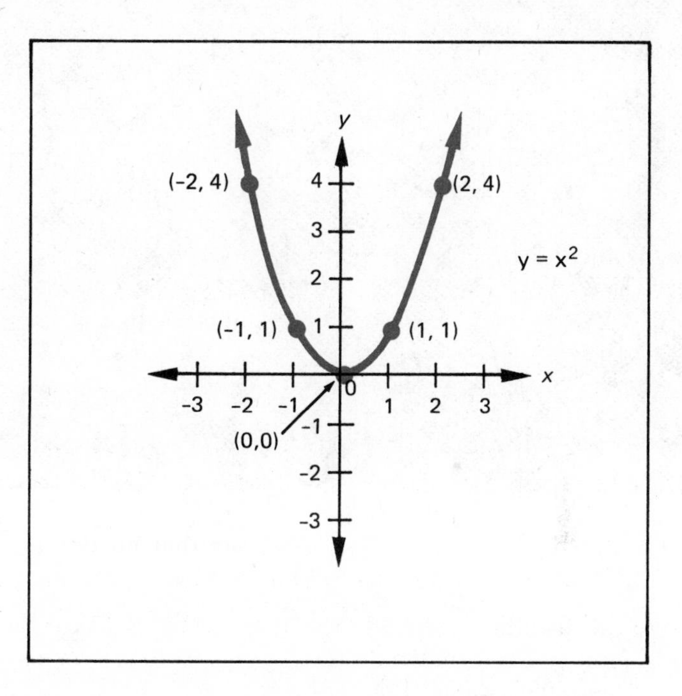

The smooth curve in figure 24 is called a **parabola.** The path of most projectiles is shaped like a parabola and the cables on suspension bridges hang in the shape of a parabola.

In figure 24, note that the point $(0, 0)$ is a turning point of the parabola $y = x^2$. That is, as we proceed from left to right along the x-axis, the values of y decrease until we reach $(0, 0)$; then the values of y begin to increase. The turning point of a parabola is called the **vertex** of the parabola.

For any parabola with equation of the form $y = ax^2 + bx + c$, the x-value of the vertex is $\frac{-b}{2a}$.

Los Angeles Times

These freeways form a pattern of intersecting parabolas and straight lines.

We can use this information to help graph a given parabola. For example, $y = x^2$ is of the form $y = ax^2 + bx + c$, with $a = 1$ and $b = 0$. Hence the x-value of the vertex is

$$\frac{-b}{2a} = \frac{-0}{2 \cdot 1} = -\frac{0}{2} = 0$$

The y-value is found by substituting zero for x in the original equation. If $x = 0$, $y = 0^2 = 0$. Therefore the vertex of $y = x^2$ is $(0, 0)$.

It should be noted that the value of a in $y = x^2$ is $+1$. If a is positive in any parabola with equation of the form $y = ax^2 + bx + c$, then the parabola opens upward. If a is negative, then the parabola opens downward. These two facts are useful when graphing parabolas.

EXAMPLE 1
Graph $y = x^2 - 4$.

Solution
The general form is $y = ax^2 + bx + c$. In this case, $a = 1$, $b = 0$, and $c = -4$. Since $a = 1$ is positive, the parabola opens upward. The x-value of the vertex is

$$\frac{-b}{2a} = \frac{-0}{2 \cdot 1} = \frac{0}{2} = 0$$

If $x = 0$, $y = 0^2 - 4 = -4$. Therefore the vertex is $(0, -4)$.

To obtain a better idea of the shape of the parabola, we choose two or three values for x on each side of the vertex and find the corresponding values of y. To the left of the vertex, we choose $x = -1$, $x = -2$, and $x = -3$, and to the right of the vertex, we choose $x = 1$, $x = 2$, and $x = 3$. The corresponding values of y are listed in the following table.

x	-3	-2	-1	0	1	2	3
y	5	0	-3	-4	-3	0	5

Using these points, we graph $y = x^2 - 4$ in figure 25.

FIGURE 25

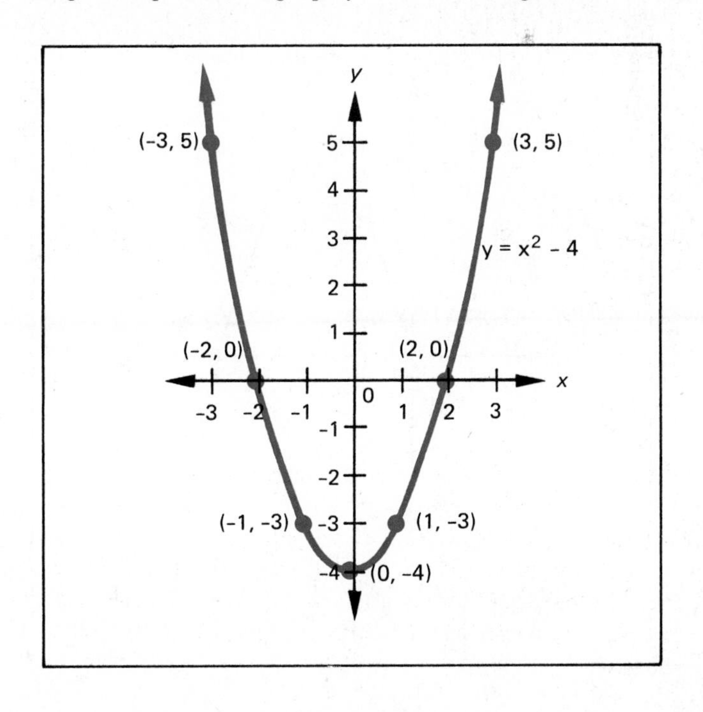

EXAMPLE 2
Graph $y = -x^2 + 4x$.

Solution
In this case, $a = -1$, $b = 4$, and $c = 0$. Since $a = -1$ is negative, the parabola opens downward. The x-value of the vertex is

$$\frac{-b}{2a} = \frac{-(4)}{2(-1)} = \frac{-4}{-2} = 2$$

If $x = 2$, $y = -2^2 + 4 \cdot 2 = -4 + 8 = 4$. Therefore the vertex is $(2, 4)$. To graph the parabola, we choose $x = 1$, $x = 0$, and $x = -1$ to the left of the vertex, and $x = 3$, $x = 4$, and $x = 5$ to the right of the vertex. The corresponding values of y are listed in the following table.

x	-1	0	1	2	3	4	5
y	-5	0	3	4	3	0	-5

Using these points, we graph $y = -x^2 + 4x$ in figure 26.

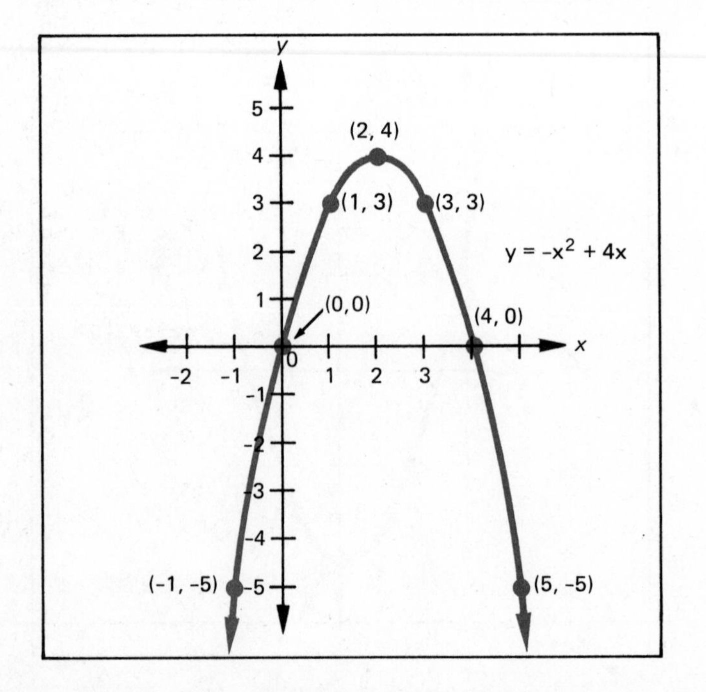

FIGURE 26

EXAMPLE 3
Graph $y = x^2 - 8x + 12$.

Solution
In this case, $a = 1$, $b = -8$, and $c = 12$. Since $a = 1$ is positive, the parabola opens upward. The x-value of the vertex is

$$\frac{-b}{2a} = \frac{-(-8)}{2 \cdot 1} = \frac{8}{2} = 4$$

If $x = 4$, $y = 4^2 - 8 \cdot 4 + 12 = 16 - 32 + 12 = -4$. Therefore the vertex is $(4, -4)$. To graph the parabola, we choose $x = 3$, $x = 2$, and $x = 1$ to the left of the vertex, and $x = 5$, $x = 6$, and $x = 7$ to the right of the vertex. The corresponding values of y are listed in the following table.

x	1	2	3	4	5	6	7
y	5	0	−3	−4	−3	0	5

Using these points, we graph $y = x^2 - 8x + 12$ in figure 27.

FIGURE 27

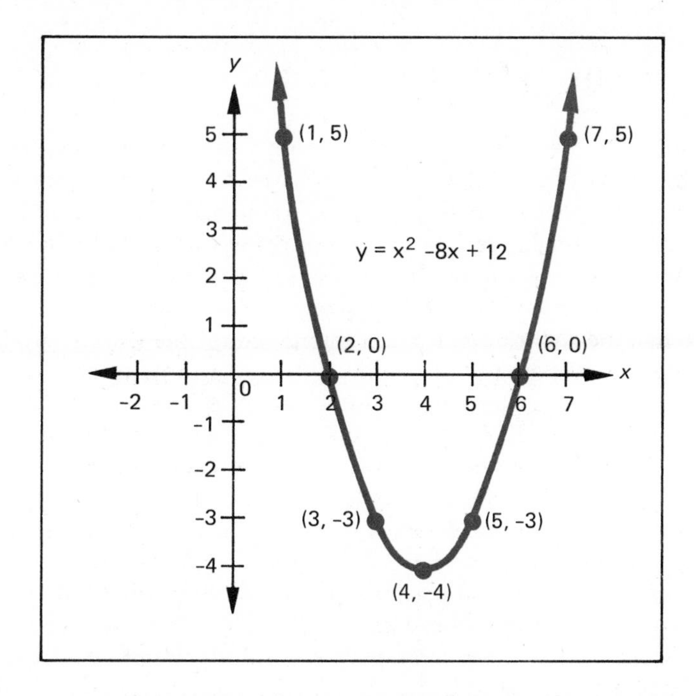

EXERCISES FOR SECTION 9.8

For exercises 1–14, graph the given equation on a Cartesian plane.

1. $y = x^2 - 1$
2. $y = -x^2 + 1$
3. $y = -x^2 + 2$
4. $y = -x^2 + 5$
5. $y = -x^2 + 6x$
6. $y = -x^2 - 4x$
7. $y = x^2 + 2x + 1$
8. $y = x^2 + 2x - 3$
9. $y = x^2 - 4x + 3$
10. $y = -2x^2 - 4x - 2$
11. $y = -x^2 + 6x - 9$
12. $y = x^2 + 2x - 8$
13. $y = x^2 - 6x + 5$
14. $y = x^2 - 4x + 2$

Just For Fun

Which is worth more, a box full of $10 gold pieces, or an identical box half full of $20 gold pieces?

9.9 INEQUALITIES IN TWO VARIABLES

In section 9.2, we graphed the solution sets of open sentences in one variable. The open sentences that we considered were either equations or inequalities, such as $x + 3 = 4$ and $x > 4$. In sections 9.7 and 9.8, we graphed the solution sets of equations in two unknowns, such as $2x - 2y = 10$ and $y = x^2 + x + 1$. In this section we shall graph the solution sets of inequalities involving two variables.

When the equation $x + y = 3$ is graphed on the Cartesian plane, we note that the points $(3, 0)$ and $(0, 3)$ are solutions of the given equation. In fact, there are infinitely many solutions to the linear equation $x + y = 3$. Each solution (ordered pair) is a point on the line. All the solutions of the equation are points on the line, and all the points on the line are solutions of the equation. But what about the points (ordered pairs) that are not on the line? What about points such as $(0, 0)$ and $(4, 0)$? Since they are not on the line, they are not solutions of the equation $x + y = 3$. But how do they compare with the solutions? Let's find out.

We can evaluate the equation $x + y = 3$ for the point $(0, 0)$. Let $x = 0$ and $y = 0$; then we have

$$0 + 0 = 3$$
$$0 = 3 \qquad \text{(not true)}$$

Substituting 0 for x and 0 for y in the equation $x + y = 3$ gives us a false statement of equality. The value 0 on the left side of the equal sign is *less than* the value 3 on the right side:

$$0 < 3 \qquad \text{(true)}$$

Let's try the other point, (4, 0). Let $x = 4$ and $y = 0$; then we have

$$4 + 0 = 3$$
$$4 = 3 \qquad \text{(not true)}$$

Again we have a false statement of equality, but note that this time the value 4 on the left side of the equal sign is *greater than* the value 3 on the right side:

$$4 > 3 \qquad \text{(true)}$$

Notice that (0, 0) is a solution of the inequality $x + y < 3$ and (4, 0) is a solution of $x + y > 3$. We have found three different types of points: those like (3, 0) and (0, 3), for which $x + y = 3$; those like (0, 0), for which $x + y < 3$; and those like (4, 0), for which $x + y > 3$.

In fact, the equation $x + y = 3$ separates the Cartesian plane into three sets. One of these sets is the graph of $x + y = 3$. Since this graph is a line, we call it "the line $x + y = 3$." The other two sets are the **half planes** that lie above and below this line. One of these half planes contains (0, 0) and is the graph of the inequality $x + y < 3$; the other contains (4, 0) and is the graph of the inequality $x + y > 3$. These two half planes are called "the half plane $x + y < 3$" and "the half plane $x + y > 3$," respectively.

How do we go about graphing an inequality such as $x + y < 3$? Recall that the line $x + y = 3$ divides the plane into two half planes. One half plane contains those points (x, y) such that $x + y < 3$, while the other half plane contains those points (x, y) such that $x + y > 3$. The set of points that satisfy the equation $x + y = 3$ are those points that lie on the line. In order to graph $x + y < 3$, we locate the boundary of this half plane, which is the line $x + y = 3$. See figure 28. Note that we draw the graph of $x + y = 3$ as a dashed line because the points on the line do not satisfy the inequality $x + y < 3$.

Which half plane is the correct half plane, the one whose points satisfy the inequality $x + y < 3$? Let's test a point on either side of the line to see which one satisfies the given inequality. One convenient point to try is the origin, (0, 0). Let $x = 0$ and $y = 0$; then we have

$$0 + 0 < 3$$
$$0 < 3 \qquad \text{(true)}$$

FIGURE 28

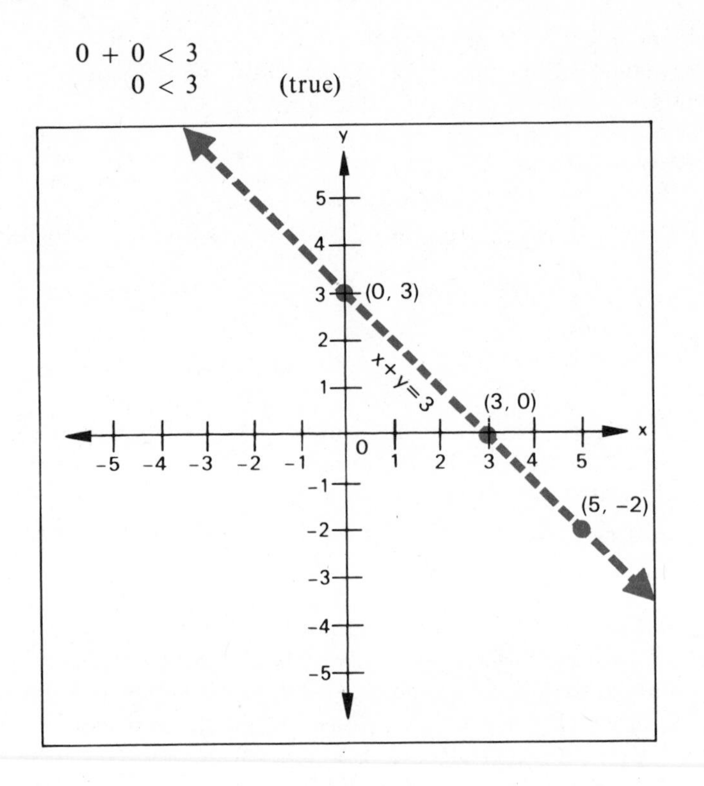

Now, if a point satisfies the given inequality, then all of the points in that half plane must also satisfy the same inequality. It should also be noted that if a point does not satisfy the given inequality, then none of the points in that half plane satisfies the given inequality and the desired half plane must be on the other side of the line.

Since the point (0, 0) does satisfy the given inequality, $x + y < 3$, we shade the half plane that contains this point. In this case we shade the half plane that is below the line $x + y = 3$ to indicate the solution. See figure 29.

If we were asked to graph $x + y \leq 3$ instead of $x + y < 3$, then we would have the same picture, with one exception. The exception would be to draw the line $x + y = 3$ as a solid line instead of a dotted line. The reason for this is that $x + y \leq 3$ means that we want the set of points that satisfy the inequality $x + y < 3$ *or* the equation $x + y = 3$. The sentence $x + y \leq 3$ is read as "$x + y$ is *less than or equal to* 3." Recall that *or* is the same as set union, and therefore we unite the half plane $x + y < 3$ and the line $x + y = 3$. Figure 30 is the graph of $x + y \leq 3$.

FIGURE 29 $x + y < 3$

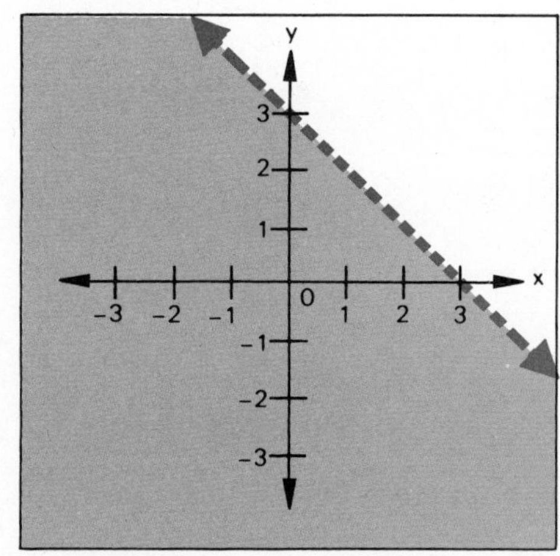

FIGURE 30 $x + y \leq 3$

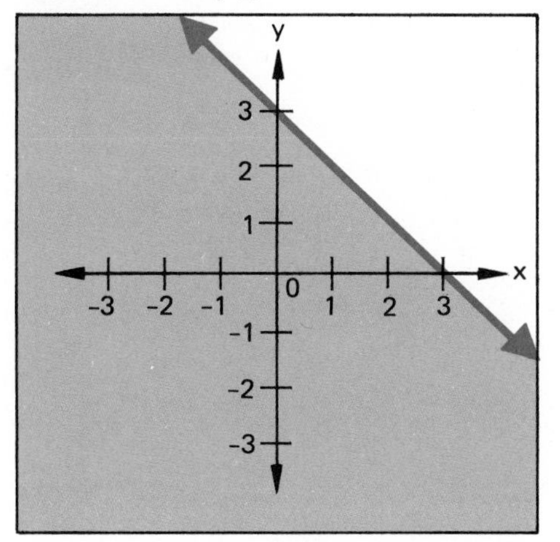

In both figure 29 and figure 30, the line $x + y = 3$ is the boundary for the solution set, a half plane. In figure 29, the boundary is not included in the solution, while in figure 30 the boundary is included in the solution.

In order to graph inequalities in two variables, we must do the following:

1. Find the boundary of the half plane. We do this by graphing the equation derived from the inequality.
2. The boundary should be a dashed line for *greater than* ($>$) or *less than* ($<$). For $\geq$ or $\leq$, the boundary should be a solid line.
3. Indicate the half plane that is the solution by shading. We do this by testing a point. If a point satisfies the given inequality, then all of the points in that half plane must also satisfy the same inequality.

EXAMPLE 1

Graph $x + 2y > 6$.

Solution

First we locate the boundary by graphing the equation $x + 2y = 6$. Note that this boundary will be a dashed line because we have a strict inequality in $x + 2y > 6$. Next we test a point to see if it satisfies the given inequality. Let's try the origin, (0, 0). Let $x = 0$ and $y = 0$; then we have

$$0 + 2(0) > 6$$
$$0 + 0 > 6$$
$$0 > 6 \qquad \text{(not true)}$$

The point (0, 0) does not satisfy the statement $x + 2y > 6$. Thus we can try a point on the other side of the line $x + 2y = 6$, (3, 3). Let $x = 3$ and $y = 3$; then we have

$$3 + 2(3) > 6$$
$$3 + 6 > 6$$
$$9 > 6 \qquad \text{(true)}$$

If a point satisfies a given inequality, then all of the points in that half plane must also satisfy the same inequality. Therefore we shade the half plane that contains the point (3, 3). In this case, we shade the half plane that is above the line $x + 2y = 6$ to indicate the solution. See figure 31.

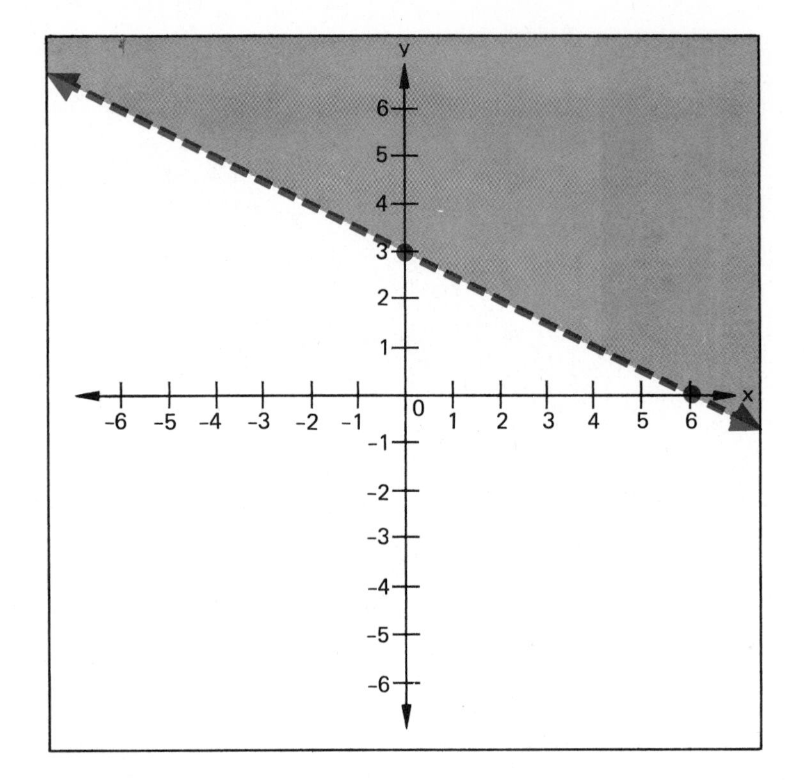

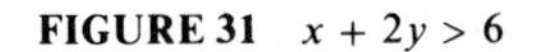
FIGURE 31 $x + 2y > 6$

EXAMPLE 2

Graph $2x - y \leq 6$.

Solution

First we locate the boundary by graphing the equation $2x - y = 6$. Note that this boundary will be a solid line because of the relationship $\leq$. Next we test a point to see if it satisfies the given inequality. Let's try the origin, $(0, 0)$. Let $x = 0$ and $y = 0$; then we have

$$\begin{aligned} 2(0) - 0 &\leq 6 \\ 0 - 0 &\leq 6 \\ 0 &\leq 6 \qquad \text{(true)} \end{aligned}$$

If a point satisfies a given inequality, then all of the points in that half plane must also satisfy the same inequality. Therefore, we shade the half plane that contains the point $(0, 0)$, that is, the half plane above the line $2x - y = 6$. See figure 32. Note that the boundary is a solid line.

EXAMPLE 3

Graph $-3x + 4y < 12$.

FIGURE 32 $2x - y \leq 6$

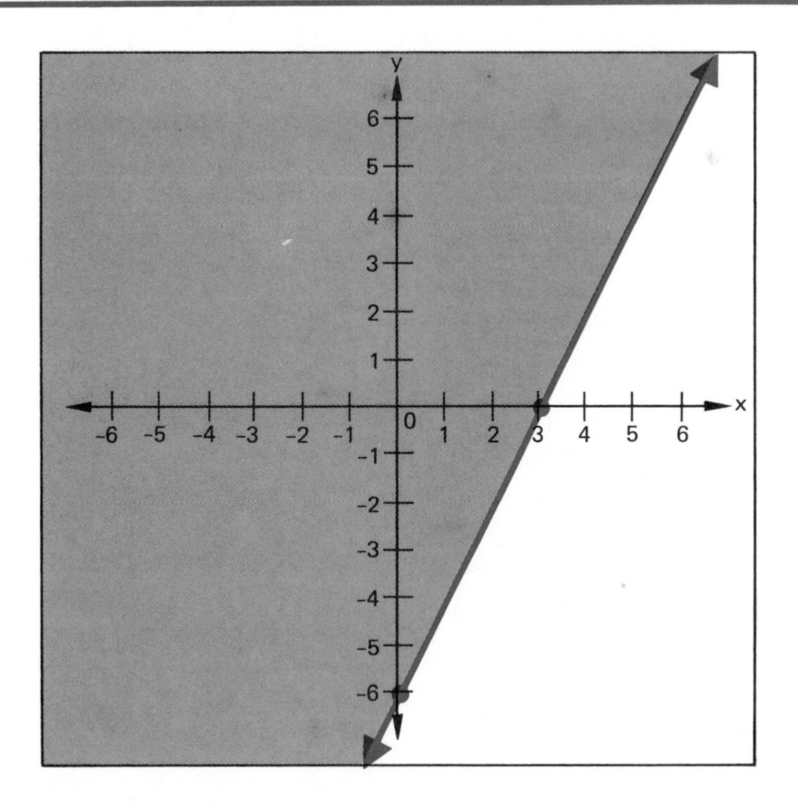

Solution

First we locate the boundary by graphing the equation $-3x + 4y = 12$. This boundary will be a dashed line because we have a strict inequality. Next we test a point to see if it satisfies the given inequality. Testing the origin, we let $x = 0$ and $y = 0$; then we have

$$-3(0) + 4(0) < 12$$
$$0 + 0 < 12$$
$$0 < 12 \qquad \text{(true)}$$

Since (0, 0) satisfies the given inequality, all of the points in that half plane will also satisfy the given inequality. Therefore we shade the half plane that contains the point (0, 0), which is the half plane below the line $-3x + 4y = 12$. See figure 33. Note that the boundary is a dotted line.

In the preceding section we graphed equations and in this section we have graphed inequalities. We can also combine equations or inequalities to get compound sentences, for example, the conjunction of $x - y \leq 2$ and $x + y < 1$. The graph of the conjunction of $x - y \leq 2$ and $x + y < 1$ is the intersection of the individual

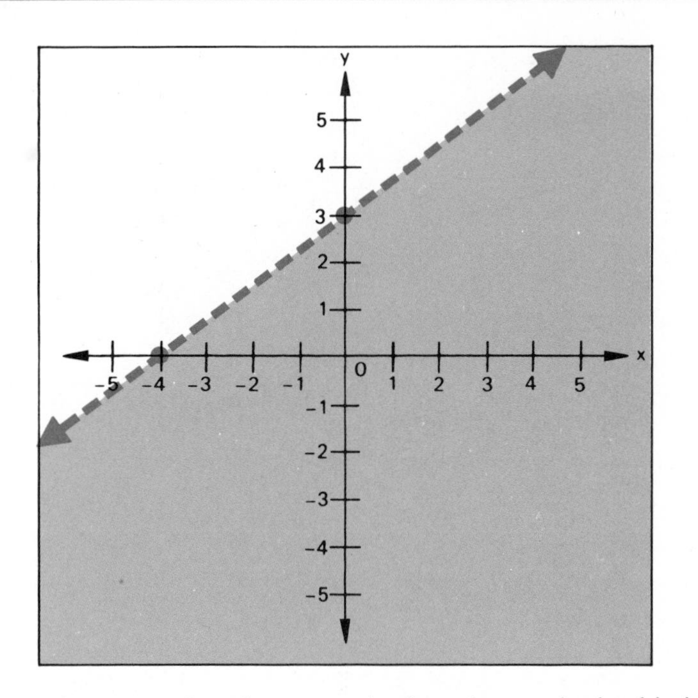

FIGURE 33 $-3x + 4y < 12$

graphs of $x - y \leq 2$ and $x + y < 1$. In order to obtain this intersection, we draw both graphs on the same Cartesian plane. The solution is the region where the two half planes intersect. The coordinates of all the points in the intersection will satisfy both of the given sentences and hence form the solution set of their conjunction. The graphs of $x - y \leq 2$ and $x + y < 1$ are shown separately in figures 34a and 34b. Figure 34c is the graph of $x - y \leq 2$ *and* $x + y < 1$. Note that figure 34c is the intersection of the graphs in figures 34a and 34b.

FIGURE 34a $x - y \leq 2$

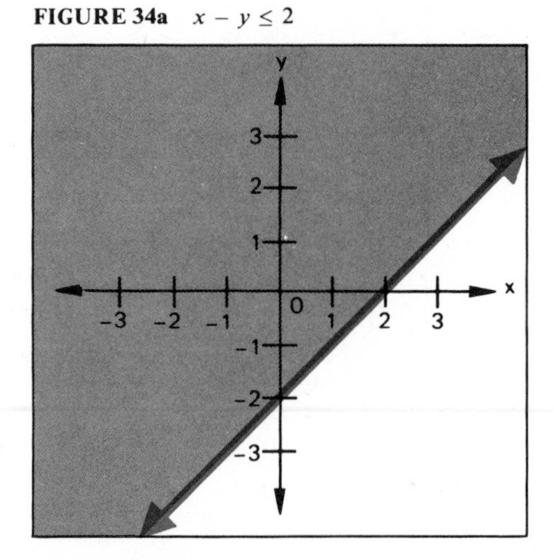

FIGURE 34b $x + y < 1$

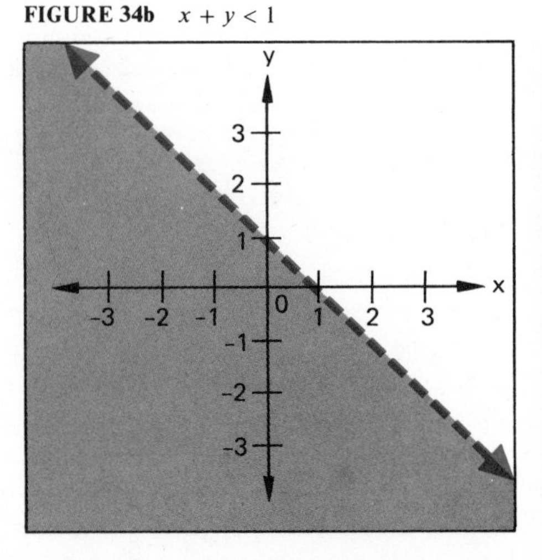

FIGURE 34c $x - y \leq 2$ and $x + y < 1$

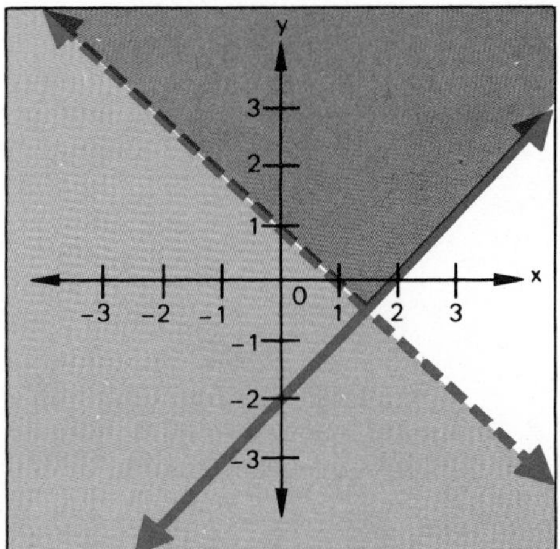

EXERCISES FOR SECTION 9.9

Graph each of the following inequalities on the Cartesian plane.

1. $x + y > 5$
2. $x + y < 8$
3. $x - y < 3$
4. $x - y > 4$
5. $x - y \leq 1$
6. $2x - y \geq 4$
7. $-2x + y \geq 4$
8. $x - y \geq 0$
9. $y - x < 0$
10. $x - 2y < 0$
11. $-2x - 2y > 0$
12. $3x - 5y \leq 15$
13. $-3x + 5y \leq 15$
14. $-3x - 5y < -15$
15. $3x + 5y > 15$
16. $-2x - 3y > -6$
17. $-3x - 2y < -6$
18. $x + y \geq 2$
19. $x + y \geq 2$ and $x - y < 2$
20. $x + y \leq 3$ and $x - y > 3$
21. $y \geq x$ and $y \geq -x$

Just For Fun

If one peacock lays three eggs in one day, then how many eggs will 33 peacocks lay in 11 days? (Be careful!)

9.10 LINEAR PROGRAMMING

In the preceding section, we graphed linear inequalities. Such graphs can be used to solve practical problems. As an example, consider a problem encountered by the E–Z Furniture Company, which manufactures tables and chairs for family rooms and patios: all of the E–Z Company's furniture undergoes some construction steps in building I and others in building II. Each table requires 3 hours' work in building I and 2 hours in building II to produce. Each chair requires 2 hours in building I and 4 hours in building II. The profit from each table is \$6, and the profit from each chair is \$5. Due to union regulations, the two buildings can only operate for at most 8 hours a day. Given these restrictions, how many tables and how many chairs should the company produce each day in order to maximize its profits?

This problem is an example of a **linear programming problem.** The theory of linear programming was developed to solve problems like these, where the goal is to maximize profits subject to a few simple restrictions. The theory has been expanded to solve similar problems of a much more complex nature. Many linear programming problems involve many more unknowns than our example and require the use of a computer to solve them. However, simple problems like that of the E–Z Furniture Company can be solved by graphing linear inequalities, as we shall see.

To solve a linear progrmming problem, we must first translate the word problem into mathematical statements. We shall now proceed to do this for the E–Z Furniture Company example. We begin by letting

x = the number of tables to be produced
y = the number of chairs to be produced

We also list all of the given information in a table:

	Time needed (hours)		
	Building I	Building II	Profit for each
Table	3	2	\$6
Chair	2	4	\$5
Time limit	8	8	

From this table, we can construct the mathematical statements needed to solve the problem. For example, if a table requires 3 hours in building I, then 2 tables would require $2 \cdot 3 = 6$ hours in building I. Therefore x tables would require $x \cdot 3 = 3x$ hours in building I. Similarly, y chairs would require $2y$ hours in building I. What do we know about the time spent in building I? It must be less than or equal to 8 hours, due to union rules. Therefore we obtain the open sentence

$$3x + 2y \leq 8$$

Similarly, x tables require $2x$ hours in building II, and y chairs require $4y$ hours in building II. Building II can also only be used for a maximum of 8 hours. Therefore, we obtain the open sentence

$$2x + 4y \leq 8$$

The same type of reasoning can be used to express the profit P in terms of x and y. A profit of \$6 is made on each table. Therefore 2 tables would yield a profit of $6 \cdot 2 = 12$ dollars. Similarly, x

tables would yield a profit of $6x$ dollars. A profit of $5 is made on each chair. Hence y chairs would yield a profit of $5y$ dollars. The total profit from both tables and chairs is $6x + 5y$. Therefore we say that

$$P = 6x + 5y$$

We now have three mathematical statements from the information in the table:

Building I:	$3x + 2y \leq 8$
Building II:	$2x + 4y \leq 8$
Profit:	$P = 6x + 5y$

The first two statements are linear inequalities; they describe the restrictions on the numbers of chairs and tables that can be manufactured each day. These inequalities are called the **constraints** of the problem. The third mathematical statement describes the relationship between the number of chairs and tables produced and the profit that can be made by selling them. This expression is called the **objective function.** Since the E–Z Furniture Company would like its profits to be as high as possible, we want to maximize the value of the objective function, $P = 6x + 5y$, subject to the constraints $3x + 2y \leq 8$ and $2x + 4y \leq 8$. In other words, we will solve the E–Z Company's problem if we can find numbers x and y such that:

(1) $3x + 2y \leq 8$
(2) $2x + 4y \leq 8$
(3) $P = 6x + 5y$ is as large as possible

Any numbers x and y that solve the problem must satisfy inequalities (1) and (2). Graphing both of these inequalities on the Cartesian plane gives us a region in which the coordinates of each point satisfy inequalities (1) and (2). Therefore the coordinates of the points in this region will be the values of x and y that we shall consider when we look for solutions to the problem.

We will now graph inequalities (1) and (2) on the same set of axes. However, we shall graph them only in the region where $x \geq 0$ and $y \geq 0$. We do this because x and y represent the numbers of tables and chairs to be manufactured, respectively, and therefore x and y cannot be negative quantities. Hence $x \geq 0$ and $y \geq 0$.

The graphs of the two inequalities are shown in figure 35. We are concerned only with the portion of this figure where the two half planes intersect. This is because the coordinates of each point in

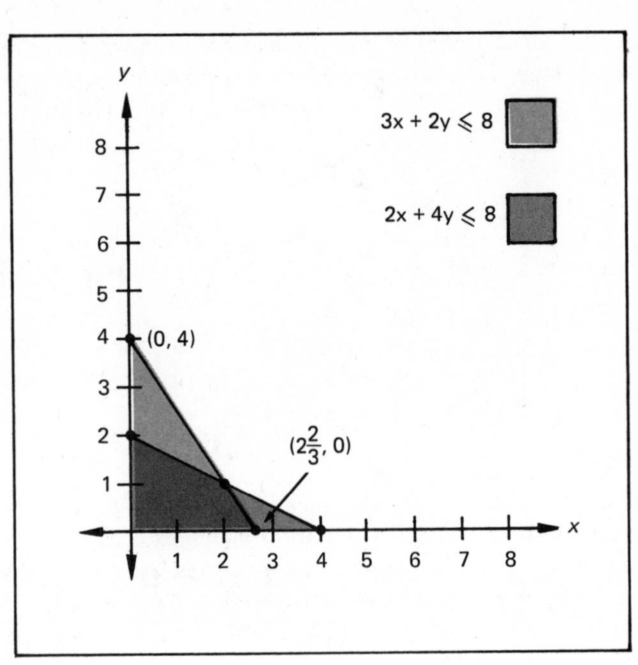

FIGURE 35 $3x + 2y \leq 8$ and $2x + 4y \leq 8$

the intersection satisfy both of the inequalities, and therefore the intersection forms the solution set of the conjunction of the two inequalities.

Let us examine this region more closely: figure 36 shows this region with its four corners labeled. The corner points are $A(0, 0)$, $B(2\frac{2}{3}, 0)$, $C(2, 1)$, and $D(0, 2)$.

It can be proved, using more advanced mathematics, that, in a linear programming problem, any maximum or minimum values of the objective function always occur at a vertex. Hence our profit expression P will have its maximum or minimum value at a corner of the region in figure 36. We can test these values by substituting them in the profit expression, $P = 6x + 5y$:

At $A(0, 0)$, we have $P = 6 \cdot 0 + 5 \cdot 0 = 0 + 0 = 0$.
At $B(2\frac{2}{3}, 0)$, we have $P = 6 \cdot \frac{8}{3} + 5 \cdot 0 = \frac{48}{3} + 0 = 16$.
At $C(2, 1)$, we have $P = 6 \cdot 2 + 5 \cdot 1 = 12 + 5 = 17$.
At $D(0, 2)$, we have $P = 6 \cdot 0 + 5 \cdot 2 = 0 + 10 = 10$.

The maximum profit occurs when $x = 2$ and $y = 1$, and it is \$17. The E–Z Furniture Company should produce 2 tables and 1 chair to obtain a maximum profit. Note that the minimum profit, \$0, occurs at $A(0, 0)$.

Keep in mind that this is an illustrative example of a linear programming problem and one method for solving such problems. More

FIGURE 36

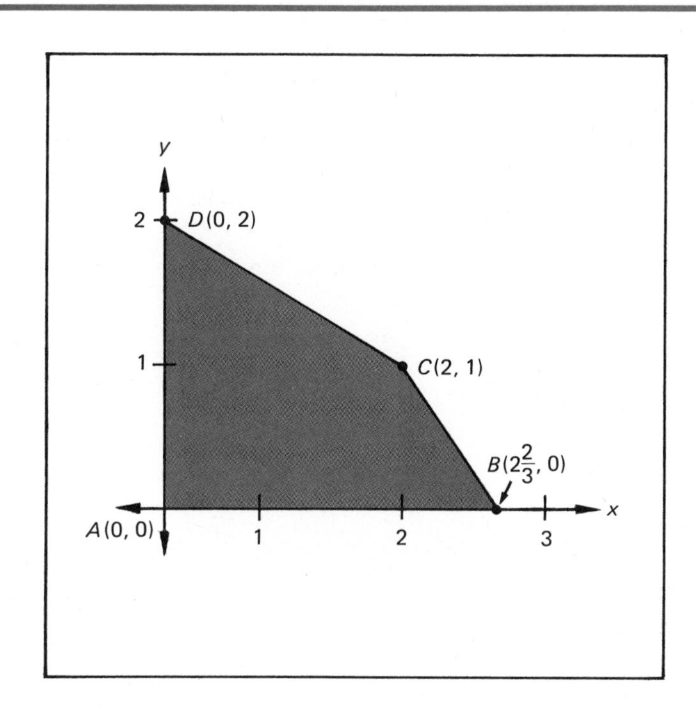

complex problems involve more variables and must be solved in a different manner.

In solving this example, we evaluated the corner points of the graph of the constraint inequalities. As frequently happens, one of these corner points—$B(2\frac{2}{3}, 0)$—does not have integer coordinates. What if the profit expression P had its maximum value at $B(2\frac{2}{3}, 0)$ instead of at $C(2, 1)$? (This could easily happen if the problem were slightly different.) In terms of the original problem, this would mean that the E–Z Furniture Company should manufacture $2\frac{2}{3}$ tables and 0 chairs. But it does not make much sense to manufacture $\frac{2}{3}$ of a table. In such a case, linear programming yields only an approximate answer to the problem. A more sophisticated mathematical technique, *integer programming,* must be used to find exact answers to problems that require whole numbers as answers.

EXAMPLE 1

The Blivit Electronic Company manufactures bleeps and peeps. Manufacturing a bleep requires 2 hours on machine A and 1 hour on machine B. Manufacturing a peep requires 1 hour on machine A and 1 hour on machine B. Machine A cannot be used more than 7 hours a day, and machine B cannot be used more than 5 hours a day. If the profit from a bleep is \$5 and that from a peep is \$4, how many of each should be produced to maximize profit?

Solution

We begin by letting

x = the number of bleeps to be manufactured
y = the number of peeps to be manufactured

Next we list all of the given information in a table:

	Time needed (hours)		
	Machine A	Machine B	Profit for each
Bleep	2	1	5
Peep	1	1	4
Time limit	7	5	

x bleeps require $2 \cdot x$, or $2x$, hours on machine A and y peeps require $1y$, or y, hours on machine A. The time limit on machine A is 7 hours. Therefore we have $2x + y \le 7$.

x bleeps require $1x$, or x, hours on machine B and y peeps require $1y$, or y, hours on machine B. The time limit on machine B is 5 hours. Therefore we have $x + y \le 5$.

The profit for x bleeps is $5x$ and the profit for y peeps is $4y$. The total profit, P, is $5x + 4y$. That is,

$$P = 5x + 4y$$

We want to maximize P.

We now have three mathematical statements from the given information on the table:

Machine A:	$2x + y \le 7$
Machine B:	$x + y \le 5$
Profit:	$P = 5x + 4y$

To find the desired region, we graph the inequalities for machines A and B on the same set of axes. We do this only where $x \ge 0$ and $y \ge 0$, as shown in figure 37.

The feasible region whose vertices will provide a maximum or minimum is that region where the half planes intersect. The region and its vertices are shown in figure 38.

FIGURE 37 $2x + y \leq 7$ and $x + y \leq 5$

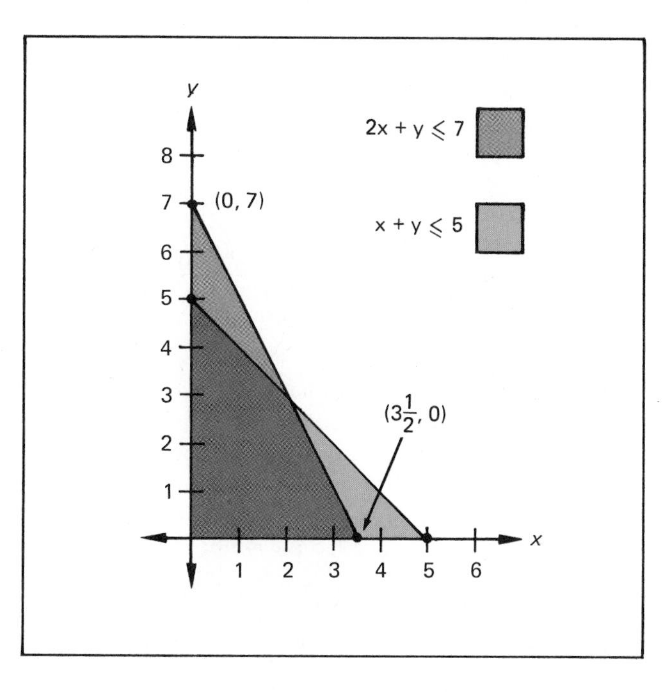

FIGURE 38

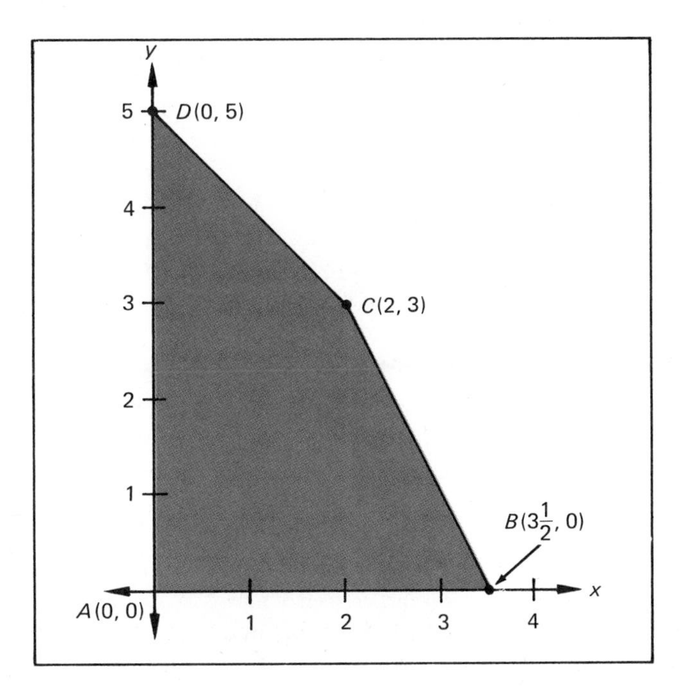

We now test the profit expression $P = 5x + 4y$ at each of these vertices:

At $A(0, 0)$, $P = 5 \cdot 0 + 4 \cdot 0 = 0 + 0 = 0$.
At $B(3\frac{1}{2}, 0)$, $P = 5 \cdot \frac{7}{2} + 4 \cdot 0 = \frac{35}{2} + 0 = 17.5$.
At $C(2, 3)$, $P = 5 \cdot 2 + 4 \cdot 3 = 10 + 12 = 22$.
At $D(0, 5)$, $P = 5 \cdot 0 + 4 \cdot 5 = 0 + 20 = 20$.

The maximum profit occurs when $x = 2$ and $y = 3$, and it is \$22. The Blivit Electronic Company should produce 2 bleeps and 3 peeps to obtain a maximum profit.

When linear programming problems involve only two variables, the technique used in example 1 and in the E–Z Furniture Company problem is convenient, and it is the only technique we will consider here. For problems containing many variables, more complicated techniques, usually executed by a computer, must be used. The most common of these techniques is called the *simplex method.*

EXERCISES FOR SECTION 9.10

1. Find the maximum value of $P = 3x + 2y$ under the following conditions:

   $x \geq 0, \quad y \geq 0, \quad x + y \leq 5, \quad x - y \leq 1$

2. Find the maximum value of $P = 5x + 8y$ under the following conditions:

   $x \geq 0, \quad y \geq 0, \quad x + y \leq 7, \quad 3x + y \leq 15$

3. Find the maximum and minimum values of $P = 4x + 3y$ under the following conditions:

   $x \geq 0, \quad y \geq 0, \quad 2x + y \leq 12, \quad x + y \leq 7$

4. Find the maximum and minimum values of $P = x + 4y$ under the following conditions:

   $x \geq 0, \quad y \geq 0, \quad x + y \leq 10, \quad 3x - y \leq 6$

5. A manufacturer makes bikes and wagons. To produce a bike requires 2 hours on machine A and 4 hours on machine B. To produce a wagon requires 3 hours on machine A and 2 hours on machine B. Machine A can operate at most 12 hours per day and machine B can operate at most 16 hours per day. If the manufacturer makes a profit of \$12 on a bike and \$10 on a wagon, how many of each should be produced in order to maximize profit?

6. A manufacturer makes lawn mowers and snow blowers. To produce a lawn mower requires 3 hours on machine A and 2 hours on machine B. To produce a snow blower requires 2 hours on machine A and 4 hours on machine B. Machine A can operate at most 18 hours per day and machine B can operate at most 20 hours per day. If the manufacturer makes a profit of \$20 on a lawn mower and \$30 on a snow blower, how many of each should be produced in order to maximize profit?

7. Frank Sloane raises pheasants and partridges and has room for at most 100 birds. It costs him \$2 to raise a pheasant and \$3 to raise a partridge, and he has \$240 to cover these costs. If he can make a profit of \$7 on each pheasant and \$8 on each partridge, how many of each bird should he raise in order to maximize his profit?

8. Betty Juarez has a 100-acre farm where she

raises two crops, potatoes and cauliflower. It costs her \$20 to raise an acre of potatoes and \$40 to raise an acre of cauliflower, and she has \$2600 to cover the costs. If she can make a profit of \$35 on each acre of potatoes and \$60 on each acre of cauliflower, how many acres of each crop should she plant in order to maximize her profit?

9. The Long Island Shellfish Company processes (cleans, sorts, opens, and freezes) oysters and clams. In a given week, the company can process 600 bushels of shellfish, of which 100 bushels of oysters and 200 bushels of clams are required by regular customers (restaurants). The profit on a bushel of oysters is \$8 and on a bushel of clams is \$10. How many bushels of oysters and of clams should the company process in order to maximize its profit?

Just For Fun

Two people have the same parents. They were born on the same day, and at the same place, but they are not twins. How are they related?

9.11 SUMMARY

In this chapter, we explored the topic of algebra and some of its basic characteristics. Algebra is the area of mathematics that generalizes the facts of arithmetic. In algebra, letters are used to denote numbers or a certain set of numbers. The expression $4 + 2$ is an arithmetic expression, while $3x + 2y$ is an algebraic expression. The letters are called *variables,* since they are used to represent an unknown member of the set. Sentences such as $x + 1 = 3$, which cannot be classified as true or false until we replace x by a number, are called *open sentences.* There are open sentences of *equality* and open sentences of *inequality* (for example, $x + 1 > 3$). We can find the solution sets for these sentences, and we can graph these solution sets on the number line.

In solving most equations, we make use of the following axioms:

1. If the same quantity or two equal quantities are added to two equal quantities, then the sums are equal.

2. If the same quantity or two equal quantities are subtracted from two equal quantities, then the remainders (differences) are equal.

3. If equal quantities are multiplied by equal quantities, then the products are equal.
4. If equal quantities are divided by equal quantities (except zero), then the quotients are equal.

Once we have found a solution to an equation, we can check our answer by replacing the variable by the solution in the original equation to see if the left side of the equation equals the right side of the equation.

One of the oldest applications of algebra is in solving word problems. Most word problems require the use of algebra in order to find the solution in a systematic manner, as opposed to trial and error. In order to solve a verbal problem, we perform the following steps: (1) let the unknown be represented by a variable such as x, y, and so on; (2) translate the given relationships with the unknown into an equation; (3) solve the resulting equation in order to find the value of the unknown. It is important to check the answer to make sure that it is correct. This is done by checking to see if the answer satisfies the original problem (not just the equation).

Linear equations in two unknowns are equations of the form $Ax + By = C$, where A, B, and C are real numbers. They are called linear equations because the graph of such an equation is a straight line. Since a linear equation such as $x + y = 1$ contains two variables, x and y, each of its solutions is an ordered pair (x, y). There are infinitely many ordered pairs that will satisfy a given linear equation. In order to find solutions for a linear equation, select a value for x and substitute it for x in the equation; then solve the resulting equation for y. In listing the solution as an ordered pair, the x value is always the first value, and the y value is listed second: (x, y).

Since each ordered pair represents a point on the Cartesian plane, we can locate several solutions for a linear equation on the plane and connect these points to obtain the graph of the equation. When we graph a linear equation, it is usually convenient to find the points where the line crosses each axis, that is, the *x-intercept* and the *y-intercept*. We should also find a third solution to the equation to check that all three points do lie on the same straight line. If one of the points is not on the line, then we must go back and check for an error in determining the solutions to the given linear equation.

The graph of an equation of the form $y = ax^2 + bx + c$ is a parabola. If $a > 0$, then the parabola opens upward, while if $a < 0$, then the graph opens downward. The turning point of a parabola is called the *vertex* of the parabola. For any parabola of the form $y = ax^2 + bx + c$, the x-value of the vertex is $\frac{-b}{2a}$. The y-value is found by substituting $\frac{-b}{2a}$ for x in the original equation. After determining the vertex of the parabola and which way it opens, the next step in graphing the parabola is to make a table of values. This is done by choosing two or three values of x on each side of the vertex and finding the corresponding values of y. Then the points are plotted and connected by a smooth curve to obtain the graph of the parabola.

In order to graph an inequality in two variables, we do the following:

1. Find the boundary of the half plane. We do this by graphing the equation derived from the inequality.
2. Make the boundary a dashed line for a strict inequality such as $>$ or $<$. But for $\geq$ or $\leq$, make the boundary a solid line.
3. Use shading to indicate the half plane that is the solution. We do this by testing a point. If a point satisfies the given inequality, then all of the points in the half plane must also satisfy the same inequality.

Graphs of linear inequalities can be used to solve some linear programming problems. To solve a linear programming problem, first translate the problem into mathematical statements. These mathematical statements can be obtained by placing all of the given information in a table. Next, we express the constraints in terms of linear inequalities in x and y, and graph this system of inequalities in the region of the Cartesian plane where $x \geq 0$ and $y \geq 0$. The solution set for the system of inequalities is a polygonal region, one of whose vertices will provide the desired maximum or minimum value for an expression of profit, cost, or some other quantity.

You should be aware that this chapter is intended only to provide an introduction to algebra. There are many topics in algebra that are beyond the scope of this text. You should be aware that algebra is one of the cornerstones of mathematics and provides methods for solving a multitude of problems.

Review Exercises for Chapter 9

1. Find the solution set for each of the following open sentences. Unless otherwise noted, the replacement set is the set of real numbers.

   a. $x - 2 = 0$
   b. $x + 4 = -2$
   c. $x - 2 = 6 - 2$
   d. $x + 4 \leq 7$ (x is a natural number)
   e. $x - 2 < 4$ (x is a whole number)
   f. $-3 < x < 3$ (x is an integer)
   g. $-4 \leq x \leq 3$ (x is a natural number)

2. Graph the solution set (on a number line) for each of the following open sentences. In each case, the replacement set is the set of real numbers.
   a. $x < 3$ b. $x - 1 \geq 1$
   c. $-1 < x < 2$ d. $-2 \leq x \leq 1$
   e. $0 < x \leq 3$ f. $x - 2 < x$

3. Solve each of the following equations. (The replacement set is the set of real numbers.)
   a. $2x - 2 = 10$ b. $2y + 4 = y + 6$
   c. $5z - 3 = 2z + 3$ d. $2y + 6 = 5y - 3$
   e. $\frac{x}{2} - 2 = 3$ f. $\frac{z}{3} = 2 - z$

4. The sum of two consecutive odd integers is 28. Find the numbers.

5. Ike is 2 years older than Bill and the sum of their ages is 50. How old is each?

6. Five less than 3 times a number is 16. Find the number.

7. The width of a rectangle is 5 metres less than its length. If the perimeter of the rectangle is 30 metres, find the dimensions of the rectangle.

8. Frank has $3.00 in dimes and quarters in his pocket. If he has twice as many quarters as dimes, how many of each type of coin does he have?

9. Find three solutions for each of the following equations, and then use the solutions to graph the equation on a Cartesian plane.
   a. $x - y = 2$ b. $2x + y = 4$
   c. $2x - y = 4$ d. $-2x + y = 6$
   e. $3x - 5y = 15$ f. $y = x$

10. Graph the given equation on a Cartesian plane.
   a. $y = x^2$ b. $y = x^2 - 4$
   c. $y = -x^2 + 2$ d. $y = x^2 - 6x$
   e. $y = x^2 + 2x - 3$
   f. $y = -x^2 + 6x - 5$

11. Graph each of the following inequalities on the Cartesian plane.
   a. $x + y > 3$ b. $x - y \leq 2$
   c. $-x + y > 2$ d. $3x + 5y > -15$
   e. $x - 2y \leq 0$ f. $-3x - 2y \leq -6$

12. Find the maximum value of $P = 2x + 3y$ under the following conditions:

$$x \geq 0, \quad y \geq 0, \quad 2x + y \leq 10, \quad x + 3y \leq 15$$

13. A manufacturer makes couches and recliners. To produce a couch requires 3 hours in the frame shop and 2 hours in the upholstery shop. To produce a recliner requires 4 hours in the frame shop and 4 hours in the upholstery shop. The frame shop can operate at most 24 hours per day, and the upholstery shop can operate at most 20 hours per day. If the manufacturer makes a profit of $30 on a couch and $45 on a recliner, how many of each should be produced in order to maximize profit?

14. The Safety-First Corporation manufactures two types of smoke alarms, a standard model and a deluxe model. To produce a standard smoke alarm requires 1 hour on machine A and 1 hour on machine B. To produce a deluxe smoke alarm requires 1 hour on machine A and 2 hours on machine B. Due to costs and safety regulations, machine A can operate at most 35 hours per week and machine B can operate at most 40 hours per week. If the manufacturer makes a profit of $7 on each standard model and $10 profit on each deluxe model, how many of each should the company produce in order to maximize profits?

Just For Fun

A train one mile long travels through a tunnel one mile long at a rate of one mile per hour. How long will it take the train to pass completely through the tunnel?

10 An Introduction to Geometry

After studying this chapter, you will be able to do the following:

1. Identify **points, lines, half lines, rays,** and **line segments**
2. Find the intersection and union of lines, rays, line segments, and half lines
3. Find the intersection and union of angles, of the **interiors** of angles, and of the **exteriors** of angles
4. Identify **acute angles, obtuse angles, right angles,** and **straight angles**
5. Use the Pythagorean theorem to find the length of a side of a right triangle when given the measures of the other two sides
6. Identify **equilateral triangles, isosceles triangles, right triangles, acute triangles, obtuse triangles,** and **equiangular triangles**
7. Identify **trapezoids, parallelograms, rhombuses, rectangles,** and **squares**
8. Determine whether a **network** is **traversable** by identifying the number of even and odd vertices in the network.

Symbols frequently used in this chapter

$\cdot C$	point C
$\overleftrightarrow{AB}$	line AB
$\overrightarrow{AB}$	ray AB
$\overline{AB}$	segment AB
$\overset{\circ\!\!\longrightarrow}{AB}$	half line AB
$\angle RST$	angle RST
$\overset{\circ\!-\!\circ}{TR}$	open line segment TR
$\overleftrightarrow{MN} \perp \overleftrightarrow{BT}$	line MN is perpendicular to line BT
$m(\overline{AB})$	measure of line segment AB

10.1 INTRODUCTION

What is geometry? To begin with, the word *geometry* is derived from two Greek words. The first part of the word *geometry* is taken from the Greek word *ge*, which means *earth,* and the second part is

Le Jeune/Stockmarket

taken from the Greek word *metron*, which means *measure*. Therefore we can safely assume that early geometry concerned itself with the measure of the earth, or earth measurement.

The Egyptians were one of the first people to use geometry. One of the ways that the Egyptians used geometry was to survey land, or "measure earth." The Egyptians had to pay taxes for their land. The kings sent workers to measure the people's land, and taxes were levied accordingly. The Egyptians used this form of taxation as early as 1300 B.C.

From this beginning, the study of geometry was carried on by the Greeks. Thales was one of the first Greeks to concern himself with proving that mathematical statements were true. Pythagoras was the next major contributor to the study of geometry (approximately 550 B.C.). You may be familiar with the Pythagorean theorem, which states that the sum of the squares of the legs of a right triangle is equal to the square of the hypotenuse.

The next great Greek scholar was Plato. He founded a school in the city of Athens. Plato thought that geometry was so important that he had the slogan "Let no one ignorant of geometry enter my doors" posted at the entrance to his school.

Euclid, another Greek, is known as the "father of geometry." In approximately 300 B.C., Euclid gathered all of the mathematical works known to exist at that time and produced a book called *The Elements*. From this book, the modern high school geometry course has evolved.

In this chapter, we will *not* concern ourselves with proofs of theorems. Instead, we will examine some of the more basic elements of geometry, such as points, lines, and planes. These are some of the same ideas that interested the ancient Greeks.

10.2 POINTS AND LINES

The most basic terms of geometry are *point, line,* and *plane*. Each of these terms presents a problem for us since we cannot define it. Granted, you can find a description for these words in the dictionary —but can you *define* them? For instance, what is a **point?** You might want to describe a point as a dot on a piece of paper, but what size dot? Other concepts of a point are of a point on the number line or a point on the Cartesian plane. The important thing to remember is that a geometric point has no dimension; that is, it has no length, breadth, or thickness. The one thing that a geometric point does possess is position. We can represent a point by a dot, just as we used the symbol "2" to represent the number two. Points are labeled or named by using capital letters. Therefore, if we wish to represent a point C, we make a dot and label it with a capital C as shown below

$\cdot\, C$

We now have an intuitive idea of what a point is and how we shall represent it. We shall not define a point, just as we did not define a set in chapter 1. (Recall that one of the ingredients of an axiomatic system is a collection of undefined terms which are used to define other terms.)

A **line** is formed by the intersection of two flat surfaces. For example, the intersection of two walls in one corner of a room can be thought of as a line. A point in geometry has only position, while a line in geometry has only one dimension, length. A geometric line has no width. Note that when we refer to a *line* we are talking about a *straight line,* unless otherwise noted.

We can think of a line as a set of points. The points are *on* the line and the line passes *through* the points. It is interesting to note that a line may be extended infinitely far in either direction. What this means is that a line has no endpoints. Normally we work with only pieces or parts of lines; these are described later in this section.

We can name a line in different ways. We can, for example, name a line by writing a lower-case letter such as a, b, c, or d near the line. Figure 1 shows three lines, line a, line b, and line c. It is more common to name a line by using two points on the line. The line shown in figure 2 can be denoted by $\overleftrightarrow{AB}$, $\overleftrightarrow{AC}$, $\overleftrightarrow{BC}$, $\overleftrightarrow{CD}$, and so on. Since all the points A, B, C, and D lie on the same line, we say that they are *collinear.* **Collinear points** are points that lie on the same line.

FIGURE 1

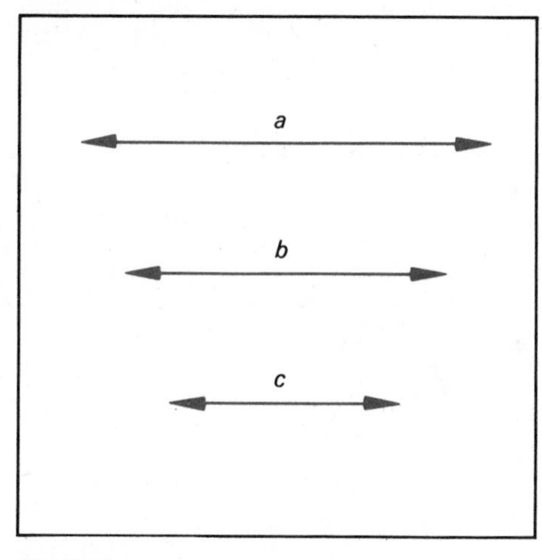

FIGURE 2

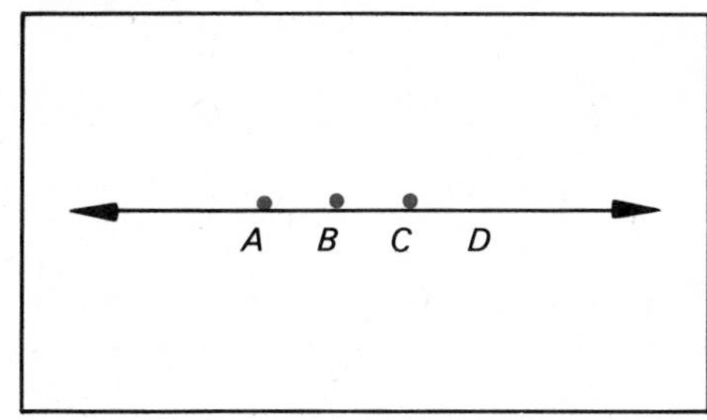

We can name a line by using only two letters, such as $\overleftrightarrow{AB}$, because one and only one straight line can be drawn through any two points. This is another way of saying that any two different points determine a unique line; or we could say that there is exactly one line containing any two different points.

A point on a line separates one part of the line from another part. In fact, when we place a point on a line as in figure 3, we separate the line into three sets: the given point and two *half lines.* The point P is not a point on either of the half lines. A half line is a set of points; if we include point P with the set of the points that constitute a half line, we get what is known as a *ray*. A **ray** has only one endpoint and may be extended indefinitely in only one direction from that endpoint. Figure 4 illustrates ray AB. The notation used to denote ray AB is $\overrightarrow{AB}$.

FIGURE 3

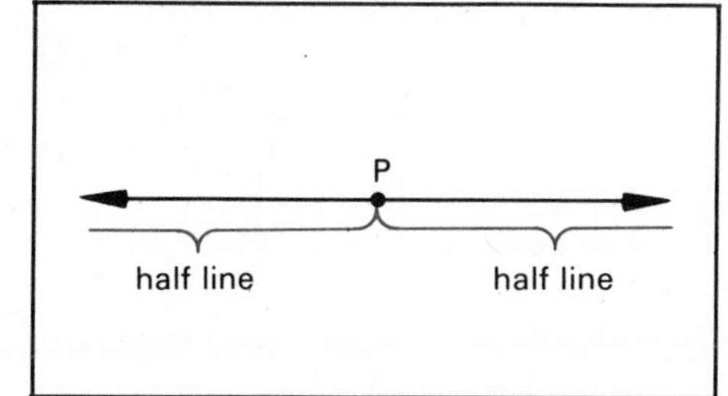

FIGURE 4

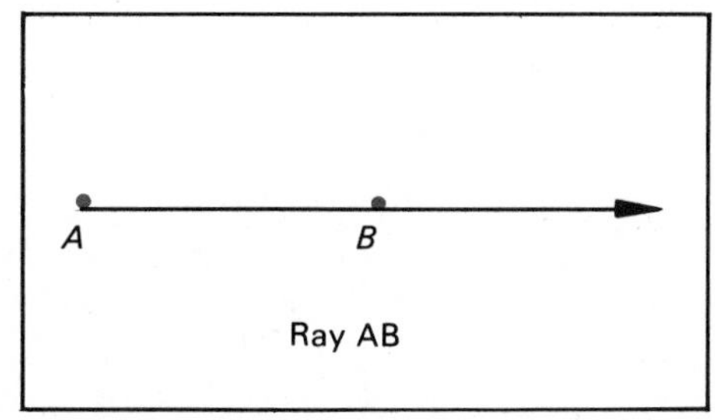

If we consider only those points between and including A and B, as in figure 5, we have a *line segment*. A **line segment** is that part of a line contained between two of its points, including the two endpoints. Line segment AB is denoted by $\overline{AB}$. Remember that a line may be extended indefinitely in either direction, a ray can be extended indefinitely in only one direction, and a line segment cannot be extended at all.

FIGURE 5

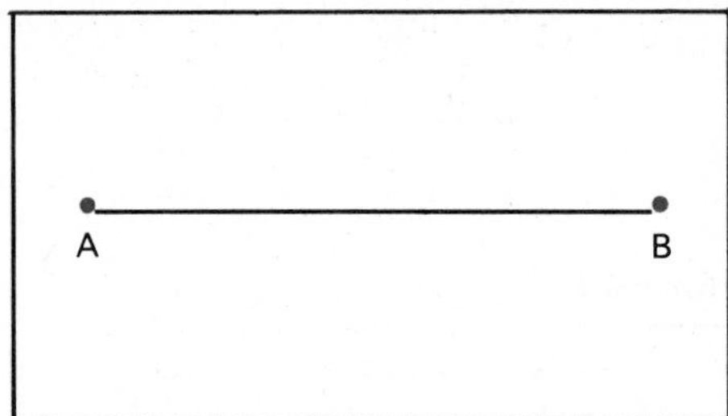

Thus far in our discussion, we have covered the concepts of point, line, half line, ray, and line segment. These terms and their corresponding notations are:

Description	Diagram	Notation
point P	•	P
line PQ	P Q	$\overleftrightarrow{PQ}$
half line PQ	P Q	$\overset{\circ\!\!\rightarrow}{PQ}$
ray PQ	P Q	$\overrightarrow{PQ}$
ray QP	P Q	$\overrightarrow{QP}$
line segment PQ	P Q	$\overline{PQ}$

Ray PQ is denoted by $\overrightarrow{PQ}$, which means that the endpoint of the ray is P and the ray is directed towards the point Q. Ray QP is denoted by $\overrightarrow{QP}$, which means that the endpoint of the ray is Q and the ray is directed towards the point P. (Refer to the chart above.) Ray PQ ($\overrightarrow{PQ}$) and ray QP ($\overrightarrow{QP}$) are distinct; they involve different sets of points.

FIGURE 6

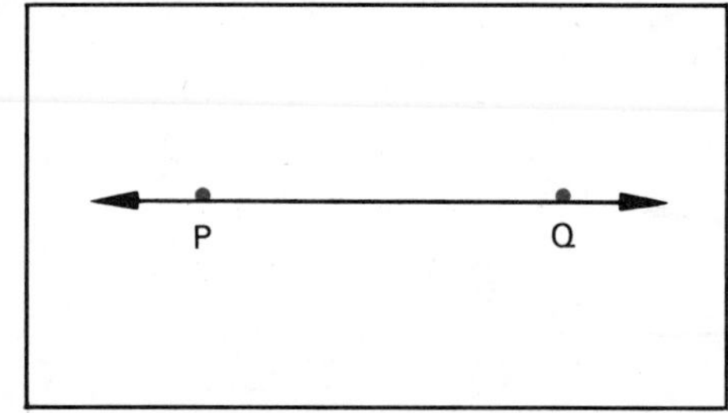

By now you have probably discovered that lines, rays, half lines, and line segments that pass through a given pair of points have some points in common and some points not in common. We can illustrate this by means of the two set operations, intersection and union. Consider line PQ in figure 6. Using this line, what is $\overrightarrow{PQ} \cap \overrightarrow{QP}$?

$\overrightarrow{PQ} \cap \overrightarrow{QP}$ is the intersection of ray PQ and ray QP. Ray PQ consists of the set of points which has the end point P and is directed towards the right through Q. Ray QP consists of the set of points which has the endpoint Q and is directed towards the left through P. Their intersection is the set of points common to both rays, that is, line segment PQ:

$$\overrightarrow{PQ} \cap \overrightarrow{QP} = \overline{PQ}$$

See figure 7.

FIGURE 7

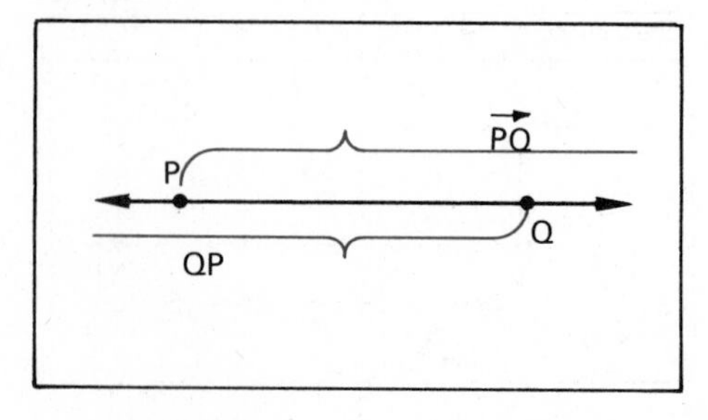

What is $\overrightarrow{PQ} \cup \overrightarrow{QP}$? We again refer to figures 6 and 7, but instead of the intersection we want the union of the two sets of points. That gives us all of the points on the line PQ. Therefore, we have

$$\overrightarrow{PQ} \cup \overrightarrow{QP} = \overleftrightarrow{PQ}$$

FIGURE 8

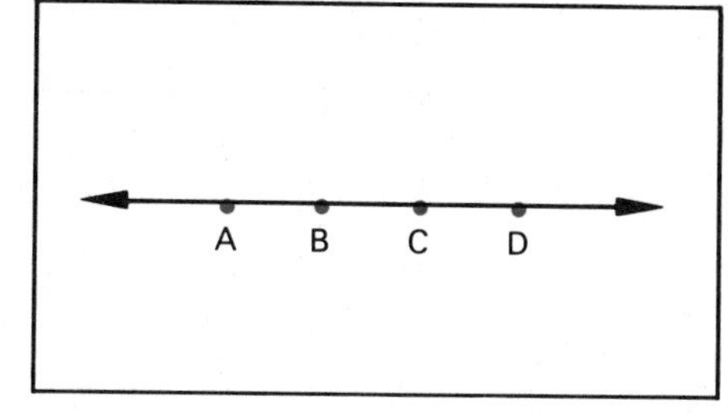

EXAMPLE 1

Use the line in figure 8 with the indicated points to find each of the following:

a. $\overrightarrow{AB} \cap \overrightarrow{CA}$ b. $\overrightarrow{AB} \cap \overrightarrow{BC}$

c. $\overrightarrow{BA} \cup \overrightarrow{BC}$ d. $\overline{AB} \cap \overline{CD}$

Solution

a. $\overrightarrow{AB} \cap \overrightarrow{CA}$ is the intersection of ray AB and ray CA. The set of points common to both of these rays is line segment AC. Therefore $\overrightarrow{AB} \cap \overrightarrow{CA} = \overline{AC}$.

b. $\overrightarrow{AB} \cap \overrightarrow{BC}$ is the intersection of ray AB and ray BC. These two rays are both directed to the right and have all of the points in ray BC in common. Therefore $\overrightarrow{AB} \cap \overrightarrow{BC} = \overrightarrow{BC}$.

FIGURE 9

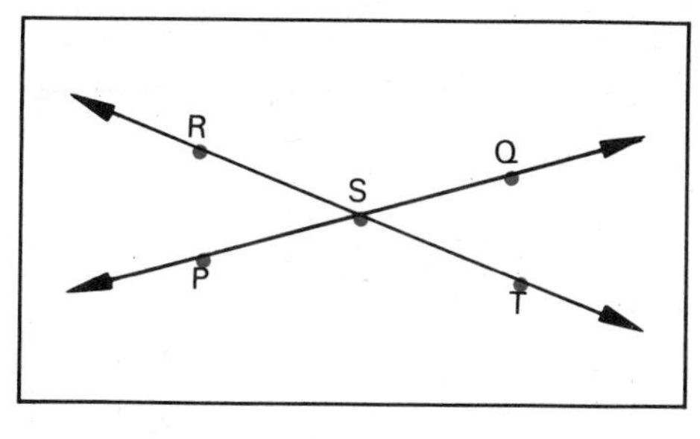

c. $\overrightarrow{BA} \cup \overrightarrow{BC}$ is the union of the two rays, and, since they are directed in opposite directions, their union will result in all of the points on the line. Therefore $\overrightarrow{BA} \cup \overrightarrow{BC} = \overleftrightarrow{AC}$. Note that it would also be correct to denote the answer as $\overleftrightarrow{AB}$, $\overleftrightarrow{BC}$, $\overleftrightarrow{CD}$, $\overleftrightarrow{BD}$, and so on.

d. $\overline{AB} \cap \overline{CD}$ is the intersection of line segment AB and line segment CD. Examining the diagram, we see that there are no points common to these two line segments. Their intersection is empty. Therefore, $\overline{AB} \cap \overline{CD} = \phi$.

When we have two different lines that contain the same point, these lines are said to **intersect** at that point. In figure 9, lines PQ and RT intersect in point S; that is, $\overleftrightarrow{PQ} \cap \overleftrightarrow{RT} = \{S\}$.

The set of points formed by two intersecting lines has many interesting subsets. For example, consider $\overrightarrow{SQ} \cup \overrightarrow{ST}$. What is the result when we unite the sets of points in ray SQ and ray ST? The result is a geometric figure formed by two rays drawn from the same point, as shown in figure 10. This figure is called an *angle*. An **angle** ($\angle$) is the union of two rays which have a common endpoint. The rays are called the **sides** of the angle and the common endpoint is called the **vertex** of the angle. Therefore

FIGURE 10 $\overrightarrow{SQ} \cup \overrightarrow{ST} = \angle QST$

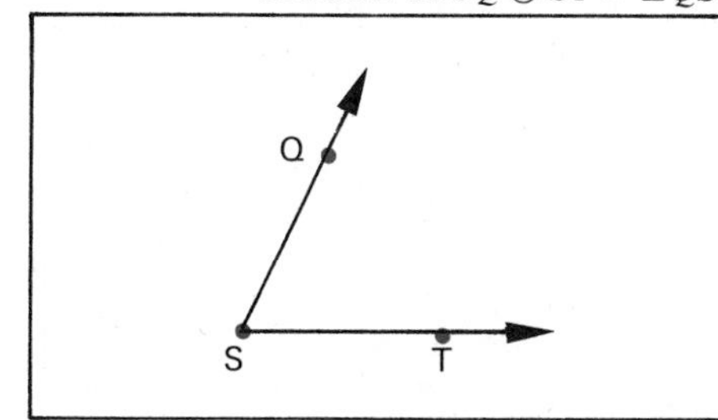

$$\overrightarrow{SQ} \cup \overrightarrow{ST} = \angle QST$$

We use capital letters to label an angle, and the name of the angle is written with the vertex letter in the middle. The first and third letters are used to designate the sides of the angle. In figure 9, we can see that

$$\overrightarrow{SR} \cup \overrightarrow{SP} = \angle RSP$$
$$\overrightarrow{SR} \cup \overrightarrow{SQ} = \angle RSQ$$
$$\overrightarrow{SP} \cup \overrightarrow{ST} = \angle PST$$

Angles RSP and RSQ are **adjacent angles** because they have the same vertex and a common side between them. Angles RSQ and QST are also adjacent angles, but angles RSP and QST are *not* adjacent angles. They do have the same vertex, but they do not have a common side. Angles RSP and QST are angles where the sides of one angle extend through the vertex and form the sides of the other. Angles of this type are called **vertical angles.** In figure 9, angles RSQ and PST are also vertical angles.

Consider the union of rays SR and ST in figure 9. We see that $\overrightarrow{SR} \cup \overrightarrow{ST} = \overleftrightarrow{RT}$. But, since an angle is the union of two rays that have a common endpoint, we can also say that $\overrightarrow{SR} \cup \overrightarrow{ST} = \angle RST$. Angle RST is a special kind of angle since its sides form a straight line. Angle RST is referred to as a **straight angle.** Therefore a line such as line PR in figure 11 can also be thought of as $\angle PQR$. (Remember that the vertex letter of an angle is always written in the middle when we label the angle.)

FIGURE 11

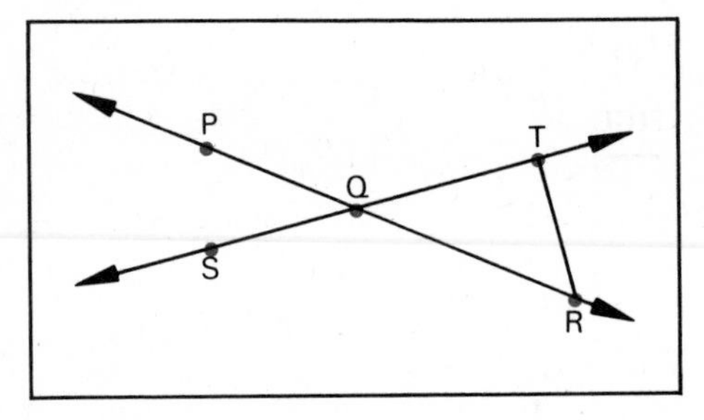

EXAMPLE 2

Use figure 11 with the indicated points to find each of the following:

a. $\overleftrightarrow{PR} \cap \overleftrightarrow{ST}$ b. $\overrightarrow{QP} \cup \overrightarrow{QT}$

c. $\overrightarrow{PR} \cap \overrightarrow{RP}$ d. $\overrightarrow{QP} \cup \overrightarrow{QR}$

Solution

a. $\overleftrightarrow{PR} \cap \overleftrightarrow{ST}$ is the intersection of lines PR and ST, and the two lines intersect at point Q. Therefore, $\overleftrightarrow{PR} \cap \overleftrightarrow{ST} = \{Q\}$.

b. $\overrightarrow{QP} \cup \overrightarrow{QT}$ is the union of rays QP and QT. The union of two rays which have a common endpoint is an angle. Therefore $\overrightarrow{QP} \cup \overrightarrow{QT} = \angle PQT$.

c. $\overrightarrow{PR} \cap \overrightarrow{RP}$ is the intersection of two rays having opposite direction. The set of points that they have in common is line segment PR. Therefore $\overrightarrow{PR} \cap \overrightarrow{RP} = \overline{PR}$.

d. $\overrightarrow{QP} \cup \overrightarrow{QR}$ is the union of two rays of opposite direction, so their union will result in all of the points on the line. Therefore $\overrightarrow{QP} \cup \overrightarrow{QR} = \overleftrightarrow{PR}$. (*Note:* Recall that we can also say that $\overrightarrow{QP} \cup \overrightarrow{QR} = \angle PQR$.)

FIGURE 12

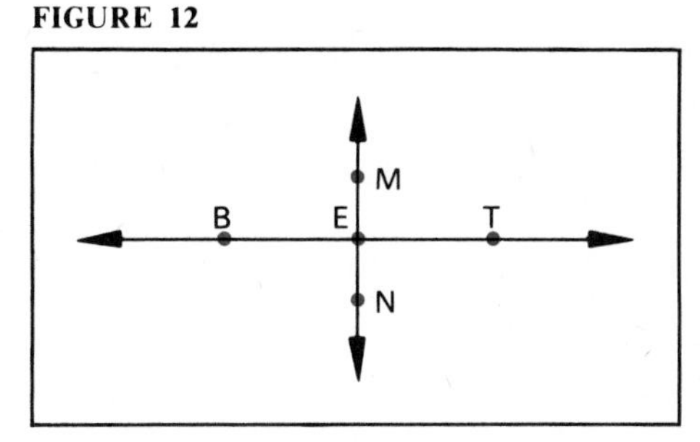

There are many other kinds of angles that can be defined. However, before we can discuss these angles, we must define *perpendicular lines*. Two lines that intersect so as to form a pair of equal adjacent angles are called **perpendicular lines.** Each line is said to be perpendicular to the other. In figure 12, $\overleftrightarrow{MN}$ is perpendicular to $\overleftrightarrow{BT}$; we denote this by $\overleftrightarrow{MN} \perp \overleftrightarrow{BT}$. Note also that $\overleftrightarrow{EM} \perp \overleftrightarrow{BT}$.

A right angle is an angle whose sides are perpendicular. In figure 12, $\angle MET$ is a **right angle,** as are $\angle BEM$, $\angle BEN$, and $\angle NET$. An angle that is wider than a right angle is called an **obtuse angle.** An angle that is narrower than a right angle is called an **acute angle.** In figure 13, $\angle$ FID is an obtuse angle, and $\angle$ DIG is an acute angle.

FIGURE 13

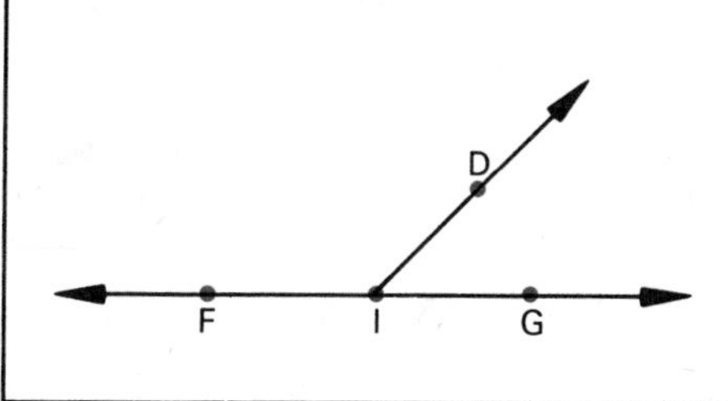

EXAMPLE 3

Use figure 13 with the indicated points to find each of the following:

a. $\overrightarrow{FG} \cap \overrightarrow{ID}$ b. $\overrightarrow{IF} \cup \overrightarrow{ID}$

c. $\overline{FI} \cap \overline{IG}$ d. $\angle FID \cap \angle DIG$

Solution

a. $\overrightarrow{FG} \cap \overrightarrow{ID}$ is the intersection of two rays; the only thing they have in common is point I. Therefore $\overrightarrow{FG} \cap \overrightarrow{ID} = \{I\}$.

b. $\overrightarrow{IF} \cup \overrightarrow{ID}$ is the union of two rays with a common endpoint. Therefore $\overrightarrow{IF} \cup \overrightarrow{ID} = \angle FID$.

c. $\overline{FI} \cap \overline{IG}$ is the intersection of two line segments with point I in common. Therefore $\overline{FI} \cap \overline{IG} = \{I\}$.

d. $\angle FID \cap \angle DIG$ is the intersection of two angles. They are adjacent angles because they have the same vertex and a common side. The intersection of these two angles is the common side, namely, $\overrightarrow{ID}$. Therefore $\angle FID \cap \angle DIG = \overrightarrow{ID}$.

Remember that any dot we place on a piece of paper will have some measurement such as height, width, and even thickness. The same can be said for a line segment that we may draw on paper; that is, any line that we draw will have some width and thickness in addition to length. But, in geometry, points and lines do not have such characteristics. The points and lines that we draw are diagrams that represent the points and lines of geometry.

EXERCISES FOR SECTION 10.2

For exercises 1–14, use figure 14 to find each of the following:

1. $\overline{QI} \cap \overline{IK}$
2. $\overline{QI} \cap \overline{CK}$
3. $\overline{QI} \cup \overline{IK}$
4. $\overline{UI} \cap \overline{CK}$
5. $\overrightarrow{QI} \cap \overline{QU}$
6. $\overrightarrow{QI} \cap \overrightarrow{CI}$
7. $\overrightarrow{IU} \cup \overrightarrow{IK}$
8. $\overleftrightarrow{UI} \cap \overline{CK}$
9. $\overrightarrow{IK} \cap \overline{IU}$
10. $\overrightarrow{IK} \cup \overline{IU}$
11. $\overrightarrow{IK} \cap \overrightarrow{IQ}$
12. $\overrightarrow{IK} \cap \overrightarrow{IC}$
13. $\overrightarrow{CI} \cap \overline{QU}$
14. $\overline{QU} \cap \overrightarrow{IC}$

FIGURE 14

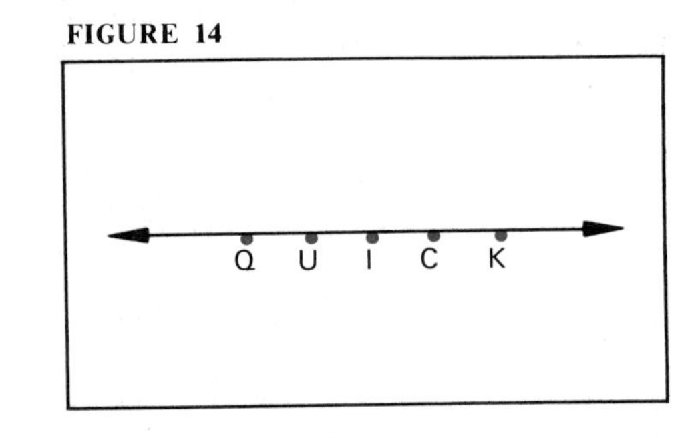

For exercises 15–28, use figure 15 to find each of the following:

15. $\overleftrightarrow{BT} \cap \overleftrightarrow{SW}$
16. $\overrightarrow{BT} \cap \overrightarrow{SW}$
17. $\overrightarrow{TB} \cap \overrightarrow{WS}$
18. $\overline{BE} \cup \overline{ET}$
19. $\overline{SE} \cap \overline{ET}$
20. $\overline{BE} \cap \overline{ET}$
21. $\overrightarrow{ET} \cup \overrightarrow{EW}$
22. $\overrightarrow{EB} \cup \overrightarrow{ES}$
23. $\overrightarrow{ET} \cup \overrightarrow{EB}$
24. $\overrightarrow{BE} \cap \overrightarrow{ET}$
25. $\overrightarrow{BE} \cup \overrightarrow{ET}$
26. $\overrightarrow{EW} \cup \overrightarrow{EB}$
27. $\angle BEW \cap \angle WET$
28. $\angle SET \cap \angle WET$

FIGURE 15

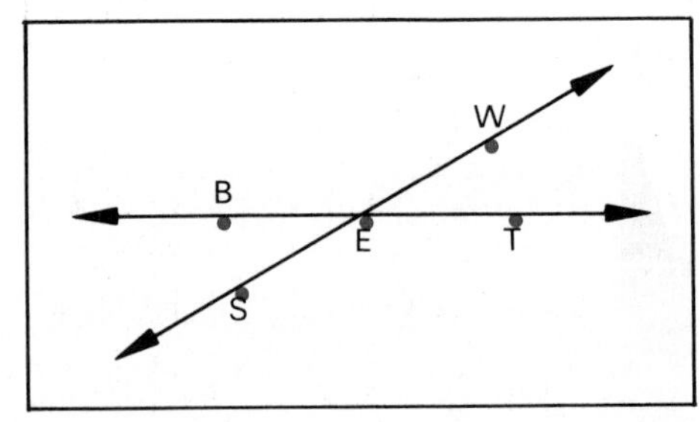

For exercises 29–41, use figure 16 to find each of the following:

29. $\overleftrightarrow{WM} \cap \overleftrightarrow{ZP}$
30. $\overrightarrow{ZA} \cap \overrightarrow{PA}$
31. $\overrightarrow{AP} \cup \overrightarrow{AZ}$
32. $\overrightarrow{AP} \cap \overrightarrow{AZ}$

33. $\overline{ZW} \cap \overline{MP}$

34. $\overrightarrow{AP} \cup \overrightarrow{AM}$

35. $\overrightarrow{AZ} \cup \overrightarrow{AW}$

36. $\overrightarrow{PA} \cap \overline{ZW}$

37. $\angle PAM \cap \angle ZAM$

38. $\angle PAW \cap \angle ZAW$

39. $\angle PAM \cap \angle ZAW$

40. $\angle ZAP \cap \angle WAM$

*41. $\overline{ZA} \cup \overline{AW} \cup \overline{WZ}$

FIGURE 16

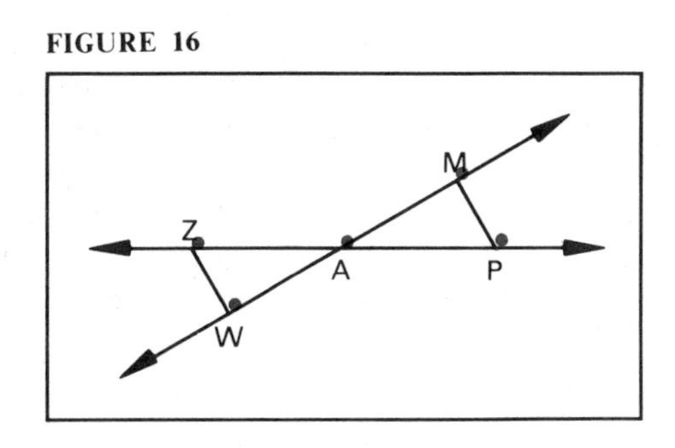

Just For Fun

In this section we discovered that one and only one straight line can be drawn through any two given points. We can also say that two points determine a unique line. How many points determine a unique circle? What is the minimum number of points required?

USDA, Soil Conservation Service

This surveyor's instrument is mounted on a three-legged stand for stability on rough terrain.

10.3 PLANES AND ANGLES

A **plane** can be represented by a flat surface. The floor of a room, a table top, or a desk top can be thought of as planes. A plane divides, or separates, one portion of space from another. A floor separates the space above the floor from the space below the floor. A wall is a plane that separates the space in one room from the space in the adjoining room. The wall of a building is a plane that separates the space inside from the space outside. In figure 17, each of the faces of the pyramid is part of a plane.

A plane has two dimensions, length and width. In geometry, a plane does not have any thickness.

Just as *point* and *line* are undefined terms in geometry, so is *plane.* Although our concept of a plane is intuitive, we can still discuss some of the properties of a plane. For example, we can think of a line as a set of points, and we can do the same for a plane. A plane is a set of points. The points are *on* the plane, and the plane *contains* the points. **Coplanar points** are points that are on the same plane, just as collinear points are points that are on the same line.

FIGURE 17

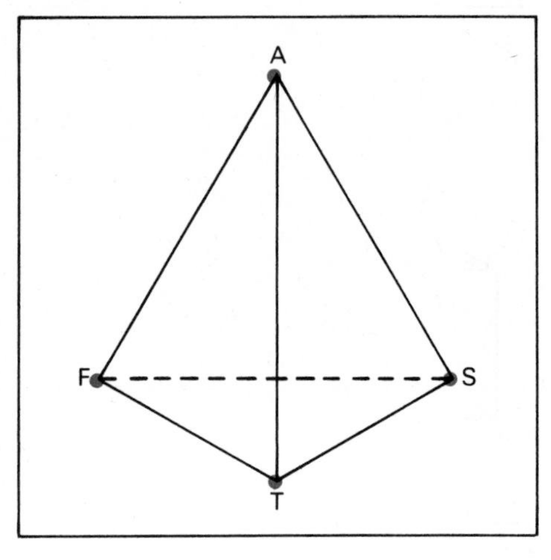

A unique plane is determined by any three noncollinear points. In other words, if we are given three distinct points which are not all on the same line, then there is one and only one plane containing all three points. For example, in figure 17, points F, S, and T are not on the same line, so they determine a unique plane, namely, plane FST. Note that none of the other planes in figure 17 contain all three of these points. Each of the other planes does contain two of these points; by using a different third point, a different plane is determined.

Have you ever noticed that easels, telescopes, cameras, and Christmas trees are all mounted on stands that have three legs? The reason that these stands, or tripods, have only three legs is that the three legs will always rest on some plane. A stand, table, or chair that has four legs tends to wobble unless all four legs are exactly the same length; this is not the case for an object with three legs.

Since lines and planes are both composed of points, we can make the observation that if two different points of a line are on a plane, then all the points on the line are also on the plane. In other words, if two points of a line are on a plane, then the line must also be on the plane, since two points determine a line. For an example, see line AB in figure 18.

FIGURE 18

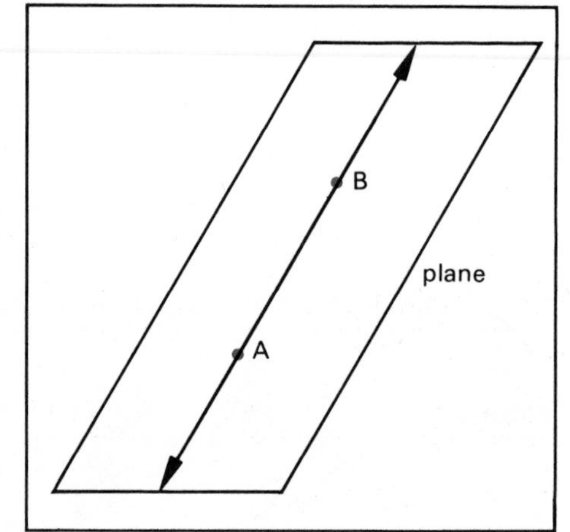

Figure 18 illustrates another important concept regarding the relationship between lines and planes. Any line on a given plane divides that plane into two half planes. Note that in figure 18, $\overleftrightarrow{AB}$ separates the points on the plane into two half planes. The points on $\overleftrightarrow{AB}$ are not on either half plane; thus a line on a plane divides the plane into two half planes, but the result is three sets of points: the points on the line, the points on one half plane, and the points on the other half plane.

Two different planes either intersect in a line or they do not intersect at all. If two distinct planes do not intersect, we call them **parallel planes.** The plane of a wall intersects the plane of the floor; that is, they meet in a line. But the floor and ceiling planes are parallel, as they will never meet in a line, no matter how far they are extended. In figure 19, planes $ABCD$ and $EFGH$ are parallel planes, but planes $ABCD$ and $BGHC$ intersect in a line.

It should also be noted that lines BC and AD in figure 19 are

FIGURE 19

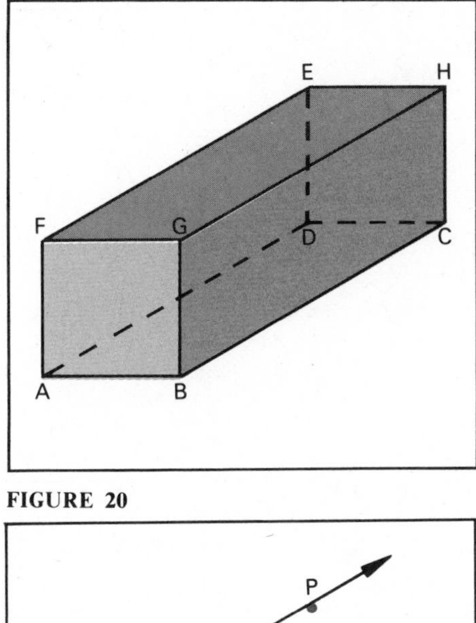

FIGURE 20

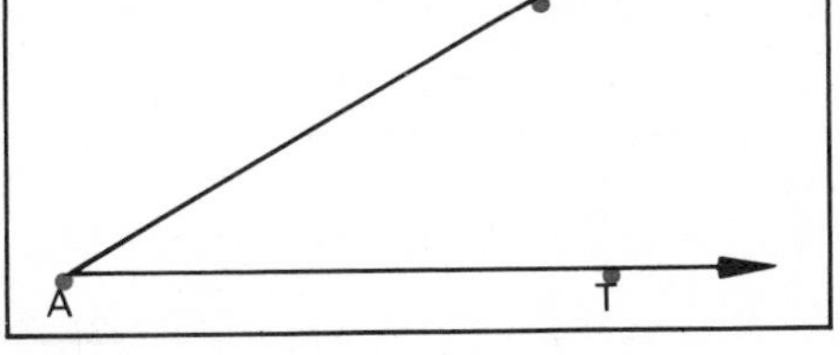

FIGURE 21

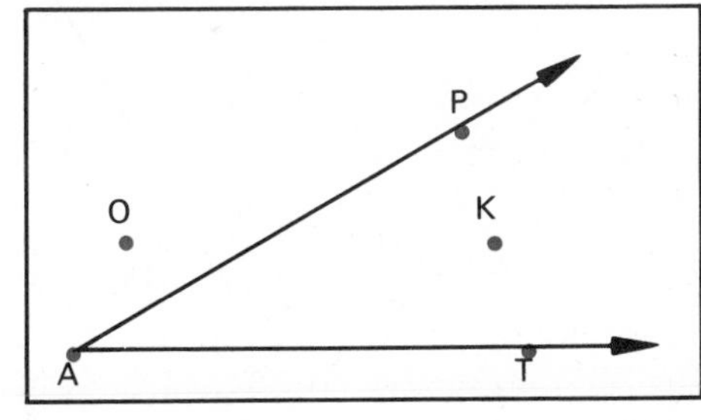

FIGURE 22

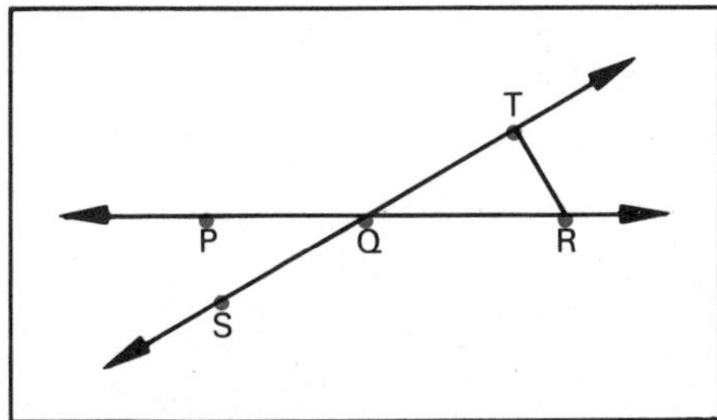

parallel lines. Two lines that are on the same plane and do not intersect however far they are extended are called **parallel lines.** In figure 19, lines AB and GH will not intersect, no matter how far we extend them, but they are not parallel lines because they are not in the same plane. If two lines do not lie in the same plane, they are called **skew lines.**

In the preceding section, we noted that an angle is the union of two rays with a common endpoint. A line on a plane divides that plane into three sets of points, the points on the line, the points on one half plane, and the points on the other half plane. Similarly, $\angle PAT$ in figure 20 divides a plane such as this page into three parts. There are those points which are *on* angle PAT, such as point P. There are also points which are inside angle PAT; that is, they are in the **interior** of angle PAT. Finally, there are points outside angle PAT; that is, in the **exterior** of angle PAT. In figure 21, point O is an exterior point and point K is an interior point. Note that points P, A, and T are neither interior nor exterior points, as they are on the angle.

We can combine these new concepts of interior and exterior points with set intersection. Consider figure 22 with the indicated points. What is the intersection of angle TQR and $\overline{TR}$? That is, what is $\angle TQR \cap \overline{TR}$? The points that $\angle TQR$ and $\overline{TR}$ have in common are points T and R; that is, $\angle TQR \cap \overline{TR} = \{T, R\}$.

Now consider the intersection of $\overline{TR}$ and the interior of $\angle TQR$, that is, $\overline{TR} \cap$ (interior $\angle TQR$). The points that these two sets have in common are those points that are in the interior of $\angle TQR$ and are also on line segment TR. Is the answer $\overline{TR}$? No. We cannot say that our answer is line segment $\overline{TR}$ because that would mean that points T and R are members of the solution, and from our previous discussion we know that points T and R are *on* $\angle TQR$; therefore, they cannot be *inside* $\angle TQR$. Our solution consists of all those points on $\overline{TR}$ except T and R. This is an **open line segment,** and we denote it by $\overset{\circ\!-\!\circ}{TR}$. Therefore $\overline{TR} \cap$ (interior $\angle TQR$) $= \overset{\circ\!-\!\circ}{TR}$.

What is the intersection of ray QP and the exterior of angle TQR? That is, what is $\overrightarrow{QP} \cap$ (exterior $\angle TQR$)? The exterior of $\angle TQR$ consists of all those points that are not on $\angle TQR$ and not inside $\angle TQR$. Now all of the points on ray QP lie outside of $\angle TQR$, except one point, namely, Q. Therefore the solution is the half line QP (excluding Q), and we denote this by $\overset{\circ\!\rightarrow}{QP}$. That is, $\overrightarrow{QP} \cap$ (exterior $\angle TQR$) $= \overset{\circ\!\rightarrow}{QR}$.

EXAMPLE 1

Use figure 23 with the indicated points to find each of the following:

FIGURE 23

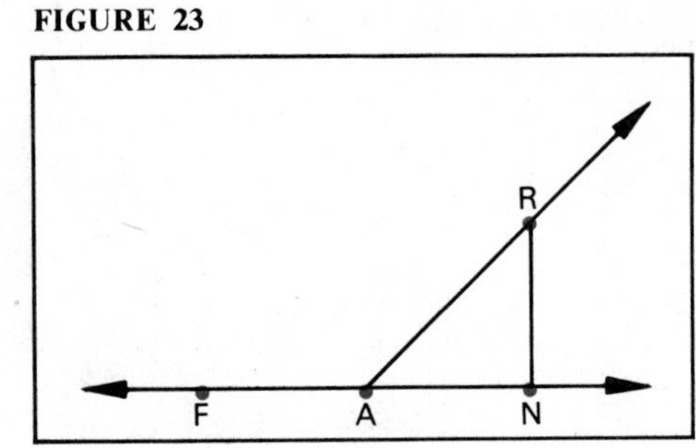

a. $\overrightarrow{AR} \cup \overrightarrow{AN}$

b. $\overrightarrow{AF} \cap$ (exterior $\angle RAN$)

c. $\overline{RN} \cap$ (interior $\angle RAN$)

d. (interior $\angle RAN$) $\cap$ (exterior $\angle RAN$)

Solution

a. $\overrightarrow{AR} \cup \overrightarrow{AN}$ is the union of two rays with a common endpoint; that is, an angle. Therefore $\overrightarrow{AR} \cup \overrightarrow{AN} = \angle RAN$.

b. $\overrightarrow{AF} \cap$ (exterior $\angle RAN$) is the intersection of those points on $\overrightarrow{AF}$ with those points that are in the exterior of $\angle RAN$. This is all the points on $\overrightarrow{AF}$ except A. Therefore $\overrightarrow{AF} \cap$ (exterior $\angle RAN$) $= \overset{\circ\!\rightarrow}{AF}$.

c. $\overline{RN} \cap$ (interior $\angle RAN$) is the intersection of those points that are on $\overline{RN}$ and also in the interior of $\angle RAN$. This is all the points on $\overline{RN}$ except R and N. Therefore $\overline{RN} \cap$ (interior $\angle RAN$) $= \overset{\circ\!-\!\circ}{RN}$.

d. The set of points that are in the interior of $\angle RAN$ and the points that are in the exterior of $\angle RAN$ do not intersect. They are separated by those points that are on $\angle RAN$. Therefore (interior $\angle RAN$) $\cap$ (exterior $\angle RAN$) $= \phi$.

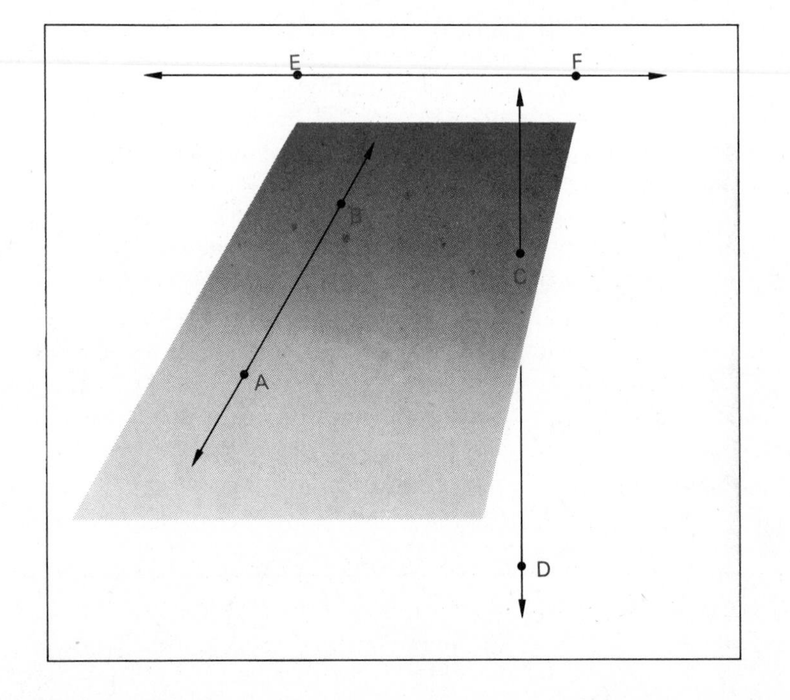

FIGURE 24

Remember that a plane is a flat surface, and that it separates one part of space from an adjoining part of space. It is also a set of points. A line on a plane divides that plane into half planes, but there are other possible relationships between lines and planes. Line AB in figure 24 is an example of a line on a plane; since two points of $\overleftrightarrow{AB}$ are on the plane, all of the points on $\overleftrightarrow{AB}$ are on the plane. Line CD intersects the plane on only one point, while line EF does not intersect the plane at all. Line EF is parallel to the given plane and skew to line AB.

EXERCISES FOR SECTION 10.3

For exercises 1–8, use figure 25 with the indicated points to find each of the following:

1. $\overrightarrow{AR} \cap \overrightarrow{RS}$
2. $\overrightarrow{RA} \cup \overrightarrow{RS}$
3. $\overrightarrow{AS} \cap$ (interior $\angle ARS$)
4. $\overrightarrow{AS} \cap$ (exterior $\angle ARS$)
5. $\overrightarrow{RA} \cup \overrightarrow{RD}$
6. $\overrightarrow{RS} \cap$ (interior $\angle DRA$)
7. $\overrightarrow{RS} \cap$ (exterior $\angle DRA$)
8. (interior $\angle DRA$) $\cap$ (exterior $\angle ARS$)

FIGURE 25

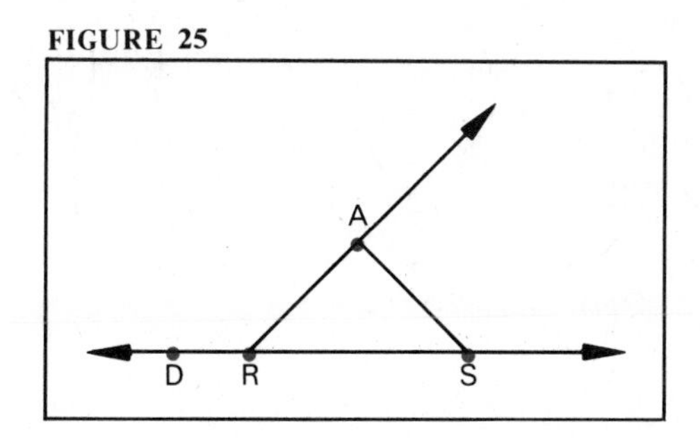

For exercises 9–18, use figure 26 with the indicated points to find each of the following:

9. $\overrightarrow{OT} \cup \overrightarrow{ON}$
10. $\overrightarrow{OT} \cap \overrightarrow{ON}$
11. $\overrightarrow{OE} \cap$ (interior $\angle TON$)
12. (interior $\angle TOE$) $\cap$ (exterior $\angle EON$)
13. (exterior $\angle TOL$) $\cap$ (interior $\angle EON$)
14. (interior $\angle TON$) $\cap$ (interior $\angle TOE$)
15. (interior $\angle TOE$) $\cap$ (interior $\angle EON$)
16. (exterior $\angle EON$) $\cap \overrightarrow{OL}$
17. (exterior $\angle TOE$) $\cap$ (interior $\angle EON$)
18. (interior $\angle TOE$) $\cap$ (exterior $\angle LOT$)

FIGURE 26

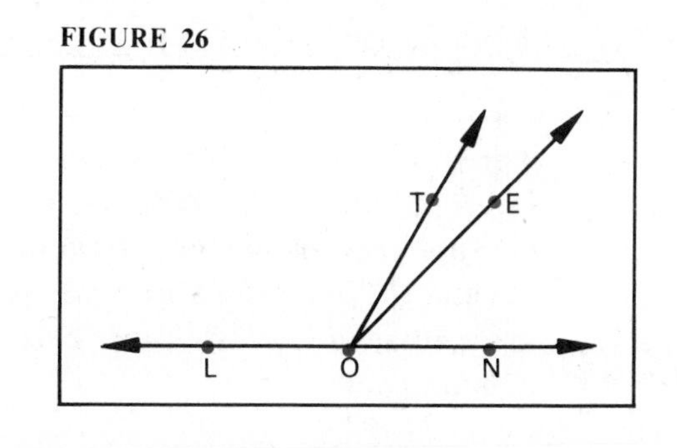

For exercises 19–30, use figure 27 with the indicated points to find each of the following:

19. $\overrightarrow{OT} \cup \overrightarrow{OC}$
20. $\overrightarrow{OC} \cup \overrightarrow{OS}$
21. $\overrightarrow{OJ} \cup \overrightarrow{OS}$
22. $\overrightarrow{OS} \cap \overrightarrow{OT}$
23. (interior $\angle SOC$) $\cap$ (interior $\angle SOT$)
24. (exterior $\angle COT$) $\cap$ (interior $\angle JOS$)
25. $\overrightarrow{OT} \cap$ (exterior $\angle JOS$)

26. $\angle COT \cap \angle SOC$
27. $\angle JOS \cap \angle COT$
28. (exterior $\angle COT$) $\cap$ (exterior $\angle COS$)
29. (exterior $\angle SOJ$) $\cap$ (exterior $\angle SOC$)
30. (interior $\angle JOS$) $\cap$ (interior $\angle JOC$)

FIGURE 27

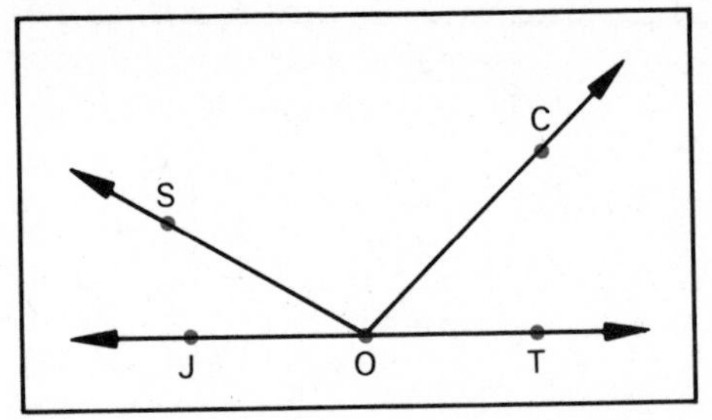

Indicate whether each statement is true or false.

31. Two parallel planes will never intersect.
32. The intersection of two planes that are not parallel is a line.
33. The intersection of a line and a plane is never a point.
34. Two lines on the same plane must intersect at some point.
35. Skew lines intersect at some point.
36. Parallel lines intersect at some point.
37. An angle is formed by the union of two rays with a common endpoint.
38. The interior of an angle is the intersection of two half planes.

Just For Fun

Three intersecting planes can have (*a*) no common intersection, (*b*) intersection in a common line, or (*c*) intersection in one common point. Conditions (*a*) and (*b*) are shown in figure 28 below. Can you draw three planes so that their intersection is one common point?

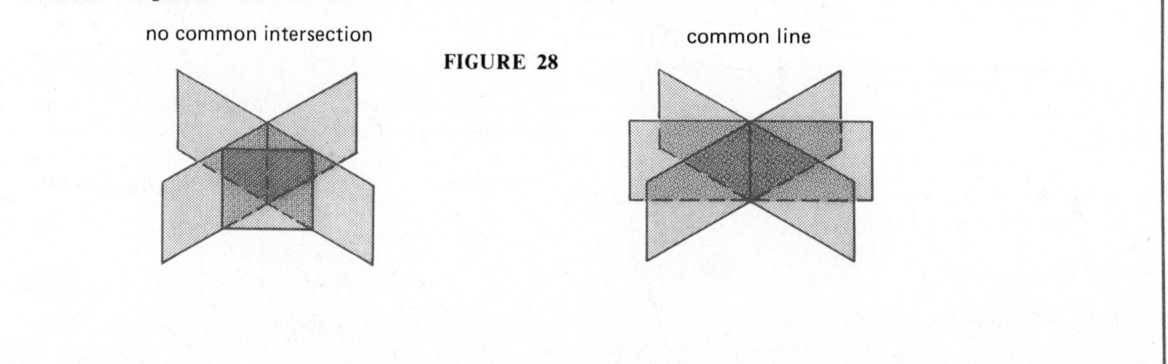

FIGURE 28

10.4 POLYGONS

A **broken line** is a set of connected line segments, that is, a set of line segments that have been placed end to end. The four figures in figure 29 are examples of broken lines.

Now let us take each of the broken lines in figure 29 and "close" it so that it appears as shown in figure 30. A **closed broken line** begins and ends at the same point. You will note that each of the figures in figure 30 is a closed broken line. But, in addition, each is also a *simple closed broken line.* A **simple closed broken line** is one that does not intersect itself. None of the closed broken line figures shown in figure 31 is simple, since in each case the broken line intersects itself.

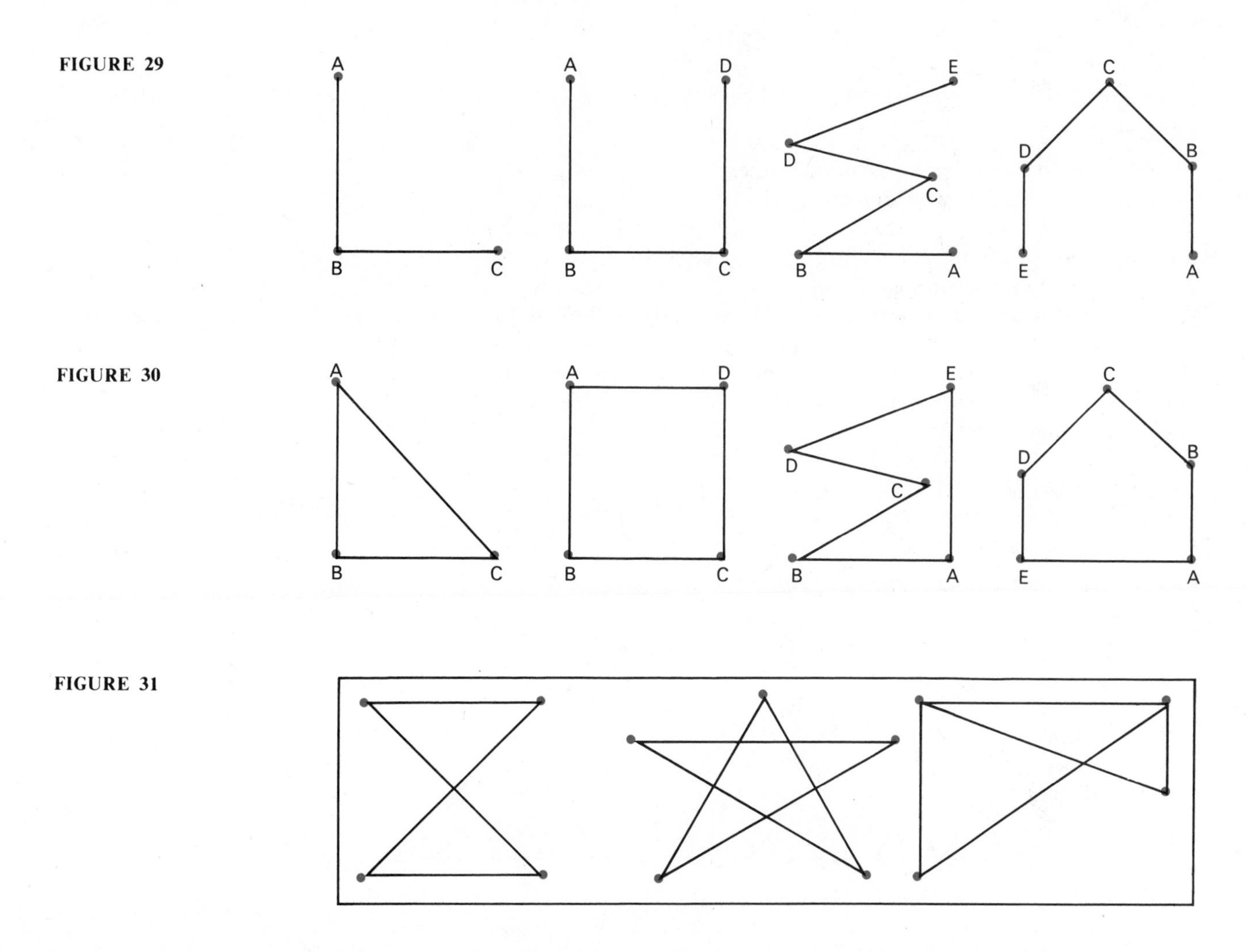

FIGURE 29

FIGURE 30

FIGURE 31

A simple closed broken line is called a **polygon.** The connected line segments are the **sides** of the polygon, and the points at which the line segments are connected are the **vertices** of the polygon. Note that any two consecutive sides of a polygon form an angle of the polygon. In figure 32, polygon $ABCD$ is a simple closed broken line. Its sides are $\overline{AB}$, $\overline{BC}$, $\overline{CD}$, and $\overline{DA}$. Its vertices are A, B, C, and D, and its angles are $\angle DAB$, $\angle ABC$, $\angle BCD$, and $\angle CDA$.

FIGURE 32

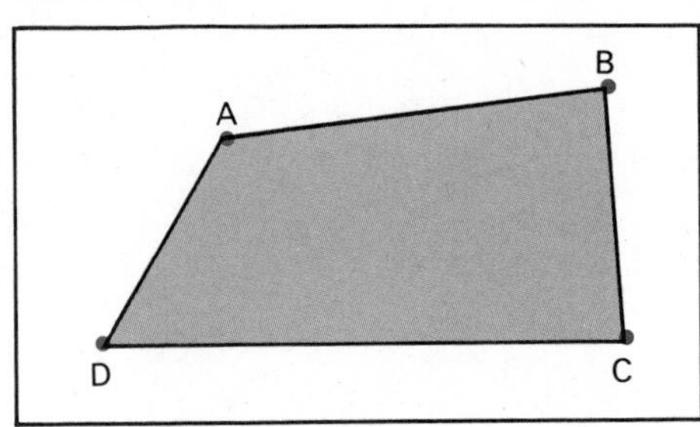

Polygons are classified according to the number of sides that they have. Below is a partial list of some types of polygons and the number of sides that each has.

Polygon	Number of sides	Polygon	Number of sides
triangle	3	octagon	8
quadrilateral	4	nonagon	9
pentagon	5	decagon	10
hexagon	6	dodecagon	12
heptagon	7	icosagon	20

In order to form a simple closed broken line, we need at least three line segments. If we form such a closed broken line, then we have a polygon with three sides, that is, a **triangle.** There are many different kinds of triangles, and they can be classified according to the characteristics of their sides or their angles. A triangle in which the measures of all three sides are equal is called an **equilateral triangle.** A triangle in which two sides are of equal length is called an **isosceles triangle.** A triangle in which no two sides are of equal length is called a **scalene triangle.** An example of each type of triangle is shown in figure 33.

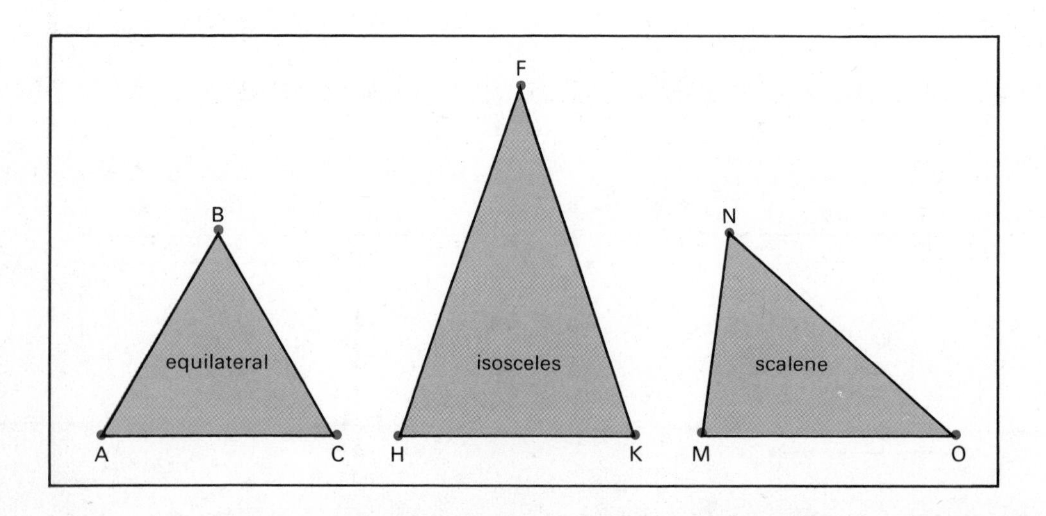

FIGURE 33

Can we form a triangle given any three line segments? The answer is no. There is a certain requirement that must be fulfilled in order to construct a triangle from three segments. For example, suppose we are given three line segments that have the following lengths: 10 centimetres, 5 centimetres, and 2.5 centimetres. Can we form a triangle with these three line segments?

10 centimetres

5 centimetres

2.5 centimetres

Norman Prince

If we attempt to construct a triangle with these segments, we have the situation illustrated in figure 34.

FIGURE 34

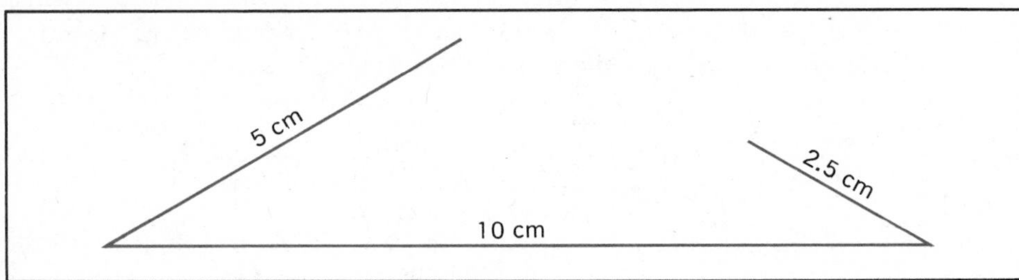
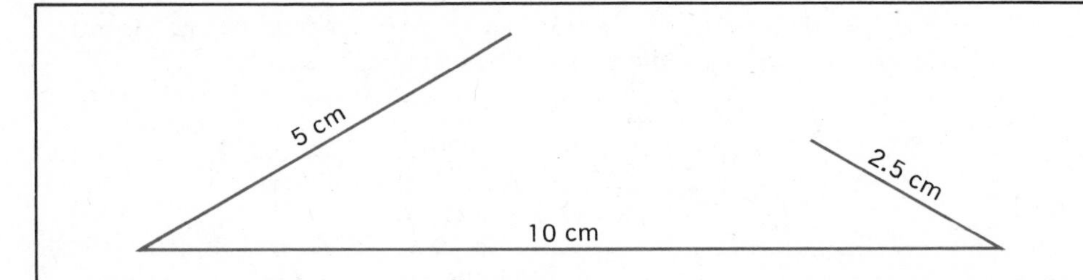

As you can see, we are not able to form a simple *closed* broken line using these three line segments. In fact, in order to construct any triangle, the sum of the measures of any two of the line segments must be greater than the measure of the third line segment. If we already have a triangle, then we know that the sum of the lengths of any two of the sides is greater than the length of the third side.

Is it possible to construct a triangle whose sides measure 4, 4, and 8? The answer is no, since $4 + 4$ is not greater than 8. Is it possible to construct a triangle whose sides measure 4, 5, 6? Yes, because $4 + 5 > 6$, $4 + 6 > 5$, and $5 + 6 > 4$.

We can also classify triangles according to the types of angles that are in the triangle. If all of the angles in the triangle are acute angles, then the triangle is called an **acute triangle.** If a triangle has an obtuse angle in it, then it is called an **obtuse triangle.** Can a triangle have more than one obtuse angle in it? We cannot form such a triangle because the figure formed is not a closed broken line, as shown in figure 35.

FIGURE 35

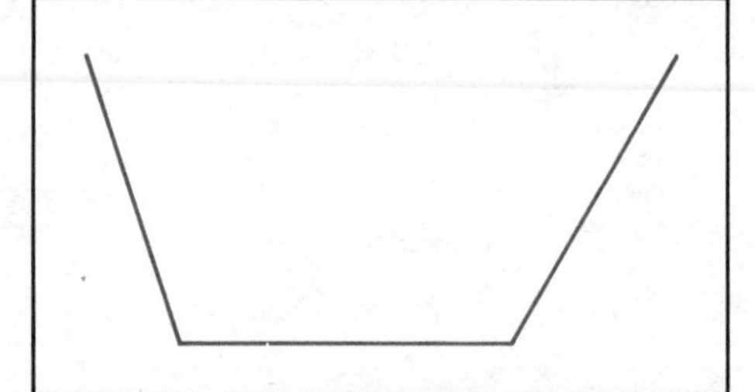

If the measures of all of the angles in a triangle are equal, then the triangle is called an **equiangular triangle.** It can be shown that the sum of the measures of the interior angles of a triangle is equal to 180°. Therefore, if a triangle is equiangular, each angle will measure exactly 60°.

If a triangle contains a right angle, then it is called a **right triangle.**

Examples of an acute triangle, an obtuse triangle, a right triangle, and an equiangular triangle are shown in figure 36.

Consider right triangle ABC in figure 37. The right angle is $\angle BCA$. The sides of the triangle that form the right angle are called the **legs** of the right triangle (a and b in figure 37) and the side opposite the right angle is called the **hypotenuse** (c in figure 37).

FIGURE 36

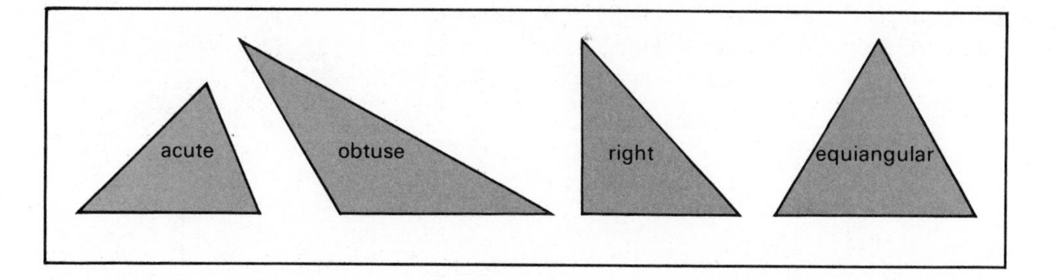

FIGURE 37

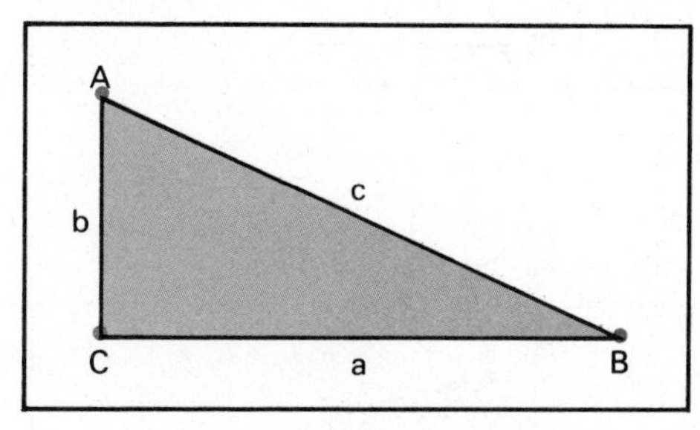

Probably one of the most famous theorems in geometry deals with the right triangle. It is called the *Pythagorean theorem* since its proof was supposedly discovered by Pythagoras (approximately 550 B.C.). The Pythagorean theorem states that:

The square of the hypotenuse of a right triangle is equal to the sum of the squares of the legs.

Using right triangle ABC in figure 37 as a reference, we can also state the Pythagorean theorem as

$$c^2 = a^2 + b^2$$

Suppose, in figure 37, we know that $m(\overline{AC}) = 3$ and $m(\overline{BC}) = 4$, where $m(\overline{AC})$ is the measure of $\overline{AC}$ and $m(\overline{BC})$ is the measure of $\overline{BC}$. What is the length of $\overline{AB}$?

We can use the Pythagorean theorem to find the answer. Using the formula $c^2 = a^2 + b^2$, we are given $a = 4$ and $b = 3$. Therefore we can substitute the appropriate values in the formula and solve it for c.

$$c^2 = a^2 + b^2$$
$$c^2 = 4^2 + 3^2 \qquad \text{substituting } a = 4 \text{ and } b = 3$$
$$c^2 = 16 + 9$$
$$c^2 = 25$$
$$c = \sqrt{25}$$
$$c = 5 \qquad \text{length of } \overline{AB}$$

FIGURE 38

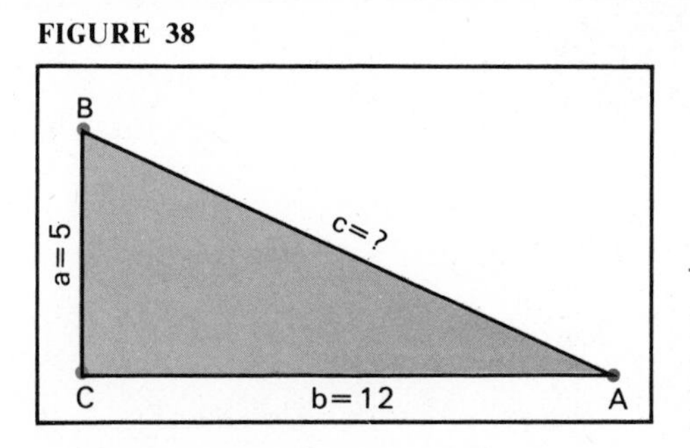

EXAMPLE 1

Find the length of hypotenuse $\overline{AB}$ in right triangle ABC, given that $m(\overline{AC}) = 12$ and $m(\overline{BC}) = 5$. See figure 38.

Solution

Using the formula $c^2 = a^2 + b^2$, we have $a = 5$ and $b = 12$. Therefore

$$
\begin{aligned}
c^2 &= a^2 + b^2 \\
c^2 &= 5^2 + 12^2 \\
c^2 &= 25 + 144 \\
c^2 &= 169 \\
c &= \sqrt{169} \\
c &= 13 \qquad \text{length of } \overline{AB}
\end{aligned}
$$

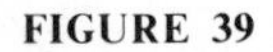

FIGURE 39

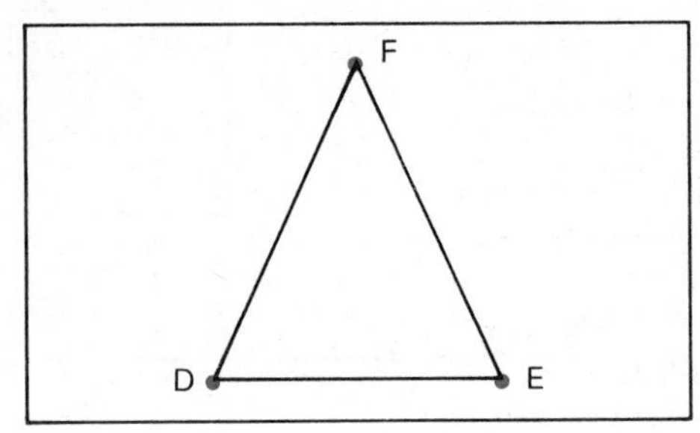

If a triangle is isosceles, then it has two sides of equal length. It can also be shown that the measures of the two angles opposite the sides of equal length are equal. For example, given isosceles triangle DEF in figure 39, with $m(\overline{DF}) = m(\overline{FE})$, then the angles opposite these sides are also equal. That is, $m(\angle FDE) = m(\angle FED)$. The measure of these two angles depends upon the size of the third angle.

If we form a simple closed broken line using four line segments, such as $ABCD$ in figure 40, we have a polygon with four sides; this polygon is called a **quadrilateral.**

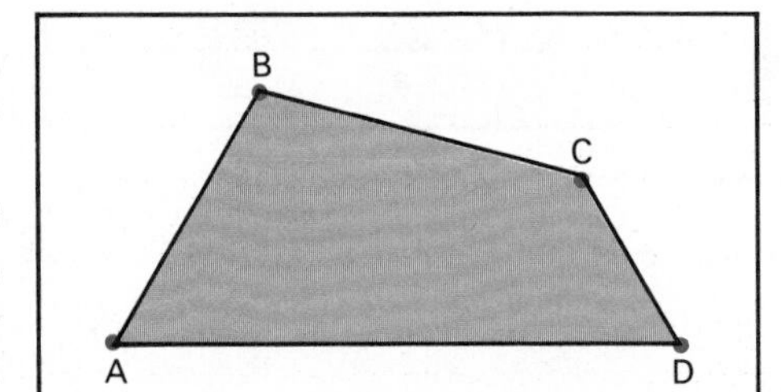

FIGURE 40 Quadrilateral $ABCD$

There are many kinds of quadrilaterals and each type has some particular properties. Any four-sided polygon is a quadrilateral, but a quadrilateral that has only two parallel sides is called a **trapezoid.** The two parallel sides are called the **bases** of the trapezoid, and if the two nonparallel sides are equal in length, then the trapezoid is an **isosceles trapezoid.** A trapezoid and an isosceles trapezoid are shown in figure 41.

FIGURE 41

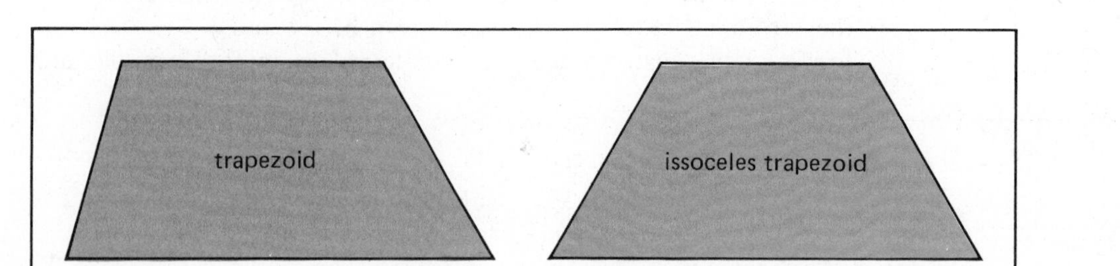

If a quadrilateral has both pairs of opposite sides parallel, then it is called a **parallelogram.** Note that we did not say anything about the angles of a parallelogram; they can be of any type, as long as both pairs of opposite sides are parallel. Figure 42 shows a parallelogram.

FIGURE 42

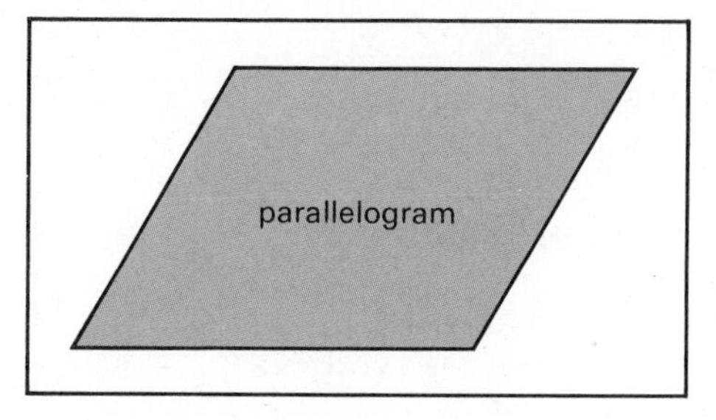

FIGURE 43

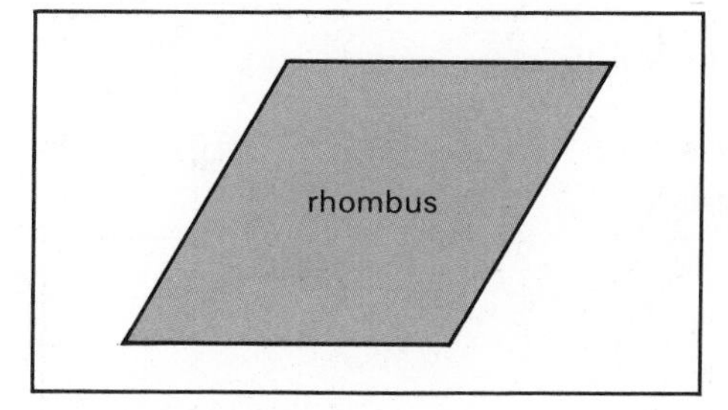

FIGURE 44

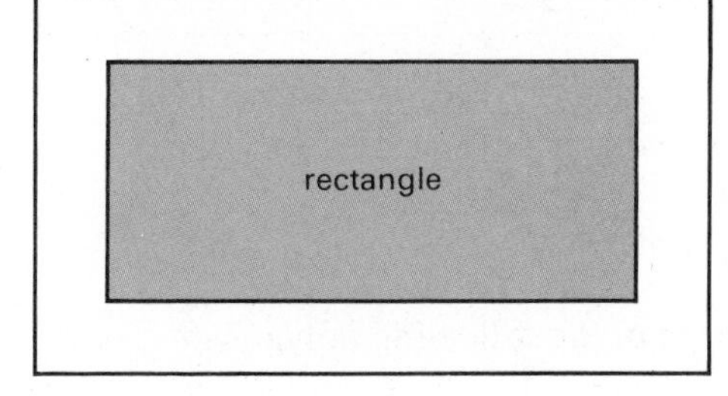

FIGURE 45

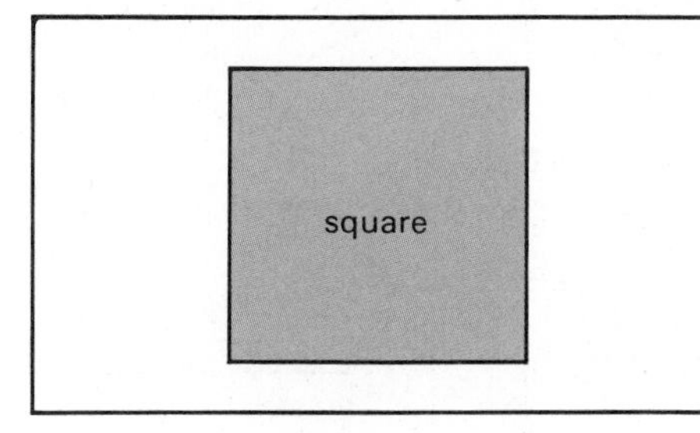

A parallelogram that has adjacent sides of equal length is called a **rhombus.** Note that if the adjacent sides of a parallelogram are of equal length, then all of its sides are of equal length. Therefore, we can also describe a rhombus as a parallelogram with four equal sides. See figure 43.

A parallelogram that contains a right angle is called a **rectangle.** If a parallelogram contains one right angle, then it must contain four right angles. Also, any quadrilateral with four right angles is a rectangle, as shown in figure 44.

A **square** can be described in a number of ways. For example, a square is a rectangle with two adjacent sides of equal length. We can also say that a square is a quadrilateral with four sides of equal length and four angles of equal measure. Note that the square can be considered to be a special case of the rhombus, since a rhombus is a parallelogram with four sides of equal length. See figure 45.

Below is a diagram that illustrates the relationships among the various quadrilaterals.

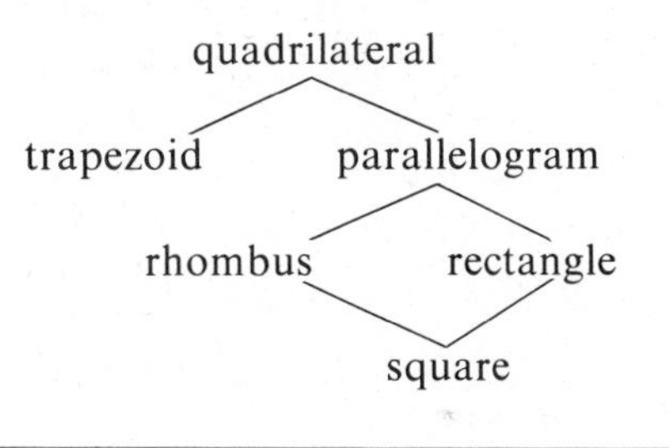

EXAMPLE 2

Determine whether each sentence is true or false.

a. Every square is a rectangle.

b. Every rectangle is a square.

c. Every square is a rhombus.

d. Every rhombus is a square.

e. Every parallelogram is a rectangle.

f. Every trapezoid is a parallelogram.

Solution

a. True, a square is a rectangle with two adjacent sides equal.

b. False, a rectangle is a quadrilateral with four right angles; the adjacent sides are not necessarily equal.

c. True, a square is a special case of the rhombus.

d. False, a rhombus is a parallelogram with four equal sides; the four angles are not necessarily equal.

e. False, a parallelogram is a quadrilateral with both pairs of opposite sides parallel. The angles of a parallelogram are not necessarily right angles.

f. False, a trapezoid is a quadrilateral that has only two sides parallel.

EXERCISES FOR SECTION 10.4

1. Which of the following are broken lines?

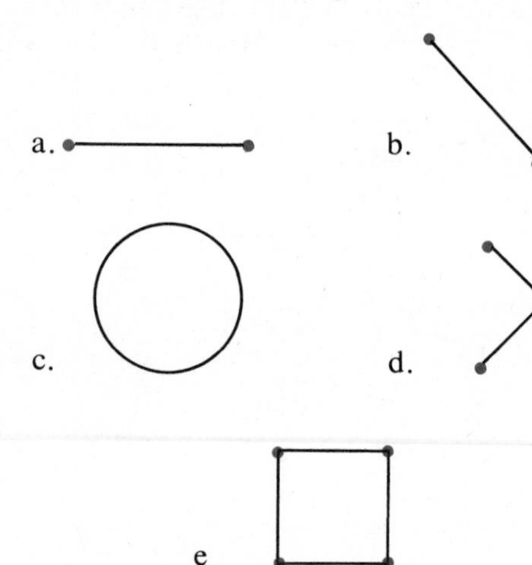

2. Which of the following are simple closed broken lines?

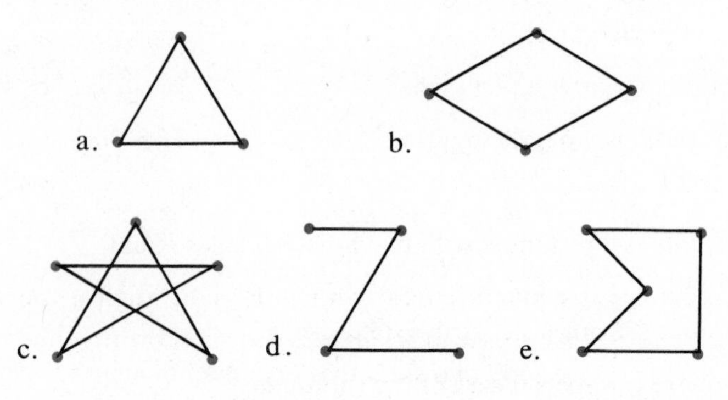

3. Which of the following are polygons?

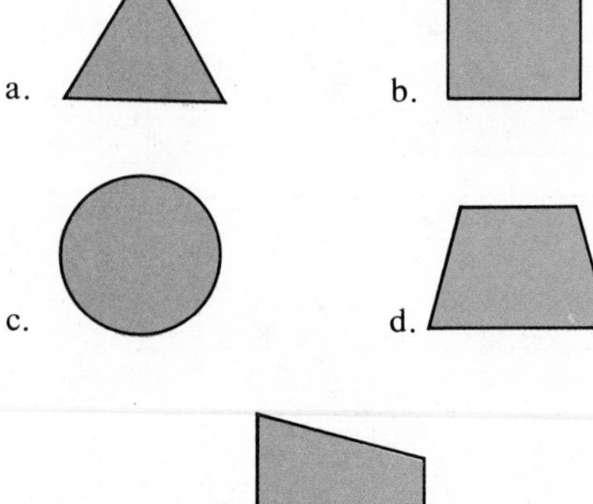

4. Identify each of the following polygons.

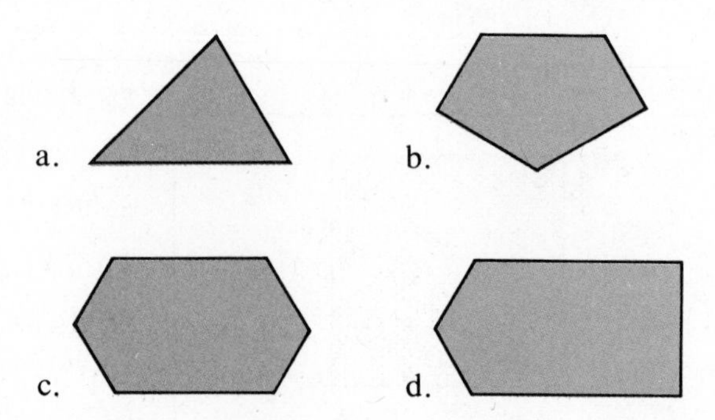

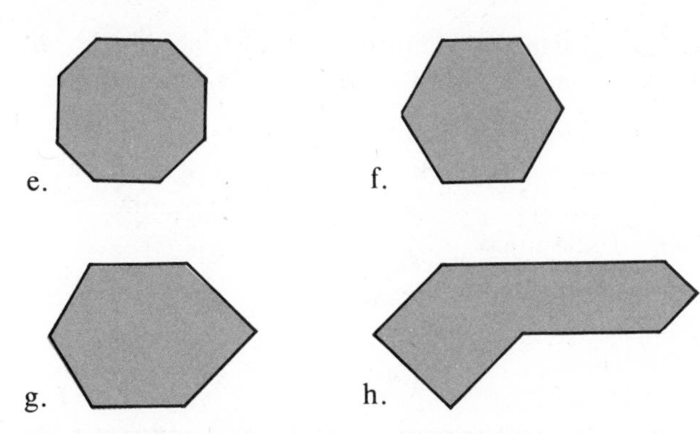

5. Tell whether each of the following is true or false.
   a. A scalene triangle has no sides equal.
   b. An isosceles triangle has all three sides equal.
   c. An equiangular triangle is an acute triangle.
   d. A triangle may contain two obtuse angles.
   e. In a right triangle, the side opposite the right angle is called the hypotenuse.

6. Tell whether each of the following is true or false.
   a. The set of numbers $\{2, 7, 10\}$ may represent the lengths of the sides of a triangle.
   b. The lengths of two sides of a triangle are 5 and 6. The third side may have length 11.
   c. The set of numbers $\{2, 3, 4\}$ may represent the lengths of the sides of a right triangle.
   d. The set of numbers $\{3, 4, 5\}$ may represent the lengths of the sides of a right triangle.
   e. In any triangle, the square of one side is equal to the sum of the squares of the other two sides.

7. Find the length of hypotenuse $\overline{AB}$ in right triangle ABC, given that $m(\overline{AC}) = 9$ and $m(\overline{BC}) = 12$.

8. Find the length of hypotenuse $\overline{AB}$ in right triangle ABC, given that $m(\overline{AC}) = 15$ and $m(\overline{BC}) = 8$.

9. Find the length of hypotenuse $\overline{AB}$ in right triangle ABC, given that $m(\overline{AC}) = 10$ and $m(\overline{BC}) = 24$.

10. Find the length of leg $\overline{AC}$ in right triangle ABC, given that $m(\overline{BC}) = 12$, $m(\overline{AB}) = 13$, and $\overline{AB}$ is the hypotenuse.

11. Find the length of leg $\overline{BC}$ in right triangle ABC, given that $m(\overline{AC}) = 4$, $m(\overline{AB}) = 5$, and $\overline{AB}$ is the hypotenuse.

12. Identify each of the following quadrilaterals.

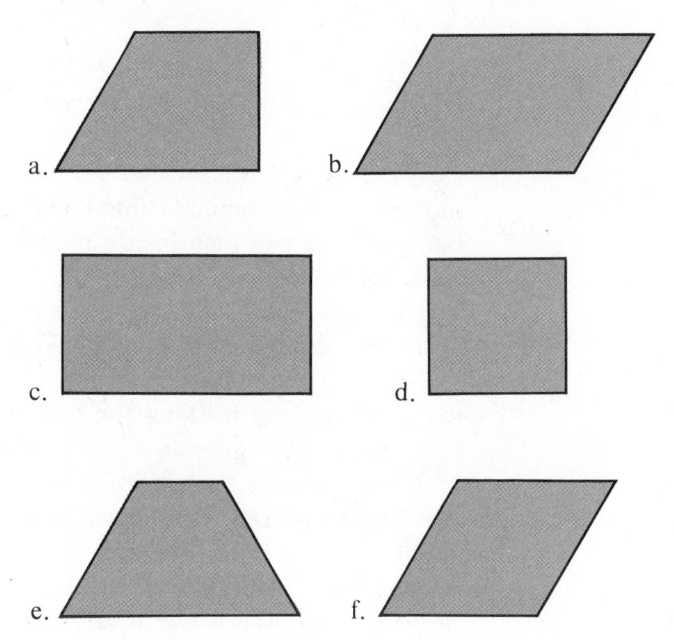

13. Tell whether each of the following is true or false.
    a. A square is a rhombus.
    b. A parallelogram is a rectangle with a right angle.
    c. A parallelogram is a polygon whose opposite sides are parallel.
    d. A rectangle is a parallelogram with four right angles.
    e. A trapezoid is a quadrilateral whose opposite sides are parallel.

14. Tell whether each of the following is true or false.
    a. An isosceles triangle is a triangle which has exactly two sides that are equal in length.

b. An acute triangle has only one acute angle.
c. A triangle can be both isosceles and obtuse.
d. An equilateral triangle can be an obtuse triangle.
e. A square is a rectangle.

15. Arrange the following terms in the order in which the definitions of each should be given: triangle, hypotenuse, polygon, right triangle.

16. Arrange the following terms in the order in which the definitions of each should be given: square, quadrilateral, parallelogram, polygon, rectangle.

Just For Fun

In chapter 9, we located natural numbers, whole numbers, integers, and rational numbers on the number line. We can also locate irrational numbers on the number line.

For example, suppose we want to locate the point that corresponds to $\sqrt{2}$. We can do this by drawing the hypotenuse of a right triangle whose legs are 1 unit in length along the number line with one vertex at 0. See figure 46.

Recall that the Pythagorean theorem tells us that the square of the hypotenuse of a right triangle equals the sum of the squares of the legs; that is, if a and b are the lengths of the legs, and c is the length of the hypotenuse, then $c^2 = a^2 + b^2$.

If $a = 1$ and $b = 1$, then

$$c^2 = 1^2 + 1^2 = 1 + 1 = 2$$

and therefore

$$c = \sqrt{2}$$

This length corresponds to a point on the number line. See figure 46.

Using the fact that $1^2 + 2^2 = 5$, can you graph $\sqrt{5}$ on the number line?

FIGURE 46

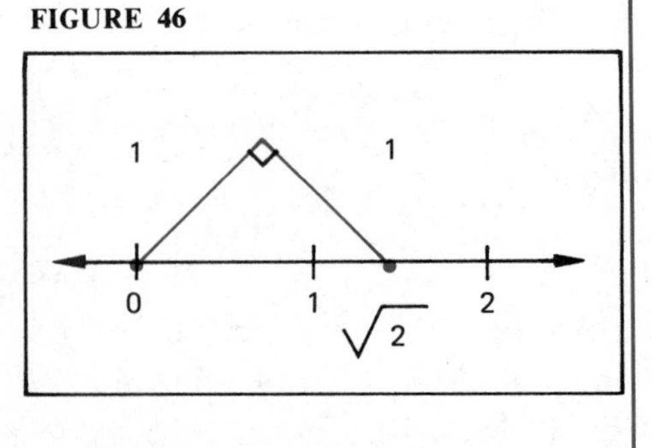

B.C. by permission of Johnny Hart and Field Enterprises, Inc.

10.5 NETWORKS

A set of line segments or arcs is called a **graph.** If it is possible to move from any point in the graph to any other point in the graph by moving along the line segments or arcs, then we say the graph is **connected.** A **network** is a connected graph.

A network is **traversable** if it can be drawn by tracing each line segment or arc exactly once without lifting the pencil from the paper. Figure 47 shows some examples of networks that are traversable. The endpoints of the arcs or line segments are called **vertices.**

FIGURE 47

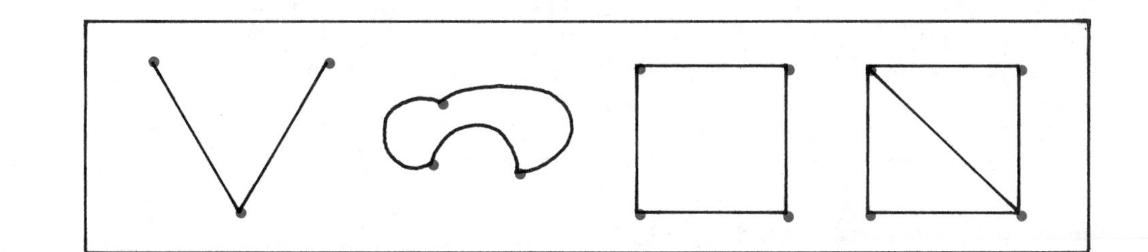

FIGURE 48a

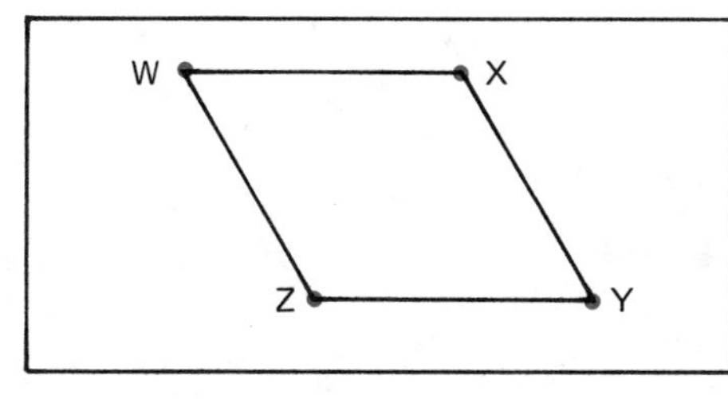

A **closed network** divides a plane into two or more regions. A **simple network** is one that does not cross itself. A network that is simple and closed is traversable. Furthermore, you can start at any vertex of a simple closed network to traverse the network. The vertex chosen for the initial point will also be the terminal point.

Consider the network in figure 48a. It is simple and closed. Therefore we can begin at any vertex and traverse the network.

The network in figure 48b is closed, but not simple. It can also be traversed, but we must begin at either U or X. If we start at U, then we end at X; if we start at X, then we end at U.

FIGURE 48b

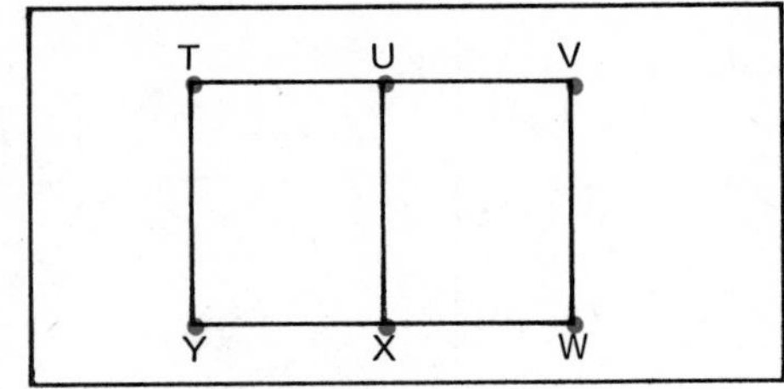

A famous puzzle is largely responsible for beginning the study of network theory. This puzzle, the "Seven Bridges of Königsberg," first attracted attention during the 1700s. Königsberg (now Kaliningrad, U.S.S.R.) was a town in Prussia built on both sides of the Pregel River. Located in the river were two islands, connected to each other and to the city by seven bridges as shown in figure 49.

The problem associated with these bridges and islands was to determine if a person could start at a given point in the town of

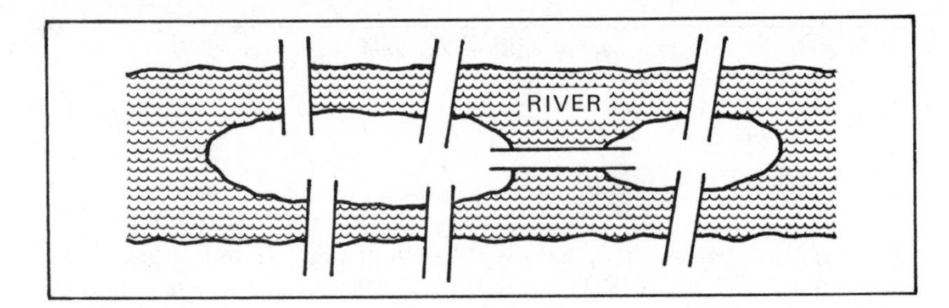

FIGURE 49

Königsberg and follow a path that would cross every bridge once and only once on a continuous walk through the town. The citizens of Königsberg tried many routes, but found that—no matter where they started, or what path they chose—they could not cross each bridge once and only once. However, it was not until Leonhard Euler (1707–1783), a Swiss mathematician, became interested in the problem that it was proved that each bridge could not be crossed once and only once on a continuous walk through the town. To prove this, Euler analyzed the problem by transforming it into a network similar to that shown in figure 50.

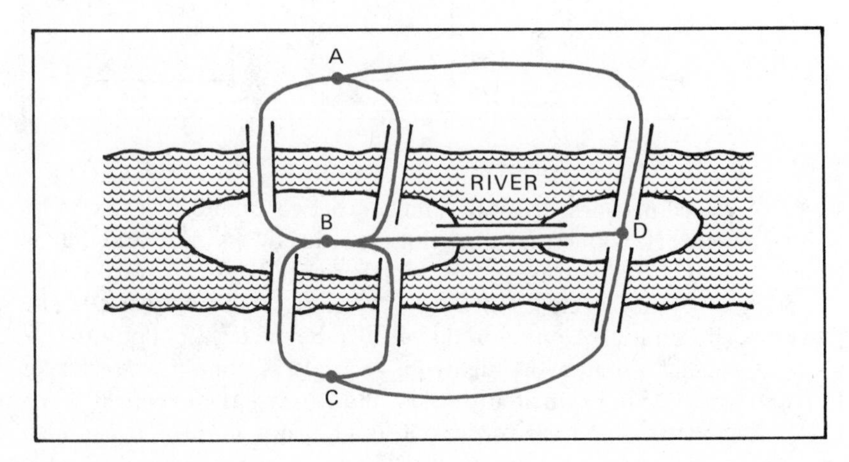

FIGURE 50

Euler called the points where the paths of the network came together *vertices*. Furthermore, he classified the vertices of a network as **odd** or **even**, depending upon whether an odd or even number of paths passed through the vertex. For example, in figure 48b, vertex T is even because two paths pass through it, and vertex X is odd because three paths pass through it. In figure 50, all of the vertices are odd.

Euler proved that any network containing only even vertices is traversable by a route beginning at any vertex and ending at the same vertex. He also showed that a network that has exactly two odd vertices is traversable, but the traversing route must start at one of the vertices and end at the other.

Finally Euler showed that if a network has more than two odd vertices, then it is not traversable. This means that the network in figure 50 is not traversable. Therefore, Euler had proved that each of the bridges of Königsberg could not be crossed once and only once on a continuous walk, because the equivalent network in figure 50 has four odd vertices.

EXAMPLE 1

For each network, identify the even and odd vertices.

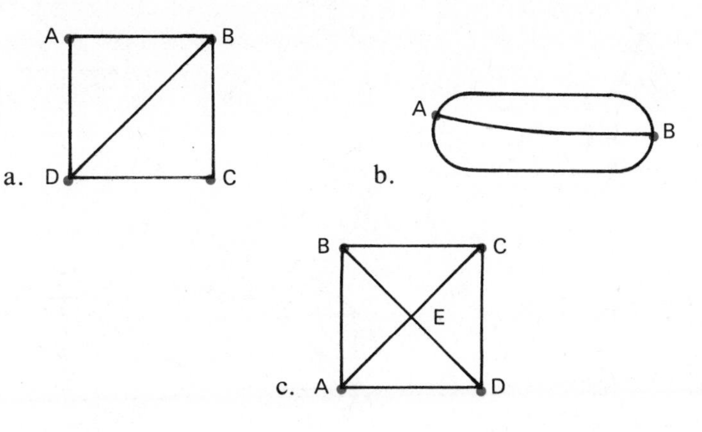

Solution

Recall that an even vertex is one that is an endpoint of an even number of arcs or line segments, and an odd vertex is one that is an endpoint of an odd number of arcs or line segments.

a. Even vertices: A and C; odd vertices: B and D

b. Even vertices: none; odd vertices: A and B

c. Even vertices: E; odd vertices: A, B, C, and D

EXAMPLE 2

Determine whether the networks in example 1 are traversable. If the network is traversable, find the possible starting points.

a. The network is traversable; *B* and *C* are the possible starting points.

b. The network is traversable; *A* and *B* are the possible starting points.

c. The network is not traversable because it has more than two odd vertices.

EXERCISES FOR SECTION 10.5

For exercises 1–10, find (a) the number of even vertices, (b) the number of odd vertices, (c) whether the network is traversable, and (d) the possible starting points if the network is traversable.

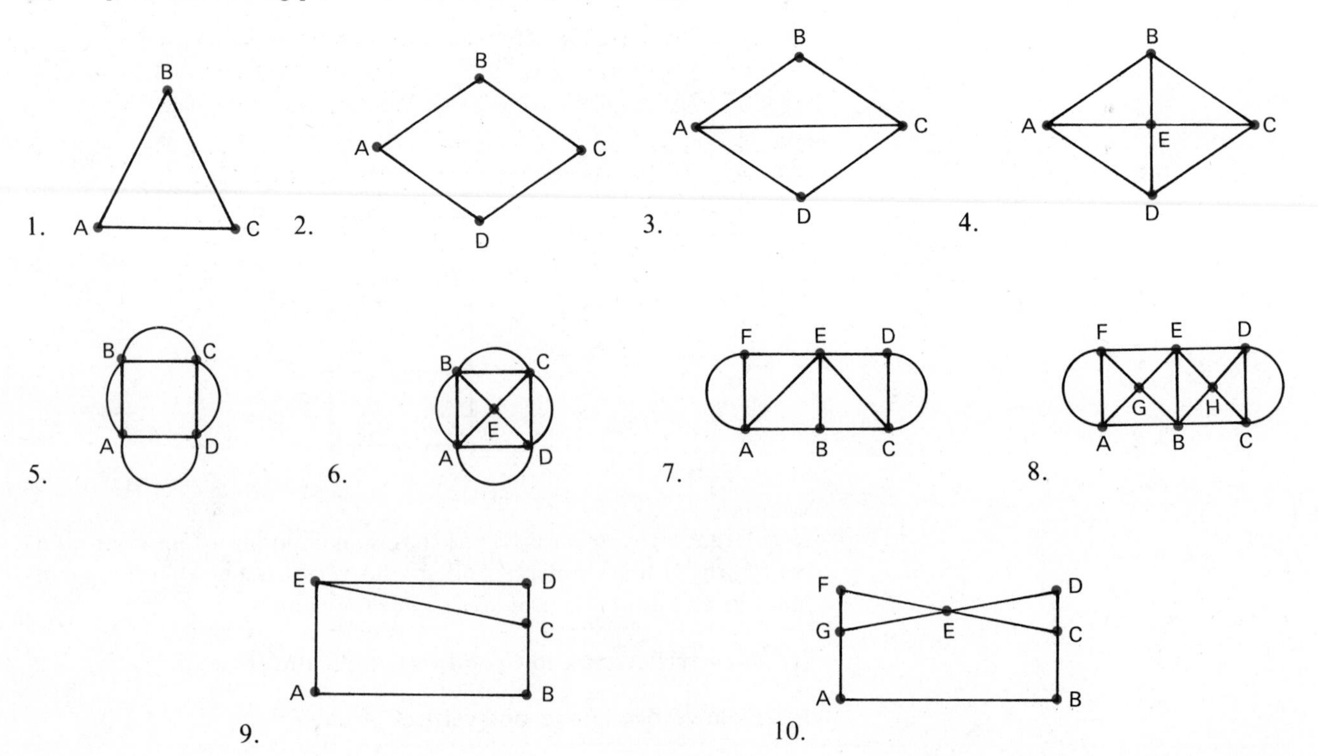

11. Is it possible to walk through a house with the floor plan given in figure 51 and pass through each doorway exactly once? (*Hint:* Use a network.)

12. Is it possible to walk through a house with the floor plan given in figure 52 and pass through each doorway exactly once? (*Hint:* Use a network.)

FIGURE 51

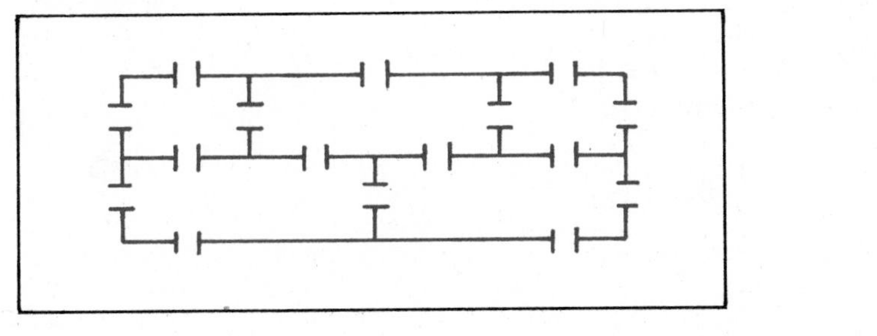

FIGURE 52

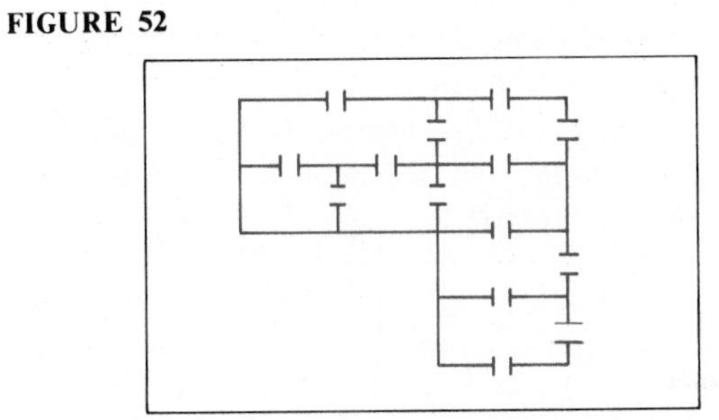

Just For Fun

Can you cut a hole in a standard-size piece of paper in such a way that your entire body can pass through it? Take a standard-size piece of paper and cut it as indicated in figure 53. If you have cut correctly, you should be able to pass your body through the hole in the paper. With some practice, you can use a smaller piece of paper and make more cuts.

FIGURE 53

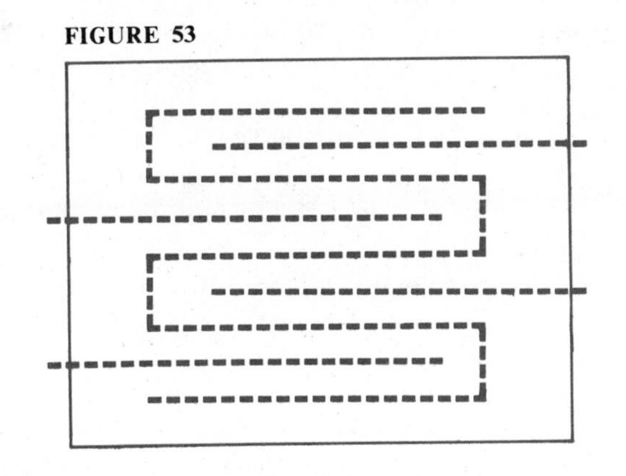

10.6 SUMMARY

Point, line, and *plane* are the most basic terms of geometry. A geometric *point* has no dimension, that is, it has no length, breadth, or thickness. The one thing that a geometric point does possess is position. A geometric *line* is formed by the intersection of two surfaces. A line in geometry has no width, but it does have length. We can think of a line as a set of points. *Collinear points* are points that lie on the same straight line. Listed below are some sets of points commonly used in geometry.

Description	Diagram	Notation
point P	•	P
line PQ	P Q	$\overleftrightarrow{PQ}$
half line PQ	P Q	$\overset{\circ\!\!\rightarrow}{PQ}$
ray PQ	P Q	$\overrightarrow{PQ}$
ray QP	P Q	$\overrightarrow{QP}$
line segment PQ	P Q	$\overline{PQ}$

A *plane* in geometry is a flat surface, such as the floor of a room. A plane separates one portion of space from another. We can think of a plane as a set of points. The points are on the plane, and the plane contains the points. *Coplanar points* are points that are on the same plane. A unique plane is determined by any three noncollinear points. Any line on a given plane divides that plane into two half planes.

An *angle* is the union of two rays that have a common endpoint. An angle on a plane divides the points on the plane into three parts: those points on the angle, those points in the *interior* of the angle, and those points in the *exterior* of the angle.

In this chapter we discussed many different kinds of angles. You should be familiar with the following types of angles: *adjacent angles, vertical angles, straight angles, right angles, acute angles,* and *obtuse angles.*

A *broken line* is a set of connected line segments. A *simple closed broken line* is one that starts and stops at the same point and does not intersect itself. A simple closed broken line is called a *polygon.* The connected line segments are the *sides* of the polygon, and the points at which the line segments are connected are the *vertices* of the polygon. Polygons are classified according to the number of sides that they have. For example, a polygon having five sides is called a *pentagon,* and a polygon having ten sides is called a *decagon.* In this chapter we discussed in detail polygons having three sides and four sides, that is, *triangles* and *quadrilaterals.* You should be familiar with the following: *equilateral triangles, isosceles triangles, scalene triangles, obtuse triangles, acute triangles,* and *right triangles.* You should also be familiar with *trapezoids, parallelograms, rhombuses, rectangles,* and *squares.*

A set of line segments or arcs is called a *network.* A network is *traversable* if it can be drawn by tracing each line segment or arc exactly once without lifting the pencil point from the paper. Any network that has only even vertices is traversable. If a network has exactly two odd vertices, it is traversable providing we begin at one of the odd vertices. A network that has more than two odd vertices is not traversable.

Review Exercises for Chapter 10

For exercises 1–10, use figure 54 with the indicated points to find each of the following:

1. $\overleftrightarrow{AR} \cap \overleftrightarrow{LJ}$
2. $\overrightarrow{AR} \cap \overrightarrow{SR}$
3. $\overrightarrow{RS} \cap \overrightarrow{RA}$
4. $\overrightarrow{RS} \cup \overrightarrow{RA}$
5. $\overline{LS} \cap \overline{AJ}$
6. $\overrightarrow{RA} \cap \overline{LS}$
7. $\overrightarrow{RA} \cup \overrightarrow{RL}$
8. $\overrightarrow{RS} \cup \overrightarrow{RJ}$
9. $\angle SRJ \cap \angle ARJ$
10. $\angle SRJ \cap \angle LRA$

FIGURE 54

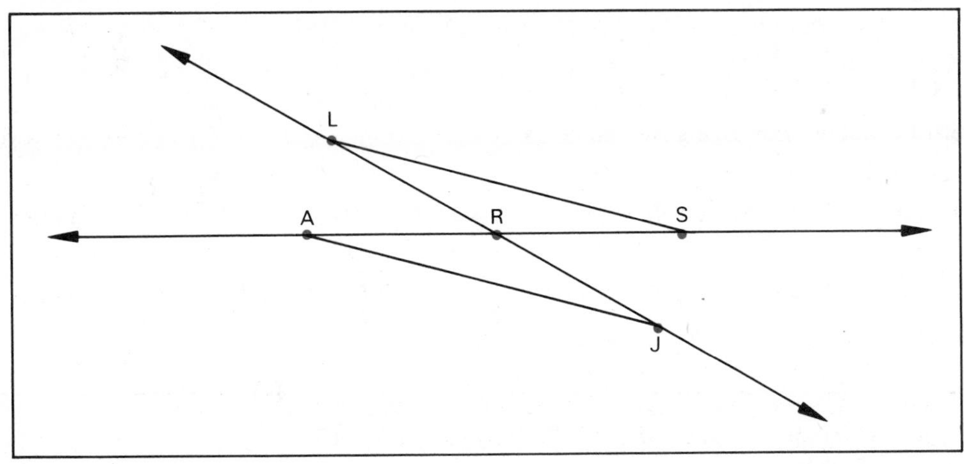

For exercises 11–20, use figure 55 with the indicated points to find each of the following:

11. $\overrightarrow{OS} \cup \overrightarrow{OJ}$
12. $\overrightarrow{OT} \cup \overrightarrow{OE}$
13. $\overrightarrow{OS} \cap \overrightarrow{OJ}$
14. $\overline{OT} \cap \overline{ST}$
15. (interior $\angle COT$) $\cap$ (interior $\angle EOT$)
16. (exterior $\angle COT$) $\cap$ (interior $\angle JOS$)
17. $\angle JOS \cap \angle EOT$
18. $\angle COT \cap \angle COS$
19. (exterior $\angle EOT$) $\cap$ (exterior $\angle JOS$)
20. (interior $\angle JOE$) $\cap$ (interior $\angle COE$)

FIGURE 55

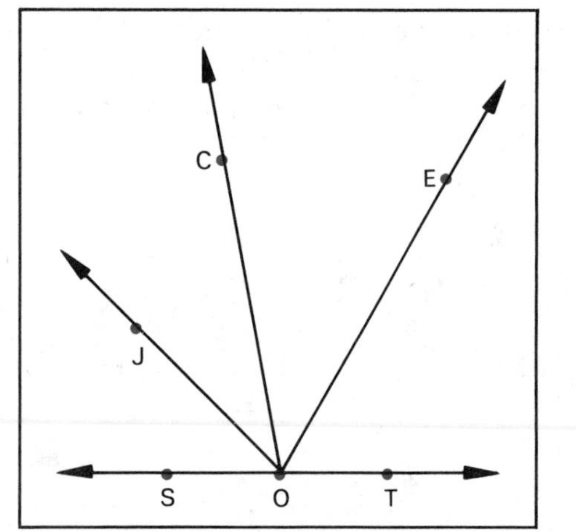

21. The lengths of two sides of a triangle are 4 and 7. The third side may be: (*a*) 1, (*b*) 2, (*c*) 3, (*d*) 4. (Choose one.)
22. Which set of numbers represents the lengths of the sides of a right triangle?

    (*a*) {2, 3, 4} (*b*) {4, 5, 6} (*c*) {7, 8, 9} (*d*) {5, 12, 13}

23. Find the length of hypotenuse $\overline{AB}$ in right triangle ABC, given that $m(\overline{AC}) = 8$ and $m(\overline{BC}) = 15$.
24. Find the length of leg $\overline{BC}$ in right triangle ABC given that $m(\overline{AC}) = 12$, $m(\overline{AB}) = 15$, and $\overline{AB}$ is the hypotenuse.

For exercises 25–28, find (a) the number of even vertices, (b) the number of odd vertices, (c) whether the network is traversable, and (d) the possible starting points if the network is traversable.

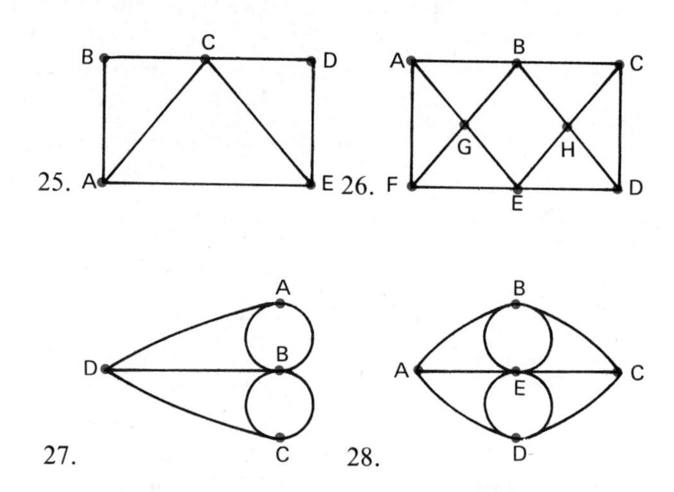

Indicate whether each sentence is true or false.

29. The sum of the squares of two sides of a triangle is equal to the square of the third side.
30. Two adjacent angles are angles which have a common vertex.
31. A parallelogram is a polygon whose opposite sides are equal in length.
32. Parallel lines are lines which lie in the same plane and do not intersect however far they are extended.
33. An acute triangle is a triangle in which one angle is acute.
34. If two angles of a triangle are equal in measure, the sides opposite these angles are of equal length.

Just For Fun

Cut a strip from a standard-size piece of paper: the strip of paper should measure approximately $\frac{1}{2}''$ by 11″. Give the paper a half-twist and then tape or glue the two ends together. You have constructed a Mobius strip. It should resemble figure 56a.

The Mobius strip has some interesting properties. For example, it has only one surface. You can demonstrate this by drawing a continuous line, or shading one side all the way around without lifting your pencil. You will note that this will mark or shade the entire surface.

Next, cut the strip in the middle, as you normally would to obtain two loops. See figure 56b. If you have done everything correctly, you should get one bigger loop.

FIGURE 56a

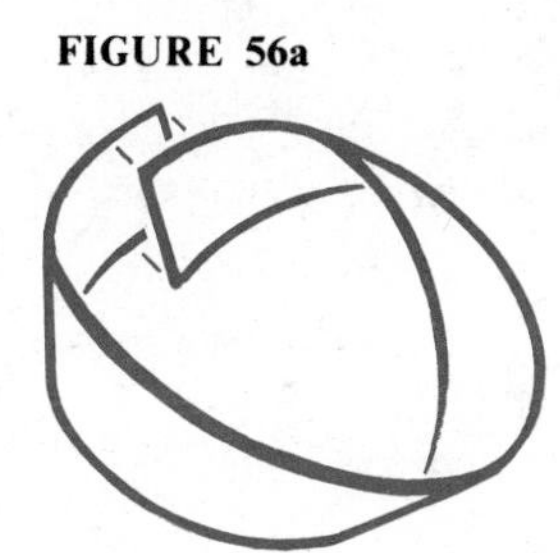

FIGURE 56b

NO MONEY DOWN
DISCOUNT HOUSE
Olympic
OLYMPIC
$1.25 A WEEK
take it easier
BUY WITH
TIME PAYMENTS
I'M BUYIN' FROM BRIAN
TRANSISTOR POCKET Radios
VICTOR
20% to 40%
DISCOUNT
others
R.C.A. VICTOR
RADIO

11 CONSUMER MATHEMATICS

After studying this chapter, you will be able to do the following:

1. Express the relationship between two quantities as a **ratio,** and solve a **proportion** for the missing term
2. Convert **percentages** to fractions or decimals, and convert fractions or decimals to percentages
3. Find the **markup** on an item when given the **cost** or **selling price** and the **percentage markup**
4. Find the amount of **markdown** and **sale price** when given the original retail price and **percentage markdown**
5. Use the formula $I = Prt$ to compute **simple interest**
6. Compute **compound interest** and **compound amounts**
7. Find the **effective annual interest rate** when money is compounded annually, semiannually, or quarterly
8. Determine the annual **premium** for the following types of life insurance policies: **term** (5-year), **straight life, limited-payment life** (20-year), and **endowment** (20-year)
9. Find the **true annual interest rate** when an item is purchased on the installment plan
10. Find the amount of interest paid when an item is purchased on the installment plan
11. Determine the monthly payments for principal and interest for mortgages of various lengths at various interest rates.

11.1 INTRODUCTION

Most of us have probably heard the phrase *caveat emptor* at one time or another. It means "let the buyer beware," which can also be interpreted as "the buyer buys at his own risk." Unfortunately, many businesses maintain this attitude when dealing with customers. As consumers, we must be wise and discerning shoppers. That is, we must look for the best buy when we purchase an item. When buying items on credit, purchasing insurance, or taking out a loan, consumers must have an understanding of *decimals, percents,*

Hays/Monkmeyer

simple interest, compound interest, and *effective rate of interest.* When faced with several alternatives—for example, buying on the installment plan or paying cash—we must make an intelligent decision based on our own particular situation. The topics in this chapter are designed to give you the information necessary to be a more intelligent shopper and a wiser consumer.

11.2 RATIO AND PROPORTION

The **ratio** of two quantities a and b is the quotient or indicated quotient obtained by dividing a by b. The ratio of a to b is written as

$$a \div b \quad \text{or} \quad \frac{a}{b} \quad \text{or} \quad a:b$$

For example, to indicate the ratio of 3 to 5, we could use

$$3 \div 5 \quad \text{or} \quad \frac{3}{5} \quad \text{or} \quad 3:5$$

A ratio provides us with a way of comparing two numbers by means of division. The ratio of one number to another number is the quotient of the first number divided by the second number. Therefore, the ratio of 12 to 3 is $12 \div 3$, or 4 to 1. That is,

$$12 \text{ to } 3 = \frac{12}{3} = \frac{4}{1} = 4 \text{ to } 1$$

EXAMPLE 1
Express each ratio in simplest form.

a. 12 to 36 b. 49:14 c. $\frac{14}{12}$

Solution

a. 12 to 36 $= \frac{12}{36} = \frac{1}{3}$

b. 49:14 $= \frac{49}{14} = \frac{7}{2}$

c. $\frac{14}{12} = \frac{7}{6}$

EXAMPLE 2
Find the ratio of 210 minutes to 3 hours.

Solution

The quantities compared by ratio must represent objects of the same kind measured in the same units. In this case, we have minutes compared to hours. However, to form a ratio, either both quantities should be in terms of minutes, or both should be in terms of hours. We choose to convert both to minutes. Since there are 60 minutes in an hour, we have

$$\frac{210}{180} = \frac{21}{18} = \frac{7}{6}$$

EXAMPLE 3

Express the ratio of $1\frac{1}{2}$ to $2\frac{1}{4}$ in simplest form.

Solution

The ratio of one number to another number is the quotient of the first number divided by the second number. Therefore, the ratio of $1\frac{1}{2}$ to $2\frac{1}{4}$ is the same as

$$1\frac{1}{2} \div 2\frac{1}{4}, \quad \text{or} \quad \frac{3}{2} \div \frac{9}{4}$$

This problem involves division of rational numbers. (See section 8.5, examples 8 and 9.) Hence

$$1\frac{1}{2}:2\frac{1}{4} = 1\frac{1}{2} \div 2\frac{1}{4} = \frac{3}{2} \div \frac{9}{4} = \frac{3}{2} \cdot \frac{4}{9} = \frac{12}{18} = \frac{2}{3}$$

EXAMPLE 4

A baseball team played 20 games and won 15 of them.

a. What is the ratio of the number of games won to the number of games played?

b. What is the ratio of the number of games lost to the number of games played?

c. What is the ratio of the number of games won to the number of games lost?

Solution

a. The ratio of games won to games played is 15:20.

$$15:20 = \frac{15}{20} = \frac{3}{4}$$

b. games lost = games played − games won
games lost = 20 − 15 = 5
The ratio of games lost to games played is 5:20.

$$5{:}20 = \frac{5}{20} = \frac{1}{4}$$

c. The ratio of games won to games lost is 15:5.

$$15{:}5 = \frac{15}{5} = \frac{3}{1}$$

A **proportion** is the equality of two ratios. That is, a proportion is a statement that says two ratios are equal. For example,

$$\frac{1}{2} = \frac{4}{8}$$

is a proportion. It states that the two ratios $\frac{1}{2}$ and $\frac{4}{8}$ are equal. If the ratios $a{:}b$ and $c{:}d$ are equal, then we can form a proportion by writing

$$a{:}b = c{:}d, \quad \text{or} \quad \frac{a}{b} = \frac{c}{d}$$

We can read this proportion by saying that "a is to b as c is to d." In this proportion, b and c are called the **means,** while a and d are called the **extremes.**

In any proportion, the product of the means equals the product of the extremes:

$$\textbf{if} \quad \frac{a}{b} = \frac{c}{d}, \quad \textbf{then} \quad ad = bc$$

For example, in the proportion $\frac{1}{2} = \frac{4}{8}$, 2 and 4 are the means, while 1 and 8 are the extremes.

$$\text{If} \quad \frac{1}{2} = \frac{4}{8}, \quad \text{then} \quad \underbrace{1 \cdot 8}_{\text{extremes}} = \underbrace{2 \cdot 4}_{\text{means}}, \quad \text{or} \quad 8 = 8$$

Many times we know three terms of a proportion and need to find the fourth term. For example, suppose that in a certain mathematics class, the ratio of the number of men to the number of women is 4:3. If there are 12 women in the class, how many men are in the class?

To solve this problem, note that the ratio 4:3 is the same as the ratio of the number of men to the number of women (12). We can, therefore, set up the proportion

$$\frac{x}{12} = \frac{4}{3}$$

where x represents the number of men in the class. By applying the rule that the product of the means equals the product of the extremes, we have

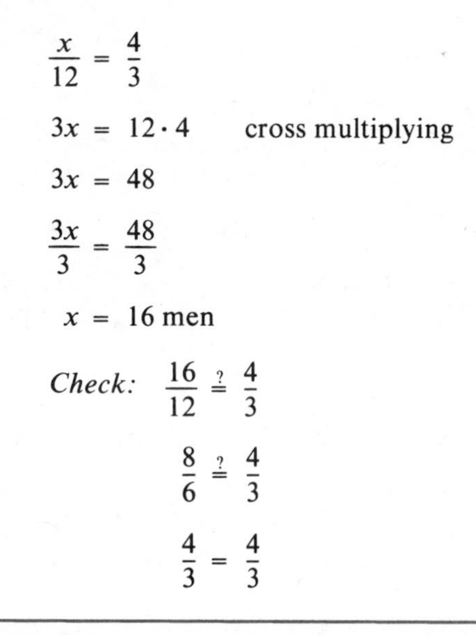

$$\frac{x}{12} = \frac{4}{3}$$

$$3x = 12 \cdot 4 \qquad \text{cross multiplying}$$

$$3x = 48$$

$$\frac{3x}{3} = \frac{48}{3}$$

$$x = 16 \text{ men}$$

Check: $\frac{16}{12} \stackrel{?}{=} \frac{4}{3}$

$$\frac{8}{6} \stackrel{?}{=} \frac{4}{3}$$

$$\frac{4}{3} = \frac{4}{3}$$

O'Brien/Stockmarket

EXAMPLE 5

A car travels 400 miles on 20 gallons of gasoline. At this rate, how many gallons of gasoline will be consumed on a trip of 900 miles?

Solution

Let x = the number of gallons used. We can then write the proportion

$$\frac{20}{x} = \frac{400}{900}$$

Solving this proportion,

$$400 \cdot x = 20 \cdot 900 \quad \text{cross multiplying}$$

$$400x = 18{,}000$$

$$\frac{400x}{400} = \frac{18{,}000}{400}$$

$$x = 45 \text{ gallons}$$

Check:

$$\frac{900}{45} \stackrel{?}{=} \frac{400}{20}$$

$$\frac{100}{5} \stackrel{?}{=} \frac{20}{1}$$

$$\frac{20}{1} = \frac{20}{1}$$

EXAMPLE 6

Find the value of x in each of the following proportions.

a. $4:8 = 3:x$ b. $3:7 = x:28$ c. $x:6 = 10:12$

Solution

To find the values of x in each proportion, we use the property that, for any proportion, the product of the means equals the product of the extremes.

a. $4:8 = 3:x$

$$\frac{4}{8} = \frac{3}{x}$$

$$4 \cdot x = 8 \cdot 3$$

$$4x = 24$$

$$x = 6$$

Check:

$$\frac{4}{8} \stackrel{?}{=} \frac{3}{6}$$

$$\frac{1}{2} = \frac{1}{2}$$

b. $3:7 = x:28$

$$\frac{3}{7} = \frac{x}{28}$$

$$3 \cdot 28 = 7 \cdot x$$

$$84 = 7x$$

$$12 = x$$

Check:

$$\frac{3}{7} \stackrel{?}{=} \frac{12}{28}$$

$$\frac{3}{7} \stackrel{?}{=} \frac{6}{14}$$

$$\frac{3}{7} = \frac{3}{7}$$

c. $x:6 = 10:12$

$$\frac{x}{6} = \frac{10}{12}$$

$$12x = 60$$

$$x = 5$$

Check: $\frac{5}{6} \stackrel{?}{=} \frac{10}{12}$

$$\frac{5}{6} = \frac{5}{6}$$

EXAMPLE 7

If $\frac{1}{2}$ inch represents 5 feet in a scale drawing, then how many inches will represent 30 feet?

Solution

We can solve this problem using a proportion. Let x be the number of inches that represent 30 feet. Since $\frac{1}{2}$ inch represents 5 feet, and x inches represent 30 feet, we have the ratios $\frac{1}{2}:x$ and 5:30. The proportion is

$$5:30 = \frac{1}{2}:x$$

$$\frac{5}{30} = \frac{\frac{1}{2}}{x}$$

$$5 \cdot x = \frac{1}{2} \cdot 30$$

$$5x = 15$$

$$x = 3$$

Check: $\frac{5}{30} \stackrel{?}{=} \frac{\frac{1}{2}}{3}$

$$\frac{1}{6} = \frac{1}{6}$$

To summarize what we have learned: The ratio of two quantities a and b is the quotient or indicated quotient obtained by dividing a by b. If the ratio of a and b, $a:b$, and the ratio of c and d, $c:d$, are equal, then we can form the proportion

$$a:b = c:d, \quad \text{or} \quad \frac{a}{b} = \frac{c}{d}$$

In any proportion, the product of the means equals the product of the extremes. That is, $a \cdot d = b \cdot c$.

EXERCISES FOR SECTION 11.2

1. Express each ratio as a fraction.
   a. 3 to 2 b. 4 to 7 c. 8:5 d. 5:6
2. Express each ratio as a fraction.
   a. 4 to 11 b. 7 to 9 c. 9:7 d. 11:7
3. Express each ratio in simplest form.
   a. 12 to 4 b. 14:28 c. $\frac{18}{12}$
   d. 36:9 e. 5 to 30 f. $1\frac{1}{2}:2\frac{1}{4}$
4. Express each ratio in simplest form.
   a. 14 to 56 b. $\frac{14}{12}$ c. 2:18
   d. 14:49 e. $3\frac{1}{3}:2\frac{2}{3}$ f. $2\frac{1}{3}:2\frac{5}{6}$
5. A mathematics class has 30 students in it. There are 18 women and 12 men in the class.
   a. What is the ratio of men to women?
   b. What is the ratio of women to men?
   c. What is the ratio of the number of men to the number of students in the class?
6. The perimeter of a rectangle is 36 metres and the width is 8 metres. Find the ratio of the length of the rectangle to its width.
7. The perimeter of a rectangle is 44 metres and the length is 19 metres. Find the ratio of the width of the rectangle to its length.
8. A final examination has 40 true–false questions and 60 multiple-choice questions.
   a. What is the ratio of true–false questions to multiple-choice questions?
   b. What is the ratio of multiple-choice questions to true–false questions?
   c. What is the ratio of true–false questions to the total number of questions on the examination?
9. Find the value of x in each of the following proportions.
   a. $x:7 = 6:21$ b. $3:x = 14:28$
   c. $5:1 = x:6$ d. $1:7 = 5:x$
10. Find the value of x in each of the following proportions.
    a. $1:2 = 4:x$ b. $5:7 = x:21$
    c. $2:x = 8:12$ d. $x:6 = 4:8$
11. On a map, a line segment 3 inches long represents a distance of 18 miles. Using the same scale, how many miles long is a road which measures 2 inches on the map?
12. Addie takes 3 minutes to read an article of 350 words. At the same rate, how many minutes will it take her to read another article of 875 words?
13. If one-half of a number is 30, then how much is two-thirds of the same number?
14. A student who is 6 feet tall stands next to a flag pole that is 50 feet tall. If the shadow cast by the student is 9 feet long, how long is the shadow cast by the flag pole?

Just For Fun

How long will it take to cut a 12-foot log into 1-foot lengths, allowing 2 minutes for each cut?

11.3 PERCENTAGES, DECIMALS, AND FRACTIONS

The concept of percentage is one that occurs daily in our lives. For example, when we read the newspaper we might find statements like these:

"The consumer price index rose one-tenth of one percent last month."
"Sibley's Department Store's fall sale on women's fashions will feature savings of 20 percent and more on all items in stock."
"First Federal Savings Bank offers loans at 9 percent."
"The sales tax is 7 percent."

We can describe a **percentage** as a ratio with a denominator of 100. That is, a percentage is the ratio of any number to 100. The symbol for percent is %. Hence "20 percent" can also be written as "20%." Since a percentage is the ratio of any number to 100, 20% means the ratio of 20 to 100; that is, $20\% = \frac{20}{100}$. Similarly,

$$18\% \text{ means } 18{:}100, \text{ or } \frac{18}{100}$$

$$6\% \text{ means } 6{:}100, \text{ or } \frac{6}{100}$$

$$\tfrac{1}{2}\% \text{ means } \tfrac{1}{2}{:}100, \text{ or } \frac{\frac{1}{2}}{100}$$

EXAMPLE 1

Express each percentage as a fraction in simplest form.

a. 20% b. $33\frac{1}{3}\%$ c. $\frac{1}{2}\%$

Solution

A percentage is the ratio of a number to 100.

a. $20\% = \dfrac{20}{100} = \dfrac{1}{5}$

b. $33\frac{1}{3}\% = \dfrac{33\frac{1}{3}}{100} = \dfrac{\frac{100}{3}}{100} = \dfrac{100}{3} \div \dfrac{100}{1} = \dfrac{100}{3} \cdot \dfrac{1}{100} = \dfrac{1}{3}$

c. $\dfrac{1}{2}\% = \dfrac{\frac{1}{2}}{100} = \dfrac{1}{2} \div \dfrac{100}{1} = \dfrac{1}{2} \cdot \dfrac{1}{100} = \dfrac{1}{200}$

It is sometimes necessary to express a percentage as a decimal, such as in finding the sales tax on a purchase of $15 if the sales tax is 7 percent. To convert a percentage to a decimal, we use a technique similar to that used for converting a percentage to a fraction. Seven percent means $\frac{7}{100}$, and this is the same as 0.07. Therefore, $7\% = 0.07$. The 7 percent sales tax on \$15 is found by multiplying 15 by 0.07.

$$\begin{array}{r} 15 \\ \times\ 0.07 \\ \hline 1.05 \end{array}$$

The sales tax is \$1.05.

EXAMPLE 2
Express 15% as a decimal.

Solution

$$15\% = \frac{15}{100} = 0.15$$

Note that, to convert a percentage to a decimal, we can drop the percent sign and move the decimal two places to the left. We can do this because a percentage is the ratio of a number to 100, and to divide a number by 100 is the same as moving the decimal point two places to the left. For example, $125 \div 100 = 1.25$. Therefore we can convert 15% to a decimal directly:

$$15\% = 0.15$$

Note that we drop the percent sign and move the decimal point two places to the left.

EXAMPLE 3
Express each percentage as a decimal.

a. 3% b. 18% c. $\frac{1}{2}\%$

Solution

a. $3\% = 0.03$

b. $18\% = 0.18$

c. $\frac{1}{2}\% = 0.5\% = 0.005$

(*Note:* To express $\frac{1}{2}\%$ as a decimal, we first had to convert $\frac{1}{2}$ to a decimal: $\frac{1}{2} = 0.5$. For a review of converting fractions to decimals, see section 8.6.)

Thus far we have converted percentages to fractions and to decimals. Next we want to consider changing fractions and decimals to percentages.

To change a fraction such as $\frac{1}{4}$ to percentage notation, we again make use of the idea that percentage is the ratio of a number to 100. We can think of $\frac{1}{4}$ as a ratio of 1:4. Thus we want to convert $\frac{1}{4}$ to an equivalent ratio with 100 as its denominator. We can use a proportion to do this. That is,

$$\begin{aligned} \frac{1}{4} &= \frac{x}{100} \\ 4 \cdot x &= 1 \cdot 100 \\ 4x &= 100 \\ x &= 25 \end{aligned}$$

Therefore $\frac{1}{4} = \frac{25}{100}$, or 25%.

EXAMPLE 4

Express each fraction as a percentage.

a. $\frac{1}{2}$ b. $\frac{1}{8}$ c. $1\frac{3}{4}$

Solution

To change a fraction to a percentage, we first find an equivalent ratio with a denominator of 100. We can do this by means of proportion. After finding this equivalent ratio, we drop the denominator (100) and add a percent sign.

a. $$\begin{aligned} \frac{1}{2} &= \frac{x}{100} \\ 2 \cdot x &= 1 \cdot 100 \\ 2x &= 100 \\ x &= 50 \end{aligned}$$

Therefore $\frac{1}{2} = \frac{50}{100}$, or 50%.

b. $$\frac{1}{8} = \frac{x}{100}$$

$$8 \cdot x = 1 \cdot 100$$

$$8x = 100$$

$$x = \frac{100}{8} = 12.5$$

Therefore $\frac{1}{8} = \frac{12.5}{100}$, or 12.5%.

c. First we rewrite $1\frac{3}{4}$ as $\frac{7}{4}$; then we set up the necessary proportion:

$$\frac{7}{4} = \frac{x}{100}$$

$$4 \cdot x = 7 \cdot 100$$

$$4x = 700$$

$$x = 175$$

Therefore $\frac{7}{4} = \frac{175}{100}$, or 175%.

When expressing a decimal as a percentage, it is important to remember that one decimal place to the right of the decimal point represents tenths, two decimal places represent hundredths, three decimal places represent thousandths, and so on. For example,

$$0.5 = \frac{5}{10}, \qquad 0.12 = \frac{12}{100}, \quad \text{and} \quad 0.125 = \frac{125}{1000}$$

A percentage is the ratio of a number to 100. Therefore, to convert a decimal to a percentage, we must obtain an equivalent expression with a denominator of 100. To express 0.15 as a percentage, we first rewrite 0.15 as $\frac{15}{100}$. Next we drop the denominator (100) and add a percent sign. Therefore $0.15 = \frac{15}{100} = 15\%$.

To express 0.5 as a percentage, we first express it as $\frac{5}{10}$. However, $\frac{5}{10}$ does not have a denominator of 100. Hence we use a proportion to find an equivalent ratio. That is,

$$\frac{5}{10} = \frac{x}{100}$$

$$10 \cdot x = 5 \cdot 100$$

$$10x = 500$$

$$x = 50$$

Therefore, $0.5 = \frac{5}{10} = \frac{50}{100}$, or 50%.

After examining these examples, you may have discovered another way to express a decimal as a percentage:

To write a decimal as a percentage, move the decimal point two places to the right and add a percent sign.

For example, 0.15 = 15% and 0.5 = 50%. In each case, the decimal point has been moved two places to the right and a percent sign has been added.

EXAMPLE 5
Express each decimal as a percentage.

a. 0.07 b. 0.1 c. 3.2 d. 0.003

Solution
To convert a decimal to a percentage, move the decimal point two places to the right and add a percent sign.

a. 0.07 = 7% b. 0.1 = 10%

c. 3.2 = 320% d. 0.003 = 0.3%

To summarize what we have learned in this section: A percentage is the ratio of a number to 100. To express a percentage as a fraction, drop the percent sign and write the given number over 100; then reduce the resulting fraction to simplest form. To express a percentage as a decimal, drop the percent sign and move the decimal point two places to the left. To express a fraction as a percentage, use a proportion to find an equivalent fraction with a denominator of 100; then drop the denominator (100) and add a percent sign. To change a decimal to a percentage, move the decimal point two places to the right and add a percent sign.

EXERCISES FOR SECTION 11.3

In exercises 1–6, express each percentage as a fraction in simplest terms.

1. a. 15% b. 25% c. 75%
2. a. 10% b. 30% c. 48%
3. a. $4\frac{1}{2}\%$ b. $2\frac{1}{3}\%$ c. $6\frac{1}{4}\%$
4. a. $7\frac{3}{4}\%$ b. $8\frac{1}{2}\%$ c. $20\frac{1}{3}\%$
5. a. 6.5% b. 2.3% c. 150%
6. a. 13.7% b. 3.6% c. 250%

In exercises 7–9, express each percentage as a decimal.

7. a. 17% b. 3% c. $4\frac{1}{2}\%$

8. a. 12% b. 4% c. $5\frac{1}{2}\%$
9. a. 6.5% b. 300% c. 0.25%

In exercises 10–15, express each fraction as a percentage.

10. a. $\frac{1}{4}$ b. $\frac{2}{5}$ c. $\frac{3}{8}$
11. a. $\frac{3}{4}$ b. $\frac{4}{5}$ c. $\frac{5}{8}$
12. a. $1\frac{1}{2}$ b. $\frac{1}{25}$ c. $\frac{3}{25}$
13. a. $2\frac{3}{5}$ b. $1\frac{1}{8}$ c. $\frac{7}{8}$
14. a. $\frac{1}{3}$ b. $1\frac{4}{5}$ c. $\frac{3}{16}$
15. a. $\frac{2}{3}$ b. $2\frac{3}{4}$ c. $\frac{1}{16}$

In exercises 16–19, express each decimal as a percentage.

16. a. 0.05 b. 0.32 c. 0.5
17. a. 0.09 b. 2.14 c. 0.9
18. a. 0.005 b. 0.314 c. 5.12
19. a. 1.125 b. 0.010 c. 3.01

Just For Fun

When it is 12 noon in New York City (eastern standard time), what time is it in Tokyo?

11.4 MARKUPS AND MARKDOWNS

The price that we pay for an item when we buy it from a retailer is the retail price, or **selling price.** The price that the retailer pays for the item and the selling price are not the same. The amount that a retailer pays for goods is called the **cost** of the item. The difference between the *selling price* of an item and the *cost* of that item is the retailer's **profit margin.** For example, if a color television has a selling price of $400 and it cost the dealer $300, then the profit margin is

$$\$400 - \$300 = \$100$$

The $100 represents a profit margin, but it is not all profit. Out of this $100, the dealer has to meet expenses such as utility costs (heat, light, phone), employees' wages, insurance premiums, taxes, rent or mortgage payments, and so on. Another term commonly used to describe this profit margin is **markup.** Markup is the difference between the selling price and the cost of an item. That is,

$$\text{markup} = \text{selling price} - \text{cost}$$

Using this equation, we can derive two other equations:

$$\text{selling price} = \text{markup} + \text{cost}$$
$$\text{cost} = \text{selling price} - \text{markup}$$

If the selling price of a ring is \$90 and the cost is \$50, then the markup is \$90 − \$50 = \$40. It is important to know the amount of markup on an item because this amount tells you how much of the price you pay goes to the retailer to cover overhead (the cost of running the business) and profit, and how much goes to the manufacturer or wholesaler.

Markup can also be given as a percentage of the cost (**percentage markup on cost**) or as a percentage of the selling price (**percentage markup on selling price**). Percentage markup on cost tells you the percentage by which the cost of the item was increased to obtain the price you pay, the selling price. Percentage markup on selling price gives the percentage of the selling price that the retailer retains for overhead and profits.

Most retailers work with percentage markups because they deal with large lots of merchandise. A markup of 50% of cost can be applied to a whole group of items whose individual selling prices might vary widely. Then the total markup for the group is easily figured as a percentage of the total cost, without any need to count the items in the group or figure individual markup costs or selling prices.

EXAMPLE 1

The pro shop at the National Golf Club sells a certain brand of golf clubs at prices based on a markup of 40% of the cost. If the cost of a set of these golf clubs is \$150, what is the selling price?

Solution

Since the markup is determined by the cost, we find 40% of \$150:

$$40\% \text{ of } \$150 = 0.40 \times \$150 = \$60.00 = \text{markup}$$
$$\text{selling price} = \text{markup} + \text{cost}$$
$$\text{selling price} = \$60 + \$150 = \$210$$

EXAMPLE 2

Another pro shop, at Southampton Golf Links, sells a different brand of golf clubs for \$300 per set. If the cost of a set of these golf clubs is \$200, what is the percentage of markup on the cost?

Solution

$$\text{markup} = \text{selling price} - \text{cost}$$

Therefore,

$$\text{markup} = \$300 - \$200 = \$100$$

Now we must find what the percentage of markup is, based on the cost. That is, we must find what percentage $100 (the markup) is of $200 (the cost). To do this, we form a ratio, markup:cost. Therefore

$$\frac{\text{markup}}{\text{cost}} = \frac{\$100}{\$200} = \frac{1}{2} = 0.5 = 50\%$$

The markup ($100) is 50% of the cost ($200).

Check: 50% of $200 = 0.50 × $200 = $100 = markup

To find what percentage one number is of another, we form a ratio between the two numbers and convert the ratio (fraction) to a percentage. For example, suppose you took a quiz and got 12 out of 15 questions correct. To find the percentage of correct answers, we form a ratio, 12:15, and convert it to a percentage:

$$\frac{12}{15} = \frac{4}{5} = 0.8 = 80\%$$

Recall that we can also do this by means of a proportion. That is,

$$\frac{12}{15} = \frac{x}{100}$$

$$15 \cdot x = 12 \cdot 100$$

$$15x = 1200$$

$$x = 80, \quad \text{and} \quad \frac{80}{100} = 80\%$$

EXAMPLE 3

a. What percentage of 108 is 27?

b. What percentage of 60 is 48?

Solution

a. $$\frac{27}{108} = \frac{x}{100}$$

$$108 \cdot x = 27 \cdot 100$$

$$108x = 2700$$

$$x = 25, \text{ and } \frac{25}{100} = 25\%$$

b. $$\frac{48}{60} = \frac{x}{100}$$

$$60 \cdot x = 48 \cdot 100$$

$$60x = 4800$$

$$x = 80, \text{ and } \frac{80}{100} = 80\%$$

Alternate Solution

a. Express $\frac{27}{108}$ as a decimal and convert the resulting decimal to a percentage:

$$\frac{27}{108} = \frac{1}{4} = 0.25 = 25\%$$

b. $\frac{48}{60} = \frac{4}{5} = 0.8 = 80\%$

EXAMPLE 4
Al's Appliance Outlet sells a particular black-and-white television for \$110. If the set costs Al \$80, what is the percentage of markup on the cost?

Solution

$$\text{markup} = \text{selling price} - \text{cost}$$

Therefore,

$$\text{markup} = \$110 - \$80 = \$30$$

Since the percentage of markup on cost is equal to the ratio of markup to cost, we have

$$\frac{\$30}{\$80} = \frac{3}{8} = 0.375 = 37.5\%$$

The markup ($30) is 37.5% of the cost ($80).

Check: 37.5% of $80 = 0.375 × $80 = $30 = markup

EXAMPLE 5

A stereo sells for $320. The markup is 60% of the cost. What is the cost of the stereo system?

Solution

$$\text{selling price} = \text{markup} + \text{cost}$$

Since the markup is 60% of the cost, we can also state that

$$\text{selling price} = (60\% \text{ of cost}) + \text{cost}$$

This means that the selling price is 60% of the cost plus the full cost. But the full cost is 100% of the cost. Therefore,

$$\text{selling price} = (60\% \text{ of cost}) + (100\% \text{ of cost})$$

Thus, the selling price is 160% of the cost, so $320 represents 160% of the cost. To find the cost, we divide $320 by 160%, or 1.60.

$$\frac{\$320}{160\%} = \frac{\$320}{1.60} = \$200 = \text{cost}$$

Check: 60% of $200 = 0.60 × $200 = $120 = markup

$$\begin{aligned} \text{selling price} &= \text{markup} + \text{cost} \\ \$320 &= \$120 + \$200 \\ \$320 &= \$320 \end{aligned}$$

EXAMPLE 6

A coat retails for $120. The markup is 25% of the cost. What is the cost of the coat?

Solution

$$\begin{aligned} \text{selling price} &= \text{markup} + \text{cost} \\ &= (25\% \text{ of cost}) + \text{cost} \\ &= (25\% \text{ of cost}) + (100\% \text{ of cost}) \\ &= 125\% \text{ of cost} \end{aligned}$$

The selling price (\$120) is 125% of the cost. To find the cost, we divide \$120 by 125%, or 1.25:

$$\frac{\$120}{125\%} = \frac{\$120}{1.25} = \$96 = \text{cost}$$

Check:

$$\begin{aligned} 25\% \text{ of } \$96 &= 0.25 \times \$96 = \$24 = \text{markup} \\ \text{selling price} &= \text{markup} + \text{cost} \\ \$120 &= \$24 + \$96 \\ \$120 &= \$120 \end{aligned}$$

Thus far we have discussed markup in terms of percentage of cost. However, there are many businesses that figure their markup on the selling price. Suppose a coat costs a retailer \$50 and the markup is 20% of the selling price. What is the selling price for this particular coat? Recall that

$$\text{cost} = \text{selling price} - \text{markup}$$

Then

$$\text{cost} = \text{selling price} - (20\% \text{ of selling price})$$

Note that the selling price is 100% of the selling price. Hence

$$\text{cost} = (100\% \text{ of selling price}) - (20\% \text{ of selling price})$$

This means that

$$\text{cost} = 80\% \text{ of selling price}$$

To find the selling price, we divide \$50 by 80%, or 0.80:

$$\frac{\$50}{80\%} = \frac{\$50}{0.80} = \$62.50 \qquad \text{selling price}$$

We can check our work. The selling price is \$62.50 and the cost is \$50. Therefore the markup is \$62.50 − \$50.00 = \$12.50. We now find 20% of \$62.50 and check to see if it is \$12.50.

$$20\% \text{ of } \$62.50 = 0.20 \times \$62.50 = \$12.50 \quad \text{markup}$$

The markup is \$12.50, which is 20% of the selling price, \$62.50.

EXAMPLE 7

A 10-speed bike costs a retailer \$90. The markup is 25% of the selling price. Find the selling price.

Solution

$$\begin{aligned} \text{selling price} &= \text{markup} + \text{cost} \\ &= (25\% \text{ of selling price}) + \text{cost} \end{aligned}$$

This means that the cost is 100% − 25% = 75% of the selling price. That is, \$90 is 75% of the selling price. Therefore to find the selling price, we divide \$90 by 75%, or 0.75:

$$\frac{\$90}{75\%} = \frac{\$90}{0.75} = \$120 \qquad \text{selling price}$$

Check: The markup is \$120 − \$90 = \$30. Hence, 25% of \$120 should be \$30.

$$25\% \text{ of } \$120 = 0.25 \times \$120 = \$30 = \text{markup}$$

The solution checks.

EXAMPLE 8

A color television costs a retailer \$300. The markup is 30% of the selling price. Find the selling price.

Solution

$$\begin{aligned} \text{selling price} &= \text{markup} + \text{cost} \\ &= (30\% \text{ of selling price}) + \text{cost} \end{aligned}$$

Therefore, the cost is 100% − 30% = 70% of the selling price. That is, \$300 is 70% of the selling price. To find the selling price, we divide \$300 by 70%, or 0.70:

$$\frac{\$300}{70\%} = \frac{\$300}{0.70} = \$428.57 \qquad \text{selling price, to the nearest cent}$$

Check: The markup is

$$\$428.57 - \$300.00 = \$128.57$$

Therefore, 30% of \$428.57 should be \$128.57.

30% of \$428.57

$$= 0.30 \times \$428.57 = \$128.57 = \text{markup, to the nearest cent}$$

The solution checks.

Many times, retailers cannot sell everything at the retail selling price. This happens with defective merchandise, overstocked items, discontinued models, unpopular styles, and so on. A retailer still wants to sell the merchandise in stock, but, since it cannot be sold at the original price, it must be reduced in price. Therefore, the merchandise is sold at a new, lower price called the **sale price.** The change, or difference, between the original price and the sale price is called the **markdown.**

$$\text{markdown} = \text{original price} - \text{sale price}$$

Markdown can be expressed in terms of a dollar amount, and it can also be expressed in terms of percentage reduction.

It is important for consumers to understand markdowns so that they can make judgments about the actual savings that sale prices represent. Percentage markdowns are usually a better indicator of savings than the dollar amount of the markdown. For example, a \$10 markdown on an item that ordinarily sells for \$100 is a savings of 10%, but a \$10 markdown on a \$20 item is a savings of 50%.

If a retailer is selling coats for \$30 that were originally priced at \$40, we can find the amount of markdown by subtracting \$30 from \$40. That is,

$$\text{markdown} = \text{original price} - \text{sale price}$$

In this case, the markdown is \$40 − \$30 = \$10. But what is the percent reduction? To find the percent reduction, we must find what percent \$10 is of \$40. Therefore we divide \$10 by \$40:

$$\frac{\$10}{\$40} = \frac{1}{4} = 0.25 = 25\%$$

The original price was reduced by 25%. That is, the markdown is 25% of the original price. If we want to find the percentage markdown based on the sale price, we divide the markdown (\$10) by the sale price (\$30):

$$\frac{\$10}{\$30} = \frac{1}{3} = 33\frac{1}{3}\%$$

The markdown of $10 is $33\frac{1}{3}\%$ of the sale price.

EXAMPLE 9

Al's Appliance Outlet had a clearance sale on last year's color television sets. A certain set originally selling for $480 was advertised at a reduction of 20%.

a. What was the dollar markdown?

b. What was the sale price?

Solution

a. We find 20% of the original price, $480.

$$20\% \text{ of } \$480 = 0.20 \times \$480 = \$96 = \text{markdown}$$

b. sale price = original price − markdown
sale price = $480 − $96 = $384

EXAMPLE 10

During a clearance sale, a pair of boots that originally sold at $50 was reduced to $40.

a. What was the percentage markdown based on the original price?

b. What was the percentage markdown based on the sale price?

Solution

$$\begin{aligned} \text{markdown} &= \text{original price} - \text{sale price} \\ &= \$50 - \$40 = \$10 \end{aligned}$$

a. To find the percentage markdown based on the original price, we divide $10 by $50:

$$\frac{\$10}{\$50} = \frac{1}{5} = 0.20 = 20\%$$

b. To find the percentage markdown based on the sale price, we divide $10 by $40:

$$\frac{\$10}{\$40} = \frac{1}{4} = 0.25 = 25\%.$$

EXERCISES FOR SECTION 11.4

For exercises 1 and 2, express your answers to the nearest tenth of a percentage.

1. a. What percentage of 48 is 24?
   b. What percentage of 48 is 12?
   c. What percentage of 39 is 13?
   d. What percentage of 24 is 16?
   e. What percentage of 216 is 54?
   f. What percentage of 170 is 54?

2. a. What percentage of 76 is 38?
   b. What percentage of 56 is 7?
   c. What percentage of 114 is 38?
   d. What percentage of 7 is 2?
   e. What percentage of 56 is 21?
   f. What percentage of 96 is 60?

3. Find the dollar amount of markup, to the nearest cent.

	Cost	Percentage markup on cost
a.	\$ 50.00	25%
b.	\$300.00	$33\frac{1}{3}\%$
c.	\$ 25.00	30%
d.	\$175.00	$12\frac{1}{2}\%$

4. Find the dollar amount of markup, to the nearest cent.

	Cost	Percentage markup on cost
a.	\$ 35.00	7%
b.	\$210.00	$12\frac{1}{2}\%$
c.	\$ 29.95	20%
d.	\$ 49.95	$18\frac{1}{2}\%$

5. Find the markup and the percentage of markup on the cost. Express each percentage to the nearest tenth.

	Selling Price	Cost
a.	\$ 50.00	\$20.00
b.	\$100.00	\$75.00
c.	\$ 13.50	\$10.50
d.	\$ 19.95	\$15.95

6. Find the markup and the percentage of markup on the cost. Express each percentage to the nearest tenth.

	Selling Price	Cost
a.	\$220.00	\$110.00
b.	\$ 8.00	\$ 6.00
c.	\$ 14.75	\$ 10.00
d.	\$ 99.99	\$ 75.49

7. A coat sells for \$125. The markup is 40% of the cost. What is the cost of the coat, to the nearest cent?

8. A pair of boots retails for \$60. The markup is $33\frac{1}{3}\%$ of the cost. What is the cost of the boots?

9. A television sells for \$358. The markup is 30% of the cost. What is the cost of the television, to the nearest cent?

10. A lawnmower costs a retailer \$100. The markup is 25% of the selling price. Find the selling price, to the nearest cent.

11. A snowblower costs a retailer \$310. The markup is 40% of the selling price. Find the selling price, to the nearest cent.

12. A sporting-goods dealer pays \$18.50 for each basketball that she buys. If she wants a markup of 35% on the selling price, what should she set as the selling price?

13. Find the markup and the cost, to the nearest cent.

	Selling Price	Percentage markup on selling price
a.	\$400.00	40%
b.	\$ 10.98	20%
c.	\$ 49.95	$12\frac{1}{2}\%$

14. Find the markup and the cost, to the nearest cent.

	Selling Price	Percentage markup on selling price
a.	$ 60.00	$33\frac{1}{3}\%$
b.	$ 99.99	25%
c.	$299.95	$12\frac{1}{2}\%$

15. Al's Appliance Outlet had a clearance sale on refrigerators. A certain model, originally priced at $395, was advertised at 20% off. What was the dollar markdown? What was the sale price?

16. A bike shop advertised one of its 10-speed bikes for $150. After some time had passed, the bike had not sold, so the dealer lowered the price to $120. What was the percentage of markdown on the original price?

17. Dinah's Donut Shop sells fresh doughnuts for $1.50 per dozen. Day-old doughnuts are sold for $1.00 per dozen. What percentage (to the nearest tenth) is the markdown on the original price?

18. A department store sells men's suits for $199.95. During a clearance sale, the price is reduced 25%. What is the dollar markdown and the sale price, to the nearest cent?

19. During an inventory clearance sale, a used-car dealer reduced the price of all the cars on his lot by $300. If a certain car was originally priced at $1195, what was the sale price? What percentage (to the nearest tenth) is the markdown on the sale price?

20. A coat that was originally priced to sell for $72.50 was reduced to $60.00. What percentage (to the nearest tenth) was the markdown on the original price? What percentage is the markdown (to the nearest tenth) on the sale price?

Just For Fun

The Bills are in first place and the Dolphins are in fifth, while the Patriots are midway between them. If the Jets are ahead of the Dolphins and the Colts are immediately behind the Patriots, then who is in second place?

11.5 SIMPLE INTEREST

Interest is a familiar term that many people have trouble understanding. If an individual borrows money from a bank, then he or she must pay a fee to the bank for the use of the money. This fee is called **interest.** Banks also pay interest to people who deposit their money in savings accounts. Whenever merchandise is bought on credit, the customer pays interest on the unpaid balance. Most department stores charge $1\frac{1}{2}$-percent interest per month on the unpaid balance, after 30 days.

If a person borrows \$1000 from a bank, then the \$1000 is called the *principal.* Similarly, if a person has \$1000 in a savings account, this is also called the principal. The amount of money on which interest is paid is always called the **principal.** The amount of interest depends on the principal; that is, the interest is a certain percentage of the principal. This percentage is called the **rate of interest,** or **rate.** Unless otherwise noted, the rate of interest is an annual one.

When money is borrowed, the borrower agrees to pay back the principal and the interest within a specified period of time. For example, an auto loan may be given for a period of 3 years. At the end of 3 years, both the principal and the interest have been paid, so the interest for such a loan is computed for a period of 3 years. This period is called the **time** of the loan. The interest due depends on three things: the principal, the rate of interest, and the time.

O'Brien/Stockmarket

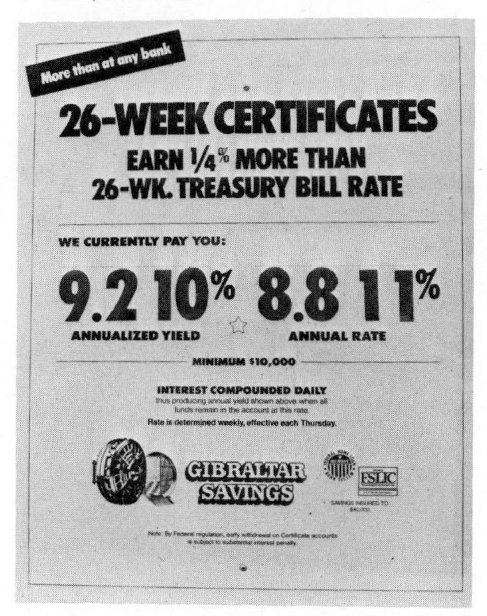

More than at any bank
26-WEEK CERTIFICATES
EARN 1/4% MORE THAN
26-WK. TREASURY BILL RATE
WE CURRENTLY PAY YOU:
9.2 10%
8.8 11%
ANNUALIZED YIELD
ANNUAL RATE
MINIMUM \$10,000
INTEREST COMPOUNDED DAILY
thus producing annual yield shown above when all
funds remain in the account at this rate
Rate is determined weekly, effective each Thursday.
GIBRALTAR
SAVINGS
FSLIC
Note: By Federal regulation, early withdrawal on Certificate accounts
is subject to substantial interest penalty.

Interest is calculated by multiplying the principal times the rate of interest times the time (**in years**). **That is,**

$$\textbf{Interest} = \textbf{Principal} \times \textbf{rate} \times \textbf{time}$$

or

$$I = Prt$$

This formula determines the *simple interest.* **Simple interest** is the cost of borrowing money computed on the original principal only. This formula can also be used to find the simple interest earned on an investment.

If a person borrows \$500 for a period of one year at 8% simple interest, we can find the amount of interest by means of the formula $I = Prt$. In this case, we have $P = \$500$, $r = 0.08$ (because $8\% = 0.08$), and $t = 1$ year.

$$\begin{aligned} I &= Prt \\ &= \$500 \times 0.08 \times 1 \\ &= \$40.00 \end{aligned}$$

Therefore the borrower would have to pay \$40 interest on the loan. The total amount that must be repaid is \$500 (the principal) plus \$40 (the interest), or \$540. The amount repaid is equal to the principal plus the interest:

$$A = P + I$$

EXAMPLE 1

A woman borrows \$2000 for 3 years at a simple interest rate of 9%. How much interest does she pay? What is the total amount that must be repaid?

Solution

We use the formula $I = Prt$ with $P = \$2000$, $r = 0.09$, and $t = 3$:

$$\begin{aligned} I &= Prt \\ &= \$2000 \times 0.09 \times 3 \\ &= \$540 \end{aligned}$$

The total amount that must be repaid is

$$\begin{aligned} A &= P + I \\ &= \$2000 + \$540 \\ &= \$2540 \end{aligned}$$

EXAMPLE 2

Carl borrowed \$1000 for 6 months at a simple interest rate of 10%. How much interest did he pay?

Solution

This problem is similar to example 1, but note that the loan was for a period of time less than one year. To use the simple-interest formula, we must express 6 months as a fraction of a year:

$$6 \text{ months} = \frac{6}{12}, \quad \text{or} \quad \frac{1}{2} \text{ year}$$

Now we have $P = \$1000$, $r = 0.10$, and $t = \frac{1}{2}$.

$$\begin{aligned} I &= Prt \\ &= \$1000 \times 0.10 \times \frac{1}{2} \\ &= \$50.00 \end{aligned}$$

EXAMPLE 3

A merchant borrowed \$2400, agreeing to repay the principal and 9% simple interest at the end of 9 months. Find the amount of interest and the total sum that must be repaid.

Solution
$P = \$2400$, $r = 0.09$, and $t = \frac{9}{12} = \frac{3}{4}$. Therefore,

$$I = Prt$$
$$= \$2400 \times 0.09 \times \frac{3}{4}$$
$$= \$162.00$$

The total amount to be repaid is

$$A = P + I$$
$$= \$2400 + \$162$$
$$= \$2562$$

EXAMPLE 4
How much will \$1000 earn in 2 years at $8\frac{1}{2}$ percent simple interest?

Solution
We use the formula for simple interest with $P = \$1000$, $r = 0.085$, and $t = 2$:

$$I = Prt$$
$$= \$1000 \times 0.085 \times 2$$
$$= \$170$$

EXERCISES FOR SECTION 11.5

1. Find the simple interest on a \$2000 loan for 3 years at 7%.
2. Find the simple interest on a \$3500 loan for 1 year at 15%.
3. Find the simple interest on a \$2000 loan for 6 months at 12%.
4. Find the simple interest on a \$3000 loan for 9 months at 15%.
5. Find the simple interest on a \$5000 loan for 6 months at $8\frac{1}{2}$ percent.
6. A merchant borrowed \$2000, agreeing to repay the principal and 12% simple interest at the end of 6 months. Find the amount of interest and the total sum that must be paid.
7. Luke borrowed \$3000 at $8\frac{1}{2}$-percent simple interest for a period of 3 years. How much interest will he pay? What is the total amount that he must pay?
8. To help pay her tuition bill, Sally borrowed \$800 at 8% simple interest for 18 months. How much interest will she pay? What is the total amount she will pay?
9. How much simple interest will \$5000 earn in

3 years at a rate of $8\frac{1}{2}$ percent? How much interest will it earn in 3 months?

10. How much simple interest will $6000 earn in 4 years if the rate of interest is 9%? How much will it earn in 4 months?

11. Connie has a balance due of $240 on her charge account. The rate is $1\frac{1}{2}$-percent simple interest *per month* on the unpaid balance. If Connie decides not to pay anything toward her balance this month, and she does not charge anything to her account next month, what will be her new balance due next month?

12. Sam has a balance due of $300 on his charge account. The rate is $1\frac{1}{2}$-percent simple interest per month on the unpaid balance. If Sam pays $180 toward his balance this month, and does not charge anything else next month, what will be the balance due on his charge account next month?

Just For Fun

True or false: A lead pencil contains lead.

11.6 COMPOUND INTEREST

If a person borrows $1000 at 8% simple interest for a period of 1 year, then at the end of that year $1080 must be repaid:

$$I = Prt = \$1000 \times 0.08 \times 1 = \$80 \qquad A = P + I = \$1000 + \$80 = \$1080$$

If the borrower did not pay back any of the loan or the interest by the end of the first year, and if he or she wanted to continue the loan for another year at the same rate, then he or she would owe $1080 plus the interest on $1080, which is $86.40. That is, $1166.40 would have to be repaid:

$$I = Prt = \$1080 \times 0.08 \times 1 = \$86.40 \qquad A = P + I = \$1080 + \$86.40 = \$1166.40$$

This is an example of *compound interest.* For this example, we would say that $1000 was loaned for a period of two years, with interest *compounded annually.* Banks pay compound interest on their savings accounts. Most banks pay interest that is compounded quarterly, and some banks pay interest compounded daily.

In a situation where the interest due at the end of a certain period is added to the principal, and then both the principal and interest earn interest for the next period, the interest paid is called **compound interest.** The interest for each succeeding period is greater than the previous one, because the principal keeps increasing.

EXAMPLE 1

a. Find the amount of simple interest when \$100 is invested for 3 years at 6%.

b. Find the amount of *compound* interest when \$100 is invested for 3 years at 6% compounded annually.

Solution

a. $I = Prt$
$= \$100 \times 0.06 \times 3$
$= \$18$

b. Since this is compound interest, we must find the interest at the end of each year and add it to the principal before computing the interest for the next year.

1st year: $I = Prt$
$= \$100 \times 0.06 \times 1$
$= \$6.00$

Amount at end of 1 year $= \$100 + \$6 = \$106$

2nd year: $I = Prt$
$= \$106 \times 0.06 \times 1$
$= \$6.36$

Amount at end of 2 years $= \$106.00 + \$6.36 = \$112.36$

3rd year: $I = Prt$
$= \$112.36 \times 0.06 \times 1$
$= \$6.74$

Amount at end of 3 years $= \$112.36 + \$6.74 = \$119.10$

The total interest is equal to the total amount due at the end of 3 years, minus the principal:

Interest $=$ amount $-$ principal $= \$119.10 - \$100.00 = \$19.10$

Note that in example 1 the interest earned in 3 years on $100 at 6% *simple* interest was $18, while the interest earned on the same amount over the same time period at 6% *compounded annually* was $19.10. The difference in the two methods of investment is $1.10. This points out the advantage of compound interest over simple interest when money is invested at a given rate over a number of interest periods. Although in this instance the difference is small—only $1.10—remember that the investment was also a small one, $100. Imagine what the difference would be if the investment were $1000, $10,000, or $100,000!

The computation in part *b* of example 1 was somewhat tedious. Fortunately, there is an easier way to compute compound interest. Banks use computers to calculate interest. However, table 1 can also be used to compute compound interest. It shows the amounts that must be paid at the end of various interest periods if $1 is invested at compound interest at one of the given interest rates.

TABLE 1 Compounded Amount of $1

Periods	Interest Rate Per Period				
	2%	3%	4%	6%	8%
1	1.020	1.030	1.040	1.060	1.080
2	1.040	1.061	1.082	1.124	1.166
3	1.061	1.093	1.125	1.191	1.260
4	1.082	1.126	1.170	1.262	1.360
6	1.126	1.194	1.265	1.419	1.587
8	1.172	1.267	1.369	1.594	1.851
10	1.219	1.344	1.480	1.791	2.159
12	1.268	1.426	1.601	2.012	2.518
20	1.486	1.806	2.191	3.207	4.661
24	1.608	2.033	2.563	4.049	6.341

Table 1 is an example of a compound-interest table. It will enable us to do the problems in this section. There are tables that are much more detailed, in which, the interest is carried out to four or five decimal places, and values are given for many more interest rates and numbers of periods.

Table 1 shows the amount that $1 will accumulate when interest is paid at the indicated rate and compounded for the indicated number of interest periods. For example, the entry in the fourth row and 8% column of table 1 is 1.360. This means that, if $1 is placed in a bank that pays 8%, compounded for each of 4 interest periods, then the total accumulation in the account will be $1.36. That is, $1 will grow to $1.36 in 4 interest periods at 8% a period. The amount accumulated in this way is called the **compound amount.** To find the compound amount on larger amounts of money, we multiply the principal by the compound amount for $1. For example, to find the compound amount if $1000 is deposited at 6% compounded annually for 10 years, we first find the compound amount for $1 using these figures for time and interest rate. Using the table, we obtain $1.791. The compound amount for $1000 is therefore

$$\$1000 \times 1.791 = \$1791$$

EXAMPLE 2

Find the compound amount on deposit when $500 is deposited for 10 years at 8% compounded annually.

Solution

Using table 1, the compound amount for $1 at 8% for 10 interest periods is $2.159. Therefore to find the compound amount for $500 at 8% for 10 periods, we multiply $500 times 2.159:

$$\text{compound amount} = \$500 \times 2.159 = \$1079.50$$

We can find the total compound interest by subtracting the principal from the compound amount. That is,

$$\text{compound interest} = \text{compound amount} - \text{principal}$$

In example 2, the compound amount is $1079.50 and the principal is $500. Therefore

$$\text{compound interest} = \$1079.50 - \$500 = \$579.50$$

We noted earlier that banks usually compound interest more often than once a year. There are some banks that compound interest semiannually, others that compound interest quarterly, and others that compound interest daily. If a bank offers 6% interest compounded semiannually, then there are two interest periods every year, and the bank pays 3% interest every 6 months.

To find the compound amount when $1000 is deposited for 4 years at 6% compounded semiannually, we must first determine the number of interest periods and the interest rate per period. Since the principal is deposited for 4 years compounded semiannually, there are 8 interest periods. The rate of 6% per year is the same as

3% per half-year. Using table 1, we find 8 under the period column and go across to the 3% column. The entry is 1.267. Now we multiply 1.267 by $1000:

$$\text{compound amount} = \$1000 \times 1.267 = \$1267$$

EXAMPLE 3

Find the compound amount on deposit when $500 is deposited for 5 years at 8% compounded quarterly.

Solution

Interest compounded quarterly means that there are 4 interest periods per year, so for 5 years there are 20 interest periods. Eight percent interest per year is the same as 2% every quarter-year. The compound amount for $1 at 2% for 20 interest periods is $1.486. Therefore,

$$\text{compound amount} = \$500 \times 1.486 = \$743$$

EXAMPLE 4

Find the compound amount and the compound interest on $2000 invested at 12% compounded quarterly for 6 years.

Solution

$$\text{number of interest periods per year} = 4$$

$$\text{total number of interest periods} = 6 \times 4 = 24$$

$$\text{interest rate each interest period} = \frac{1}{4} \text{ of } 12\% = 3\%$$

The compound amount for $1 at 3% for 24 interest periods is $2.033. Therefore,

$$\text{compound amount} = \$2000 \times 2.033 = \$4066$$
$$\text{compound interest} = \$4066 - \$2000 = \$2066$$

EXAMPLE 5

If $1000 were invested at 12% compounded semiannually, how long would it take for the investment to double itself?

Solution

The rate per interest period is $\frac{1}{2}$ of 12%, or 6%. Examining the column headed 6% in table 1, we see that $1 will double itself in 12 periods; that is, the compound amount for $1 at 6 percent for 12 periods is $2.012. The interest is compounded semiannually, so 12 periods is $12 \div 2 = 6$ years. Therefore, $1000 will double itself in

6 years when invested at 12% compounded semiannually. The compounded amount would be $1000 × 2.012 = $2012. (*Note:* This compound amount is a little more than twice $1000, but the answer to the original question is correct to the nearest year.)

EXERCISES FOR SECTION 11.6

For exercises 1–6, use table 1 to find the compound amount for each investment if interest is compounded annually.

1. $500 at 8% for 10 years
2. $800 at 6% for 6 years
3. $1000 at 4% for 3 years
4. $2500 at 3% for 2 years
5. $5000 at 6% for 8 years
6. $10,000 at 8% for 10 years

For exercises 7–12, use table 1 to find the compound amount for each investment if interest is compounded semiannually.

7. $500 at 8% for 10 years
8. $800 at 6% for 6 years
9. $1000 at 4% for 3 years
10. $2500 at 12% for 3 years
11. $5000 at 12% for 5 years
12. $10,000 at 16% for 2 years

For exercises 13–18, use table 1 to find the compound amount for each investment if interest is compounded quarterly.

13. $500 at 8% for 2 years
14. $750 at 12% for 3 years
15. $1000 at 16% for 5 years
16. $2500 at 8% for 6 years
17. $5000 at 12% for 6 years
18. $10,000 at 16% for 5 years
19. Find the compound amount and the compound interest on $3000 invested for 5 years at 12% compounded quarterly.
20. Find the compound amount and the compound interest on $2500 invested for 10 years at 16% compounded semiannually.
21. If $100 were invested at 6% compounded semiannually, how long would it take for the investment to double itself?
22. If $100 were invested at 24% compounded quarterly, how long would it take for the investment to quadruple itself?

Just For Fun

There are eleven denominations of Federal Reserve notes now in circulation. Two of these are the one-dollar bill and the two-dollar bill. Can you name the others?

11.7 EFFECTIVE RATE OF INTEREST

In our previous discussions pertaining to interest, we have seen that interest rates are usually stated on a yearly basis even though the period of payment is not a year. For example, an interest rate of 8% compounded quarterly means that we use an interest rate of 2% for each interest period. Similarly, an interest rate of 6% compounded semiannually means that we use an interest rate of 3% for each period.

To find the compound amount on deposit when $1 is deposited for 1 year at 8% compounded quarterly, we can use table 1 from section 11.6 to find that the amount is $1.082. That is, $1 will compound to $1.082 in 1 year with interest compounded quarterly. The compound amount on deposit when $1 is deposited for 1 year at 6% compounded semiannually is $1.061. That is, $1 will yield $1.061 in one year with interest compounded semiannually.

We have chosen these two examples to help illustrate the *effective annual interest rate,* also referred to as *effective rate.* An **effective annual interest rate** is the annual interest rate which gives the same yield as a nominal interest rate compounded several times a year. For example, an interest rate of 8% compounded quarterly is equivalent to an effective rate of 8.2% compounded annually. This is because, as noted in the preceding paragraph, $1 compounds to $1.082 in one year at 8% compounded quarterly. Therefore, the total interest earned in one year is

$$\$1.082 - \$1.00 = \$0.082$$

This is an effective rate of 8.2% since

$$8.2\% \text{ of } \$1.00 = \$0.082$$

Thus the interest earned at 8% compounded quarterly is the same as the interest earned at an annual interest rate of 8.2%.

Similarly, a 6% interest rate compounded semiannually is equivalent to an effective rate of 6.1% compounded annually. (Note that $1 compounds to $1.061 in one year at 6% compounded semiannually.)

The effective rate is usually greater than the stated annual rate (often called the **nominal rate**). If interest is paid more than once a year, the effective annual rate is greater than the stated annual rate. If interest is paid annually, then the effective rate is the same as the stated annual rate. In the preceding examples, the effective rate was greater than the stated annual rate.

Effective rate is often used to compare interest rates which are compounded at different intervals. For example, if bank A offers 5% interest on its deposits compounded quarterly and bank B offers $5\frac{1}{2}$% interest on its deposits compounded semiannually, we can find the effective annual rate for each bank to determine which offers the best investment. We can also compare effective rates when borrowing money. In this case, we would select the lowest effective rate.

We can determine the effective annual rate by means of the formula

$$E = (1 + r)^n - 1$$

where

$$\begin{aligned} E &= \text{effective rate} \\ n &= \text{number of payment periods per year} \\ r &= \text{interest rate per period} \end{aligned}$$

We shall now use this formula to find the effective annual rate for 6% compounded semiannually:

$$E = (1 + r)^n - 1$$

The interest rate is compounded semiannually, so there are $n = 2$ payment periods per year. Therefore, $r = 6\% \div 2 = 3\% = 0.03$. (Note that r is expressed without a percentage sign.) Substituting these values in the formula, we have

$$\begin{aligned} E &= (1 + 0.03)^2 - 1 \\ E &= (1.03)^2 - 1 \\ E &= 1.0609 - 1 \\ E &= 0.0609, \text{ or } 6.1\% \end{aligned}$$

EXAMPLE 1

What is the effective rate, if money is invested at 8% compounded quarterly?

Solution

The number of payment periods is 4, because interest is compounded quarterly. The interest rate per period is $8\% \div 4 = 2\%$. Therefore $r = 0.02$. Substituting these values in the formula, we have

$$\begin{aligned} E &= (1 + r)^n - 1 \\ E &= (1 + 0.02)^4 - 1 \\ E &= (1.02)^4 - 1 \\ E &= 1.0824 - 1 \\ E &= 0.0824, \text{ or } 8.2\% \qquad \text{(to the nearest tenth)} \end{aligned}$$

This answer checks with the one we obtained earlier using the compound-interest table.

In example 1 we rounded our answer to the nearest tenth so that it could be compared with the first illustrative example in this section. However, all other problems in this section will be rounded to the nearest hundredth of a percentage. Also, it is recommended that all calculation involving the use of this formula be done on a pocket calculator (to save time and relieve tension).

EXAMPLE 2

What is the effective rate, if money is invested at 7% compounded semiannually?

Solution

$n = 2, \quad r = 7\% \div 2 = 3.5\% = 0.035$

$$\begin{aligned} E &= (1 + r)^n - 1 \\ E &= (1 + 0.035)^2 - 1 \\ E &= (1.035)^2 - 1 \\ E &= 1.071225 - 1 = 0.071225 \\ E &= 7.12\% \end{aligned}$$

EXAMPLE 3

Bank A offers its depositors an interest rate of 5% compounded quarterly and bank B offers its depositors a rate of $5\frac{1}{2}\%$ compounded semiannually. Which bank makes the better offer?

Solution

We must compare the effective annual rate for each bank.

bank A: $n = 4, r = 5\% \div 4 = 1.25\% = 0.0125$

$$\begin{aligned} E &= (1 + r)^n - 1 \\ E &= (1 + 0.0125)^4 - 1 \\ E &= (1.0125)^4 - 1 \\ E &= 1.0509 - 1 = 0.0509 \\ E &= 5.09\% \end{aligned}$$

bank B: $n = 2, \quad r = 5\frac{1}{2}\% \div 2 = 2.75\% = 0.0275$

$$\begin{aligned} E &= (1 + r)^n - 1 \\ E &= (1 + 0.0275)^2 - 1 \\ E &= (1.0275)^2 - 1 \\ E &= 1.0558 - 1 = 0.0558 \\ E &= 5.58\% \end{aligned}$$

The effective rate of bank B is greater than that of bank A by 5.58% − 5.09% = 0.49%. Therefore bank B offers a better interest rate to its depositors.

EXERCISES FOR SECTION 11.7

Answers involving fractional parts of a percentage should be rounded to the nearest hundredth of a percentage.

1. What is *effective annual interest rate?*
2. What is the effective rate if money is invested at 8% compounded semiannually?
3. What is the effective rate if money is invested at 9% compounded semiannually?
4. What is the effective rate if money is invested at 10% compounded quarterly?
5. What is the effective rate if money is invested at 12% compounded quarterly?
6. What is the effective rate if money is invested at 5% compounded (a) annually? (b) semiannually? (c) quarterly?
7. What is the effective rate if money is invested at 6% compounded (a) annually? (b) semiannually? (c) quarterly?
8. Which is the higher interest rate: 5% compounded quarterly or $5\frac{1}{2}$% compounded semiannually?
9. Bob invested his money at $5\frac{1}{2}$% compounded quarterly while Larry invested his money at $5\frac{3}{4}$% compounded semiannually. Who receives the better interest rate?
10. Ruth invested her money at $6\frac{3}{4}$% compounded quarterly while Julia invested her money at 6.9% compounded semiannually. Who receives the better interest rate?

Just For Fun

Mario and Roy arranged an unusual race. They agreed that the man whose car crossed the finish line first would be the loser. The man whose car crossed the finish line second would be the winner. How should Mario and Roy drive to have such a race?

11.8 LIFE INSURANCE

Almost all consumers purchase insurance at some point in their lifetimes. The types of insurance that a person may obtain include car insurance, health insurance, fire insurance, and life insurance. Many people obtain all of these types of insurance to protect themselves, their homes, and their families.

The basic concept of insurance is the sharing of risks or losses. That is, a person who buys an insurance policy is agreeing to share the risks involved with other people buying a similar policy. A policy is a contract between the insured person, or **policyholder,** and the insurer, or **underwriter.** The fee that an insured person pays to the insurer is called a **premium.** Premiums can be paid monthly, quarterly, semiannually, or annually.

In this section we shall consider only one type of insurance, life insurance. There are three basic types of life insurance policies: *whole life, term,* and *endowment.* A **whole life insurance policy** is one in which the insurer will pay the face value of the policy to a designated person called the **beneficiary** when the insured person dies, regardless of when this occurs. Premiums for a whole life policy can be paid in two different ways. A **straight life** policy is one in which the policyholder pays premiums until death. A limited-payment life policy is one in which the policyholder pays premiums for a certain number of years. Whether the policy is a straight life policy or a limited-payment policy, it is in effect for the insured person's lifetime. Premiums for a limited-payment life policy are higher than those for a straight-life policy.

A **term insurance policy** is one in which the insurer will pay the face value of the policy to a beneficiary if the insured person dies during the **term** or period of time stated in the policy. A common term is 5 years. For example, if a 20-year-old college student purchased a 5-year term life insurance policy, then the insurance company is liable for payment of the face value of the policy if the insured dies within the 5-year period. Once the term of the policy expires, the insurance company is no longer liable. Usually term insurance can be renewed for another term, but at a higher rate.

An **endowment insurance policy** is similar to a term policy in that it insures the life of the insured for a specified term or number of years. But if the insured is living at the end of the term of the policy, then the face value of the policy will be paid to the policyholder. Hence the use of the word *endowment.* A common term for an endowment policy is 20 years.

It should be noted that the premiums are different on these different types of life insurance policies. Different rates are also applied in each case. Typical annual premium rates per $1000 of life insurance are given in table 2. These premiums are usually lower for women because women live longer than men. (For the sake of convenience in making computations, we shall assume that the rates are the same for males and females.)

TABLE 2 Annual Premium Rate per $1000 of Life Insurance

Age	5-year Term	Straight life	Limited-Payment (20-year)	Endowment (20-year)
20	$6.35	$14.07	$23.60	$44.84
25	$6.88	$15.91	$26.21	$45.21
30	$7.82	$18.32	$29.18	$45.61
35	$9.07	$21.67	$32.65	$46.48
40	$11.31	$25.79	$37.04	$47.90
45	$14.17	$30.82	$42.37	$50.30
50	$18.08	$37.48	$48.19	$53.64
55	$24.63	$46.25	$54.30	$58.68
60	—	$58.14	$63.64	$67.55

Insurance rates are determined by using data about large samples of the population. The most important pieces of information used are the average death rates for different age groups. It stands to reason that a 20-year-old male should not be charged the same premium as a 65-year-old male, as the 65-year-old will pay fewer premiums. The average death rate of 20-year-old males is about 2 per 1000, and they can expect to live for 50 years more. The average death rate of 65-year-old males is about 32 per 1000, and they can expect to live for 13 years more. This kind of information can be found in tables called **mortality tables.** People who help construct these tables determining what premiums an insurance company should charge are called **actuaries.**

EXAMPLE 1

Tom Jones is 20 years old and wishes to purchase $10,000 worth of life insurance. Determine the annual cost of a straight life policy.

Solution

Using table 2, we see that the premium for a 20-year-old person is $14.07 per $1000 of straight life insurance. The policy is worth $10,000; hence we have

$$10 \times \$14.07 = \$140.70 \quad \text{annual premium}$$

EXAMPLE 2
Sue Smith is 25 years old and wishes to purchase $15,000 worth of 5-year term life insurance. Determine the annual cost of this policy.

Solution
Using table 2, we see that the premium for a 25-year-old person for a 5-year term policy is $6.88 per $1000 of life insurance. The policy is worth $15,000; hence, we have

$$15 \times \$6.88 = \$103.20 \qquad \text{annual premium}$$

EXAMPLE 3
John Jackson is 40 years old and wishes to purchase $20,000 of life insurance. Determine the annual cost of a straight life policy and of a limited-payment life policy for 20 years.

Solution
From the table, the straight life premium per $1000 for a 40-year-old is $25.79. Therefore, the annual cost on a $20,000 policy is

$$20 \times \$25.79 = \$515.80$$

The limited-payment life (also called *20-payment life*) premium per $1000 for a 40-year-old is $37.04. Therefore, the annual cost on a $20,000 policy is

$$20 \times \$37.04 = \$740.80.$$

EXAMPLE 4
If John Jackson (see example 3) lives for 25 years after he purchases his insurance policy, which policy (straight or limited payment) would cost more? How much more?

Solution
If John Jackson lives for 25 years after purchasing a straight life policy, he would pay 25 annual premiums. Hence, the cost is 25 times $515.80, the annual straight life premium.

$$25 \times \$515.80 = \$12,895.00$$

On a 20-year limited-payment life policy, John Jackson would pay 20 annual premiums. Therefore, the cost is 20 times $740.80, the annual limited-payment life premium.

$$20 \times \$740.80 = \$14,816.00$$

The limited-payment life policy would cost more. It would cost

$14,816 − $12,895 = $1921 more

EXERCISES FOR SECTION 11.8

Use table 2 for the exercises.

1. Find the annual premium for each of the given insurance policies.

	Face Value of Policy	Age at Issue	Type of Policy
a.	$12,000	20	5-year term
b.	$15,000	25	straight life
c.	$25,000	30	limited-payment life (20-year)
d.	$20,000	25	endowment (20-year)
e.	$15,000	45	straight life

2. Find the annual premium for each of the given insurance policies.

	Face Value of Policy	Age at Issue	Type of Policy
a.	$20,000	20	limited-payment life (20-year)
b.	$15,000	30	5-year term
c.	$25,000	25	straight life
d.	$10,000	35	limited-payment life (20-year)
e.	$50,000	40	endowment (20-year)

3. Stan Simon is 20 years old and wishes to purchase $10,000 worth of life insurance. Determine the annual cost of a 5-year term policy. If Stan renews his policy when it expires, what is the difference in total cost between the first and second policy?

4. Jan Peters is 25 years old and wishes to purchase $12,000 worth of life insurance. Determine the annual cost of a 5-year term policy. If Jan renews her policy when it expires, what is the difference in total cost between the first and second policy?

5. Jane Lark is 30 years old and wishes to purchase $20,000 worth of life insurance. Determine the annual cost of a straight life policy and also of a limited-payment life policy for 20 years. If Jane lives for 30 years after she purchases her policy, which policy would cost more? How much more?

6. Carl Ronold is 35 years old and wishes to purchase $25,000 worth of life insurance. Determine the annual cost of a straight life policy and also of a limited-payment life policy for 20 years. If Carl lives for 20 years after he purchases his policy, which policy would cost more? How much more?

7. Rich Nelson obtained a $20,000 limited-payment life policy for 20 years when he was 40 years old. Determine the annual cost of this policy. What is the total cost if Rich Nelson lives to be 65 years old? If he had obtained the same policy when he was 25 years old, what would have been the total he paid for the insurance?

8. Lois Carson obtained a $20,000 straight life insurance policy at age 40. Determine the annual cost for this policy. If she died 18 years later, how much more did her beneficiary collect than Lois Carson paid in premiums?

9. If you pay your premiums in quarterly payments, each payment is 26% of the annual premium. Find the quarterly payment for each of the following:

	Face Value of Policy	Age at Issue	Type of Policy
a.	$15,000	25	endowment (20-year)
b.	$25,000	30	straight life
c.	$10,000	20	5-year term
d.	$30,000	45	limited-payment life (20-year)
e.	$50,000	25	straight life

Just For Fun

An Amtrak train leaves Miami for Boston traveling at the rate of 80 miles per hour. Two hours later, another train leaves Boston for Miami traveling at the rate of 60 miles per hour. When the two trains meet, which train is closer to Miami?

11.9 INSTALLMENT PLANS AND MORTGAGES

Many consumers purchase items on an installment plan. That is, they obtain possession of an item immediately and agree to pay the purchase price plus an additional charge in a series of regular payments, usually monthly.

If a stereo set that sells for $1000 can be purchased on an installment plan for $200 down and 12 monthly payments of $70 each, we find the total cost of the set when it is purchased on credit by multiplying the monthly payment times the number of payments to obtain the total amount paid in monthly payments, and then we add the down payment.

$70 × 12 = $840	total amount of monthly payments
$840 + $200 = $1040	total cost

To find the amount of interest paid (also called *service charge* or *finance charge*), we subtract the cash price from the total installment cost:

$$\$1040 - \$1000 = \$40 \qquad \text{interest}$$

As we shall see, this amount does not reflect the true interest rate, but because the actual dollar amount involved is relatively small, consumers are more than willing to pay it.

Consider the case of Sam Larson, who borrows $1200 from the Friendly Finance Company. The finance company advertised an interest rate of 10% simple interest. Sam wanted to borrow the money for 6 months and repay the loan in 6 monthly payments. Hence we have

$$I = Prt$$

$$I = \$1200 \times 0.10 \times \frac{6}{12}$$

$$= \$60$$

The total amount that Sam has to repay is $1200 + $60, or $1260. If he is going to repay the loan in six monthly payments, he will pay $210 each month.

It should be noted that the Friendly Finance Company computed the interest as though Sam Larson owed the $1200 for the entire six months. But, he did not! Sam owed $1200 for only one month. At the end of the first month, he paid the finance company $210. Of this amount, $200 was applied toward the principal (the other $10 is interest). Since each payment includes $200 of the principal, the amount that Sam actually owes decreases by $200 each month. The following list shows what Sam owes each month:

$1200	—original amount	(owed for first month only)
$1000	= $1200 − $200	(amount owed for second month only)
$800	= $1000 − $200	(amount owed for third month only)
$600	= $800 − $200	(amount owed for fourth month only)
$400	= $600 − $200	(amount owed for fifth month only)
$200	= $400 − $200	(amount owed for sixth month only)
$4200	Total	

Thus, Sam's debt to the finance company is equivalent to owing $4200 for one month only. The interest that Sam paid was $60. Using the formula $I = Prt$, we can find the value of r:

$$\$60 = \$4200 \times r \times \frac{1}{12}$$

$$60 = 350 \times r$$

Solving this equation for r (correct to the nearest thousandth),

$$\begin{aligned} r &= 0.171 \\ &= 17.1\% \quad \text{(to the nearest tenth of a percentage)} \end{aligned}$$

Therefore, Sam paid about 17.1% interest on his loan, not 10% as advertised.

In 1969, Congress passed a Truth in Lending Act which requires all sellers to reveal the true annual interest rate that they charge. However, the Act does not establish any maximums on interest rates or finance charges.

There is a formula that can be used to determine the **true annual interest rate** on installment loans. It is:

$$i = \frac{2nr}{n + 1} \quad \text{where} \quad \begin{aligned} i &= \text{the true interest} \\ n &= \text{number of payments} \\ r &= \text{the nominal rate of interest} \end{aligned}$$

Applying this formula to the previous example, where $n = 6$ and $r = 10\% = 0.10$, we have

$$\begin{aligned} i &= \frac{2nr}{n + 1} \\ &= \frac{2 \times 6 \times (0.10)}{6 + 1} = \frac{12(0.10)}{7} \\ &= \frac{1.2}{7} \\ &= 0.171 \quad \text{(to the nearest thousandth)} \\ &= 17.1\% \quad \text{(to the nearest tenth of a percentage)} \end{aligned}$$

EXAMPLE 1

Stella Frisco purchased a stereo set advertised for \$1000. She bought the set on the installment plan, paying \$200 down and agreeing to pay the balance in 12 monthly payments. The finance charge was 5% simple interest on the balance. What was the true annual interest rate?

Solution

Using the formula $i = \dfrac{2nr}{n + 1}$, we have $n = 12$ and $r = 5\% = 0.05$.

Therefore

$$
\begin{aligned}
i &= \frac{2nr}{n+1} \\
&= \frac{2 \times 12 \times (0.05)}{12 + 1} = \frac{24(0.05)}{13} \\
&= \frac{1.2}{13} = 0.092 \qquad \text{(to the nearest thousandth)} \\
&= 9.2\% \qquad \text{(to the nearest tenth of a percentage)}
\end{aligned}
$$

Stephen McBrady

With most charge accounts and credit cards, there is an interest rate of $1\frac{1}{2}\%$ per month on the unpaid balance. For example, if a person purchases an item that costs \$90 and charges it, then at the end of the month he has the choice of paying the full \$90 or a portion of the \$90. If he chooses to pay a portion of the \$90 (say \$30), then the next monthly statement would be for \$60 plus $1\frac{1}{2}\%$ of \$60, the unpaid balance, or \$0.90. Thus the next monthly statement would be for \$60.90 (assuming no other charges). If another \$30 payment is made, then the following monthly statement would be for \$60.90 − \$30.00 = \$30.90, plus $1\frac{1}{2}\%$ of \$30.90, or \$0.46. Thus, the statement would be for \$31.36, and a payment for this amount would clear the account.

EXAMPLE 2

Fred Worth purchased a chair for \$300 and charged it. The store uses an interest rate of $1\frac{1}{2}\%$ per month on the unpaid balance. If Fred decides to make payments of \$100 per month, how much interest does he pay?

Solution

The first bill will be for \$300 (assuming no other charges).

1. Fred makes a payment of \$100. The balance is \$200. The interest is $1\frac{1}{2}\%$ of \$200, or \$3. Hence the next bill is for \$203.
2. Fred makes another payment of \$100. The balance is \$103. The interest is $1\frac{1}{2}\%$ of \$103, or \$1.55. Hence the next bill is for \$104.55.
3. Fred makes another payment of \$100. The balance is \$4.55. The interest is $1\frac{1}{2}\%$ of \$4.55, or \$0.07. Hence the next bill is for \$4.62.

4. Fred makes a payment of $4.62.

Fred has paid a total of $304.62, so he has paid a total of $304.62 − $300 = $4.62 in interest.

A home mortgage is probably the largest loan that most people assume. Consumers must borrow money to make such a large purchase: the average price for a new home today is $45,000, and it is predicted that the price of a new home will increase to $90,000 by 1986.

Typically, most people purchase a home by making a specified down payment and borrowing the balance from a bank or other lending institution. The borrower agrees to make regular payments on the principal and interest until the loan is paid off. This process is called **amortizing** the loan. Mortgages on new homes are usually for a period of 20, 25, or 30 years. Mortgages on older homes sometimes run for shorter periods of time.

The monthly payments necessary to amortize a loan (pay off the principal and interest) are compiled in tables which are called **amortization tables.** Listed in table 3 are typical monthly payments per $1000 of mortgage. The amount that must be paid monthly for principal and interest is given for different interest rates and different periods of time.

TABLE 3. Monthly Mortgage Payment per $1000

	Length of Mortgage		
Rate	10 years	20 years	30 years
7%	$11.61	$7.75	$6.65
$7\frac{1}{2}$%	$11.87	$8.06	$6.99
8%	$12.13	$8.36	$7.34
$8\frac{1}{2}$%	$12.40	$8.68	$7.69
9%	$12.67	$9.00	$8.05

If a person assumes a $15,000 mortgage for 20 years at $8\frac{1}{2}$%, then according to table 3 the monthly payment for each $1000 is $8.68. The mortgage is for $15,000, so we multiply $8.68 by 15 to find the total monthly payment. Therefore the monthly payment on this mortgage is 15 × $8.68, or $130.20. It should be noted that the payments could be higher if the mortgagee also has to make monthly payments on property taxes or insurance. For example, if property taxes were $1200 a year, then an additional $100 ($1200 ÷ 12) would have to be paid each month. Also, fire insurance might cost about $8 per month. Hence, the total monthly payment would be $130.20 + $100 + $8 = $238.20.

EXAMPLE 3

The Carsons assumed a $30,000 mortgage for 30 years at 9%. What is their monthly payment?

Solution

The monthly payment per $1000 at 9% for a 30-year mortgage is $8.05. The mortgage is for $30,000. Hence, we multiply $8.05 by 30:

30 × $8.05 = $241.50 (monthly payment)

EXAMPLE 4

How much total interest will the Carsons (example 3) pay on their mortgage?

Solution

Their monthly payment is $241.50 and the mortgage is for 30 years. Hence there will be 30 × 12, or 360 payments.

360 × $241.50 = $86,940 (total payment)
$86,940 − $30,000 = $56,940 (total interest)

EXAMPLE 5

How much interest would the Carsons save if they assume the same $30,000 mortgage at 9% for 20 years instead of 30 years?

Solution

The monthly payment per $1000 at 9% for a 20-year mortgage is $9.00. The mortgage is for $30,000. Therefore

monthly payment = 30 × $9.00 = $270

The mortgage is for 20 years. Hence, there will be 12 × 20, or 240 payments.

240 × $270.00 = $64,800 (total payment)
$64,800 − $30,000 = $34,800 (total interest)

Interest on 30-year mortgage = $56,940
Interest on 20-year mortgage = $34,800
$22,140 (interest saved)

Note that in both cases in example 5, the interest is more than the face value of the loan. This is the reason that banks lend money—to make money! Homebuyers should be aware of the costs involved in assuming a mortgage. One disadvantage of a mortgage is the total amount of interest to be paid on such a loan. Advantages to consider include: the satisfaction of owning a home, the fact that the loan will be repaid with cheaper dollars because of inflation, and the protection against inflation that a home provides. The value of a home is sure to increase, but the monthly payment for principal and interest will remain constant throughout the term of the mortgage.

EXERCISES FOR SECTION 11.9

1. Carl Thomas purchased a television set advertised for $500. He bought the television on the installment plan, paying $100 down and agreeing to pay the balance in 12 monthly payments. The store charged a finance charge of 8% simple interest on the balance. What was the true annual interest rate (correct to the nearest tenth of a percentage)?

2. A used car is advertised for $2000. It may be purchased on the installment plan by paying $200 down and agreeing to pay the balance plus 12% simple interest on the balance in 24 monthly payments. What is the true annual interest rate (correct to the nearest tenth of a percentage)?

3. An easy chair is advertised for $450. It may be purchased on the installment plan for $100 down and $60.67 per month for 6 months. The store advertised a finance charge of 8%. What is the true annual interest rate (correct to the nearest tenth of a percentage)?

4. The Jaxsons purchased a new couch costing $500. They put no money down and agreed to pay $25 per month for 24 months. The store stated a finance charge of 10%. What was the true annual interest rate (correct to the nearest tenth of a percent)? How much more will the Jaxsons pay for the couch by buying it on the installment plan?

5. Sandra West purchased a coat for $100 and charged it. The store charges an interest rate of $1\frac{1}{2}$% per month on the unpaid balance. If Sandra decides to make payments of $30 per month, how much interest will she pay by the time the coat is paid for? (Assume that there are no additional purchases and that the first payment is made before the end of the month, so no interest is charged on the first payment.)

6. Bill North received his statement from Stone's Department Store. The amount due is $220. If Bill decides to make payments of $50 per month and the interest rate is $1\frac{1}{2}$% per month on the unpaid balance, then how many months will it be before the account is paid in full? What is the total amount to be paid in interest? (Assume that there are no additional purchases and that the first payment is made before the end of the month, so no interest is charged on the first payment.)

7. Julia Swanson used her credit card to charge purchases totaling $180 at the Super Discount House. If Julia decides to make payments of $40 per month, and the interest rate is $1\frac{1}{2}$% per month on the unpaid balance, then how much interest will she pay? (Assume that there are no additional purchases and that the first payment is made before the end of the month, so no interest is charged on the first payment.)

8. The Smiths assumed a $20,000 mortgage for 20 years at $8\frac{1}{2}\%$. What is their monthly payment? How much total interest will the Smiths pay on their mortgage?

9. The Garcias assumed a $30,000 mortgage for 30 years at 8%. What is their monthly payment? How much total interest will the Garcias pay on their mortgage?

10. The Donovans need to borrow $20,000 to buy a house. Bank A will give them a 30-year mortgage at $7\frac{1}{2}\%$. Bank B will give them a 20-year mortgage at 9%. Which mortgage should the Donovans assume so that they will pay the smallest amount of total interest?

For exercises 11–14, find the monthly payment for principal and interest for each mortgage.

	Amount of Mortgage	Interest Rate	Term of Mortgage
11.	$12,000	9%	10 years
12.	$25,000	$8\frac{1}{2}\%$	20 years
13.	$40,000	8%	30 years
14.	$22,000	$7\frac{1}{2}\%$	30 years

Just For Fun

Whose picture is on the front of a $10 bill? What is pictured on the back of it?

11.10 SUMMARY

The *ratio* of two quantities a and b is the quotient or indicated quotient obtained by dividing a by b. The ratio of a to b is written as $a \div b$, $\frac{a}{b}$, or $a:b$. A *proportion* is defined to be two equal ratios. In any proportion, the product of the means equals the product of the extremes. That is,

$$\text{if } \frac{a}{b} = \frac{c}{d}, \text{ then } ad = bc$$

A *percentage* is a ratio with a denominator of 100. We can also say that a percentage is the ratio of any number to 100. The symbol for percentage is %. For example, 6% means 6:100, or $\frac{6}{100}$. We can also express a percent as a decimal; that is,

$$6\% = \frac{6}{100} = 0.06$$

To convert a percentage to a decimal, drop the percentage sign and move the decimal point two places to the left. To change a fraction to a percentage, find an equivalent fraction with a denominator of 100; then drop the denominator and add a percentage sign. To change a decimal to a percentage, move the decimal point two places to the right and add a percentage sign.

Markup is the difference between the selling price and cost of an item. *Markdown* is the difference between the original selling price and the sale price. Both of these terms can be expressed in terms of a dollar amount or in terms of a percentage.

Simple interest is found by means of the formula

$$I = Prt$$

where I = interest, P = principal, r = rate of interest, and t = time in years. *Compound interest* occurs when interest due at the end of a certain period of time is added to the principal, and both the principal and the interest from the first period earn interest for the next period. Table 1 in section 11.6 shows the amount that $1 will accumulate when interest is paid at the indicated rate for the indicated number of interest periods.

An *effective annual interest rate* is an interest rate which is compounded annually. For example, an 8% interest rate compounded quarterly is equivalent to an effective rate of 8.2% compounded annually. We can determine the effective annual rate by means of the formula

$$E = (1 + r)^n - 1$$

where E = effective rate, n = number of payment periods per year, and r = interest rate per period.

There are three basic kinds of life insurance policies: *whole life, term,* and *endowment.* A *whole life insurance policy* is one in which the insurance company will pay the face value of the policy to the beneficiary when the insured person dies, regardless of when this happens. Premiums for a whole life policy can be paid in two different ways. A *straight life* policy is one in which the policyholder pays premiums until death. A *limited-payment life* policy is one in which the policyholder pays premiums for a certain number of years. A *term insurance policy* is one where the insurance company will pay the face value of the policy to a beneficiary when the insured person dies, providing this occurs during the period of time (the *term*) stated in the policy. An *endowment policy* is similar to a term policy in that it insures the life of the insured for a specific term. However, if the insured is living at the end of the term, then the face value of the policy will be paid to the policyholder. Typical annual premium rates per \$1000 of life insurance for various policies and different age groups are listed in table 2, section 11.8.

Installment buying allows a consumer to obtain possession of an article immediately by agreeing to pay the purchase price plus an additional charge in a series of regular payments. When a person buys an article on the installment plan, it is important that he or she know the true annual interest rate that is being charged. A formula that can be used to determine this true annual interest rate is

$$i = \frac{2nr}{n + 1}$$

where i = the true interest rate, n = number of payments, and r = the stated interest rate.

A *home mortgage* is probably the largest loan that most people assume. The monthly payments necessary to amortize a loan are compiled in tables called *amortization tables.* Table 3, section 11.9, lists typical monthly payments per \$1000 of mortgage for various interest rates and periods of time.

Review Exercises for Chapter 11

1. Express each ratio in simplest form.
   a. 36:54 b. $\frac{19}{38}$
   c. 5 to 30 d. $1\frac{1}{2}:2\frac{1}{4}$
2. Find the value of x in each proportion.
   a. $4:x = 5:10$ b. $x:6 = 4:8$
   c. $5:2 = x:6$ d. $2:7 = 6:x$
3. Express each percentage as a fraction in simplest form.
   a. 54% b. $4\frac{1}{2}\%$
   c. 2.3% d. 125%
4. Express each percentage as a decimal.
   a. 19% b. $4\frac{1}{2}\%$
   c. 6.5% d. 0.25%
5. Express each fraction as a percentage.
   a. $\frac{2}{5}$ b. $\frac{3}{8}$
   c. $2\frac{3}{5}$ d. $\frac{1}{16}$
6. Express each decimal as a percentage.
   a. 0.09 b. 2.14 c. 0.003 d. 2.125
7. Twenty-three percent of a certain group of mathematics students had previously studied algebra. If 46 students in this group had studied algebra, how many students are in the group?
8. A suit sells for \$125. The markup is 30% of the cost. What is the cost of the suit, correct to the nearest cent?
9. A portable television costs a retailer \$60. If the markup is 25% of the selling price, find the selling price, correct to the nearest cent.
10. Find the markup and the cost, correct to the nearest cent, of a new couch which retails for \$400, if the percentage markup on the selling price is 40%.
11. A clothing store sells topcoats for \$124.99.

During a clearance sale the price is reduced by 25%. What is the dollar markdown and what is the sale price, correct to the nearest cent?

12. Find the amount of simple interest on a $2000 loan for 3 years at 9%.

13. Find the amount of simple interest on a $5000 loan for 9 months at 12%.

14. How much interest will $4000 earn in 4 years at 8% simple interest? How much in 4 months?

15. Use table 1 to find the compound amount and the compound interest on $4000 invested at 16% compounded semiannually for 10 years.

16. Use table 1 to find the compound amount and the compound interest on $5000 invested at 12% compounded quarterly for 6 years.

17. What is the effective rate if money is invested at 5% compounded (a) annually, (b) semiannually, (c) quarterly?

18. Use table 2 to find the annual premium for each type of insurance policy listed.

	Face Value of Policy	Age at Issue	Type of Policy
a.	$20,000	30	5-year term
b.	$15,000	25	straight life
c.	$30,000	35	limited-payment life (20-year)
d.	$40,000	40	endowment (20-year)

19. A videotape recorder is advertised for $2000. It may be purchased on the installment plan by paying $200 down and agreeing to pay the balance plus 12% simple interest on the balance in 24 monthly payments. What is the true annual interest rate (correct to the nearest tenth of a percentage)?

20. Tom Downs purchased a tape recorder for $125 and charged it. The store charges an interest rate of $1\frac{1}{2}$% per month on any unpaid balance. If Tom decides to make payments of $40 per month, how much interest will he pay before the tape recorder is paid for? (Assume that there are no additional purchases and that the first payment is made before the end of the month, so no interest is charged on the first payment.)

21. The Stones assumed a $25,000 mortgage for 30 years at 8%. Use table 3 to find their monthly payment? How much total interest will the Stones pay on their mortgage?

22. Find the monthly payment for the principal and interest for each of the following mortgages. (Use table 3, section 11.9)

	Amount of Mortgage	Interest Rate	Term of Mortgage
a.	$30,000	$7\frac{1}{2}$%	30 years
b.	$25,000	8%	20 years
c.	$15,000	$8\frac{1}{2}$%	10 years
d.	$50,000	9%	30 years

Just For Fun

True or false: Arabic numerals were invented by the Arabs.

APPENDIX A
TABLES

TABLE 1 Factorials

n	$n!$
0	1
1	1
2	2
3	6
4	24
5	120
6	720
7	5,040
8	40,320
9	362,880
10	3,628,800
11	39,916,800
12	479,001,600
13	6,227,020,800
14	87,178,291,200
15	1,307,674,368,000

TABLE 2 Squares, square roots, and prime factors for the numbers 1 through 100

No.	Square	Square Root	Prime Factors	No.	Square	Square Root	Prime Factors
1	1	1.000		51	2,601	7.141	$3 \cdot 17$
2	4	1.414	2	52	2,704	7.211	$2^2 \cdot 13$
3	9	1.732	3	53	2,809	7.280	53
4	16	2.000	2^2	54	2,916	7.348	$2 \cdot 3^3$
5	25	2.236	5	55	3,025	7.416	$5 \cdot 11$
6	36	2.449	$2 \cdot 3$	56	3,136	7.483	$2^3 \cdot 7$
7	49	2.646	7	57	3,249	7.550	$3 \cdot 19$
8	64	2.828	2^3	58	3,364	7.616	$2 \cdot 29$
9	81	3.000	3^2	59	3,481	7.681	59
10	100	3.162	$2 \cdot 5$	60	3,600	7.746	$2^2 \cdot 3 \cdot 5$
11	121	3.317	11	61	3,721	7.810	61
12	144	3.464	$2^2 \cdot 3$	62	3,844	7.874	$2 \cdot 31$
13	169	3.606	13	63	3,969	7.937	$3^2 \cdot 7$
14	196	3.742	$2 \cdot 7$	64	4,096	8.000	2^6
15	225	3.873	$3 \cdot 5$	65	4,225	8.062	$5 \cdot 13$
16	256	4.000	2^4	66	4,356	8.124	$2 \cdot 3 \cdot 11$
17	289	4.123	17	67	4,489	8.185	67
18	324	4.243	$2 \cdot 3^2$	68	4,624	8.246	$2^2 \cdot 17$
19	361	4.359	19	69	4,761	8.307	$3 \cdot 23$
20	400	4.472	$2^2 \cdot 5$	70	4,900	8.367	$2 \cdot 5 \cdot 7$
21	441	4.583	$3 \cdot 7$	71	5,041	8.426	71
22	484	4.690	$2 \cdot 11$	72	5,184	8.485	$2^3 \cdot 3^2$
23	529	4.796	23	73	5,329	8.544	73
24	576	4.899	$2^3 \cdot 3$	74	5,476	8.602	$2 \cdot 37$
25	625	5.000	5^2	75	5,625	8.660	$3 \cdot 5^2$
26	676	5.099	$2 \cdot 13$	76	5,776	8.718	$2^2 \cdot 19$
27	729	5.196	3^3	77	5,929	8.775	$7 \cdot 11$
28	784	5.292	$2^2 \cdot 7$	78	6,084	8.832	$2 \cdot 3 \cdot 13$
29	841	5.385	29	79	6,241	8.888	79
30	900	5.477	$2 \cdot 3 \cdot 5$	80	6,400	8.944	$2^4 \cdot 5$
31	961	5.568	31	81	6,561	9.000	3^4
32	1,024	5.657	2^5	82	6,724	9.055	$2 \cdot 41$
33	1,089	5.745	$3 \cdot 11$	83	6,889	9.110	83
34	1,156	5.831	$2 \cdot 17$	84	7,056	9.165	$2^2 \cdot 3 \cdot 7$
35	1,225	5.916	$5 \cdot 7$	85	7,225	9.220	$5 \cdot 17$
36	1,296	6.000	$2^2 \cdot 3^2$	86	7,396	9.274	$2 \cdot 43$
37	1,369	6.083	37	87	7,569	9.327	$3 \cdot 29$
38	1,444	6.164	$2 \cdot 19$	88	7,744	9.381	$2^3 \cdot 11$
39	1,521	6.245	$3 \cdot 13$	89	7,921	9.434	89
40	1,600	6.325	$2^3 \cdot 5$	90	8,100	9.487	$2 \cdot 3^2 \cdot 5$
41	1,681	6.403	41	91	8,281	9.539	$7 \cdot 13$
42	1,764	6.481	$2 \cdot 3 \cdot 7$	92	8,464	9.592	$2^2 \cdot 23$
43	1,849	6.557	43	93	8,649	9.644	$3 \cdot 31$
44	1,936	6.633	$2^2 \cdot 11$	94	8,836	9.695	$2 \cdot 47$
45	2,025	6.708	$3^2 \cdot 5$	95	9,025	9.747	$5 \cdot 19$
46	2,116	6.782	$2 \cdot 23$	96	9,216	9.798	$2^5 \cdot 3$
47	2,209	6.856	47	97	9,409	9.849	97
48	2,304	6.928	$2^4 \cdot 3$	98	9,604	9.899	$2 \cdot 7^2$
49	2,401	7.000	7^2	99	9,801	9.950	$3^2 \cdot 11$
50	2,500	7.071	$2 \cdot 5^2$	100	10,000	10.000	$2^2 \cdot 5^2$

APPENDIX B
ANSWERS TO ODD-NUMBERED EXERCISES AND ALL REVIEW EXERCISES

CHAPTER 1

SECTION 1.2

1. a. true b. true c. true
   d. true e. true f. false

3. a. true b. false c. false
   d. true e. false f. false

5. a. {Sunday, Monday, Tuesday, Wednesday, Thursday, Friday, Saturday}
   b. {Sunday, Tuesday, Wednesday, Thursday, Saturday}
   c. $\emptyset$
   d. {January, March, May, July, August, October, December}
   e. $\emptyset$

7. a. $\{1, 2, 3, 4, 5, 6, 7, 8, 9\}$
   b. $\{6, 7, 8, \ldots\}$
   c. $\{6, 7, 8, 9\}$
   d. {January, February, March, April, September, October, November, December}
   e. {May, June, July, August}

9. a. $\{x \mid x$ is a letter in the English alphabet$\}$
   b. $\{x \mid x$ is a counting number$\}$
   c. $\{x \mid x$ is an even counting number$\}$
   d. $\{x \mid x$ is a counting number divisible by 5$\}$

JUST FOR FUN

facetious

SECTION 1.3

1. a. true b. false c. false
   d. true e. false f. false

3. a. false b. false c. false
   d. true e. true f. false

5. a. $\{10, 4\}, \{10\}, \{4\}, \emptyset$
   b. $\{m, a, t, h\}, \{m, a, t\}, \{m, a, h\}, \{m, t, h\}, \{a, t, h\}, \{m, a\}, \{m, t\}, \{m, h\}, \{a, t\}, \{a, h\}, \{t, h\}, \{m\}, \{a\}, \{t\}, \{h\}, \emptyset$
   c. $\{i, o, u\}, \{i, o\}, \{i, u\}, \{o, u\}, \{i\}, \{o\}, \{u\}, \emptyset$
   d. $\emptyset$

7. a. $\{y\}$ b. $\{e, i, g\}$ c. $\{e, i, g, h\}$
   d. $\{g, h\}$ e. U f. $\emptyset$

JUST FOR FUN

$\frac{4 + 4 + 4}{4} = 3, \quad \sqrt{4 + 4 + 4 + 4} = 4$

$\frac{4 \times 4 + 4}{4} = 5, \quad 4 + \frac{4 + 4}{4} = 6$, and so on.

SECTION 1.4

1. a. $A \cap B = \{4, 6\}; A \cup B = \{1, 2, 3, 4, 6, 7, 8\}$
   b. $A \cap B = \emptyset; A \cup B = \{a, b, c, d, e, f\}$

c. $A \cap B = B; A \cup B = A$
d. $A \cap B = \{t, s\}; A \cup B = \{g, i, a, n, t, s, j, e\}$
e. $A \cap B = B; A \cup B = A$
f. $A \cap B = \emptyset; A \cup B = \{1, 2, 3, 4, \ldots\}$

3. a. $\{1, 3, 5\}$ b. $\{2, 4\}$
c. $\emptyset$ d. $\{1, 2, 3, 4, 5\}$
e. $\{1, 2, 3, 4, 5\}$ f. $\emptyset$

5. a. $\{1, 9\}$ b. $\{0, 1, 2, 3, 5, 9\}$
c. $\{4\}$ d. $\{6, 7, 8, 9\}$
e. $\{0, 1, 2, 3, 4, 6, 7, 8, 9\}$
f. $\{0, 2, 3, 4, 5, 6, 7, 8\}$
g. $\{0, 2, 3, 4, 5, 6, 7, 8\}$
h. $\{0, 2, 3, 4, 5, 6, 7, 8\}$

7. a. false b. false c. false
d. false e. true f. true

JUST FOR FUN

29

SECTION 1.5

1. a. region 2 b. region 4
c. regions 1, 2, and 3 d. regions 1, 3, and 4
e. regions 1, 2, and 3 f. region 2

3. a. region V
b. regions I, II, III, IV, V, VI, and VII
c. regions II, IV, V, VI, and VII
d. regions II, IV, and V
e. regions I, II, IV, V, and VI
f. regions IV, V, VI, VII, and VIII

5. a. $A \cap B$ $(A' \cup B')'$

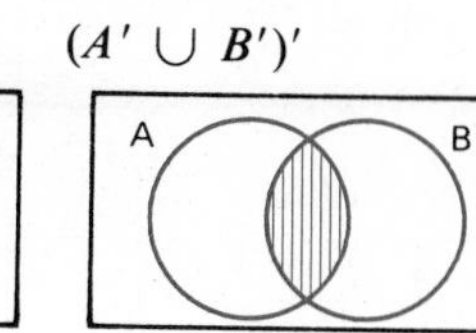

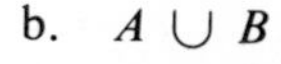

b. $A \cup B$ $(A' \cap B')'$

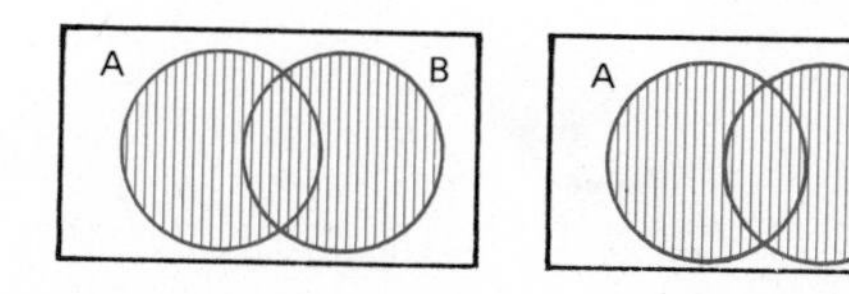

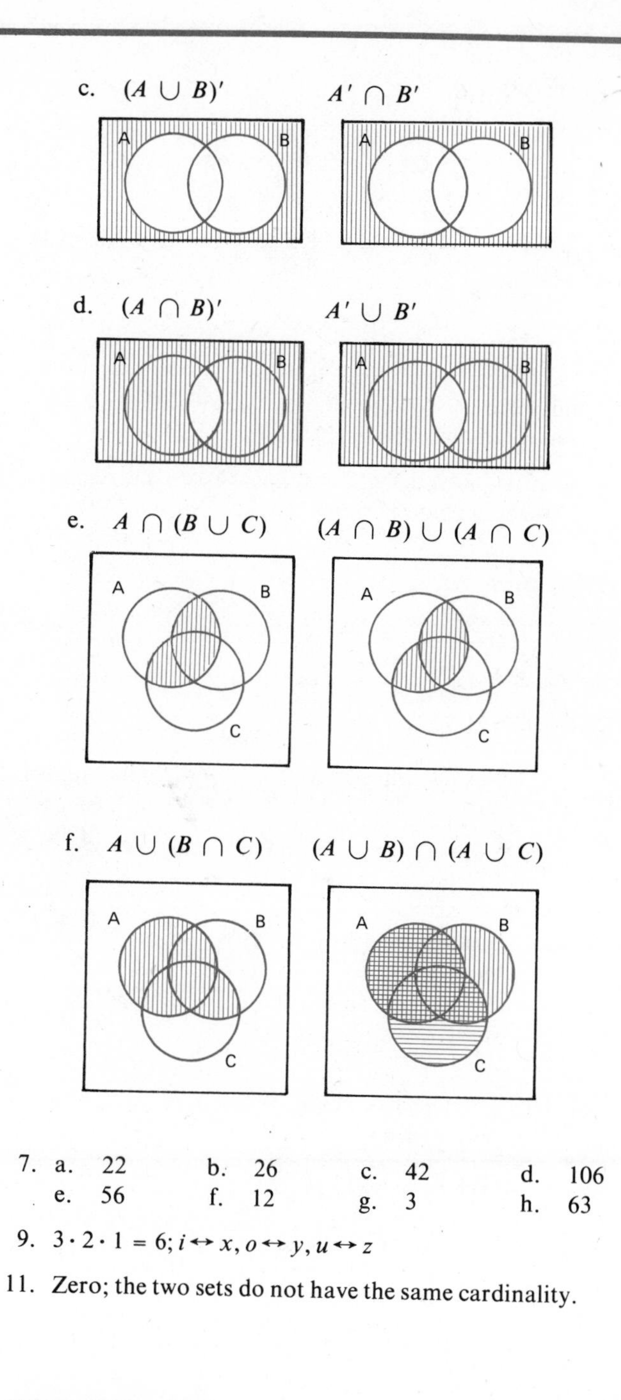

c. $(A \cup B)'$ $A' \cap B'$

d. $(A \cap B)'$ $A' \cup B'$

e. $A \cap (B \cup C)$ $(A \cap B) \cup (A \cap C)$

f. $A \cup (B \cap C)$ $(A \cup B) \cap (A \cup C)$

7. a. 22 b. 26 c. 42 d. 106
e. 56 f. 12 g. 3 h. 63

9. $3 \cdot 2 \cdot 1 = 6; i \leftrightarrow x, o \leftrightarrow y, u \leftrightarrow z$

11. Zero; the two sets do not have the same cardinality.

JUST FOR FUN

All of them have 28 days.

SECTION 1.6

1. a. 1 b. 19 c. 9 d. 22

3. a. 18 b. 45 c. 12 d. 43

5. a. 43 b. 12 c. 100 d. 11

7. a. 89 b. 6 c. 8 d. 22

9. a. 82 b. 19 c. 69 d. 49

11. According to the given information, the agency did not interview 1000 commuters. They used a universal set containing at least 1233 commuters.

13. UCLA—Kentucky; St. Johns—USC; Maryland—Ohio State; Notre Dame—NCS

JUST FOR FUN

no

SECTION 1.7

1. $A \times B = \{(a, 10), (a, 20), (b, 10), (b, 20), (c, 10), (c, 20)\}$
   $B \times A = \{(10, a), (10, b), (10, c), (20, a), (20, b), (20, c)\}$
   $n(A \times B) = 6$

3. $C \times D = \{(2, 1), (2, 3), (2, 5), (4, 1), (4, 3), (4, 5), (6, 1), (6, 3), (6, 5)\}$
   $D \times C = \{(1, 2), (1, 4), (1, 6), (3, 2), (3, 4), (3, 6), (5, 2), (5, 4), (5, 6)\}$
   $n(C \times D) = 9$

5. $T \times T = \{(t, t), (t, f), (f, t), (f, f)\}$,
   $n(T \times T) = 4$

7.

B			
y	$(4, y)$	$(5, y)$	$(6, y)$
x	$(4, x)$	$(5, x)$	$(6, x)$
	4	5	6
		A	

9.

11. a. $\{(a, b), (a, c), (a, d), (b, b), (b, c), (b, d)\}$
    b. $\{(a, c), (a, d), (a, e), (b, c), (b, d), (b, e)\}$
    c. 9
    d. $\{(b, c), (b, d), (b, e)\}$
    e. $\{(c, a), (c, b), (c, c), (c, d), (d, a), (d, b), (d, c), (d, d), (e, a), (e, b), (e, c), (e, d)\}$
    f. $\{(a, c), (a, d), (b, c), (b, d)\}$

13. No, the ordered pairs are different.

JUST FOR FUN

1	8	6
10	5	0
4	2	9

REVIEW EXERCISES FOR CHAPTER 1

1. A set is any collection of objects.

2. a. $\{a, b, c, \ldots, z\}$;
   $\{x \mid x$ is a letter in the English alphabet$\}$
   b. {Huron, Erie, Michigan, Superior, Ontario};
   $\{x \mid x$ is a Great Lake$\}$
   c. $\{1, 2, 3, 4, 5, 6, 7\}$;
   $\{x \mid x$ is a counting number less than 8$\}$
   d. $\{3, 4, 5, \ldots\}$;
   $\{x \mid x$ is a counting number greater than 2$\}$
   e. $\{2, 4, 6, 8, \ldots\}$; $\{x \mid x$ is an even counting number$\}$

3. a. false b. true c. false
   d. true e. false f. true

4. a. true b. true c. true
   d. false e. true f. false

5. a. true b. true
   c. true d. true

6. $\{i, o, u\}, \{i, o\}, \{i, u\}, \{o, u\}, \{i\}, \{o\}, \{u\}, \emptyset$

7. a. $\{1, 3, 5\}$ b. $\{0, 1, 2, 3, 4, 5\}$
   c. $\{2\}$ d. $\{0, 1, 2, 3\}$
   e. U f. $\{0, 1, 3, 4, 5\}$
   g. C h. $\{0, 1, 2, 3, 4\}$

8. a. regions 1, 2, and 3 b. regions 1, 3, and 4
   c. region 4 d. regions II, IV, and V
   e. regions I, II, IV, V, and VI
   f. regions I, IV, V, VII, and VIII

9. For example: Bob—5, Ted—10, Joe—15; $3 \cdot 2 \cdot 1 = 6$

10. a. 11 b. 24 c. 31 d. 5
    e. 34 f. 38 g. 17 h. 47

11. a. 33 b. 6 c. 93 d. 90

12. Don—Bob, Ted—Joe, Cal—Jon, Jim—Al

13. a. $A \times B = \{(1, a), (1, b), (1, c), (2, a), (2, b), (2, c), (3, a), (3, b), (3, c)\}$
    b. $n(A \times B) = 9$
    c.

B			
c	$(1, c)$	$(2, c)$	$(3, c)$
b	$(1, b)$	$(2, b)$	$(3, b)$
a	$(1, a)$	$(2, a)$	$(3, a)$
	1	2	3
		A	

    d. 1 → a, b, c 2 → a, b, c 3 → a, b, c

14. a. For example: m—e, a—a, t—s, h—y
    b. $4 \cdot 3 \cdot 2 \cdot 1 = 24$
    c. $\{m, a, t, h\}$, $\{m, a, t\}$, $\{m, a, h\}$, $\{m, t, h\}$, $\{a, t, h\}$, $\{m, a\}$, $\{m, t\}$, $\{m, h\}$, $\{a, t\}$, $\{a, h\}$, $\{t, h\}$, $\{m\}$, $\{a\}$, $\{t\}$, $\{h\}$, $\emptyset$
    d. $\{(m, e), (m, a), (m, s), (m, y), (a, e), (a, a), (a, s), (a, y), (t, e), (t, a), (t, s), (t, y), (h, e), (h, a), (h, s), (h, y)\}$

15. a. when $A \subseteq B$ b. when $A = B$
    c. always d. when $B = \emptyset$
    e. always

JUST FOR FUN

	Step	Amount Left After Each Step 8 gallon	5 gallon	3 gallon
1.	Fill 5 gal from 8 gal.	3	5	0
2.	Fill 3 gal from 5 gal.	3	2	3
3.	Empty 3 gal into 8 gal.	6	2	0
4.	Empty 5 gal into 3 gal.	6	0	2
5.	Fill 5 gal from 8 gal.	1	5	2
6.	Fill 3 gal from 5 gal.	1	4	3
7.	Empty 3 gal into 8 gal.	4	4	0

CHAPTER 2

SECTION 2.2

1. a. simple
   b. compound; negation
   c. compound; biconditional
   d. compound; negation
   e. compound; conditional
   f. neither
   g. compound; conjunction
   h. compound; disjunction

3. a. $P \wedge Q$ b. $Q \vee P$
   c. $\sim P \vee \sim P$ d. $Q \rightarrow \sim P$
   e. $\sim(\sim P)$ f. $P \leftrightarrow Q$

5. a. $P \vee Q$ b. $\sim Q \wedge P$
   c. $\sim(\sim Q)$ d. $Q \leftrightarrow P$
   e. $P \rightarrow Q$ f. $\sim P \wedge \sim Q$

7. a. $P \vee M$ b. $G \wedge P$
   c. $\sim A \rightarrow D$ d. $B \vee \sim W$
   e. $T \leftrightarrow F$ f. $E \rightarrow R$

9. a. I like algebra and I like geometry.
   b. If I like algebra then I do not like geometry.
   c. I like algebra or I like geometry.
   d. I like algebra or I do not like geometry.
   e. I do not like algebra and I do not like geometry.
   f. I like algebra iff I like geometry.

JUST FOR FUN

Let 13 = XIII; then XIII = VIII.

SECTION 2.3

1. a. $\sim(P \wedge Q \rightarrow R)$ b. none
   c. $\sim P \wedge (Q \rightarrow R)$ d. none
   e. $\sim P \vee (Q \wedge R)$ f. $P \wedge (Q \leftrightarrow R)$

3. a. $Q \wedge P \rightarrow R$ b. $(R \wedge P) \vee Q$
   c. $\sim(Q \wedge P)$ d. $(Q \wedge R) \vee P$
   e. $P \leftrightarrow R \wedge Q$ f. $\sim P \wedge \sim R$

5. a. $E \rightarrow C \wedge G$ b. $S \vee (J \wedge I)$
   c. $C \rightarrow E \wedge Z$ d. $\sim(\sim F)$
   e. $\sim S \wedge \sim L$

7. a. Algebra is difficult, and logic is easy or Latin is interesting.
   b. If algebra is difficult and logic easy, then Latin is interesting.
   c. Algebra is difficult, or logic is easy and Latin is interesting.
   d. Algebra is difficult, and if logic is easy then Latin is interesting.
   e. It is not the case that algebra is difficult and logic is easy.
   f. Algebra is not difficult iff logic is easy and Latin is not interesting.

JUST FOR FUN

160 ft

SECTION 2.4

1.

~	(P	∧	Q)
F	T	T	T
T	T	F	F
T	F	F	T
T	F	F	F

3.

~	P	∧	~	Q
F		F	F	
F		F	T	
T		F	F	
T		T	T	

5.

P	∧	~	P
T	F	F	
F	F	T	

7.

P	∨	~	Q
T	T	F	
T	T	T	
F	F	F	
F	T	T	

9.

~	(P	∨	~	Q)
F	T	T	F	
F	T	T	T	
T	F	F	F	
F	F	T	T	

11.

~	P	∨	(P	∧	~	Q)
F		F	T	F	F	
F		T	T	T	T	
T		T	F	F	F	
T		T	F	F	T	

13. a.

(P	∨	Q)	∧	(~	P	∨	~	Q)
T	T	T	F	F		F	F	
T	T	F	T	F		T	T	
F	T	T	T	T		T	F	
F	F	F	F	T		T	T	

b.

(P	∧	~	Q)	∨	(~	P	∧	Q)
T	F	F		F	F		F	T
T	T	T		T	F		F	F
F	F	F		T	T		T	T
F	F	T		F	T		F	F

JUST FOR FUN

336

SECTION 2.5

1.

P	$\to$	Q
T	T	T
T	F	F
F	T	T
F	T	F

3.

$\sim$	P	$\to$	$\sim$	Q
F		T	F	
F		T	T	
T		F	F	
T		T	T	

5.

$\sim$	P	$\to$	Q
F		T	T
F		T	F
T		T	T
T		F	F

7.

$\sim$	P	$\leftrightarrow$	$\sim$	Q
F		T	F	
F		F	T	
T		F	F	
T		T	T	

9.

P	$\vee$	Q	$\to$	$\sim$	Q
T	T	T	F	F	
T	T	F	T	T	
F	T	T	F	F	
F	F	F	T	T	

11.

$(P$	$\to$	$Q)$	$\vee$	P	$\to$	Q
T	T	T	T	T	T	T
T	F	F	T	T	F	F
F	T	T	T	F	T	T
F	T	F	T	F	F	F

13.

P	$\wedge$	Q	$\leftrightarrow$	P	$\vee$	Q
T	T	T	T	T	T	T
T	F	F	F	T	T	F
F	F	T	F	F	T	T
F	F	F	T	F	F	F

15.

$(P$	$\vee$	$Q)$	$\wedge$	R
T	T	T	T	T
T	T	T	F	F
T	T	F	T	T
T	T	F	F	F
F	T	T	T	T
F	T	T	F	F
F	F	F	F	T
F	F	F	F	F

17.

$(P$	$\wedge$	$Q)$	$\vee$	$(P$	$\wedge$	$R)$
T	T	T	T	T	T	T
T	T	T	T	T	F	F
T	F	F	T	T	T	T
T	F	F	F	T	F	F
F	F	T	F	F	F	T
F	F	T	F	F	F	F
F	F	F	F	F	F	T
F	F	F	F	F	F	F

19.

P	$\leftrightarrow$	Q	$\vee$	R
T	T	T	T	T
T	T	T	T	F
T	T	F	T	T
T	F	F	F	F
F	F	T	T	T
F	F	T	T	F
F	F	F	T	T
F	T	F	F	F

21. yes

$\sim$	$(P$	$\vee$	$Q)$	$\leftrightarrow$	$\sim$	P	$\wedge$	$\sim$	Q
F	T	T	T	T	F		F	F	
F	T	T	F	T	F		F	T	
F	F	T	T	T	T		F	F	
T	F	F	F	T	T		T	T	

23. yes

P	$\wedge$	$\sim$	Q	$\leftrightarrow$	$\sim$	$(\sim$	P	$\vee$	$Q)$
T	F	F		T	F	F		T	T
T	T	T		T	T	F		F	F
F	F	F		T	F	T		T	T
F	F	T		T	F	T		T	F

25. a. The tide is not out, or we can go clamming.
b. Bill did not drive his van, or he brought the packages.

c. Today is not Wednesday, or tomorrow is not Friday.
d. If two does equal three, then four equals six.
e. If Bob passed the test, then he is unhappy about something else.

JUST FOR FUN

3

SECTION 2.6

1. $P \vee \sim Q$
3. $\sim(\sim P \vee \sim Q)$
5. $\sim(\sim P \vee Q)$
7. $P \wedge Q$
9. $P \wedge Q$
11. John did not go to the party and Janie did not go to the party.
13. It is false that I passed the test and I did not study too much.
15. Logic is not dull or it is interesting.
17. It is false that the bus is not late and my watch is working correctly.
19. Either x is not greater than zero or x is not negative.
21. It is false that the wind did not come up and we sailed.
23. $A' \cup B'$
25. $(A \cup B)'$
27. $(A' \cup B')'$
29. $A' \cap B$

JUST FOR FUN

They are the same length.

SECTION 2.7

1. invalid
3. invalid
5. valid
7. valid
9. valid
11. invalid
13. invalid
15. valid
17. valid

JUST FOR FUN

SECTION 2.8

1. Universal Negative

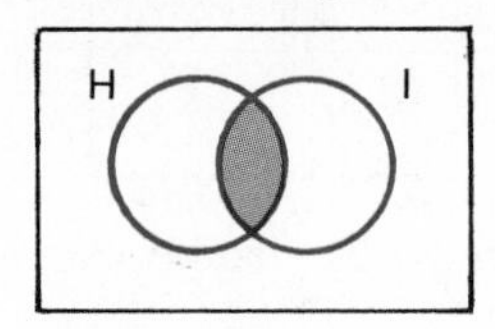

3. Particular Negative

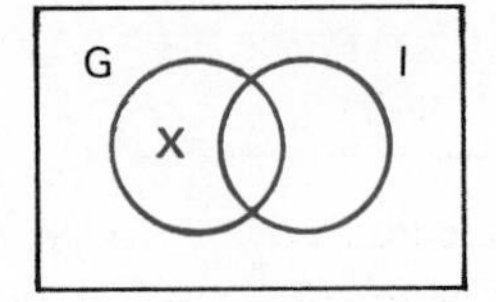

5. Particular Affirmative

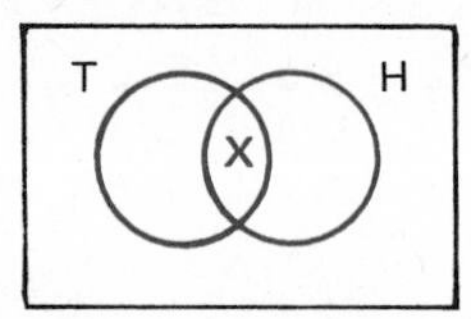

7. Particular Negative

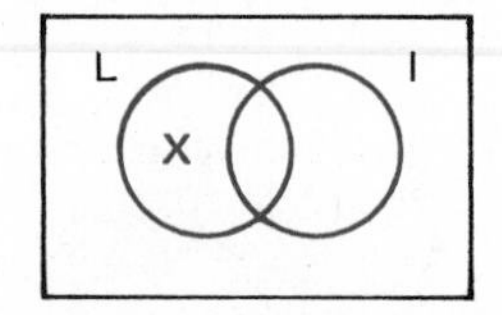

9. Particular Affirmative

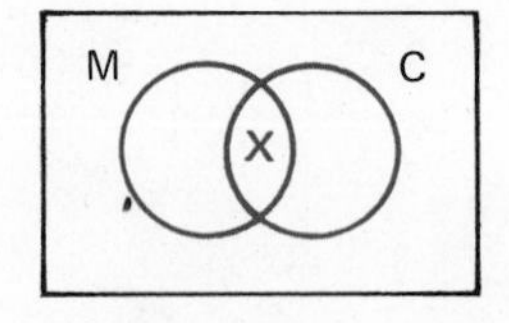

11. Particular Negative

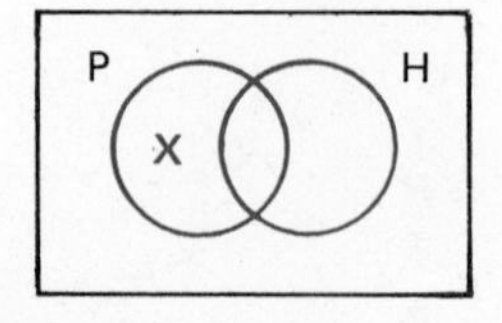

13. Universal Negative

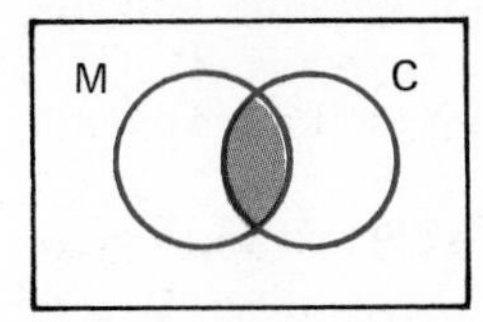

15. Particular Negative

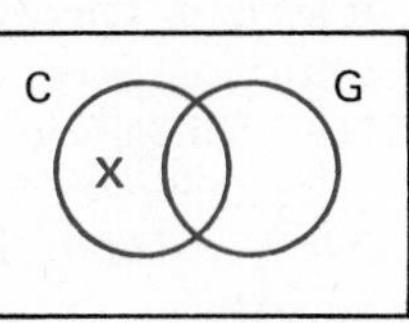

17. Particular Negative

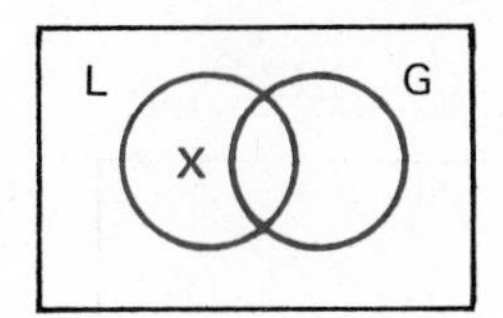

19. Universal Negative

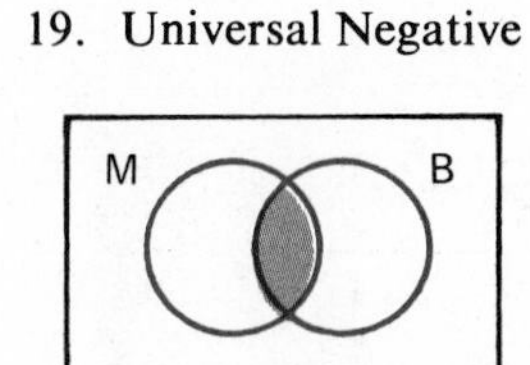
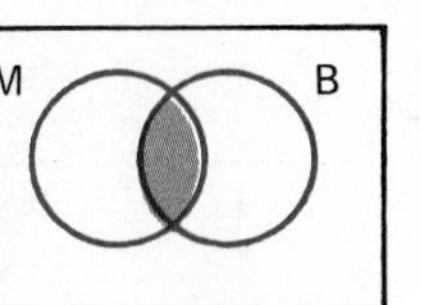

21. consistent 23. consistent 25. consistent
27. consistent 29. consistent 31. consistent
33. consistent 35. consistent

JUST FOR FUN

11.25 seconds

SECTION 2.9

1. invalid 3. valid 5. valid
7. invalid 9. valid 11. invalid
13. valid 15. invalid 17. invalid
19. valid 21. valid 23. invalid
25. invalid

JUST FOR FUN

Q and Z

SECTION 2.10

1. *converse:* a. $P \to S$
   inverse: $\sim S \to \sim P$
   contrapositive: $\sim P \to \sim S$
   b. $S \to F$ c. $F \to \sim S$
   $\sim F \to \sim S$ $S \to \sim F$
   $\sim S \to \sim F$ $\sim F \to S$

   d. $\sim G \to L$ e. $\sim W \to \sim T$
   $\sim L \to G$ $T \to W$
   $G \to \sim L$ $W \to T$

3. a. $P \to Q$
   b. $\sim Q \to \sim P$
   c. $P \to Q$
   d. $P \to Q$
   e. $Q \leftrightarrow P$

5. a. $P \to Q$
   b. $\sim P \to \sim Q$
   c. $Q \to P$
   d. $P \leftrightarrow Q$
   e. $Q \to P$

7. No, the statement is equivalent to "If I pass you, then you will come to class every day."

JUST FOR FUN

1: special agent; 2: spy; 3: spy

REVIEW EXERCISES FOR CHAPTER 2

1. a. compound; disjunction
   b. neither
   c. simple
   d. compound; conjunction
   e. compound; negation
   f. compound; conditional

2. a. $P \wedge Q$ b. $\sim Q \to R \vee P$
   c. $\sim P \wedge \sim Q$ d. $R \to P \wedge Q$
   e. $\sim(P \wedge Q) \vee R$ f. $P \leftrightarrow Q \vee R$

3. a. Sam is sulky, and Tom is tense or Freddy is ready.
   b. Sam is sulky and Tom is tense, or Freddy is ready.
   c. If Sam is sulky or Tom is tense, then Freddy is ready.
   d. Sam is sulky, or if Tom is tense then Freddy is ready.
   e. If Freddy is ready and Tom is tense, then Sam is not sulky.
   f. Sam is not sulky iff Tom is tense and Freddy is not ready.

4. a.

$\sim$	P	$\to$	Q
F		T	T
F		T	F
T		T	T
T		F	F

b.

P	$\vee$	$\sim$	Q
T	T	F	
T	T	T	
F	F	F	
F	T	T	

c.

P	$\vee$	Q	$\leftrightarrow$	P	$\wedge$	$\sim$	Q
T	T	T	F	T	F	F	
T	T	F	T	T	T	T	
F	T	T	F	F	F	F	
F	F	F	T	F	F	T	

d.

P	$\vee$	Q	$\rightarrow$	R
T	T	T	T	T
T	T	T	F	F
T	T	F	T	T
T	T	F	F	F
F	T	T	T	T
F	T	T	F	F
F	F	F	T	T
F	F	F	T	F

e.

P	$\wedge$	$\sim$	Q	$\rightarrow$	Q	$\vee$	R
T	F	F		T	T	T	T
T	F	F		T	T	T	F
T	T	T		T	F	T	T
T	T	T		F	F	F	F
F	F	F		T	T	T	T
F	F	F		T	T	T	F
F	F	T		T	F	T	T
F	F	T		T	F	F	F

f.

$\sim$	$(P$	$\wedge$	$\sim$	$Q)$	$\rightarrow$	$\sim$	$(\sim$	P	$\vee$	$Q)$
T	T	F			F	F	F		T	T
F	T	T			T	T	F		F	F
T	F	F			F	F	T		T	T
T	F	F			F	F	T		T	F

5. a. Today is not Monday or tomorrow is not Sunday.
   b. It is false that Scott is first and Joe is not second.
   c. It is false that Hugh is not painting and not cutting the grass.
   d. Norma did not go to the store and Laurie did not go swimming.
   e. It is false that mathematics is difficult and logic is not easy.

6. yes

$\sim$	P	$\vee$	Q	$\leftrightarrow$	$\sim$	$(P$	$\wedge$	$\sim$	$Q)$
F		T	T	T	T	T	F	F	
F		F	F	T	F	T	T	T	
T		T	T	T	T	F	F	F	
T		T	F	T	T	F	F	T	

7. a. valid b. valid

8. a. inconsistent b. consistent
   c. consistent d. inconsistent

9. a. valid b. invalid
   c. valid d. valid
   e. invalid

10. a. *converse:* $G \rightarrow P$
    inverse: $\sim P \rightarrow \sim G$
    contrapositive: $\sim G \rightarrow \sim P$
    b. $\sim C \rightarrow P$, $\sim P \rightarrow C$, $C \rightarrow \sim P$
    c. $\sim C \rightarrow \sim A$, $A \rightarrow C$, $C \rightarrow A$

11. No, statement is equivalent to "If I marry you, then I will get a job."

12. a. Nixon did not know about Watergate or he is telling the truth.
    b. It is not the case that if Nixon knew about Watergate then he is telling the truth.

13. a. $P \rightarrow Q$ b. $P \rightarrow Q$
    c. $\sim Q \rightarrow \sim P$ d. $\sim Q \rightarrow \sim P$
    e. $\sim P \vee Q$

JUST FOR FUN

Barry—painter; Bob—mason; Bart—carpenter

CHAPTER 3

SECTION 3.2

1. a. $\frac{1}{6}$ b. $\frac{1}{2}$ c. $\frac{1}{2}$
   d. $\frac{1}{2}$ e. $\frac{1}{3}$ f. 1

3. a. $\frac{1}{52}$ b. $\frac{1}{52}$ c. $\frac{1}{13}$
   d. $\frac{1}{2}$ e. $\frac{1}{4}$ f. $\frac{1}{26}$

5. a. $\frac{3}{13}$ b. $\frac{11}{26}$ c. $\frac{4}{13}$
   d. $\frac{1}{52}$ e. $\frac{1}{26}$ f. $\frac{1}{52}$

7. a. $\frac{5}{7}$ b. $\frac{4}{7}$ c. 1
   d. $\frac{2}{7}$ e. $\frac{3}{7}$

JUST FOR FUN

a. $\frac{0}{1} = 0$ b. $\frac{1}{0}$ is undefined
c. $\frac{0}{0}$ is meaningless

SECTION 3.3

1. a. $\frac{1}{12}$ b. 0 c. $\frac{7}{12}$
   d. $\frac{1}{2}$ e. $\frac{1}{4}$ f. $\frac{3}{4}$

3. a. $5 \times 5 = 25$ outcomes in sample space:

(1, 1)	(2, 1)	(3, 1)	(4, 1)	(5, 1)
(1, 2)	(2, 2)	(3, 2)	(4, 2)	(5, 2)
(1, 3)	(2, 3)	(3, 3)	(4, 3)	(5, 3)
(1, 4)	(2, 4)	(3, 4)	(4, 4)	(5, 4)
(1, 5)	(2, 5)	(3, 5)	(4, 5)	(5, 5)

   b. $\frac{3}{5}$ c. $\frac{3}{5}$ d. $\frac{9}{25}$
   e. $\frac{21}{25}$ f. $\frac{4}{25}$

5. a. $\frac{1}{6}$ b. $\frac{1}{18}$ c. 0
   d. $\frac{1}{2}$ e. 1 f. $\frac{7}{36}$

7. a.

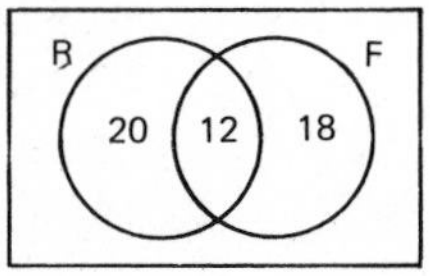

   b. $\frac{2}{5}$ c. $\frac{9}{25}$

9. a. $\frac{2}{25}$ b. $\frac{1}{5}$ c. $\frac{3}{25}$
   d. $\frac{11}{15}$ e. $\frac{19}{75}$ f. $\frac{22}{75}$

JUST FOR FUN

1 + 23 + 4 + 5 + 67 = 100

SECTION 3.4

1.

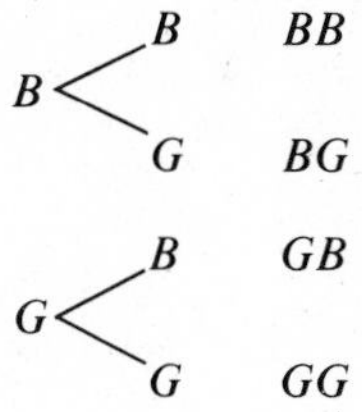

   a. $\frac{1}{4}$ b. $\frac{1}{4}$ c. $\frac{1}{4}$
   d. $\frac{3}{4}$ e. $\frac{1}{4}$

3.

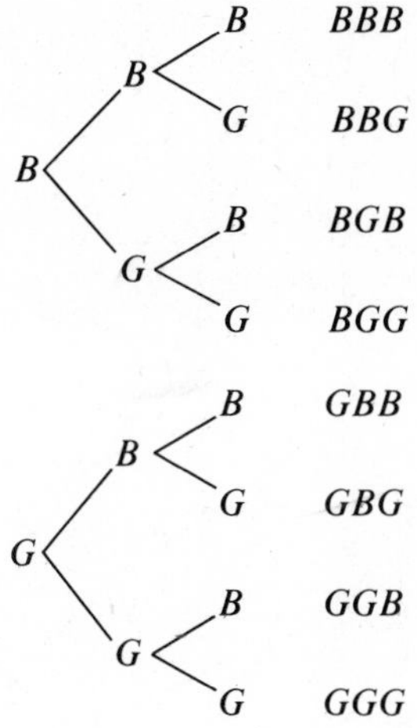

   a. $\frac{1}{8}$ b. $\frac{3}{8}$ c. $\frac{7}{8}$
   d. $\frac{1}{8}$ e. $\frac{1}{4}$ f. $\frac{1}{2}$

5.

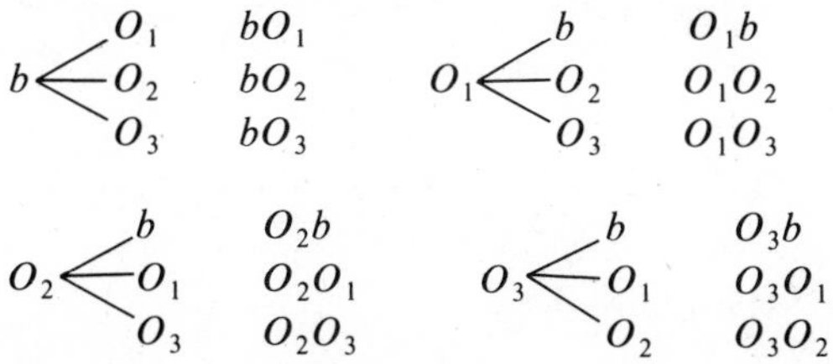

   a. $\frac{1}{2}$ b. 1 c. $\frac{1}{4}$
   d. $\frac{1}{4}$ e. $\frac{1}{2}$ f. $\frac{1}{2}$

JUST FOR FUN

One—it is one continuous groove.

SECTION 3.5

1. a. 1:5 b. 1:17 c. 1:35
 d. 35:1 e. 31:5 f. 11:1

3. a. 1:12 b. 1:3 c. 1:1
 d. 12:1 e. 3:10 f. 15:11

5. $\frac{2}{9}$

7. a. $\frac{7}{12}$ b. $\frac{1}{2}$ c. $\frac{2}{5}$
 d. $\frac{9}{14}$

9. 1:5; $3

11. $1, or 20¢ per ticket

13. $.14

JUST FOR FUN

Optical illusion; 3 or 5

SECTION 3.6

1. a. $\frac{3}{13}$ b. $\frac{3}{26}$ c. $\frac{2}{13}$
 d. $\frac{4}{13}$ e. $\frac{1}{52}$ f. $\frac{1}{52}$

3. a. $\frac{11}{221}$ b. $\frac{4}{221}$ c. $\frac{1}{221}$
 d. $\frac{1}{221}$ e. $\frac{1}{17}$ f. $\frac{1}{17}$

5. a. $\frac{1}{55}$ b. $\frac{1}{220}$ c. $\frac{1}{22}$
 d. $\frac{1}{22}$ e. $\frac{1}{22}$ f. $\frac{14}{55}$

7. $\frac{1}{36}$

9. a. $\frac{1}{2}$ b. $\frac{1}{100}$ c. 0
 d. $\frac{9}{100}$ e. $\frac{5}{18}$

JUST FOR FUN

Probability this will occur is greater than $\frac{1}{2}$.

SECTION 3.7

1. 120; 216

3. 657,720

5. 12

7. 3,276,000; 2,948,400; 2,835,000

9. 64; 56

11. a. 6 b. 120 c. 1
 d. 210 e. 20 f. 24

13. 90

15. 362,880

17. 132

19. a. 2520 b. 50,400 c. 5040
 d. 1260

21. 625

23. 1,000,000,000

JUST FOR FUN

Hawaii

SECTION 3.8

1. a. 10 b. 10 c. 35
 d. 35 e. 1 f. 1

3. 210

5. 10; 85¢

7. 3003

9. 21

11. 1960

13. 2970

15. 5400

17. 108,900

JUST FOR FUN

No, order is important.

SECTION 3.9

1. $\frac{1}{221}$

3. $\frac{143}{11,050}$

5. a. $\frac{7}{306}$ b. $\frac{7}{102}$ c. $\frac{7}{17}$
 d. $\frac{28}{153}$ e. $\frac{95}{102}$

7. a. $\frac{1}{208}$ b. $\frac{3}{104}$ c. $\frac{21}{52}$
 d. $\frac{15}{52}$ e. $\frac{15}{208}$ f. $\frac{21}{104}$

9. ${}_{52}C_{13}$; $\frac{{}_{13}C_{13}}{{}_{52}C_{13}}$

11. $\frac{1}{35}$

13. a. $\frac{{}_6C_3}{{}_{13}C_3}$ b. $\frac{{}_4C_3}{{}_{13}C_3}$ c. $\frac{{}_3C_3}{{}_{13}C_3}$
d. $\frac{{}_6C_2 \cdot {}_4C_1}{{}_{13}C_3}$ e. $\frac{{}_4C_2 \cdot {}_3C_1}{{}_{13}C_3}$ f. $\frac{{}_3C_2 \cdot {}_6C_1}{{}_{13}C_3}$

15. $\frac{1287}{2,598,960}$ 17. $\frac{1}{108,290}$ 19. $\frac{6}{4165}$

JUST FOR FUN

$1N + 3D + 2Q + 15P = \$1.00$

REVIEW EXERCISES FOR CHAPTER 3

1. a. $\frac{4}{13}$ b. $\frac{6}{13}$ c. $\frac{7}{13}$
d. $\frac{9}{13}$ e. $\frac{10}{13}$ f. $\frac{3}{13}$

2. a. $\frac{1}{6}$ b. $\frac{5}{6}$ c. $\frac{1}{6}$
d. $\frac{2}{9}$ e. $\frac{5}{12}$ f. $\frac{1}{2}$

3. a. 12
b. ($1, $2) ($2, $1) ($5, $1) ($10, $1)
($1, $5) ($2, $5) ($5, $2) ($10, $2)
($1, $10) ($2, $10) ($5, $10) ($10, $5)
c. $\frac{1}{6}$

4.

Start → H → H → H: H H H
Start → H → H → T: H H T
Start → H → T → H: H T H
Start → H → T → T: H T T
Start → T → H → H: T H H
Start → T → H → T: T H T
Start → T → T → H: T T H
Start → T → T → T: T T T

a. $\frac{1}{8}$ b. $\frac{7}{8}$ c. $\frac{1}{8}$

5. a. 1:12 b. 12:1 c. 3:10
d. 4:9 e. 1:51 f. 51:1

6. $\frac{1}{9}$ 7. 3:8

8. $2 9. $4, or 80¢ per ticket

10. Independent; it does not matter what ball is chosen first because there is replacement. The occurrence of one event does not affect the occurrence of a second event.

11. Yes; mutually exclusive events cannot happen at the same time. Either a seven or an eleven may occur, but they cannot occur at the same time.

12. a. $\frac{25}{102}$ b. $\frac{1}{221}$ c. $\frac{13}{51}$
d. $\frac{13}{204}$ e. $\frac{40}{221}$ f. $\frac{4}{663}$

13. a. $\frac{4}{312}$ b. $\frac{72}{312}$ c. $\frac{1}{12}$
d. $\frac{7}{12}$ e. $\frac{1}{156}$

14. a. 7776 b. $\frac{1}{1296}$

15. a. 468,000 b. 405,000 c. 302,400

16. 143,640 17. 90 18. 2520

19. a. 24 b. 1 c. 60
d. 20 e. 30 f. 1

20. a. 10 b. 15 c. 35
d. 35 e. 1 f. 1

21. 196,000 22. 3600

23. a. $\frac{1}{30}$ b. $\frac{1}{30}$ c. $\frac{1}{30}$
d. $\frac{1}{5}$ e. $\frac{1}{10}$

24. a. $\frac{{}_{26}C_5}{{}_{52}C_5}$ b. $\frac{{}_{26}C_5}{{}_{52}C_5}$ c. $\frac{{}_{12}C_5}{{}_{52}C_5}$
d. $\frac{{}_{40}C_5}{{}_{52}C_5}$ e. $\frac{{}_{12}C_3 \cdot {}_{40}C_2}{{}_{52}C_5}$

25. a. $\frac{624}{2,598,960}$ b. $\frac{3744}{2,598,960}$
c. $\frac{5148}{2,598,960}$ (counting straight flushes) or
$\frac{5108}{2,598,960}$ (not counting straight flushes)

CHAPTER 4

SECTION 4.2

1. mean = 5; median = 4; mode = 4
3. mean = 5.5; median = 5.5; no mode
5. mean = 7; median = 7; no mode
7. mean = 55; median = 55; no mode
9. mean = 1755.3; median = 1794; no mode
11. mode
13. No; he actually needed 10 points.
15. 710
17. 390; 78
19. Benny is right; his average is 3.27 while Larry's is 2.73.

JUST FOR FUN

Put 1 penny in one cup, 4 in another, and 5 in the third, then stack the first cup inside the second.

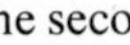

SECTION 4.3

1. $\sigma = 3$
3. $\sigma = \sqrt{6}$ or 2.4
5. $\sigma = \sqrt{13.7}$ or 3.7
7. $\sigma = \sqrt{26.8}$ or 5.2
9. $\sigma = \sqrt{21.6}$ or 4.6
11. a. 4 b. 3.5 c. 3 d. 5
    e. 4.5 f. $\sigma = \sqrt{2.6}$ or 1.6
13. a. 345 b. 370 c. none d. 340
    e. 320 f. $\sigma = \sqrt{11055.6}$ or 105.1
15. a. Rudy, Maureen, Jeff, Mark
    b. Maureen c. Eric, Maria
    d. Maureen e. Maria
    f. Mark

JUST FOR FUN

35?

SECTION 4.4

1. 94 3. 24 5. 8th 7. 15th
9. a. 3rd b. 75 c. 83 d. 75
11. Larry 13. Peter
15. a. 3 b. 92.5 c. 68
    d. 65 e. 25 f. 55

JUST FOR FUN

qoph, qiviut

SECTION 4.5

1. Bus = 180°
   Car = 120°
   Walk = 60°

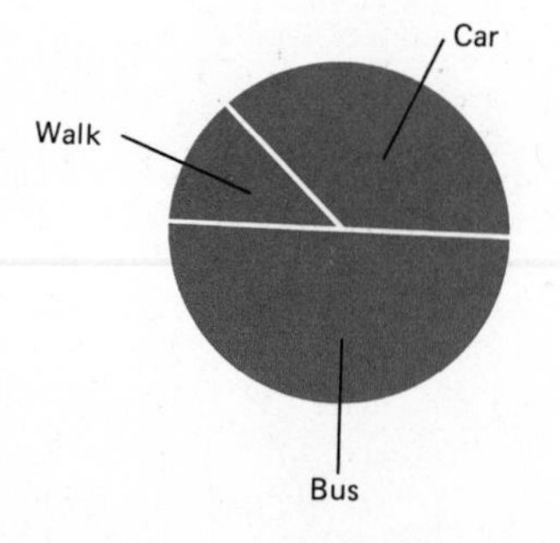

3. Food = 108°
   Household = 90°
   Transportation = 72°
   Savings = 36°
   Entertainment = 18°
   Unexpected expense = 36°

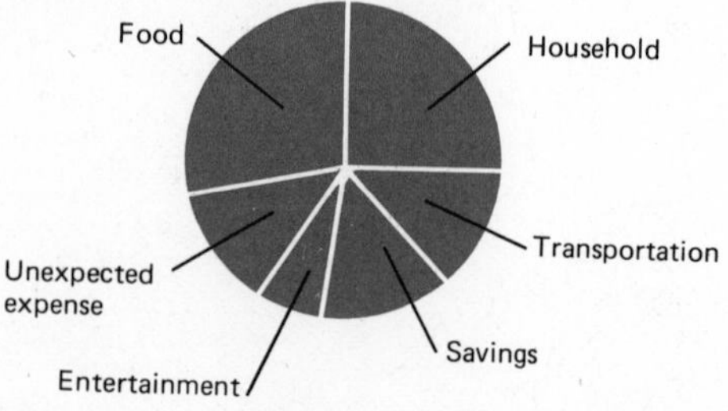

5. Personal income = 90°
   Corporate income = 90°
   Excise = 54°
   Sales = 72°
   Highway = 36°
   Miscellaneous = 18°

7.

Number	Tally	Frequency
6	~~////~~ ////	9
5	~~////~~/	6
4	~~////~~ ///	8
3	///	3
2	~~////~~ //	7
1	~~////~~ //	7

9. a.

Mark	Tally	Frequency
95–99	//	2
90–94	///	3
85–89	~~////~~	5
80–84	///	3
75–79	//	2
70–74	~~////~~ ////	9
65–69	//	2
60–64	////	4
55–59	///	3

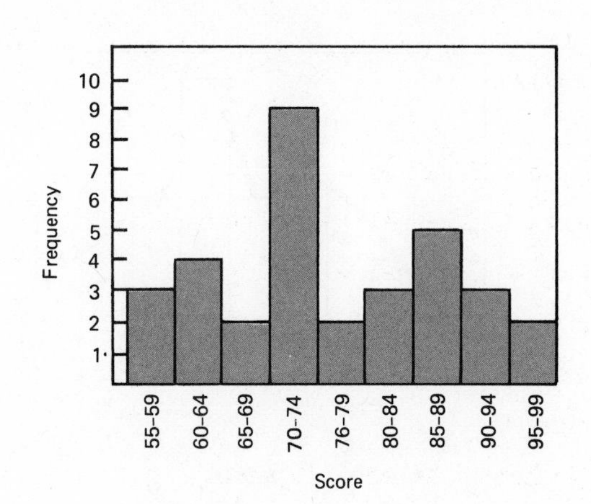

11.

Amount	Tally	Frequency
93	////	4
92	~~////~~ ///	8
91	//	2
90	//	2
89	/	1
88	~~////~~	5
87	///	3
86	~~////~~ //	7
85	////	4
84	////	4

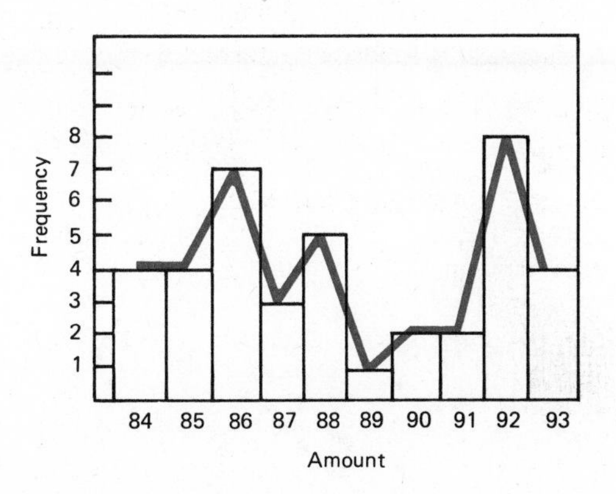

13. a.

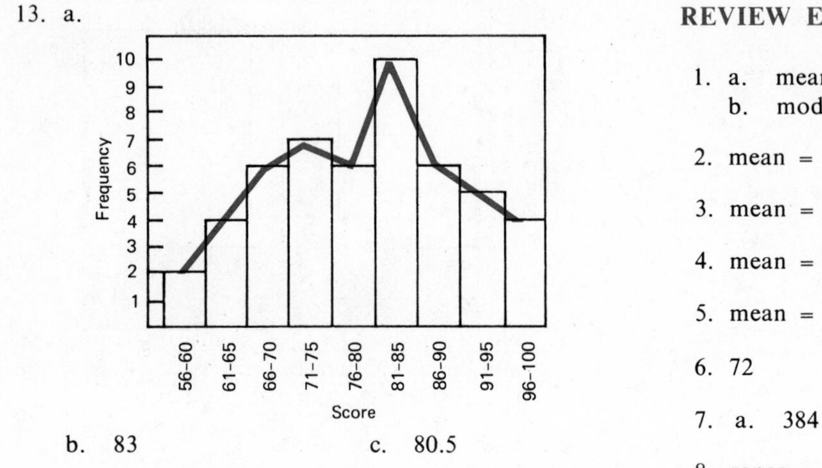

b. 83 c. 80.5

15. More men than women are involved in auto accidents.

17. Answers may vary.

JUST FOR FUN

VII + III = 10

SECTION 4.6

1. a. 68.2% b. 95.4% c. 99.7%

3. a. 50,000 b. 12,000 c. 84% d. 320

5. a. 795 b. 115 c. 6 d. 4090

7. a. 2.3% b. 15.9%
   c. 81.8% d. 97.6%

9. a. 4.5% b. 9.1% c. 74.9%
   d. 7.9% e. 3.6%

11. a. 38.4% b. 62.5% c. 2.3%
    d. 69.2% e. 28.5%

13. a. 0.6% b. 6.7%

15. a. 38.4% b. 61.6%

17. a. 0.50 or $\frac{50}{100}$ b. 0.308 c. 0.48

JUST FOR FUN

REVIEW EXERCISES FOR CHAPTER 4

1. a. mean, median, mode
   b. mode c. median d. mean

2. mean = 15.4; mode = 12; median = 16

3. mean = 6; mode = 5; median = 5

4. mean = 9.6; no mode; median = 9

5. mean = 55; no mode; median = 55

6. 72

7. a. 384 b. 76.8

8. range = 10; midrange = 65; mean = 65

9. range = 42; midrange = 79; mean = 79.7

10. $\sigma = \sqrt{5}$ or 2.2

11. $\sigma = \sqrt{24}$ or 4.9

12. 85 13. Julie

14. a.

Number	Tally	Frequency
6	卌 卌 /	11
5	卌 ////	9
4	卌 卌	10
3	////	4
2	卌 ///	8
1	卌 ///	8

b, c.

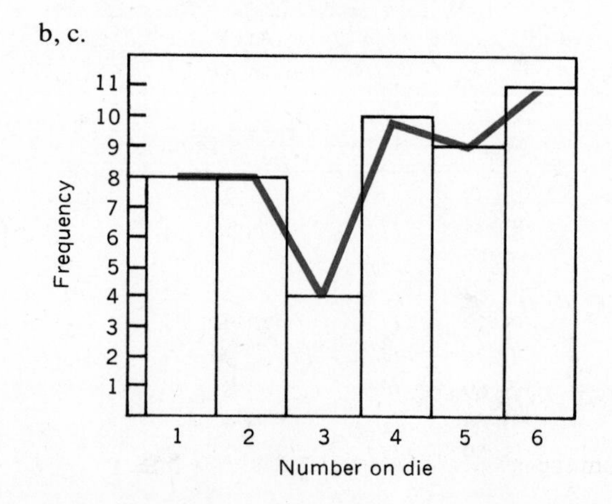

15. Individual income taxes: 137°
Social insurance receipts: 101°
Corporation income taxes: 47°
Borrowing: 47°
Excise taxes: 14°
Other: 14°

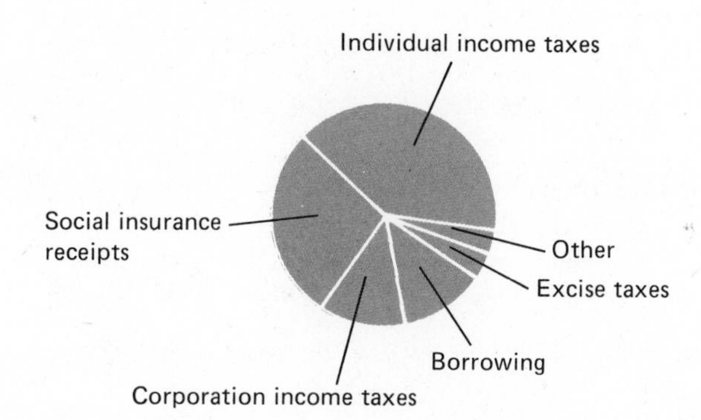

16. a.

Height	Tally	Frequency
75	//	2
74		0
73		0
72	//	2
71	//	2
70	~~////~~ ///	8
69	////	4
68	//	2
67	///	3
66	//	2
65	//	2
64	///	3

b, c.

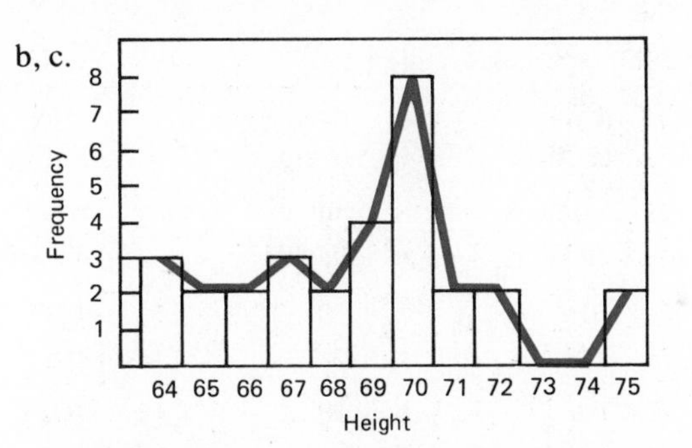

d. No, a normal curve has a normal distribution and the mean, median, and mode all have the same value at the center of distribution.

17. a. 15.9% b. 2.3% c. 15.9%
d. 81.8% e. 682

18. a. everybody but Steve
b. Steve c. no one d. Steve

19. a. 23.7 b. 22
c. 22 and 20 (bimodal)
d. 25 e. 26.5

20. a. 26 b. 26 c. 26
d. 14 e. 27
f. $\sqrt{20}$ or 4.5

21. a. third b. 75 c. 27
d. 22 e. bimodal, 22 and 20

JUST FOR FUN

8	13	12
15	11	7
10	9	14

CHAPTER 5

SECTION 5.1

JUST FOR FUN

From the initials *U.S.*, which became Ʉ or $.

SECTION 5.2

1. (*i*) To reduce the number of different sizes; (*ii*) to make conversion from one unit to another easier; (*iii*) to create a potential increase in export; (*iv*) because it is the international system of measurement.

3. cubit, span, palm, digit

5. a. deka b. deci c. kilo
 d. centi e. hecto f. milli

7. a. 3.937 b. 393.7 c. 0.3937
 d. 3937 e. 0.03937 f. 39,370

9. a. 10 b. 10 c. 10
 d. 0.001 e. 0.1 f. 10

11. a. 1 b. 10 c. 100
 d. 1 e. 1 f. 1

JUST FOR FUN

nautical mile = 6076 feet; statute mile = 5280 feet

SECTION 5.3

1. a. 26 mm b. 60 mm c. 35 mm
 d. 71 mm e. 46 mm f. 34 mm
 g. 96 mm

3. a. 1.8 b. 3.5 c. 300,000
 d. 3.5 e. 4 f. 40

5. 4200 7. 2814

9. yes, 50 mi/hr = 80 km/hr

11. 34,800,000 13. 4 cm^2 15. 2.56

17. a. 1 in^2 b. 1 m^2
 c. 1 m^2 d. 1 mi^2
 e. 1 ha^2 f. 1 square are

JUST FOR FUN

151,476

SECTION 5.4

1. a. 3000 b. 0.5 c. 1000
 d. 10 e. 200 f. 2

3. kilolitre, hectolitre, dekalitre, litre, decilitre, centilitre, millilitre

5. 0.2 m^3 7. 0.36 m^3

9. Clam Chowder

5 ml salt	120 ml water
60 ml butter	470 ml clams
480 ml milk	480 ml potatoes
60 ml onions	

JUST FOR FUN

3692

SECTION 5.5

1. a. 5000 b. 3000 c. 2000
 d. 5 e. 60 f. 10,000

3. a. 224 b. 9 c. 896
 d. 63 e. 9 f. 2.7

5. 0.5 g

7. 360 ℓ; 360 kg

9. a. 1 kg b. 2 oz c. 5 kg
 d. 3 t e. 2 kg f. 280 g

JUST FOR FUN

13,120

SECTION 5.6

1. a. 40 °C b. 10 °C c. 5 °C
 d. 35 °C e. 30 °C f. −5 °C

3. a. 68 °F b. 122 °F c. 185 °F
 d. 50 °F e. 32 °F f. 149 °F

5. a. −10 °C b. 59 °F
 c. 77 °F d. 30 °C

7. 104 °F 9. 3000 °C

11. −60 °C 13. 57 °C

JUST FOR FUN

247

REVIEW EXERCISES FOR CHAPTER 5

1. The metric system is a system of measurement with basic units of measure for length, area, volume, and weight in decimal relationship to each other. The basic unit is international and relates to units of length, volume, and weight. It is a well-planned logical system with uniformity, allowing for easier and more precise calculations.

2. a. kilo b. deka c. deci
   d. hecto e. centi f. milli

3. One ten-millionth of the distance from the north pole to the equator, along the meridian that passes through Dunkirk and Paris.

4. a. 100 b. 2 c. 4
   d. 350 e. 32,000 f. 1.5

5. a. 39,370 b. 0.03937 c. 3937
   d. 0.3937 e. 393.7 f. 3.937

6. 2803.2 7. 2532.8

8. a. 60 b. 90 c. 8
   d. 11 e. 6 f. 65

9. A litre is defined as the volume of a cube that is 1 decimetre long, 1 decimetre wide, and 1 decimetre high.

10. a. 2000 b. 30 c. 50
    d. 1 e. 200 f. 2000

11. 90,000 cm^3 or 0.09 m^3

12. 240 litres

13. The weight of one gram equals the weight of one millilitre of very cold water.

14. a. 4000 b. 8000 c. 3
    d. 200 e. 4 f. 1800

15. -88 °C 16. 136 °F

17. c 18. c 19. b

20. d 21. c 22. a

23. b 24. c

25. a. 1000 b. 1 000 000 c. 10
    d. 3800 e. 950 f. 1410

26. a. 10 b. 0.95 c. 106
    d. 56 e. 90 f. 35

27. Sour Cream Cookies

120 ml sour cream	224 g butter
5 ml vanilla	5 ml baking soda
240 ml brown sugar	600 ml flour
2 eggs	Bake at 177 °C

JUST FOR FUN

A	K	E	I	B	C	U	B	I	C	M	E	T	R	E
M	I	L	L	I	G	R	A	M	A	B	C	D	E	R
I	L	O	U	L	A	E	I	O	U	H	F	M	G	T
L	O	C	E	L	S	I	U	S	H	E	I	E	J	E
L	L	E	K	L	M	N	O	P	Q	C	R	T	R	M
I	I	N	J	S	T	U	S	C	O	T	T	R	V	A
M	T	T	O	N	N	E	W	X	E	A	Y	I	Z	K
E	R	I	E	A	G	R	A	M	M	R	M	C	C	E
T	E	M	B	C	D	E	O	A	F	E	G	H	I	D
R	J	E	K	L	M	L	R	D	S	H	T	A	M	E
E	O	T	P	Q	I	G	I	R	N	E	T	R	S	R
T	U	R	V	K	A	W	X	T	Y	A	T	A	E	A
Z	T	E	E	K	L	F	O	U	R	B	L	E	M	U
O	H	T	E	N	O	A	D	D	I	E	E	S	K	Q
T	A	D	C	U	B	I	C	M	E	T	R	E	I	S

CHAPTER 6

SECTION 6.2

1. a. 6 b. 1 c. 11
   d. 10 e. 9 f. 7

3. a. 9 b. 8 c. 7
   d. 6 e. 1 f. 8

5. a. 10 b. 10 c. 8
   d. 11 e. 9 f. 8

7. a. 12 b. 12 c. 9
   d. 4 e. 7 f. 9

9. a. associative property for addition
   b. closure for addition

c. identity element for addition
d. commutative property for addition
e. inverse element for multiplication
f. closure for multiplication

11. a. true b. false c. true
d. false e. false f. true

13. a. 7 b. 1 c. 5
d. 3 e. 3 f. 2

JUST FOR FUN

yes

SECTION 6.3

1. spring 3. summer 5. fall

7. winter 9. winter 11. spring

13. spring 15. fall

17. Monday 19. Wednesday 21. Monday

23. Friday 25. Thursday 27. Thursday

29. Saturday 31. Tuesday 33. Saturday

35. a. Thursday
b. Wednesday
c. Tuesday

37. Sunday × (Tuesday × Friday)
$\stackrel{?}{=}$ (Sunday × Tuesday) × Friday
Sunday × Wednesday
$\stackrel{?}{=}$ Tuesday × Friday
Wednesday = Wednesday

39. yes

JUST FOR FUN

It is practically impossible.

SECTION 6.4

1. a. 4 b. 0 c. 1
d. 2 e. 4 f. 1

3. a. 3 b. 4 c. 0
d. 2 e. 0 f. 2

5. a. 3 b. 2 c. 3
d. 4 e. 0 f. 3

7. a. 2 b. 3 c. 1
d. 4 e. 3 f. 3

9.

+	0	1	2	3	4	5	6
0	0	1	2	3	4	5	6
1	1	2	3	4	5	6	0
2	2	3	4	5	6	0	1
3	3	4	5	6	0	1	2
4	4	5	6	0	1	2	3
5	5	6	0	1	2	3	4
6	6	0	1	2	3	4	5

a. yes b. 0 c. yes
d. $(1 + 3) + 5 \stackrel{?}{=} 1 + (3 + 5)$; $4 + 5 \stackrel{?}{=} 1 + 1$; $2 = 2$
e. element: 0, 1, 2, 3, 4, 5, 6
inverse: 0, 6, 5, 4, 3, 2, 1
f. yes

11. a. true b. false c. true
d. true e. true f. false

13. a. 2 b. 4 c. 4
d. 6 e. 4 f. 6

15. 81

JUST FOR FUN

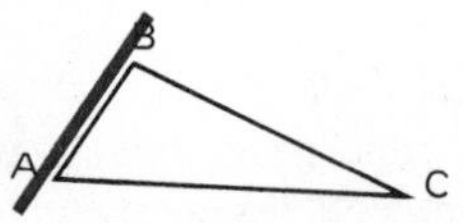

SECTION 6.5

1. Q 3. R 5. S

7. R 9. S

11. Yes; there are no new elements.

13. All of them: the inverse of P is P; the inverse of Q is S; the inverse of R is R; and the inverse of S is Q.

15. ! 17. ! 19. ?

21. ! 23. ?

25. No; there is a new element in the table, namely, !.

27. none 29. no 31. a

33. d 35. b 37. e

39. e 41. e 43. a

45. Yes; the symmetric table indicates commutativity.

47. a. yes b. none c. none
 d. no

49. a.

*	a	b	c
a	b	c	a
b	c	a	b
c	a	b	c

 b. yes c. c
 d. All; the inverse of a is b; the inverse of b is a; and the inverse of c is c.
 e. yes

JUST FOR FUN

6 dozen dozen

SECTION 6.6

1. a, d

3. An axiomatic system consists of four main parts:
 1. undefined terms
 2. defined terms (definitions)
 3. axioms
 4. theorems

5.

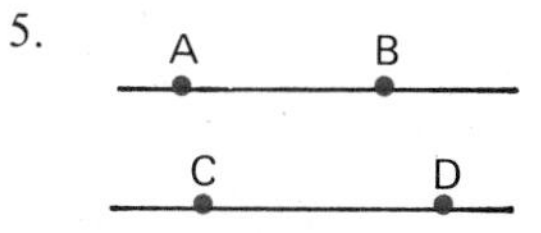

Axiom II tells us there is a line, say line AB. Axiom I tells us there are exactly 2 points on the line, A and B. Axiom IV tells us there is another line (say line CD) that has no points in common with line AB. Axiom I tells us there are exactly 2 points on the line, C and D. Axiom III tells us that the lines are distinct since for each pair of points there is one and only one line containing them. We have two pairs of points, or at least four points.

7.

	Statement	Reason
a.	1. $\overline{AC} = \overline{BC}, \overline{DC} = \overline{EC}$	1. Given
	2. $\overline{AC} + \overline{CE} = \overline{BC} + \overline{CD}$	2. Axiom I
	3. $\overline{AE} = \overline{BD}$	3. Axiom V
b.	1. $m \angle EBC = m \angle DBA$	1. Given
	2. $m \angle EBC - m \angle EBD = m \angle DBA - m \angle EBD$	2. Axiom II
	3. $m \angle 1 = m \angle 2$	3. Axiom V
c.	1. $m \angle ABC = m \angle EFG$, $m \angle 1 = m \angle 3$	1. Given
	2. $m \angle ABC - m \angle 1 = m \angle EFG - m \angle 3$	2. Axiom II
	3. $m \angle 2 = m \angle 4$	3. Axiom V
d.	1. $m \angle 1 = m \angle 2$, $m \angle 1 = m \angle 3$, $m \angle 3 = m \angle 4$	1. Given
	2. $m \angle 2 = m \angle 3$	2. Axiom V
	3. $m \angle 2 = m \angle 4$	3. Axiom V

JUST FOR FUN

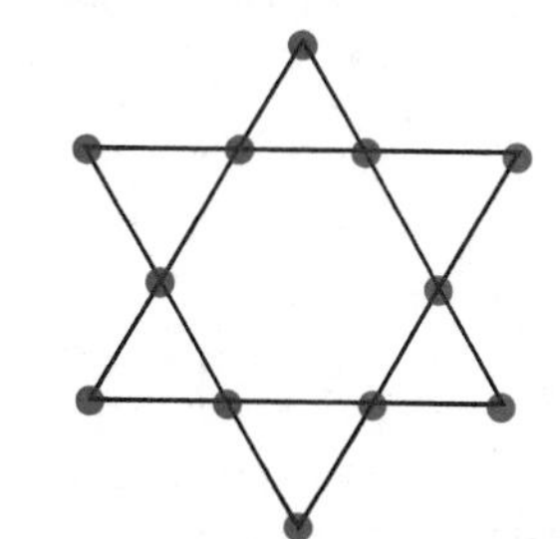

REVIEW EXERCISES FOR CHAPTER 6

1. a. 2 b. 1 c. 3
 d. 11 e. 8 f. 5

2. a. 10 b. 8 c. 8
 d. 2 e. 8 f. 11

3. a. 12 b. 8 c. 6
 d. 3 e. 6 f. 12

4. a. 2 b. 12 c. 6
 d. 2 e. 8 f. 4

5. A *mathematical system* is a set of elements together with one or more operations (rules) for combining those elements.

6. a. true b. false c. true d. false

7. a. true b. false c. true d. false

8. a. true b. true

9. a. false b. false

10. a. winter b. summer c. fall
d. fall e. fall f. fall

11. a. spring b. summer c. spring
d. fall e. winter f. summer

12. a. 2 b. 1 c. 4
d. 3 e. 2 f. 3

13. a. 3 b. 5 c. 3
d. 5 e. 1 f. 1

14. 57

15. The odometer recycles after 100,000 miles, modulo 100,000.

16. a. \$ b. ? c. π d. π
e. & f. \$ g. ? h. π
i. No; there is a new element in the table, namely π.
j. ¢ k. \$, ¢ l. no

17. An axiomatic system consists of four main parts:
   1. undefined terms
   2. defined terms (definitions)
   3. axioms
   4. theorems

18. We must have three squirrels (axiom I). They must be in a tree (axiom II). If they are all in the same tree, then there must be another squirrel in another tree (axioms IV and II). Therefore there are at least four squirrels.

If only two of the three squirrels are in a given tree (axioms II and III), then the third squirrel is in another tree (axiom II) together with a fourth squirrel (axiom III).

19. $3x + 4 + 2x + 8 + 5x - 4 = 48$
$10x + 8 = 48$
$10x = 40$
$x = 4$
If $x = 4$, $3x + 4 = 12 + 4 = 16$
$2x + 8 = 8 + 8 = 16$
$5x - 4 = 20 - 4 = 16$
and all three sides are equal.

JUST FOR FUN

The error occurs when we divide by $(a - b)$. If $a = b$, then $(a - b) = 0$, and we cannot divide by zero.

CHAPTER 7

SECTION 7.2

1. a. 𓎆𓏺𓏺𓏺𓏺𓏺𓏺𓏺𓏺𓏺 b. 𓎆𓎆𓏺𓏺𓏺 c. 𓎆𓎆𓎆𓏺𓏺𓏺𓏺
d. 𓍢𓏺𓏺 e. 𓍢𓍢𓏺
f. 𓆼𓍢𓎆𓎆𓎆𓏺𓏺

3. a. 22 b. 1213 c. 1101
d. 12,212 e. 1222 f. 2212

5. a. 𓎆𓎆𓎆𓎆𓎆𓏺𓏺𓏺𓏺
b. 𓍢𓍢𓍢𓎆𓎆𓎆𓎆𓎆
c. 𓍢𓍢𓍢𓎆𓎆𓎆𓎆𓎆𓎆𓎆𓏺𓏺𓏺𓏺
d. 𓆼𓆼𓆼

7. a. 𓏺𓏺𓏺𓏺𓏺𓏺 b. 𓎆𓎆𓎆𓎆𓎆𓎆𓎆𓎆𓏺𓏺𓏺𓏺𓏺𓏺𓏺𓏺𓏺𓏺
c. 𓍢𓎆𓎆𓏺
d. 𓆼𓆼𓎆𓎆𓎆𓎆𓎆𓎆𓎆𓏺𓏺𓏺𓏺𓏺𓏺𓏺

JUST FOR FUN

1,000,000 seconds, or approximately $11\frac{1}{2}$ days.

SECTION 7.3

1. a. ΔΓIII b. ΔΔIII
   c. ΔΔΔIIII d. ΔΔΔΔIIII
   e. H𐅄ΔΔΔΓII f. 𐅅𐅄ΔΔΔΔΓIII

3. a. 7 b. 11 c. 17
   d. 127 e. 156 f. 566

5. a. 十八 b. 二十三 c. 三十四
   d. 四十六 e. 二百三十四 f. 四百七十七

7. a. 20 b. 705 c. 13 d. 49

JUST FOR FUN

no

SECTION 7.4

1. a. b. c.
   d.
   e.
   f.

3. a. 13 b. 22 c. 31 d. 83

5. a. $(2 \times 10^2) + (4 \times 10^1) + (3 \times 10^0)$
   b. $(3 \times 10^2) + (7 \times 10^1) + (8 \times 10^0)$
   c. $(1 \times 10^3) + (2 \times 10^2) + (3 \times 10^1) + (4 \times 10^0)$
   d. $(2 \times 10^3) + (5 \times 10^1) + (1 \times 10^0)$
   e. $(1 \times 10^4) + (4 \times 10^2) + (1 \times 10^0)$

7. a. 240 b. 2311 c. 1776
   d. 4213 e. 204 f. 40,301

JUST FOR FUN

Turn the page upside down.

SECTION 7.5

1. a. 7 years 4 months b. 2 years 5 months
   c. 25 feet 9 inches d. 3 feet 11 inches
   e. 7 gross 3 dozen 2 units f. 2 gross 9 dozen 8 units

3. a. 17 b. 7 c. 9
   d. 38 e. 53 f. 266

5. a. 11_{five} b. 34_{five} c. 123_{five}
   d. 3_{five} e. 441_{five} f. 3442_{five}

7. a. 55 b. 32 c. 69
   d. 172 e. 487 f. 1729

9. a. 36_{twelve} b. 45_{twelve} c. 50_{twelve}
   d. $E5_{\text{twelve}}$ e. 176_{twelve} f. 610_{twelve}

11. a. 35 b. 131 c. 1475 d. 1618

13. 1915

15. a. 210_5 b. 17_{12}

JUST FOR FUN

1 fathom = 6 feet; 1 hand = 4 inches; 3 barleycorns = 1 inch

SECTION 7.6

1. a. 41_5 b. 41_5 c. 110_5
   d. 302_5 e. 1003_5 f. 1010_5

3. a. 4_5 b. 4_5 c. 3_5
   d. 40_5 e. 44_5 f. 234_5

5. a. 1243_5 b. 1414_5 c. 3333_5
   d. 10401_5 e. 14112_5 f. 32343_5

7. a. 3_5 b. 4_5 c. 4_5
   d. $13_5; r = 1_5$ e. 34_5 f. 12_5

JUST FOR FUN

for, four, fore

SECTION 7.7

1. a. 2 b. 3 c. 5
   d. 7 e. 11 f. 27

3. a. 101_2 b. 111_2 c. 1000_2
   d. 1011_2 e. 10001_2 f. 10111_2

5. a. 101_2 b. 110_2 c. 100_2
   d. 1000_2 e. 1001_2 f. 1100_2

7. a. 1_2 b. 1_2 c. 11_2
   d. 10_2 e. 10_2 f. 110_2

9. a. 1001_2 b. 110_2 c. 1111_2
   d. 10010_2 e. 100011_2 f. 11001_2

11. 441_5 13. 63_7 15. 11111_2

17. a. true b. true c. true d. false

JUST FOR FUN

Let the Yankee fans be Y_1, Y_2, Y_3, and let the Dodger fans be D_1, D_2, D_3.

1. D_1, D_2 go up, leaving Y_1, Y_2, Y_3, D_3.
2. Elevator returns to 1st floor empty.
3. Y_1, Y_2 go up, leaving Y_3, D_3 and joining D_1, D_2.
4. Elevator returns empty.
5. Y_3 and D_3 go up, joining Y_1, Y_2, D_1, D_2.

REVIEW EXERCISES FOR CHAPTER 7

1. a. ∩∩∩|| b. 𓍢𓍢∩| c. 𓆼𓍢∩|
2. a. 1233 b. 12,022
3. a. ∩∩∩∩∩||||| b. 𓍢𓍢∩∩∩∩∩
   c. ∩∩∩∩∩∩∩∩|||||||||
   d. 𓍢𓍢𓍢𓍢𓍢𓍢𓍢𓍢∩∩∩∩∩∩∩∩
4. a. *Simple grouping:* The position of a symbol does not affect the number represented, and a new symbol is used to indicate a certain number or group of things.
   b. *Multiplicative grouping:* Symbols are used for numbers in a basic group, together with a second symbol or notation to represent multiples of the basic group.
   c. *Place-value:* The position of a symbol matters.
5. a. Egyptian
   b. Greek, Chinese-Japanese
   c. Babylonian, Hindu-Arabic
6. a. $(3 \times 10^2) + (4 \times 10^1) + (5 \times 10^0)$
   b. $(3 \times 5^2) + (4 \times 5^1) + (2 \times 5^0)$
   c. $(1 \times 2^4) + (0 \times 2^3) + (1 \times 2^2) + (1 \times 2^1) + (1 \times 2^0)$
7. a. 55 b. 19 c. 71
   d. 30 e. 21 f. 419
8. a. 140_5 b. 201_5 c. 28_{12}
   d. 100001_2 e. 111111_2 f. $2T_{12}$
9. Base twelve is used when buying many items; for example, we use dozen (12^1), gross (12^2) and great gross (12^3). Modern computers work in base two, eight, or sixteen.

10. a. 1040_5 b. 44_5
    c. 20004_5 d. $34_5, r = 1_5$

11. a. 100_5 b. 1243_5
    c. 4_5 d. 1102_5

12. a. 100_2 b. 110_2
    c. 1110_2 d. 110111_2

13. a. 1011_2 b. 1010_2
    c. 11010_2 d. 11_2

14. a. 222_6 b. 211_3
    c. 43_5 d. 111111_2

JUST FOR FUN

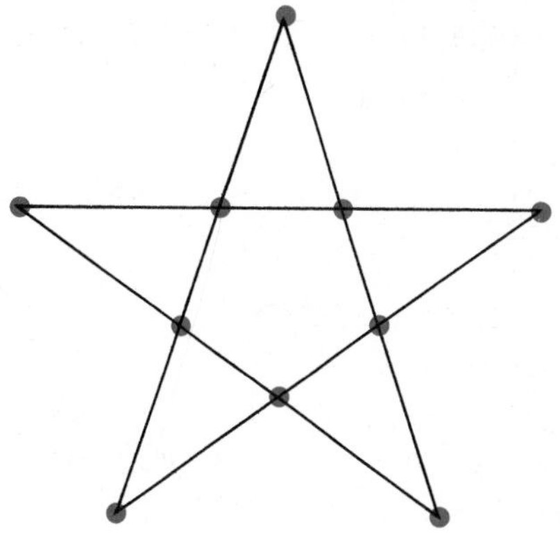

CHAPTER 8

SECTION 8.2

1. a. cardinal b. ordinal
   c. identification d. identification
   e. ordinal f. cardinal

3. a. prime b. prime c. composite
   d. composite e. composite f. composite

5. a. $2 \times 3^2 \times 13$
   b. 3×71 c. $3^4 \times 11$ d. 11^3
   e. $2 \times 11 \times 41$ f. $17 \times 19 \times 23$

JUST FOR FUN

28; 496; 8128

SECTION 8.3

1. a. 2 b. 14 c. 3
   d. 26 e. 3 f. 12

3. a. $\frac{5}{6}$ b. $\frac{7}{9}$ c. $\frac{3}{5}$
   d. $\frac{2}{3}$ e. $\frac{147}{152}$ f. $\frac{1}{2}$

5. a. 56 b. 28 c. 120
   d. 156 e. 990 f. 9879

7. a. $\frac{35}{36}$ b. $\frac{23}{36}$ c. $\frac{11}{12}$
   d. $\frac{19}{24}$ e. $\frac{6}{55}$ f. $\frac{43}{60}$

JUST FOR FUN

$4 = 2 + 2; 6 = 3 + 3; 8 = 5 + 3;$
$10 = 5 + 5; 12 = 7 + 5; 14 = 7 + 7;$
$16 = 13 + 3; 18 = 13 + 5; 20 = 13 + 7;$
$22 = 11 + 11; 24 = 19 + 5; 26 = 19 + 7;$
$28 = 23 + 5; 30 = 23 + 7$

SECTION 8.4

1. a. -1 b. -3 c. -3
   d. 2 e. 1 f. 3

3. a. -5 b. 0 c. 25
   d. 0 e. -35 f. 9

5. a. $>$ b. $<$ c. $>$
   d. $=$ e. $=$ f. $<$

7. a. even; $10 = 2 \cdot 5$ b. odd; $15 = 2 \cdot 7 + 1$
   c. odd; $21 = 2 \cdot 10 + 1$
   d. odd; $-5 = 2(-3) + 1$

9. $11 = 3 + 3 + 5$; $13 = 3 + 5 + 5$; $15 = 3 + 7 + 5$; $17 = 11 + 3 + 3$; $19 = 11 + 3 + 5$; $21 = 3 + 7 + 11$; $23 = 3 + 7 + 13$; $25 = 5 + 7 + 13$; $27 = 7 + 7 + 13$; $29 = 11 + 13 + 5$

11. Aristotle

13. $(2k + 1) + 2n = (2k + 2n) + 1 = 2(k + n) + 1$, which is odd.

JUST FOR FUN

It's a fake! No authentic coins would be dated B.C.

SECTION 8.5

1. a. $\frac{3}{8}$ b. $\frac{1}{9}$ c. $\frac{27}{43}$ d. $\frac{27}{224}$

3. a. $\frac{33}{35}$ b. $\frac{11}{12}$ c. $\frac{17}{72}$ d. $\frac{5}{33}$ e. $\frac{1}{80}$ f. $\frac{5}{9}$

5. a. $\frac{8}{35}$ b. $\frac{12}{55}$ c. $\frac{32}{91}$ d. $\frac{14}{5}$ e. $\frac{27}{44}$ f. $\frac{14}{13}$

7. a. false b. true c. false d. false e. false f. true

9. a. false b. true c. false d. true e. false

JUST FOR FUN

$\frac{1}{2}$

SECTION 8.6

1. a. 0.375 b. 0.3125 c. $0.\overline{6}$ d. $0.\overline{21}$ e. $0.\overline{09}$ f. $0.\overline{405}$

3. a. $\frac{9}{20}$ b. $\frac{7}{200}$ c. $\frac{2}{3}$ d. $\frac{4}{33}$ e. $\frac{134}{999}$ f. $\frac{719}{330}$

5. a. $\frac{5}{12}$ b. $\frac{7}{24}$ c. $\frac{9}{40}$ d. $\frac{37}{48}$ e. $\frac{69}{88}$ f. $\frac{277}{352}$

JUST FOR FUN

deny

SECTION 8.7

1. a. rational b. rational c. rational d. irrational e. irrational f. rational

3. a. terminating decimal
   b. repeating decimal
   c. nonterminating nonrepeating decimal
   d. terminating decimal
   e. repeating decimal
   f. nonterminating nonrepeating decimal

5. a. true b. true c. true d. false e. false

7. a. false b. false c. true d. false e. false f. true

JUST FOR FUN

We use $\frac{22}{7}$ only for sake of convenience. It is a close approximation of π.

REVIEW EXERCISES FOR CHAPTER 8

1. a. composite b. prime c. prime d. prime e. composite f. composite

2. a. $2 \times 3 \times 13$ b. 3×37 c. $5^2 \times 19$ d. 3×7^2 e. $3 \times 7 \times 43$ f. 11×101

3. a. 6 b. 1 c. 24 d. 6 e. 3 f. 18

4. a. 240 b. 2310 c. 144 d. 990 e. 8547 f. 11,628

5. a. -4 b. -1 c. -1 d. -4 e. 16 f. -13

6. a. 6 b. -16 c. -24 d. -9 e. -8 f. 40

7. a. $\frac{11}{15}$ b. $\frac{55}{63}$ c. $\frac{37}{143}$
   d. $\frac{5}{21}$ e. $\frac{1}{10}$ f. $\frac{19}{40}$

8. a. $\frac{2}{15}$ b. $\frac{4}{21}$ c. $\frac{6}{55}$
   d. $\frac{7}{6}$ e. $\frac{6}{5}$ f. $\frac{10}{11}$

9. a. 0.875 b. 0.4375 c. $0.\overline{18}$
   d. $0.\overline{351}$ e. $0.\overline{384615}$ f. $0.\overline{428571}$

10. a. $\frac{3}{4}$ b. $\frac{213}{1000}$ c. $\frac{157}{50}$
    d. $\frac{46}{99}$ e. $\frac{247}{99}$ f. $\frac{1373}{333}$

11. a. rational b. rational c. irrational
    d. irrational e. irrational f. rational

12. a. false b. false c. true
    d. true e. true f. true

13.

	a.	b.	c.	d.	e.	f.
I:	yes	no	no	no	no	no
II:	yes	no	yes	no	no	no
III:	yes	yes	yes	no	no	no
IV:	yes	yes	yes	yes	yes	no
V:	no	no	no	no	no	yes
VI:	yes	yes	yes	yes	yes	yes

	g.	h.	i.
I:	yes	no	no
II:	yes	no	no
III:	yes	no	no
IV:	yes	no	yes
V:	no	yes	no
VI:	yes	yes	yes

JUST FOR FUN

The number of letters in the word equals the digit.

CHAPTER 9

SECTION 9.2

1. {3} 3. {−1} 5. ∅
7. {1, 2, 3, 4} 9. $\{x \mid x > 1\}$ 11. {1, 2}
13. {0, 1, 2, 3, 4} 15. {−1, 0, 1, 2, 3}

17.

19.

21.

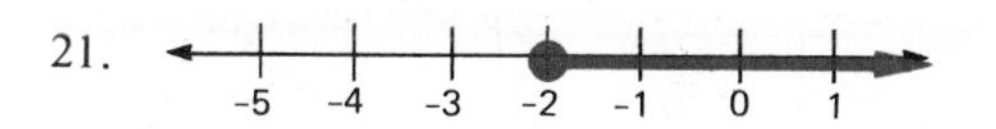

23.

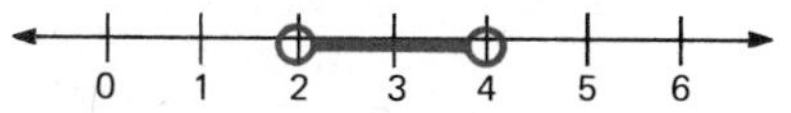

25.

27.

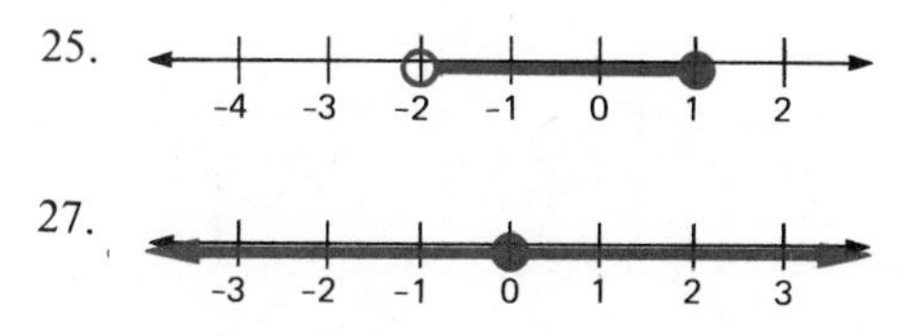

JUST FOR FUN

Write down the names of the numbers, that is, one, two, three, four, and so on.

SECTION 9.3

1. $5x$ 3. $3x$ 5. $2y$
7. $8x$ 9. $14y$ 11. $5x$
13. $2x$ 15. $5x$ 17. $5y + 2$
19. $10y + 4$ 21. $z + 2$ 23. $3y + 1$
25. $12x + 2$ 27. $3x - 6$ 29. $9z + 3$
31. 1 33. 0

JUST FOR FUN

37

SECTION 9.4

1. $x = 2$
3. $x = 8$
5. $y = 6$
7. $y = 6$
9. $x = 5$
11. $y = 4$
13. $x = 1$
15. $x = 3$
17. $y = 3$
19. $x = 10$
21. $x = 56$
23. $x = 2$
25. $z = 4$
27. $x = 4$
29. $x = 2$

JUST FOR FUN

Cut the pie in half. Place one half on top of the other, and cut these in half, resulting in four quarters. Now place two of the quarter pieces on top of the other two, and make the third cut.

SECTION 9.5

1. 7
3. 7
5. 7, 8
7. Lewis is 30, Julia is 23
9. 11
11. $l = 14$ ft, $w = 11$ ft
13. $l = 34$ m, $w = 16$ m
15. 19, 21
17. 30, 31, 32
19. 5 quarters, 10 dimes
21. 21 dimes, 7 quarters, 32 nickels
23. 10 quarters, 10 dimes, 15 nickels
25. 50°, 50°, 80°

JUST FOR FUN

A quarter and a dime; the quarter is the coin that is not a dime.

SECTION 9.6

Note: Answers may vary. The following are some possible solutions.

1. (5, 0)(0, 5)(3, 2)
3. (3, 0)(0, −3)(4, 1)
5. (0, −6)(−3, 0)(2, −10)
7. (0, 10)(2, 16)(−4, −2)
9. (0, −4)(4, 2)(2, −1)
11. (0, −3)(−5, 0)(5, −6)
13. (0, 0)(1, 1)(3, 3)
15. (0, 0)(2, −1)(4, −2)
17. (2, 5)(5, 7)(8, 9)
19. (1, −3)(4, −5)(7, −7)
21. (0, 3)(4, −2)(8, −7)

JUST FOR FUN

Tierce, hogshead, pipe, butt, and tun are names of casks (barrels). They were originally used in measuring amounts of beer and ale.

SECTION 9.7

1.

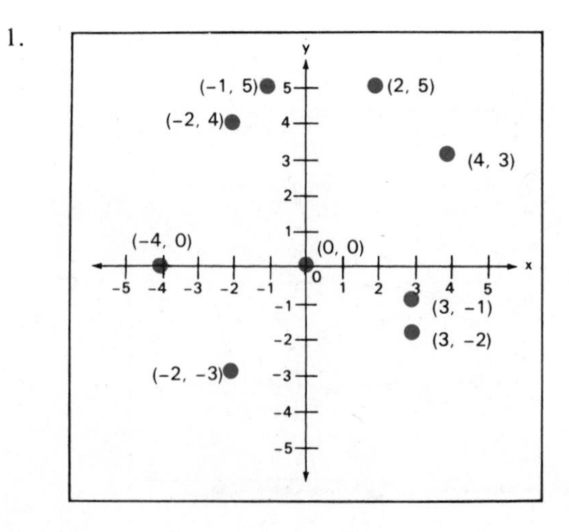

3.

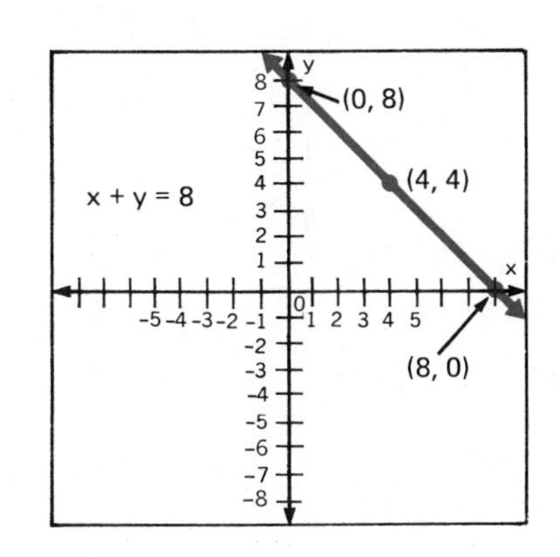
x + y = 8
(0, 8)
(4, 4)
(8, 0)
x
y

5.

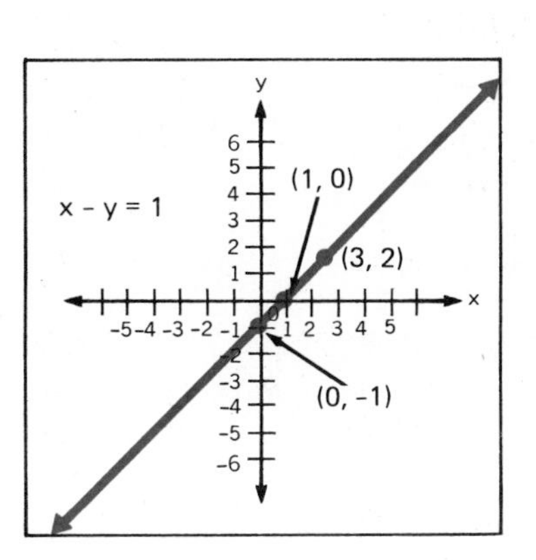
x - y = 1
(1, 0)
(3, 2)
(0, -1)
x
y

7.

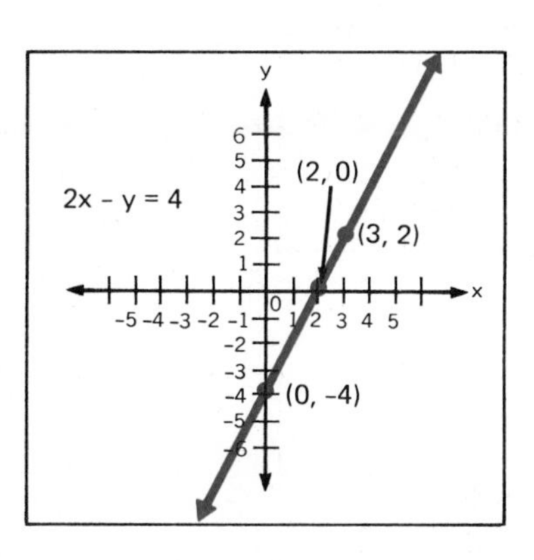
2x - y = 4
(2, 0)
(3, 2)
(0, -4)
x
y

9.

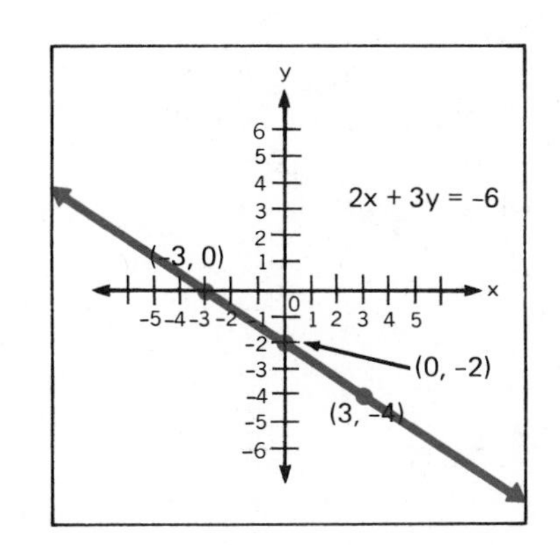
2x + 3y = -6
(-3, 0)
(0, -2)
(3, -4)
x
y

11.

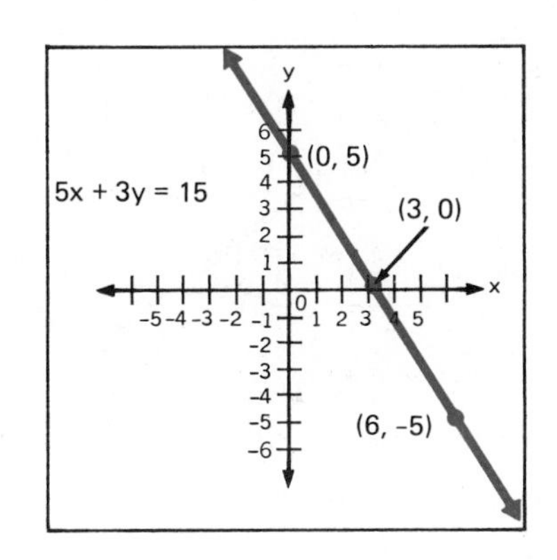
5x + 3y = 15
(0, 5)
(3, 0)
(6, -5)
x
y

13.

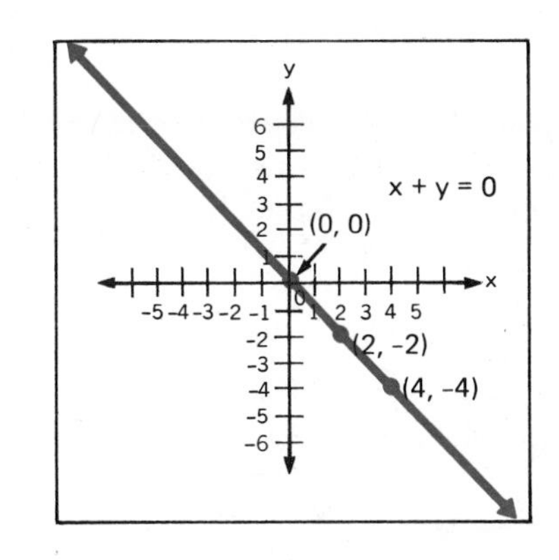
x + y = 0
(0, 0)
(2, -2)
(4, -4)
x
y

15.

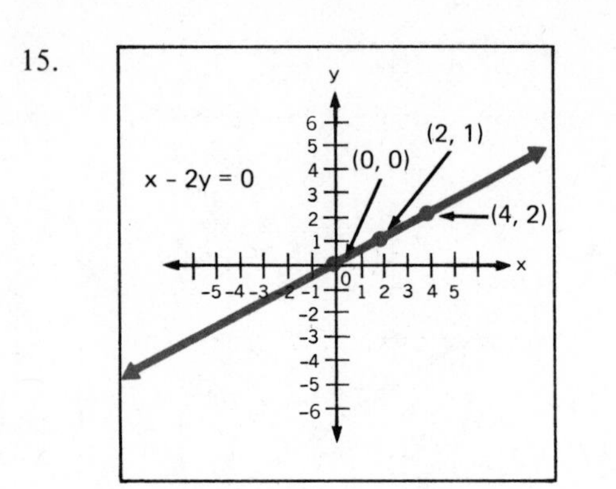

21.

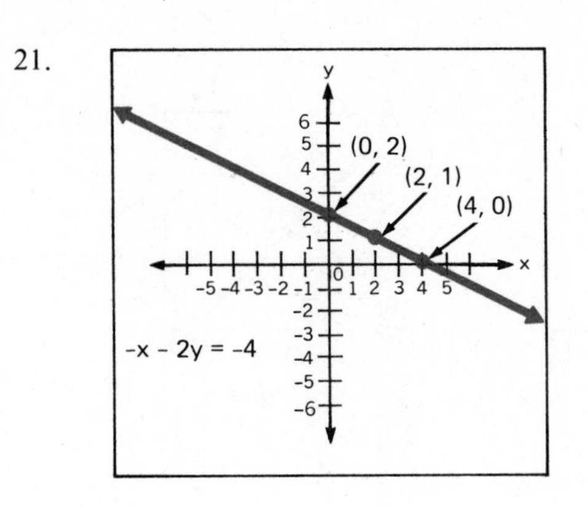

17.

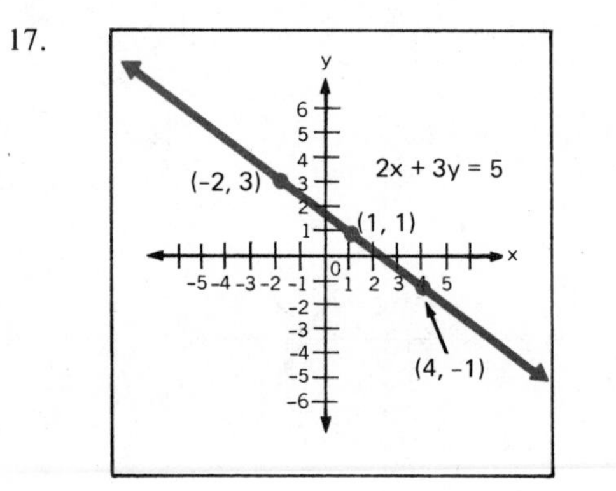

JUST FOR FUN

12

SECTION 9.8

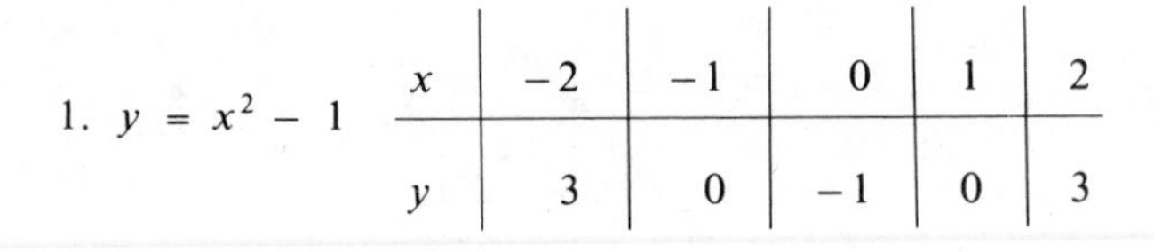

1. $y = x^2 - 1$

x	-2	-1	0	1	2
y	3	0	-1	0	3

19.

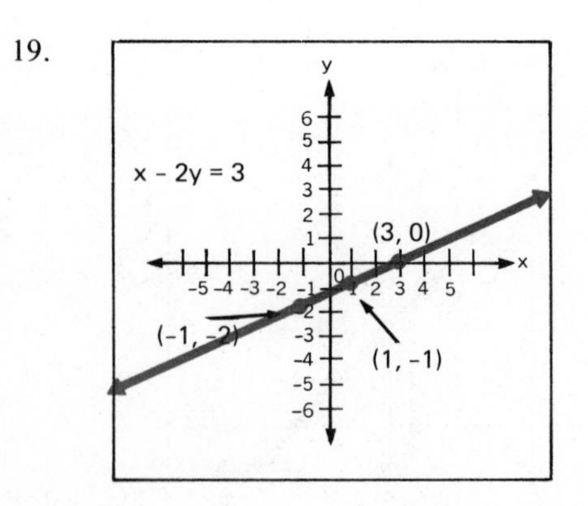

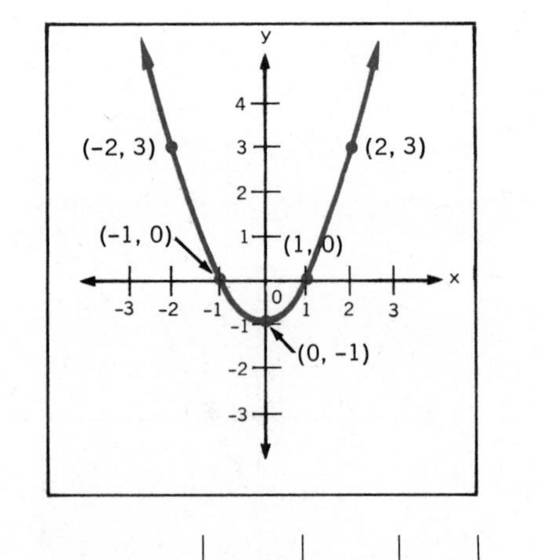

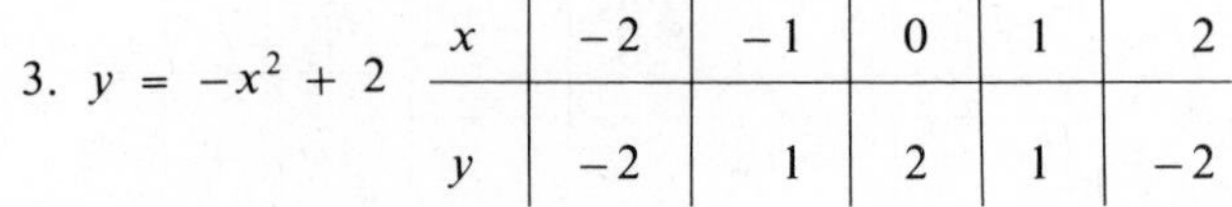

3. $y = -x^2 + 2$

x	-2	-1	0	1	2
y	-2	1	2	1	-2

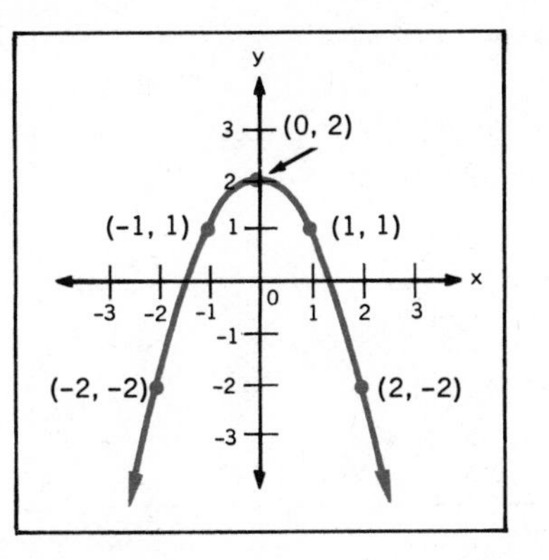

5. $y = -x^2 + 6x$

x	0	1	2	3	4	5	6
y	0	5	8	9	8	5	0

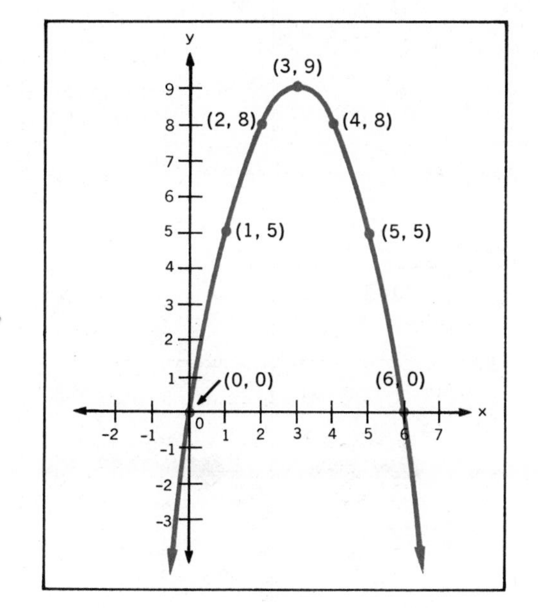

7. $y = x^2 + 2x + 1$

x	−4	−3	−2	−1	0	1	2
y	9	4	1	0	1	4	9

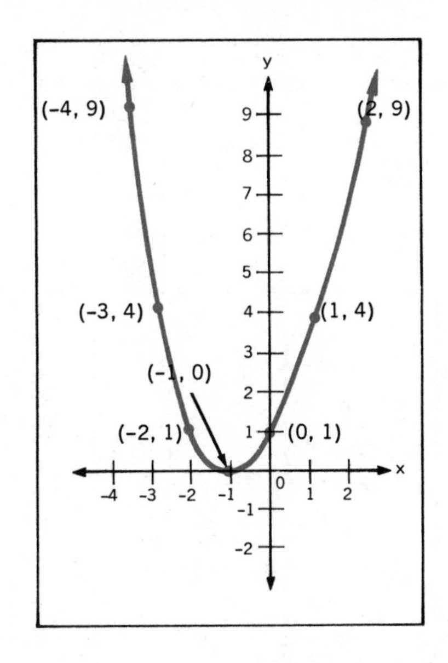

9. $y = x^2 - 4x + 3$

x	−1	0	1	2	3	4	5
y	8	3	0	−1	0	3	8

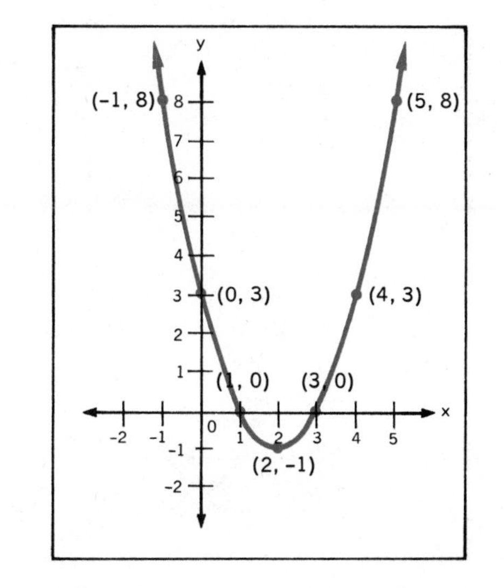

11. $y = -x^2 + 6x - 9$

x	1	2	3	4	5
y	-4	-1	0	-1	-4

13. $y = x^2 - 6x + 5$

x	0	1	2	3	4	5	6
y	5	0	-3	-4	-3	0	5

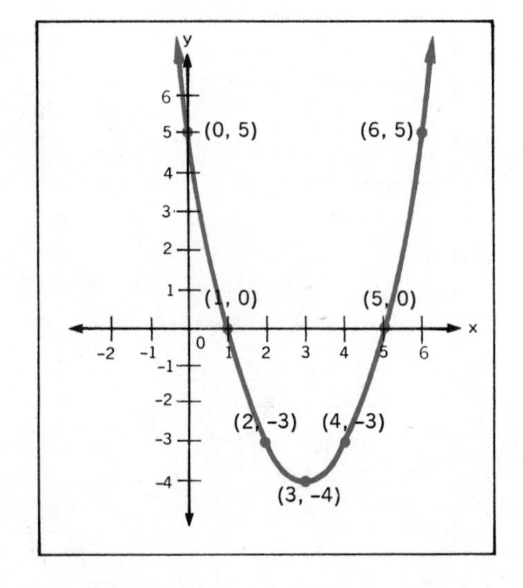

JUST FOR FUN

The box full of \$10 gold pieces; the value of gold is determined by weight, not denomination.

SECTION 9.9

1. $x + y > 5$

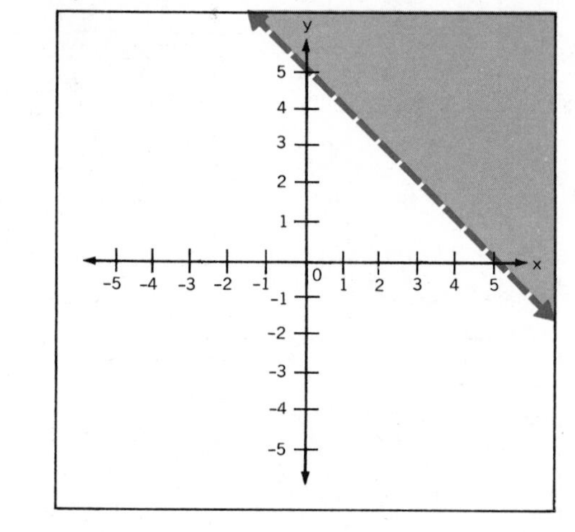

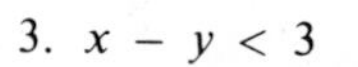

3. $x - y < 3$

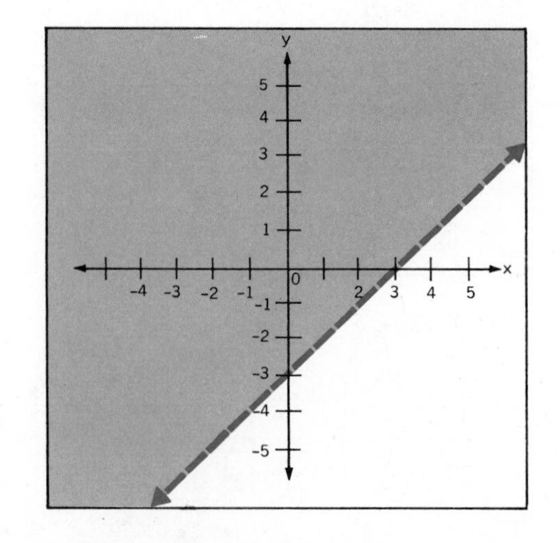

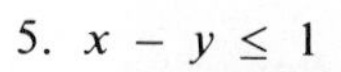
5. $x - y \leq 1$

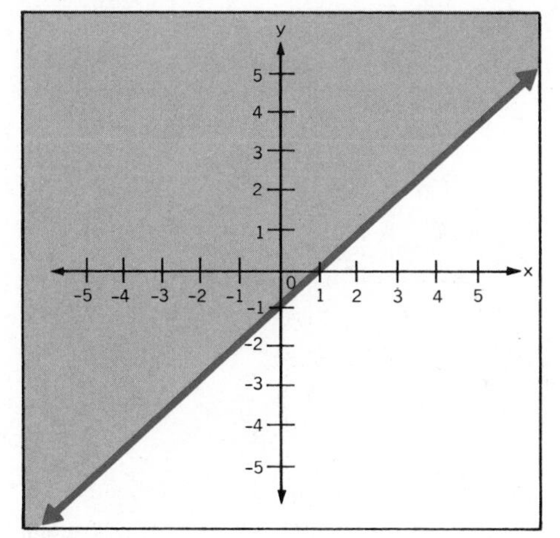

9. $y - x < 0$

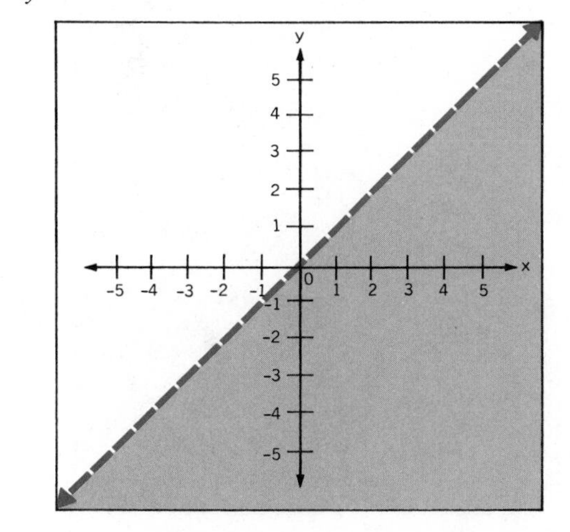

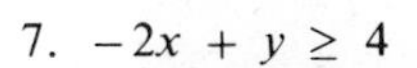
7. $-2x + y \geq 4$

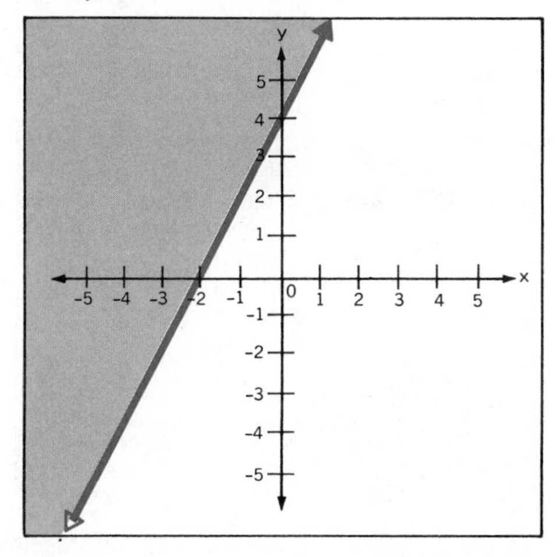

11. $-2x - 2y > 0$

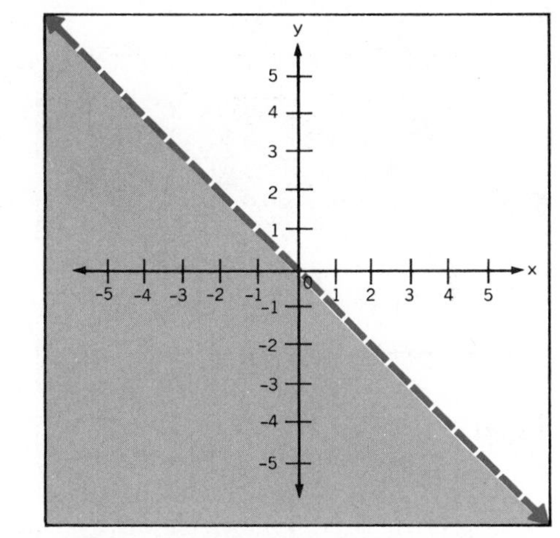

13. $-3x + 5y \leq 15$

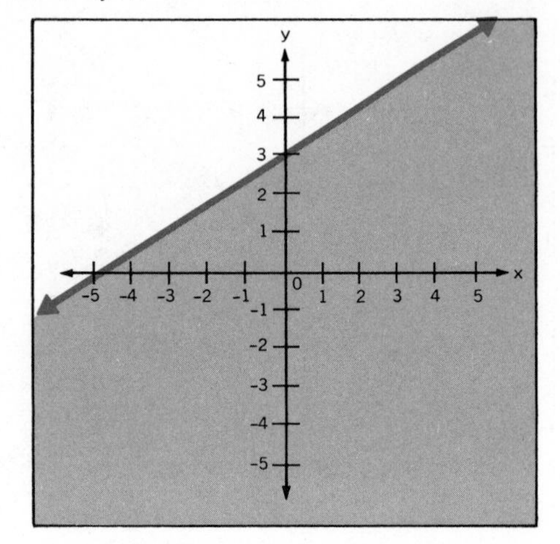

15. $3x + 5y > 15$

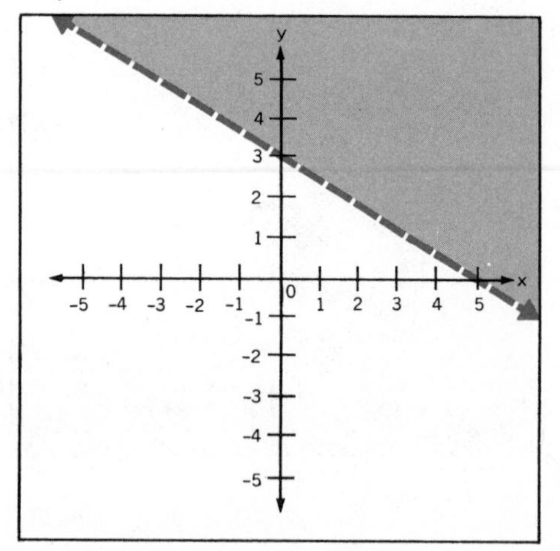

17. $-3x - 2y < -6$

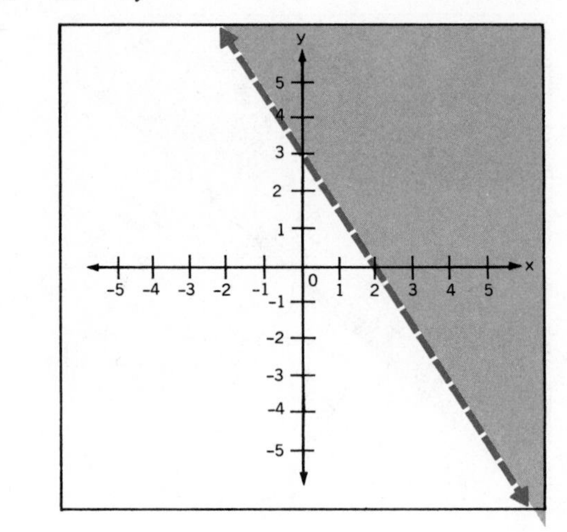

19. $x + y \geq 2$ and
$x - y < 2$

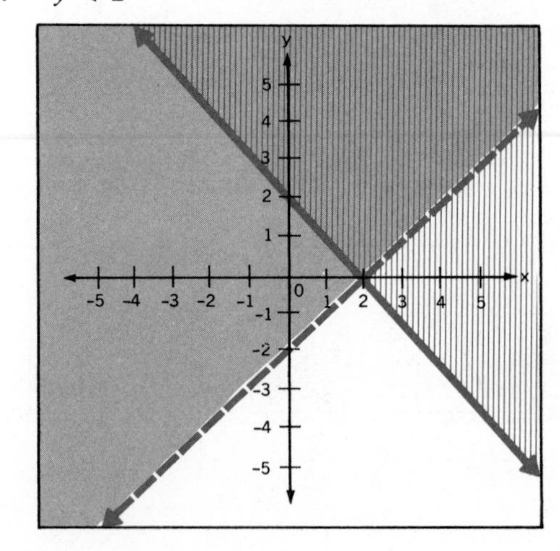

21. $y \geq x$ and $y \geq -x$

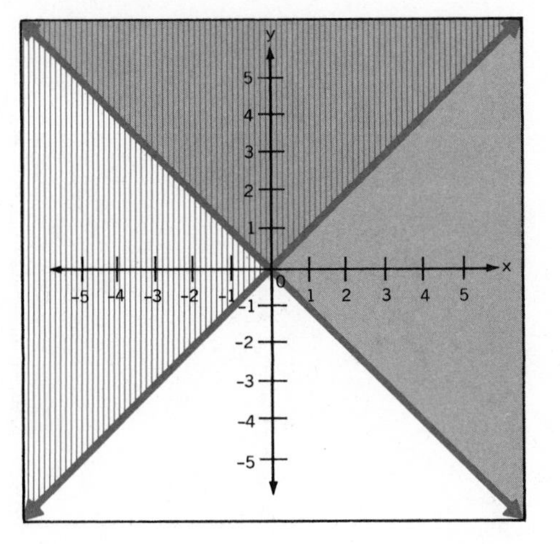

JUST FOR FUN

None; peacocks do not lay eggs.

SECTION 9.10

1. maximum value of P is 13 at (3, 2)
3. maximum value of P is 26 at (5, 2)
5. 3 bikes and 2 wagons; maximum profit is $56
7. 60 pheasants and 40 partridges; maximum profit is $740
9. 100 bushels of oysters and 500 bushels of clams; maximum profit is $5800

JUST FOR FUN

Two of a set of triplets.

REVIEW EXERCISES FOR CHAPTER 9

1. a. {2} b. {−6}
   c. {6} d. {1, 2, 3}
   e. {0, 1, 2, 3, 4, 5} f. {−2, −1, 0, 1, 2}
   g. {1, 2, 3}

2. a. b. c. d. e. f.

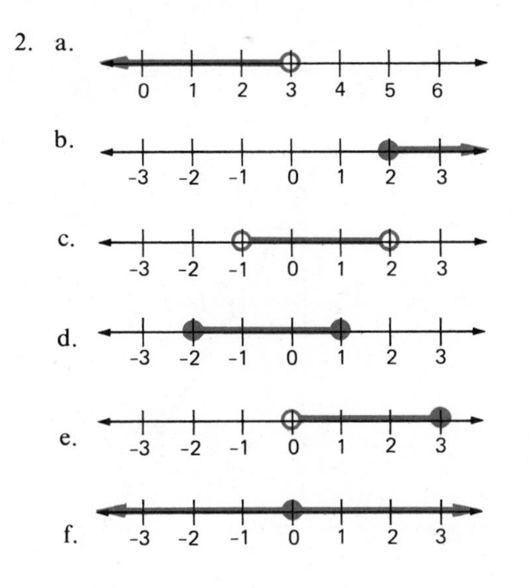

3. a. $x = 6$ b. $y = 2$ c. $z = 2$
   d. $y = 3$ e. $x = 10$ f. $z = \frac{6}{4}$

4. 13, 15

5. Bill is 24, Ike is 26

6. n = 7

7. $w = 5$ m; $l = 10$ m

8. 10 quarters; 5 dimes

9. a. $x - y = 2$

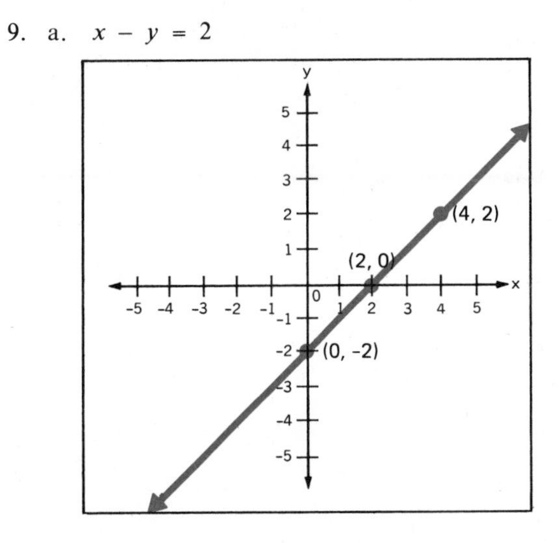

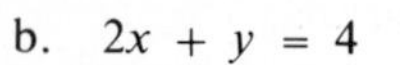

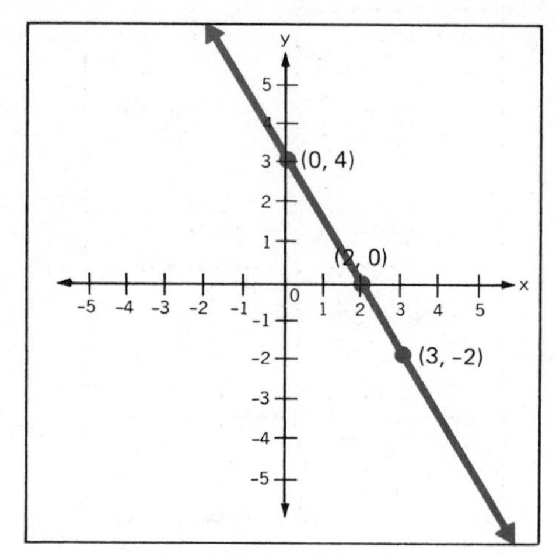

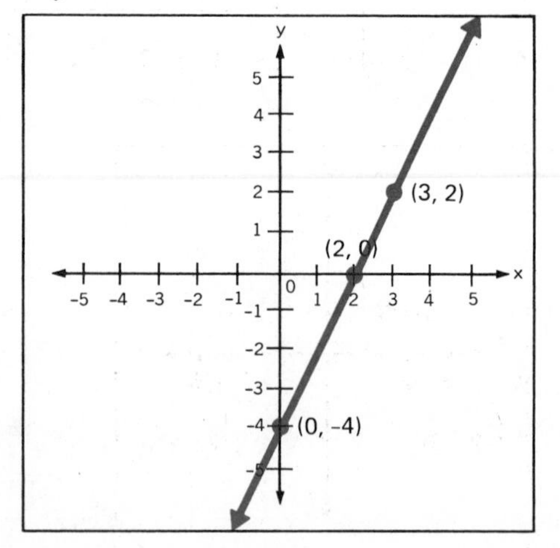

d. $-2x + y = 6$

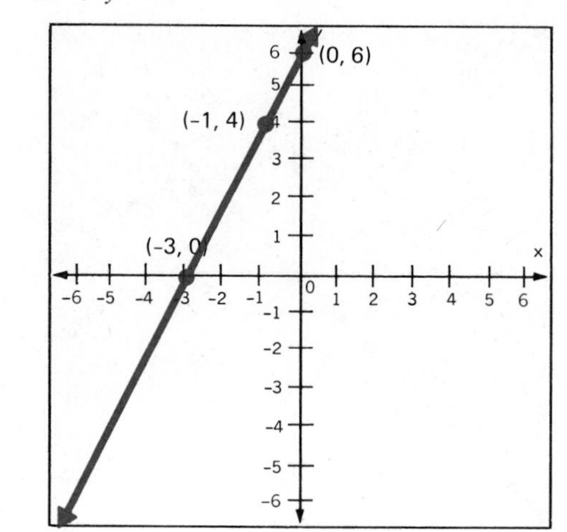

e. $3x - 5y = 15$

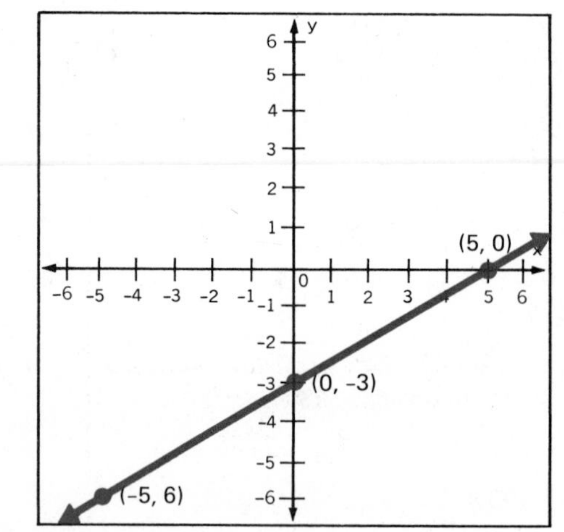

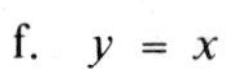
f. $y = x$

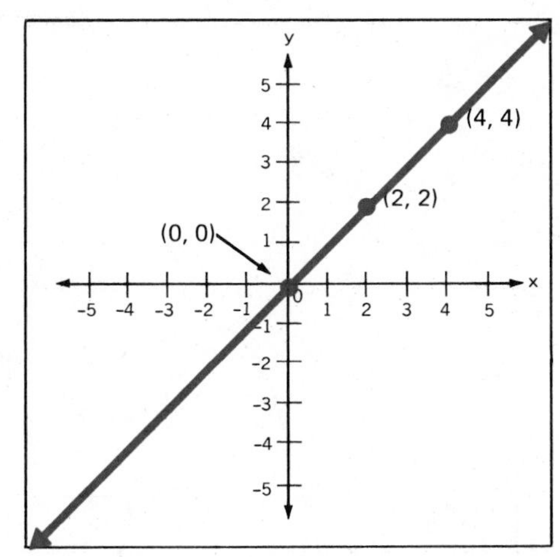

10. a. $y = x^2$

x	-2	-1	0	1	2
y	4	1	0	1	4

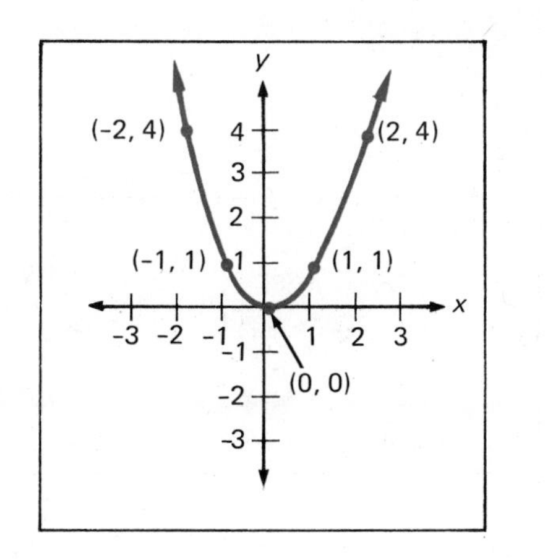

b. $y = x^2 - 4$

x	-3	-2	-1	0	1	2	3
y	5	0	-3	-4	-3	0	5

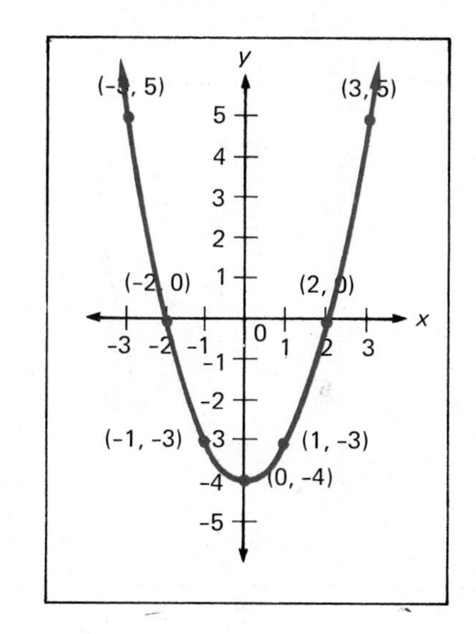

c. $y = -x^2 + 2$

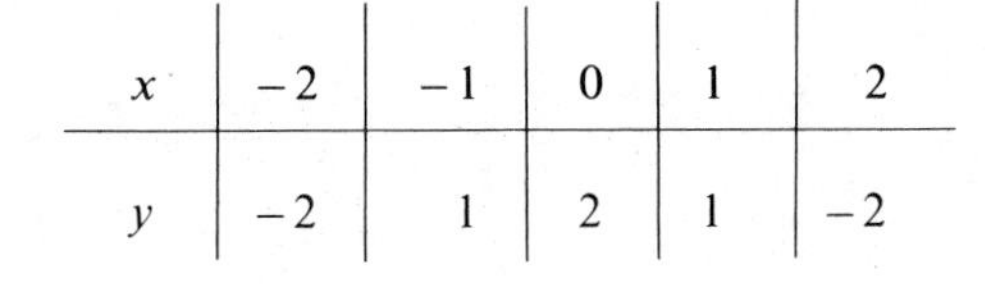

x	-2	-1	0	1	2
y	-2	1	2	1	-2

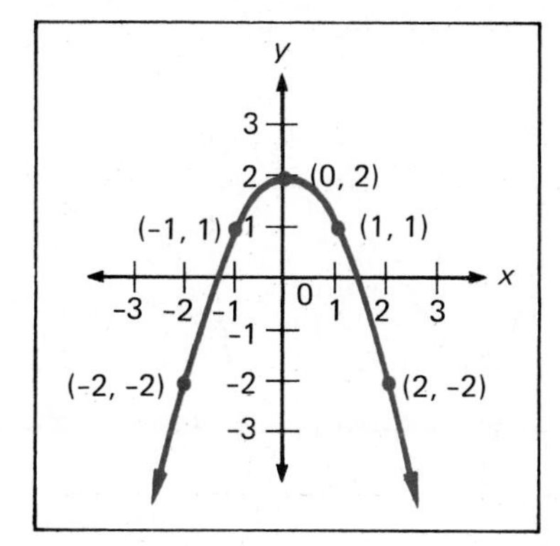

d. $y = x^2 - 6x$

x	0	1	2	3	4	5	6
y	0	-5	-8	-9	-8	-5	0

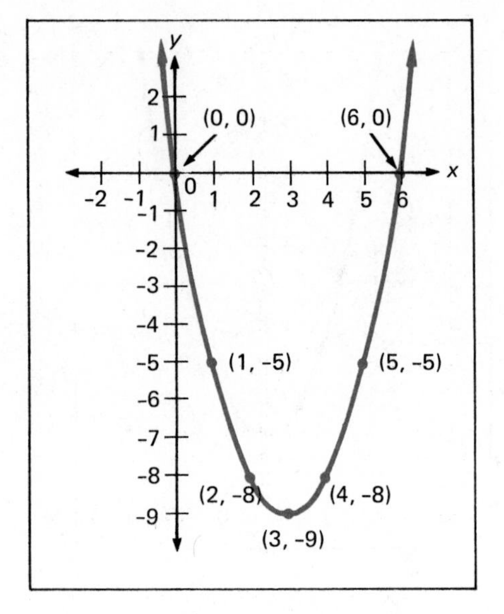

e. $y = x^2 + 2x - 3$

x	-4	-3	-2	-1	0	1	2
y	5	0	-3	-4	-3	0	5

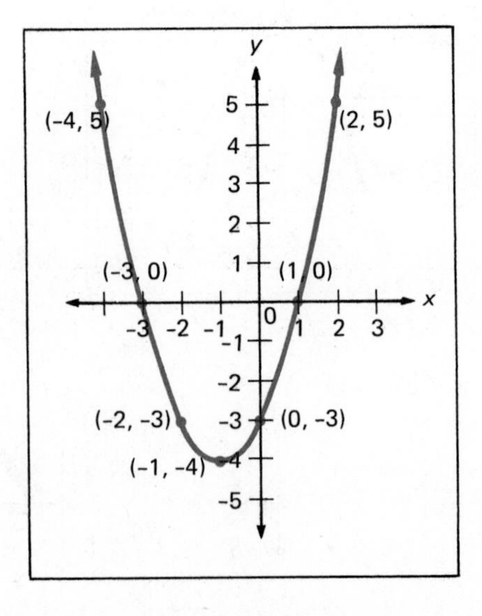

f. $y = -x^2 + 6x - 5$

x	0	1	2	3	4	5	6
y	-5	0	3	4	3	0	-5

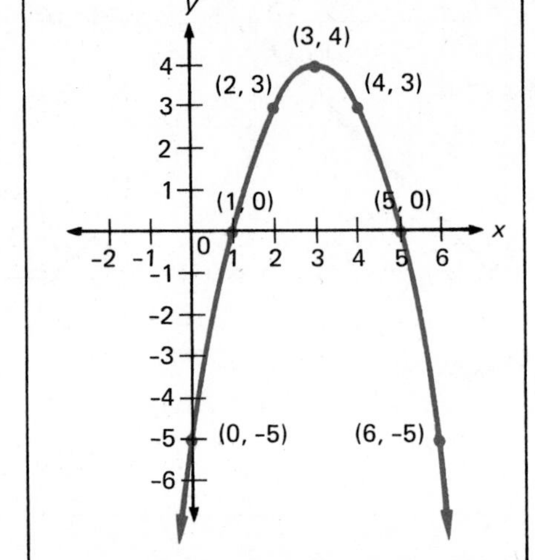

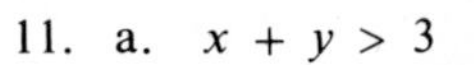

11. a. $x + y > 3$

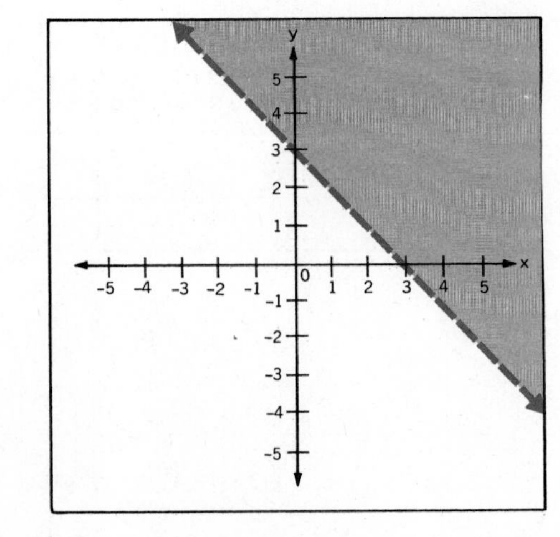

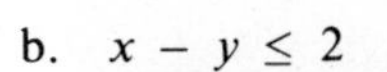

b. $x - y \leq 2$

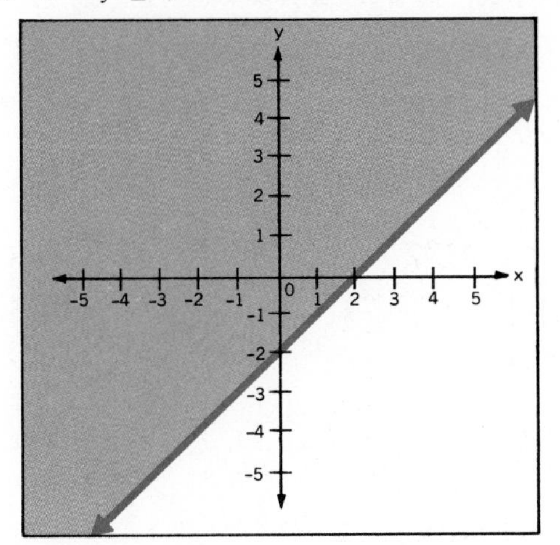

c. $-x + y > 2$

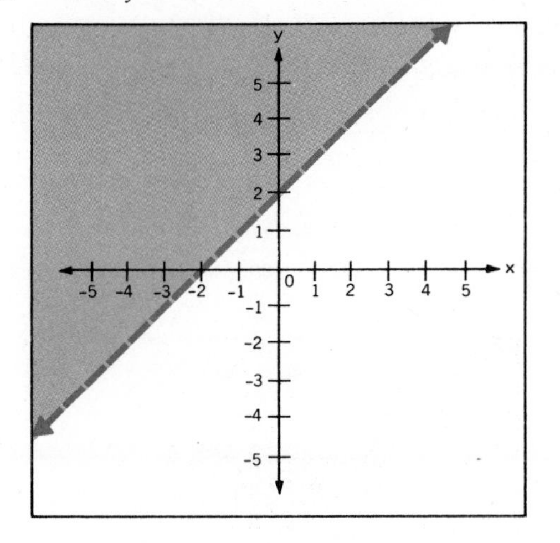

d. $3x + 5y > -15$

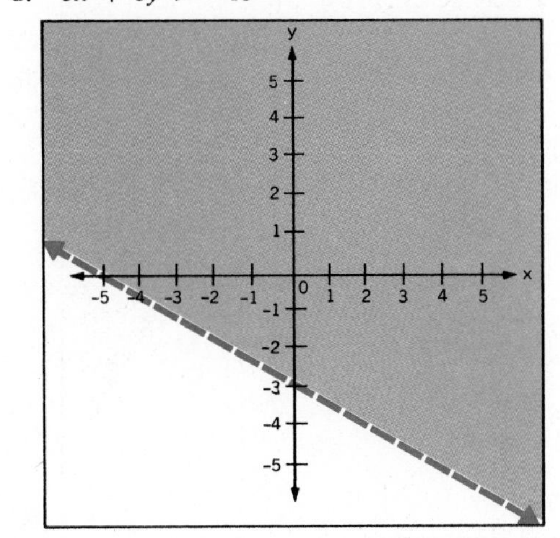

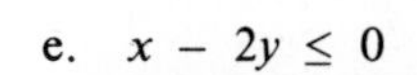

e. $x - 2y \leq 0$

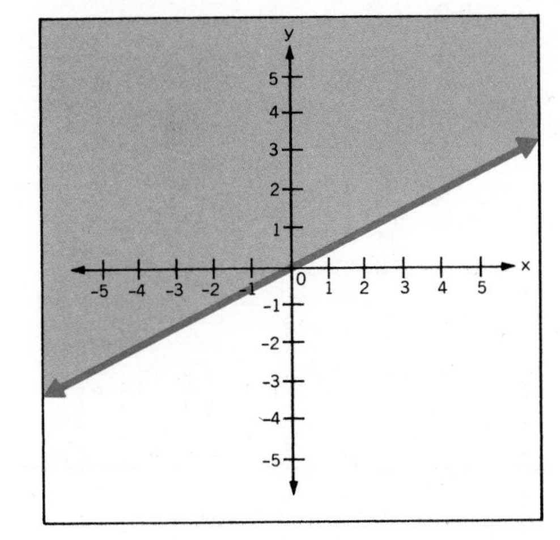

f. $-3x - 2y < -6$

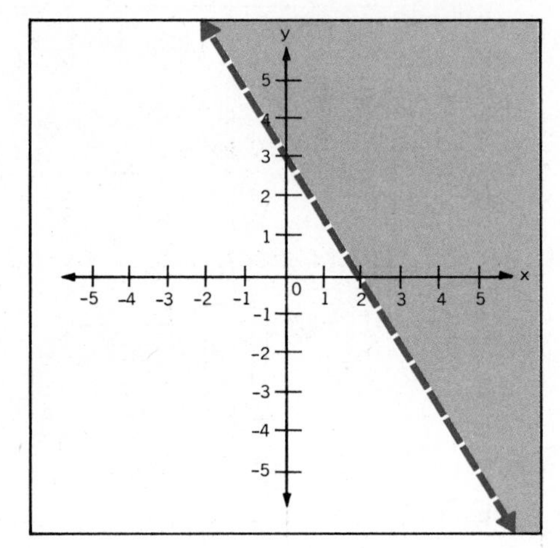

12. maximum value of P is 18 at (3, 4)

13. 4 couches and 3 recliners; maximum profit is $255

14. 30 standard and 5 deluxe models; maximum profit is $260

JUST FOR FUN

two hours

CHAPTER 10

SECTION 10.2

1. $\{I\}$ 3. $\overline{QK}$ 5. $\overline{QU}$
7. $\overleftrightarrow{UK}$ 9. $\{I\}$ 11. $\{I\}$
13. $\overline{QU}$ 15. $\{E\}$ 17. $\{E\}$
19. $\{E\}$ 21. $\angle WET$ 23. $\overleftrightarrow{BT}$ or $\angle BET$
25. $\overrightarrow{BT}$
27. $\overrightarrow{EW}$ 29. $\{A\}$ 31. $\overleftrightarrow{ZP}$ or $\angle ZAP$
33. ϕ 35. $\angle ZAW$ 37. $\overrightarrow{AM}$
39. $\{A\}$ 41. triangle WAZ

JUST FOR FUN

3

SECTION 10.3

1. R 3. $\overset{\circ\!-\!\circ}{AS}$ 5. $\angle DRA$
7. $\overset{\circ\!\rightarrow}{RS}$ 9. $\angle TON$ 11. $\overset{\circ\!\rightarrow}{OE}$
13. interior $\angle EON$ 15. ϕ
17. interior $\angle EON$ 19. $\angle COT$
21. $\angle SOJ$ 23. interior $\angle SOC$
25. $\overset{\circ\!\rightarrow}{OT}$ 27. $\{O\}$
29. exterior $\angle JOC$
31. true 33. false 35. false 37. true

JUST FOR FUN

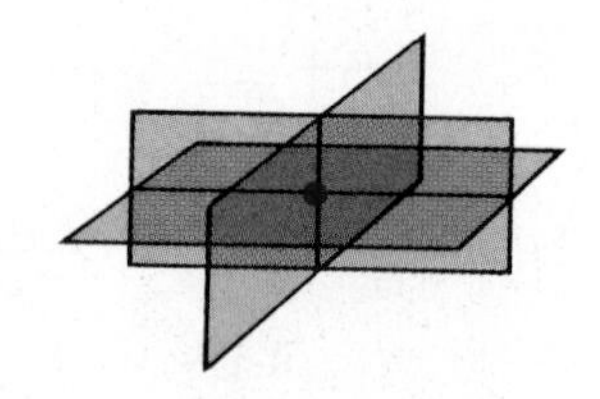

SECTION 10.4

1. b, d, e 3. a, b, d, e

5. a. true b. false c. true
d. false e. true

7. 15   9. 26   11. 3

13. a. true   b. false   c. false
d. true   e. false

15. polygon, triangle, right triangle, hypotenuse

JUST FOR FUN

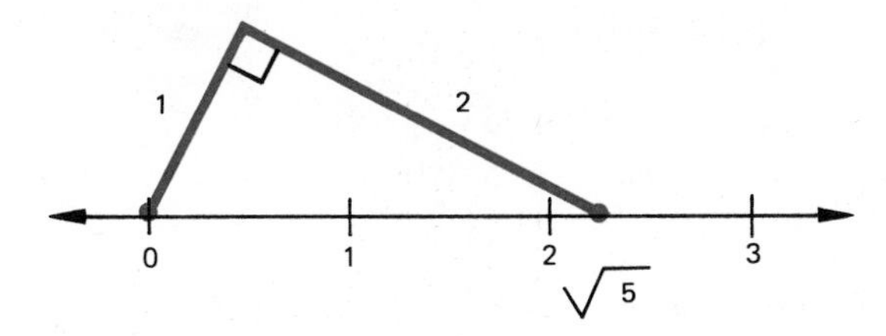

SECTION 10.5

1. a. 3   b. 0   c. yes   d. *A*, *B*, or *C*
3. a. 2   b. 2   c. yes   d. *A* or *C*
5. a. 4   b. 0   c. yes   d. *A*, *B*, *C*, or *D*
7. a. 2   b. 4   c. no   d. none
9. a. 3   b. 2   c. yes   d. *E* or *C*
11. No, there are 4 vertices of odd order.

REVIEW EXERCISES FOR CHAPTER 10

1. $\{R\}$ 2. $\overline{AS}$ 3. $\{R\}$
4. $\overleftrightarrow{AR}$ or $\angle ARS$ 5. $\emptyset$
6. $\emptyset$ 7. $\angle LRA$ 8. $\angle SRJ$
9. $\overrightarrow{RJ}$ 10. $\{R\}$ 11. $\angle SOJ$
12. $\angle TOE$ 13. $\{O\}$ 14. $\overline{OT}$
15. interior $\angle EOT$
16. interior $\angle JOS$
17. $\{O\}$ 18. $\overrightarrow{OC}$
19. interior $\angle JOE \cup$ exterior $\angle SOT$
20. interior $\angle COE$
21. d   22. d   23. 17   24. 9
25. a. 3   b. 2   c. yes   d. *A* or *E*
26. a. 4   b. 4   c. no   d. none
27. a. 0   b. 4   c. no   d. none
28. a. 3   b. 2   c. yes   d. *A* or *C*
29. false   30. false   31. false
32. true   33. false   34. true

CHAPTER 11

SECTION 11.2

1. a. $\frac{3}{2}$ b. $\frac{4}{7}$ c. $\frac{8}{5}$ d. $\frac{5}{6}$
3. a. $\frac{3}{1}$ b. $\frac{1}{2}$ c. $\frac{3}{2}$
d. $\frac{4}{1}$ e. $\frac{1}{6}$ f. $\frac{2}{3}$
5. a. $\frac{2}{3}$ b. $\frac{3}{2}$ c. $\frac{2}{5}$
7. $\frac{3}{19}$
9. a. 2   b. 6   c. 30   d. 35
11. 12 miles   13. 40

JUST FOR FUN

22 minutes

SECTION 11.3

1. a. $\frac{3}{20}$ b. $\frac{1}{4}$ c. $\frac{3}{4}$
3. a. $\frac{9}{200}$ b. $\frac{7}{300}$ c. $\frac{1}{16}$
5. a. $\frac{13}{200}$ b. $\frac{23}{1000}$ c. $\frac{3}{2}$
7. a. 0.17   b. 0.03   c. 0.045

9. a. 0.065 b. 3.0 c. 0.0025
11. a. 75% b. 80% c. 62.5%
13. a. 260% b. 112.5% c. 87.5%
15. a. $66\frac{2}{3}$% b. 275% c. 6.25%
17. a. 9% b. 214% c. 90%
19. a. 112.5% b. 1% c. 301%

JUST FOR FUN

2 A.M. the next day

SECTION 11.4

1. a. 50% b. 25% c. 33.3%
   d. 66.7% e. 25% f. 31.8%
3. a. $12.50 b. $100
   c. $7.50 d. $21.88
5.

	Markup	*Percent Markup*
a.	$30	150%
b.	$25	33.3%
c.	$3	28.6%
d.	$4	25.1%

7. $89.29 9. $275.38 11. $516.67

13.

	Markup	*Cost*
a.	$160	$240
b.	$2.20	$8.78
c.	$6.24	$43.71

15. $79; $316
17. 33.3% 19. $895; 33.5%

JUST FOR FUN

jets

SECTION 11.5

1. $420 3. $120 5. $212.50
7. $765; $3765 9. $1275; $106.25
11. $243.60

JUST FOR FUN

False, it contains graphite.

SECTION 11.6

1. $1079.50 3. $1125 5. $7970
7. $1095.50 9. $1126 11. $8955
13. $586 15. $2191 17. $10,165
19. $5418; $2418 21. 12 years

JUST FOR FUN

$5, $10, $20, $50, $100, $500, $1000, $5000, $10,000

SECTION 11.7

1. The *effective annual interest rate* is the annual interest rate which gives the same yield as the nominal interest rate compounded several times a year.
3. 9.20% 5. 12.55%
7. a. 6% b. 6.09% c. 6.14%
9. Larry; effective rate in his case is 5.83%.

JUST FOR FUN

Mario should drive Roy's car and Roy should drive Mario's car.

SECTION 11.8

1. a. \$76.20 b. \$238.65 c. \$729.50 d. \$904.20 e. \$462.30
3. \$63.50; \$26.50
5. straight life, \$366.40; limited payment life, \$583.60; limited payment life policy would cost \$680 more.
7. \$740.80; \$14,816; \$10,484
9. a. \$176.32 b. \$119.08 c. \$16.51 d. \$330.49 e. \$206.83

JUST FOR FUN

They are the same distance from Miami.

SECTION 11.9

1. 14.8% 3. 13.7% 5. \$1.85 7. \$4.95
9. \$220.20; \$49,272 11. \$152.04
13. \$293.60

JUST FOR FUN

Alexander Hamilton; U.S. Treasury Building

REVIEW EXERCISES FOR CHAPTER 11

1. a. $\frac{2}{3}$ b. $\frac{1}{2}$ c. $\frac{1}{6}$ d. $\frac{2}{3}$
2. a. 8 b. 3 c. 15 d. 21
3. a. $\frac{27}{50}$ b. $\frac{9}{200}$ c. $\frac{23}{1000}$ d. $\frac{5}{4}$
4. a. 0.19 b. 0.045 c. 0.065 d. 0.0025
5. a. 40% b. 37.5% c. 260% d. 6.25%
6. a. 9% b. 214% c. 0.3% d. 212.5%
7. 200 8. \$96.15 9. \$80
10. markup = \$160; cost = \$240
11. \$31.25; \$93.74 12. \$540
13. \$450 14. \$1280; \$106.67
15. \$18,644; \$14,644 16. \$10,165; \$5,165
17. a. 5% b. 5.06% c. 5.09%
18. a. \$156.40 b. \$238.65 c. \$979.50 d. \$1916
19. 23.0% 20. \$2.07
21. \$183.50; \$41,060
22. a. \$209.70 b. \$209 c. \$186 d. \$402.50

JUST FOR FUN

False, they were invented by people in India.

INDEX

32
33
34